Texts and
Monographs
in Physics

Norbert Straumann

# General Relativity and Relativistic Astrophysics

With 81 Figures

Springer-Verlag
Berlin Heidelberg New York Tokyo 1984

Professor Dr. Norbert Straumann

Institut für Theoretische Physik, Universität Zürich
CH-8001 Zürich, Switzerland



ISBN 3-540-13010-1 Springer-Verlag Berlin Heidelberg New York Tokyo
ISBN 0-387-13010-1 Springer-Verlag New York Heidelberg Berlin Tokyo

Library of Congress Cataloging in Publication Data. Straumann, Norbert, 1936–. General relativity and relativistic astrophysics. (Texts and monographs in physics) Revised translation of: Allgemeine Relativitätstheorie und relativistische Astrophysik. Includes bibliographical references and index. 1. General relativity (Physics) 2. Astrophysics. I. Title. II. Series. QC173.6.S7713 1984 530.1'1 84-5374



Typesetting: Konrad Triltsch, Graphischer Betrieb, Würzburg
Offset printing: Brüder Hartmann, Berlin · Bookbinding: Lüderitz & Bauer, Berlin
2153/3020-543210

*To my wife and sons,*
*Maria, Dominik, Felix, and Tobias*

# Preface

In 1979 I gave graduate courses at the University of Zurich and lectured in the 'Troisième Cycle de la Suisse Romande' (a consortium of four universities in the french-speaking part of Switzerland), and these lectures were the basis of the 'Springer Lecture Notes in Physics', Volume 150, published in 1981. This text appeared in German, because there have been few modern expositions of the general theory of relativity in the mother tongue of its only begetter. Soon after the book appeared, W. Thirring asked me to prepare an English edition for the 'Texts and Monographs in Physics'. Fortunately E. Borie agreed to translate the original German text into English. An excellent collaboration allowed me to revise and add to the contents of the book. I have updated and improved the original text and have added a number of new sections, mostly on astrophysical topics. In particular, in collaboration with M. Camenzind I have included a chapter on spherical and disk accretion onto compact objects.

This book divides into three parts. Part I develops the mathematical tools used in the general theory of relativity. Since I wanted to keep this part short, but reasonably self-contained, I have adopted the dry style of most modern mathematical texts. Readers who have never before been confronted with differential geometry will find the exposition too abstract and will miss motivations of the basic concepts and constructions. In this case, one of the suggested books in the reference list should help to absorb the material. I have used notations as standard as possible. A collection of important formulae is given at the end of Part I. Many readers should start there and go backwards, if necessary.

In the second part, the general theory of relativity is developed along rather traditional lines. The coordinate-free language is emphasized in order to avoid unnecessary confusions. We make full use of Cartan's calculus of differential forms which is often far superior computationally. The tests of general relativity are discussed in detail and the binary pulsar PSR 1913+16 is fully treated.

The last part of the book treats important aspects of the physics of compact objects. Some topics, for example the cooling of neutron stars, are discussed in great detail in order to illustrate how astrophysical problems require the simultaneous application of several different disciplines.

A text-book on a field as developed and extensive as general relativity and relativistic astrophysics must make painful omissions. Since the emphasis throughout is on direct physical applications of the theory, there is little discussion of more abstract topics such as causal spacetime structure or singularities. Cosmology, which formed part of the original lectures, has been omitted entirely. This field has grown so much in recent years that an entire book should be devoted to it. Furthermore, Weinberg's book still gives an excellent introduction to the more established parts of the subject.

The reference list near the end of the book is confined to works actually cited in the text. It is certainly much too short. In particular, we have not cited the early literature of the founders. This is quoted in the classic article by W. Pauli and in the wonderful recent book of A. Pais 'Subtle is the Lord', which gives also a historical account of Einstein's struggle to general relativity.

The physics of compact objects ist treated more fully in a book by S. L. Shapiro and S. A. Teukolsky, which just appeared when the final pages of the present English edition were typed.

I thank E. Borie for the difficult job of translating the original German text and her fine collaboration. I am particularly grateful to M. Camenzind for much help in writing the chapter on accretion and to J. Ehlers for criticism and suggestions for improvements. I profited from discussions and the writings of many colleagues. Among others, I am indepted to G. Boerner and W. Hillebrandt. R. Durrer, M. Schweizer, A. Wipf and R. Wallner helped me to prepare the final draft. I thank D. Oeschger for her careful typing of the German and English manuscripts.

For assistance in the research that went into this book, I thank the Swiss National Science Foundation for financial support.

Finally I thank my wife Maria for her patience.

Zurich, July 1984 *N. Straumann*

# Contents

# Part I

# Differential Geometry

In this purely mathematical part, we develop the most important concepts and results of differential geometry which are needed for general relativity theory.

The presentation differs little from that in many contemporary mathematical text books (however, some topics, such as fiber bundles, will be omitted). The language of modern differential geometry and the "intrinsic" calculus on manifolds are now frequently used by workers in the field of general relativity and are beginning to appear in textbooks on the subject. This has a number of advantages, such as:

(i) It enables one to read the mathematical literature and make use of the results to attack physical problems.

(ii) The fundamental concepts, such as differentiable manifolds, tensor fields, affine connection, and so on, adopt a clear and intrinsic formulation.

(iii) Physical statements and conceptual problems are not confused by the dependence on the choice of coordinates. At the same time, the role of distinguished coordinates in physical applications is clarified. For example, these can be adapted to symmetry properties of the system.

(iv) The exterior calculus of differential forms is a very powerful method for practical calculations; one often finds the results faster than with older methods.

Space does not allow us always to give complete proofs and sufficient motivation. In these cases, we give detailed references to the literature (Refs. [1]–[8]) where these can be found. Many readers will have the requisite mathematical knowledge to skip this part after familiarizing themselves with our notation (which is quite standard). This is best done by looking at the collection of important formulae at the end (p. 70).

# 1. Differentiable Manifolds

A manifold is a topological space which locally looks like the space $\mathbb{R}^n$ with the usual topology.

**Definition 1.1:** An *n-dimensional topological manifold* $M$ is a topological Hausdorff space with a countable base, which is locally homeomorphic to $\mathbb{R}^n$. This means that for every point $p \in M$ there is an open neighborhood $U$ of $p$ and a homeomorphism

$$h: U \to U'$$

which maps $U$ onto an open set $U' \subset \mathbb{R}^n$.

As an aside, we note that a topological manifold $M$ also has the following properties:

(i) $M$ is $\sigma$-compact;

(ii) $M$ is paracompact and the number of connected components is at most denumerable.

The second of these properties is particularly important for the theory of integration. For a proof, see e.g. [2], Chap. II, Sect. 15.

**Definition 1.2.:** If $M$ is a topological manifold and $h: U \to U'$ is a homeomorphism which maps an open subset $U \subset M$ onto an open subset $U' \subset \mathbb{R}^n$, then $h$ is a *chart* of $M$ and $U$ is called the *domain of the chart* or *local coordinate neighborhood.* The coodinates $(x^1, \ldots, x^n)$ of the image $h(p) \in \mathbb{R}^n$ of a point $p \in U$ are called the *coordinates* of $p$ in the chart. A set of charts $\{h_\alpha \mid \alpha \in I\}$ with domains $U_\alpha$ is called an *atlas* of $M$, if $\bigcup_{\alpha \in I} U_\alpha = M$. If $h_\alpha$ and $h_\beta$ are two charts, then both define homeomorphisms on the intersection of their domains $U_{\alpha\beta} := U_\alpha \cap U_\beta$; one thus obtains a homeomorphism $h_{\alpha\beta}$ between two open sets in $\mathbb{R}^n$ via the commutative diagram:

$$\begin{array}{ccc} & U_{\alpha\beta} & \\ {\scriptstyle h_\alpha} \swarrow & & \searrow {\scriptstyle h_\beta} \\ U'_\alpha \supset h_\alpha(U_{\alpha\beta}) & \xrightarrow{\ h_{\alpha\beta}\ } & h_\beta(U_{\alpha\beta}) \subset U'_\beta . \end{array}$$

Thus $h_{\alpha\beta} = h_\beta \circ h_\alpha^{-1}$ on the domain where the mapping is defined (the reader should draw a figure). This mapping gives a relation between the coordinates in the two charts and is called a *change of coordinates*, or *coordinate transformation*. Sometimes, particularly in the case of charts, it is useful to include the domain of a mapping in the notation; thus, we write $(h, U)$ for the mapping $h\colon U \to U'$.

**Definition 1.3:** An atlas defined on a manifold is said to be *differentiable* if all of its coordinate changes are differentiable mappings. For simplicity, unless otherwise stated, we shall always mean differentiable mappings of class $C^\infty$ on $\mathbb{R}^n$ (the derivatives of all orders exist and are continuous). Obviously, for all coordinate transformations one has (on the domains for which the mappings are defined):

$$h_{\alpha\alpha} = \text{identity and } h_{\beta\gamma} \circ h_{\alpha\beta} = h_{\alpha\gamma}, \text{ so that } h_{\alpha\beta}^{-1} = h_{\beta\alpha},$$

and hence the inverses of the coordinate transformations are also differentiable. They are thus diffeomorphisms.

If $\mathscr{A}$ is a differentiable atlas defined on a manifold $M$, then the atlas $\mathscr{D}(\mathscr{A})$ contains those charts for which all coordinate changes among charts from $\mathscr{A}$ are differentiable. The atlas $\mathscr{D}(\mathscr{A})$ is then also differentiable since, locally, a coordinate change $h_{\beta\gamma}$ in $\mathscr{D}(\mathscr{A})$ can be written as a composition $h_{\beta\gamma} = h_{\alpha\gamma} \circ h_{\beta\alpha}$ of two other coordinate changes with a chart $h_\alpha \in \mathscr{A}$ and differentiability is a local property. The atlas $\mathscr{D}(\mathscr{A})$ is clearly maximal. In other words, $\mathscr{D}(\mathscr{A})$ cannot be enlarged by the addition of further charts, and is the largest atlas which contains $\mathscr{A}$. Thus, every differentiable atlas $\mathscr{A}$ determines uniquely a maximal differentiable atlas $\mathscr{D}(\mathscr{A})$ such that $\mathscr{A} \subset \mathscr{D}(\mathscr{A})$. Furthermore, $\mathscr{D}(\mathscr{A}) = \mathscr{D}(\mathscr{B})$ if and only if the atlas $\mathscr{A} \cup \mathscr{B}$ is differentiable.

**Definition 1.4:** A *differentiable structure* on a topological manifold is a maximal differentiable atlas. A *differentiable manifold* is a topological manifold, together with a differentiable structure.

In order to define a differentiable structure on a manifold, one must specify a differentiable atlas. In general, one specifies as small an atlas as possible, rather than a maximal one, which is then obtained as described above. We shall tacitly assume that all the charts and atlases of a manifold having a differentiable structure $\mathscr{D}$ are contained in $\mathscr{D}$. As a shorthand notation, we write $M$, rather than $(M, \mathscr{D})$ to denote a differentiable manifold.

**Examples:** (a) $M = \mathbb{R}^n$. The atlas is formed by the single chart: $(\mathbb{R}^n, \text{identity})$. (b) Any open subset of a differentiable manifold has an obvious differentiable structure. It may have others as well.

**Definition 1.5:** A continuous mapping $\varphi\colon M \to N$ from one differentiable manifold to another is said to be *differentiable at the point*

$p \in M$ if for some (and hence for every) pair of charts $h\colon U \to U'$ of $M$ and $k\colon V \to V'$ of $N$ with $p \in U$ and $\varphi(p) \in V$, the composite mapping $k \circ \varphi \circ h^{-1}$ is differentiable at the point $h(p) \in U'$. Note that this mapping is defined in the neighborhood $h(\varphi^{-1}(V) \cap U)$ of $h(p)$ (if necessary, draw a sketch). The mapping $\varphi$ is *differentiable* if it is differentiable at every point $p \in M$. One can regard $k \circ \varphi \circ h^{-1}$ as a coordinate representation of $\varphi$ in the charts $h$ and $k$; for such a representation, the concept of differentiability is clear.

**Remark:** The identity and the composites of differentiable mappings are differentiable. Hence differentiable manifolds form a category.

**Notation:** Let $C^\infty(M, N)$ be the set of differentiable mappings from $M$ to $N$. We write

$$\mathscr{F}(M) = C^\infty(M) := C^\infty(M, \mathbb{R}).$$

**Definition 1.6:** A bijective differentiable mapping $\varphi$, whose inverse $\varphi^{-1}$ is also differentiable is called a *diffeomorphism.*

Every chart $h\colon U \to U'$ of $M$ is a diffeomorphism between $U$ and $U'$, provided $U'$ is taken to have the standard differentiable structure as an open set in $\mathbb{R}^n$. Differential topology is the study of properties which remain invariant under diffeomorphisms.

It is a nontrivial question as to whether two different differentiable structures can be introduced on a given topological manifold in such a way that the resulting differentiable manifolds are not diffeomorphic. For example, one can show (Kervaire and Milnor, 1963), that the topological 7-sphere has exactly 15 different structures which are not mutually diffeomorphic. Recently it has been shown that even $\mathbb{R}^4$ has more than one differentiable structure.

**Definition 1.7:** A differentiable mapping $\varphi\colon M \to N$ is called an *immersion* if (with the notation of Def. 1.5) the charts $h\colon U \to U' \subset \mathbb{R}^m$ and $k\colon V \to V' \subset \mathbb{R}^n$ can be chosen such that $k \circ \varphi \circ h^{-1}\colon h(U) \to k(V)$ is the inclusion, when we regard $\mathbb{R}^m$ as $\mathbb{R}^m \times 0 \subset \mathbb{R}^n$ (it is assumed that $m < n$).

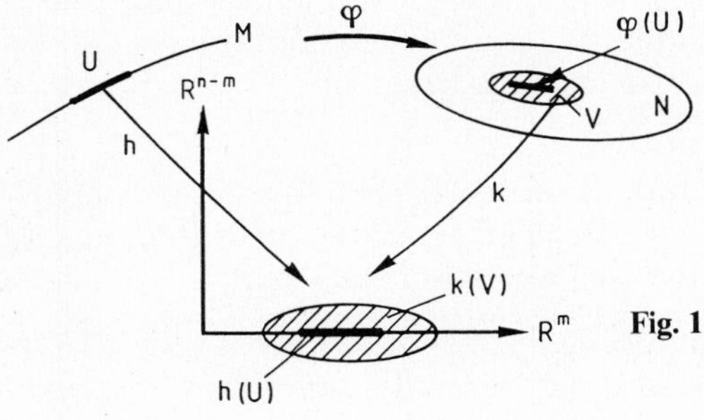


**Fig. 1**

In other words, the local coordinate representation of $\varphi$ is given by

$$(x^1, \ldots, x^m) \mapsto (x^1, \ldots, x^m, 0, \ldots, 0).$$

(See also Fig. 1.)

**Remarks:**

(i) An immersion is locally injective, but not necessarily globally injective.

(ii) If $\varphi: M \to N$ is an injective immersion, the mapping

$$\varphi: M \to \varphi(M) \subset N,$$

where $\varphi(M)$ has the induced topology, is not necessarily a homeomorphism. If, in addition, $\varphi$ is in fact a homeomorphism, then $\varphi$ is called an *embedding*.

**Definition 1.8:** If $M$ and $N$ are differentiable manifolds, then $M$ is said to be a *submanifold* of $N$ provided that

(i) $M \subset N$ (as sets).

(ii) The inclusion $M \to N$ is an embedding.

The reader is cautioned that the concepts of embedding and submanifold are not defined in a unified manner in the literature. For example, what we call submanifold is sometimes called a proper or regular submanifold.

Since the inclusion $i$ of Def. 1.8 is an immersion, one can, according to Def. 1.7, choose charts $h: U \to U' \subset \mathbb{R}^m$ and $k: V \to V' \subset \mathbb{R}^n$ whose domains are neighborhoods of the point $p \in M \subset N$ such that locally $i$ has the representation $i: (x^1, \ldots, x^m) \mapsto (x^1, \ldots, x^m, 0, \ldots, 0)$. Since $M$ has the induced topology ($i$ is an embedding, hence also a homeomorphism), it follows that $U$ has the form $U = \tilde{V} \cap M$, where $\tilde{V}$ is an open neighborhood of $p$ in $N$. If we restrict the chart $k$ to the domain $W := \tilde{V} \cap V$ then we see that

$$\mathrm{W} \cap M = \{q \in W \mid k(q) \in \mathbb{R}^m \times 0 \subset \mathbb{R}^n\}.$$

One describes this situation by saying that the submanifold $M$ lies locally in $N$ as $\mathbb{R}^m$ does in $\mathbb{R}^n$. (See Fig. 2.)

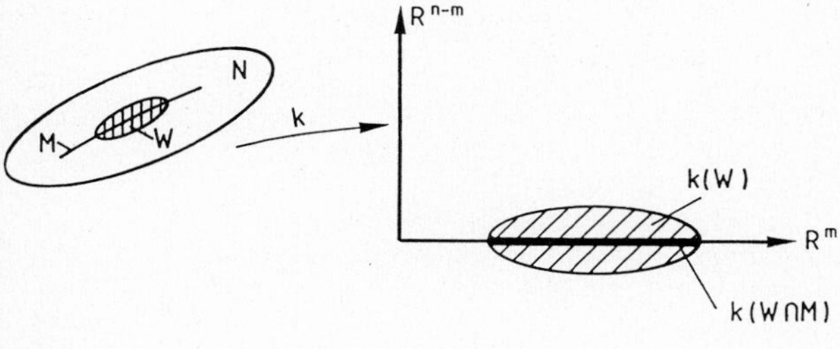


**Fig. 2**

Let $M^m$ and $N^n$ be two differentiable manifolds. We can define a product manifold by taking the underlying topological space to be the cartesian product of the two topological spaces. The differentiable structure is defined as follows: if $h: U \to U'$ and $k: V \to V'$ are charts of $M$ and $N$, then

$$h \times k: U \times V \to U' \times V' \subset \mathbb{R}^m \times \mathbb{R}^n = \mathbb{R}^{m+n}$$

is a chart of $M \times N$ and the set of all such charts defines the differentiable structure of $M \times N$.

# 2. Tangent Vectors, Vector and Tensor Fields

At every point $p$ of a differentiable manifold $M$, one can introduce a linear space, called the tangent space $T_p(M)$. A tensor field is a (smooth) map which assigns to each point $p \in M$ a tensor of a given type on $T_p(M)$.

## 2.1 The Tangent Space

Before defining the tangent space, let us introduce some basic concepts. Consider two differentiable manifolds $M$ and $N$, and the set of differentiable mappings

$$\{\varphi \mid \varphi\colon U \to N \text{ for a neighborhood } U \text{ of } p \in M\}.$$

Two such mappings $\varphi$, $\psi$ are called equivalent, $\varphi \sim \psi$, if and only if there is a neighborhood $V$ of $p$ such that $\varphi \mid V = \psi \mid V$. Here $\varphi \mid V$ denotes the restriction of $\varphi$ to the domain $V$. In other words, $\varphi$ and $\psi$ are equivalent if they coincide on some neighborhood $V$ of $p$.

**Definition 2.1:** An equivalence class of this relation is called a *germ* of smooth mappings $M \to N$ at the point $p \in M$. A germ with representative $\varphi$ is denoted by

$$\bar{\varphi}\colon (M, p) \to N \quad \text{or} \quad \bar{\varphi}\colon (M, p) \to (N, q) \quad \text{if} \quad q = \varphi(p).$$

Compositions of germs are defined naturally via their representatives. A germ of a function is a germ $(M, p) \to \mathbb{R}$. The set of all germs of functions at a point $p \in M$ is denoted by $\mathscr{F}(p)$.

$\mathscr{F}(p)$ has the structure of a real algebra, provided the operations are defined using representatives. A differentiable germ $\bar{\varphi}\colon (M, p) \to (N, q)$ defines, through composition, the following homomorphism of algebras:

$$\varphi^*\colon \mathscr{F}(q) \to \mathscr{F}(p), \quad \bar{f} \mapsto \bar{f} \circ \bar{\varphi}. \tag{2.1}$$

Obviously

$$\mathrm{Id}^* = \mathrm{Id}, \quad (\psi \circ \varphi)^* = \varphi^* \circ \psi^* . \tag{2.2}$$

If $\bar{\varphi}$ is a germ having an inverse $\bar{\varphi}^{-1}$, then $\varphi^* \circ (\varphi^{-1})^* = \mathrm{Id}$ and $\varphi^*$ is thus an isomorphism.

For every point $p \in M$ of a $n$-dimensional differentiable manifold, a chart $h$, having a neighborhood of $p$ as its domain defines an invertible germ $\bar{h}$: $(M, p) \to (\mathbb{R}^n, 0)$ and hence an isomorphism

$$h^* : \mathscr{F}_n \to \mathscr{F}(p), \; \mathscr{F}_n = \text{set of germs } (\mathbb{R}^n, 0) \to \mathbb{R} .$$

We now give three equivalent definitions of the tangent space at a point $p \in M$. One should be able to switch freely among these definitions. The "algebraic definition" is particularly handy.

**Definition 2.2:** The *tangent space* $T_p(M)$ of a differentiable manifold $M$ at a point $p$ is the set of derivations of $\mathscr{F}(p)$. A *derivation* of $\mathscr{F}(p)$ is a linear mapping $X: \mathscr{F}(p) \to \mathbb{R}$ which satisfies the Leibniz rule (product rule)

$$X(\bar{f}\bar{g}) = X(\bar{f})\, \bar{g}(p) + \bar{f}(p)\, X(\bar{g}) . \tag{2.3}$$

A differentiable germ $\bar{\varphi}$: $(M, p) \to (N, q)$ (and thus a differentiable mapping $\varphi$: $M \to N$) induces an algebra homomorphism $\varphi^* : \mathscr{F}(q) \to \mathscr{F}(p)$ and hence also a linear mapping $T_p\,\varphi$: $T_p(M) \to T_q(N)$ by

$$X \mapsto X \circ \varphi^* . \tag{2.4}$$

$T_p\,\varphi$ is called the *differential* (or *tangent map*) of $\varphi$ at $p$.

The set of derivations obviously forms a vector space.

The Leibniz rule gives

$$X(1) = X(1 \cdot 1) = X(1) + X(1), \quad \text{i.e.} \quad X(1) = 0$$

where 1 is the constant function having value unity. Linearity then implies that $X(c) = 0$ for any constant $c$.

The definition of the differential says in particular that for a given germ $\bar{f}$: $(N, q) \to \mathbb{R}$

$$T_p\,\varphi(x)(\bar{f}) = X \circ \varphi^*(\bar{f}) = X(\bar{f} \circ \bar{\varphi}), \qquad X \in T_p(M) . \tag{2.5}$$

Hence, or from (2.2), we have for a composition $(M, p) \overset{\bar{\varphi}}{\to} (N, q) \overset{\bar{\psi}}{\to} (L, r)$ the chain rule

$$T_p(\bar{\psi} \circ \bar{\varphi}) = T_q\,\bar{\psi} \circ T_p\,\bar{\varphi} \tag{2.6}$$

for differentials.

If $\bar{h}$: $(N, p) \to (\mathbb{R}^n, 0)$ is a germ of some chart, then the induced mapping $h^* : \mathscr{F}_n \to \mathscr{F}(p)$ is an isomorphism. This is then also true for the differential $T_p\,h$: $T_p\,N \to T_0\,\mathbb{R}^n$. In order to describe the latter space we use

**Lemma 2.1:** If $U$ is an open ball about the origin of $\mathbb{R}^n$, or $\mathbb{R}^n$ itself, and $f\colon U \to \mathbb{R}^n$ is a differentiable function, then there exist differentiable functions $f_1, \ldots, f_n\colon U \to \mathbb{R}$, such that

$$f(x) = f(0) + \sum_{i=1}^{n} f_i(x)\, x^i.$$

*Proof:* We have

$$\begin{aligned} f(x) - f(0) &= \int_0^1 \frac{d}{dt} f(t\,x^1, \ldots, t\,x^n)\, dt \\ &= \sum_{i=1}^{n} x^i \int_0^1 D_i f(t\,x^1, \ldots, t\,x^n)\, dt, \end{aligned}$$

where $D_i$ denotes the partial derivative with respect to the $i$th variable. It suffices to set

$$f_i(x) = \int_0^1 D_i\, f(t\,x^1, \ldots, t\,x^n)\, dt.$$

Particular derivations of $\mathscr{F}_n$ are the partial derivatives at the origin

$$\frac{\partial}{\partial x^i}\colon \mathscr{F}_n \to \mathbb{R}, \quad \bar{f} \mapsto \frac{\partial}{\partial x^i} f(0).$$

**Corollary:** The $\partial/\partial x^i$, $i = 1, \ldots, n$ form a basis of $T_0\,\mathbb{R}^n$, the vector space of derivations of $\mathscr{F}_n$.

*Proof:* a) Linear independence: If the derivation $\sum_i a^i\, \partial/\partial x^i = 0$, then applying it to the germ $\bar{x}^j$ of the $j$th coordinate function gives

$$a^j = \sum_i a^i \frac{\partial}{\partial x^i}(\bar{x}^j) = 0$$

for all $j$. Hence the $\partial/\partial x^i$ are linearly independent. b) Let $X \in T_0\,\mathbb{R}^n$ and $a^i = X(\bar{x}^i)$. We now show that

$$X = \sum_i a^i\, \partial/\partial x^i.$$

For this purpose, construct the derivation $Y = X - \sum_i a^i\, \partial/\partial x^i$, and observe that, by construction, $Y(\bar{x}^i) = 0$ for every coordinate function $x^i$. If $\bar{f} \in \mathscr{F}_n$ is any germ, then, by Lemma 2.1, it is of the form

$$\bar{f} = \bar{f}(0) + \sum_i f_i\, \bar{x}^i$$

and hence, according to the product rule

$$Y(\bar{f}) = Y(f(0)) + \sum_i f_i(0)\; Y(\bar{x}^i) = 0.$$

**Remark:** According to what has just been discussed, the tangent spaces of an $n$-dimensional differentiable manifold $M$ also have dimension $n$. Thus, the dimension of $M$ is unambiguously defined. This can also be shown to be true for topological manifolds.

The use of local coordinates $(x^1, \ldots, x^n)$ in a neighborhood of a point $p \in N^n$ enables us to write vectors in $T_p(N)$ explicitly as linear combinations of the $\partial/\partial x^i$. As a result of the isomorphism $T_p h: T_p N \to T_0 \mathbb{R}^n$, we can regard the $\partial/\partial x^i$ also as elements of $T_p N$. According to (2.5) we then have for $\bar{f} \in \mathscr{F}(p)$

$$\left(\frac{\partial}{\partial x^i}\right)_p \cdot \bar{f} = \frac{\partial}{\partial x^i} \underbrace{(f \circ h^{-1})}_{\text{coordinate representation of } f} [h(p)] .$$

If $\bar{\varphi}: (N^n, p) \to (M^m, q)$ is a differentiable germ and if we introduce local coordinates $y^1, \ldots, y^m$ in a neighborhood of $q$, then $\bar{\varphi}$ can be expressed as a germ $(\mathbb{R}^n, 0) \to (\mathbb{R}^m, 0)$ which we denote for simplicity also by $\bar{\varphi}$:

$$\begin{array}{ccc} (N, p) & \xrightarrow{\bar{\varphi}} & (M, q) \\ \text{chart} \downarrow & & \downarrow \text{chart} \\ (\mathbb{R}^n, 0) & \xrightarrow{\text{“}\bar{\varphi}\text{”}} & (\mathbb{R}^m, 0) . \end{array}$$

The tangent map $T_0 \bar{\varphi}$ is obtained as follows: Let $\bar{f} \in \mathscr{F}_m$. From (2.5) and the chain rule [we set $\varphi(x^1, \ldots, x^n) = \varphi^1(x^1, \ldots, x^n), \ldots, \varphi^m(x^1, \ldots, x^n)$]

$$T_0 \bar{\varphi}(\partial/\partial x^i)(\bar{f}) = \frac{\partial}{\partial x^i}(\bar{f} \circ \bar{\varphi}) = \frac{\partial \bar{f}}{\partial y^j}(0) \frac{\partial \varphi^j}{\partial x^i}(0) .$$

Hence

$$T_0 \bar{\varphi}(\partial/\partial x^i) = \frac{\partial \varphi^j}{\partial x^i} \frac{\partial}{\partial y^j} . \tag{2.7}$$

The matrix

$$D\varphi := \left(\frac{\partial \varphi^i}{\partial x^j}\right) \tag{2.8}$$

is called the *Jacobian* of the mapping $\varphi$.

We can also express the result (2.7) as follows: if $v = \sum a^i \, \partial/\partial x^i$, then $T_0 \bar{\varphi}(v) = \sum b^j \, \partial/\partial y^j$, where

$$b = D\varphi_0 \cdot a . \tag{2.9}$$

We summarize:

**Theorem 2.1:** If one introduces local coordinates $(x^1, \ldots, x^n)$ in a neighborhood of $p \in N^n$ and $(y^1, \ldots, y^m)$ in a neighborhood of $q \in M^m$, then the derivations $\partial/\partial x^i$ and $\partial/\partial y^i$ form bases of the vector spaces

$T_p N$ and $T_q M$, respectively. The tangent mapping of a germ $\bar{\varphi}: (N, p) \to (M, q)$ is given by

$$D\varphi_0: \mathbb{R}^n \to \mathbb{R}^m$$

with respect to these bases.

We now give the "physicist's definition" of the tangent space which starts from the preceding result. Briefly, one often says that a contravariant vector is a real $n$-tuple which transforms according to the Jacobian matrix. We now wish to express this more precisely.

If $\bar{h}$ and $\bar{k}: (N, p) \to (\mathbb{R}^n, 0)$ are germs of charts, then the coordinate transformation $\bar{g} = \bar{k} \circ \bar{h}^{-1}: (\mathbb{R}^n, 0) \to (\mathbb{R}^n, 0)$ is an invertible differentiable germ. The set of all invertible germs of coordinate transformations forms a group $G$ under composition. For two germs $\bar{h}$ and $\bar{k}$ there is exactly one $\bar{g} \in G$ such that $\bar{g} \circ \bar{h} = \bar{k}$. To each $\bar{g} \in G$ we assign the Jacobian at the origin $Dg_0$. This defines a homomorphism

$$G \to GL(n, R), \qquad \bar{g} \mapsto Dg_0$$

between $G$ and the linear group of nonsingular $n \times n$ matrices. This enables us to formulate the "physicist's definition" more precisely.

**Definition 2.3:** A tangent vector at a point $p \in N^n$ is an assignment which associates with every germ of a chart $\bar{h}: (N, p) \to (\mathbb{R}^n, 0)$, a vector $v = (v^1, \ldots, v^n) \in \mathbb{R}^n$ such that the germ $\bar{g} \circ \bar{h}$ corresponds to the vector $Dg_0 \cdot v$.

Thus if we denote by $K_p$ the set of germs of charts $\bar{h}: (N, p) \to (\mathbb{R}^n, 0)$, the "physicist's" tangent space $T_p(N)_{\text{phys}}$ is the set of mappings

$$v: K_p \to \mathbb{R}^n \quad \text{for which} \quad v(\bar{g} \circ \bar{h}) = Dg_0 \cdot v(\bar{h}) \quad \text{for all} \quad \bar{g} \in G.$$

The set of all such mappings forms a vector space, since $Dg_0$ is a linear mapping. One can choose the vector $v \in \mathbb{R}^n$ arbitrarily for a given chart $h$, and the choice for all other charts is then fixed. The vector space $T_p(N)_{\text{phys}}$ is isomorphic to $\mathbb{R}^n$. An isomorphism is given via the choice of a local coordinate system. The canonical isomorphism

$$T_p(N) \to T_p(N)_{\text{phys}}$$

with the algebraically defined tangent space (Def. 2.2) assigns to the derivation $X \in T_p(N)$ the vector $(X(\bar{h}^1), \ldots, X(\bar{h}^n)) \in \mathbb{R}^n$ for a given germ $\bar{h} = (\bar{h}^1, \ldots, \bar{h}^n): (N, p) \to (\mathbb{R}^n, 0)$. The components of this vector are precisely the coefficients of $X$ with respect to the basis $\partial/\partial x^i$. They transform according to the Jacobian under a change of coordinates.

The "geometric" definition is the most intuitive one. It identifies tangent vectors with "velocity vectors" of curves through the point $p$ at that point. (A curve on $N$ passing through $p$ is given by a differentiable mapping $w$ which maps an open interval $I \subset \mathbb{R}$ into $N$ with $0 \in I$ and $w(0) = p$.)

**Definition 2.4:** We introduce an equivalence relation on the set $W_p$ of germs of paths through $p$: $\bar{w} \sim \bar{v}$ if and only if for every germ $\bar{f} \in \mathscr{F}(p)$

$$\frac{d}{dt}(\bar{f} \circ \bar{w})(0) = \frac{d}{dt}(\bar{f} \circ \bar{v})(0).$$

An equivalence class $[w]$ of this relation is a *tangent vector* at the point $p$.

Two germs of paths define the same tangent vector if they define the same "derivative along the curve". To every equivalence class $[w]$ we can thus associate a derivation $X_w$ of $\mathscr{F}(p)$ by

$$X_w(\bar{f}) := \frac{d}{dt} \bar{f} \circ \bar{w}(0).$$

This relation defines the injective mapping

$$W_p/\sim =: T_p(M)_{\text{geom}} \to T_p(N), \qquad w \mapsto X_w$$

of the set of equivalence classes of germs of paths into the tangent space. This mapping is also surjective, since if $w(t)$ is given in local coordinates by $w(t) = (t\,a^1, \ldots, t\,a^n)$ then $X_w = \sum a^i\, \partial/\partial x^i$.

Obviously $X_w = X_v$ in precisely those cases for which $\frac{d}{dt} w^i(0) = \frac{d}{dt} v^i(0)$ for any local coordinate system. In this geometrical definition the tangent map is easily visualized. A germ $\bar{\varphi}$: $(N, p) \to (M, q)$ induces the mapping

$$T_p(N)_{\text{geom}} \to T_q(M)_{\text{geom}}, \qquad [w] \mapsto [\varphi \circ w].$$

The following equation [use (2.5)] shows that this definition of the tangent map is consistent with that in Def. 2.2:

$$X_{\varphi \circ w}(\bar{f}) = \frac{d}{dt} \bar{f} \circ (\varphi \circ w)(0) = X_w(\bar{f} \circ \bar{\varphi}) \overset{(2.5)}{=} T_p\,\varphi(X_w)(\bar{f})$$

in other words,

$$X_{\varphi \circ w} = T_p\,\varphi \cdot X_w.$$

In the following, we shall regard the three different definitions of a tangent space as equivalent and use whichever of them is most convenient.

**Definition 2.5:** The *rank* of a differentiable mapping $\varphi$: $M \to N$ at the point $p \in M$ is the number

$$rk_p\,\varphi := \text{Rank of } T_p\,\varphi,$$

i.e. the rank of the matrix $D\varphi_p$.

The rank of a mapping is bounded from below; if $rk_p\,\varphi = r$, there is a neighborhood $U$ of $p$ such that $rk_q\,\varphi \geqq r$ for all $q \in U$.

*Proof:* We choose local coordinates and consider the Jacobian $D\varphi$ corresponding to $\varphi$ in the neighborhood of $p \in V \subset \mathbb{R}^m$. The elements of this matrix describe a differentiable mapping $V \to \mathbb{R}^{m \cdot n}$ such that $q \mapsto \partial\varphi^i/\partial x^j(q)$. Since $rk_p\,\varphi = r$, there exists an $r \times r$ submatrix of $D\varphi_p$ (without loss of generality, we may take this to be the first $r$ rows and $r$ columns) which has a nonvanishing determinant at $p$. Therefore the mapping

$$V \to \mathbb{R}^{m \cdot n} \to \mathbb{R}^{r \cdot r} \to \mathbb{R},$$

$$q \mapsto D_q\,\varphi \mapsto \text{submatrix} \mapsto \text{determinant}$$

does not vanish at $p$, and hence also not in some neighborhood of $p$. The rank cannot decrease there.

Usually one defines an immersion as follows:

**Definition 2.6:** A differentiable mapping $\varphi: M \to N$ is called an *immersion* if its rank is equal to the dimension of $M$ for all $p \in M$.

One can show that this definition is equivalent to the one given in Sect. 1 (Def. 1.7). This follows easily from the inverse function theorem (see, for example, [6], p. 55).

## 2.2 Vector Fields

If we assign to every point $p$ of a differentiable manifold $M$ a tangent vector $X_p \in T_p(M)$, then we call the assignment $X: p \mapsto X_p$ a *vector field* on $M$.

If $(x^1, \ldots, x^n)$ are local coordinates in an open set $U \subset M$, then for every point $p \in U$, $X_p$ has a unique representation of the form

$$X_p = \xi^i(p)\left(\frac{\partial}{\partial x^i}\right)_p. \tag{2.11}$$

The $n$ functions $\xi^i$ $(i = 1, \ldots, n)$ defined on $U$ are the *components* of $X$ with respect to the local coordinate system $(x^1, \ldots, x^n)$. If we now consider a second local coordinate system on $U$ and let $\bar{\xi}^i$ be the components of $X$ relative to $(\bar{x}^1, \ldots, \bar{x}^n)$, then we have, according to the results of Sect. 2.1

$$\bar{\xi}^i(p) = \frac{\partial \bar{x}^i}{\partial x^j}(p)\,\xi^j(p), \qquad p \in U. \tag{2.12}$$

Hence, the property that the components of $X$ are continuous, or of class $C^r$ at a point does not depend on the choice of the local coordinate system. If $X$ is continuous or of class $C^r$ at every point of $M$, we say that the vector field is continuous, or of class $C^r$ on $M$.

In the following, we shall consider only vector fields of class $C^\infty$ (unless stated otherwise) and denote the set of such fields by $\mathscr{X}(M)$. $\mathscr{F}(M)$ or $C^\infty(M)$ denote the class of $C^\infty$ functions on $M$, as before.

If $X, Y \in \mathscr{X}(M)$ and $f \in \mathscr{F}(M)$, the assignments $p \mapsto f(p)\, X_p$ and $p \mapsto X_p + Y_p$ define new vector fields on $M$. These are denoted by $fX$ and $X + Y$, respectively. The following rules apply: If $f, g \in \mathscr{F}(M)$ and $X, Y \in \mathscr{X}(M)$, then

$$f(g\,X) = (f\,g)\,X$$
$$f(X + Y) = fX + fY$$
$$(f + g)\,X = fX + g\,X.$$

We also define the functions $Xf$ on $M$ by

$$(Xf)(p) = X_p f, \qquad (p \in M).$$

In local coordinates, we have

$$X = \xi^i\, \partial/\partial x^i,$$

where the $\xi^i$ are $C^\infty$ functions. Since

$$(Xf)(p) = \xi^i(p) \frac{\partial f}{\partial x^i}(p)$$

it follows that $Xf$ is also a $C^\infty$ function. $Xf$ is called the *derivative of $f$ with respect to the vector field $X$*. The following rules hold:

$$X(f + g) = Xf + Xg$$
$$X(fg) \quad = (Xf)\,g + f(Xg).$$

In algebraic language, $\mathscr{X}(M)$ is a module over the associative algebra $\mathscr{F}(M)$.

If we set $D_X f = Xf$ ($f \in \mathscr{F}(M), X \in \mathscr{X}(M)$), then $D_X$ is a *derivation* of the algebra $\mathscr{F}(M)$.

One can prove (see [6], p. 73) that conversely every derivation of $\mathscr{F}(M)$ is of the form $D_X$ for some vector field $X$.

The commutator of two derivations $D_1$ and $D_2$ of an algebra $\mathscr{A}$,

$$[D_1, D_2]\, a := D_1(D_2\, a) - D_2(D_1\, a), \qquad a \in \mathscr{A},$$

is also a derivation, as one can easily show by direct computation.

The commutator is antisymmetric

$$[D_1, D_2] = -[D_2, D_1]$$

and satisfies the Jacobi identity

$$[D_1, [D_2, D_3]] + [D_2, [D_3, D_1]] + [D_3, [D_1, D_2]] = 0.$$

As a result, one is led to define the commutator of two vector fields $X$ and $Y$ according to

$$[X, Y]f = X(Yf) - Y(Xf).$$

One can easily prove the following properties:

$$\begin{aligned}
&[X+Y, Z] = [X, Z] + [Y, Z]\\
&[X, Y] = -[Y, X]\\
&[fX, gY] = fg[X, Y] + f(Xg)Y - g(Yf)X, \quad f, g \in \mathscr{F}(M)\\
&[X, [Y, Z]] + [Z, [X, Y]] + [Y, [Z, X]] = 0.
\end{aligned} \tag{2.13}$$

The set of vector fields $\mathscr{X}(M)$ is an $\mathbb{R}$-*Lie-algebra* with respect to the commutator product.

If, in local coordinates,

$$X = \xi^i \frac{\partial}{\partial x^i}, \qquad Y = \eta^i \frac{\partial}{\partial x^i}$$

then one easily finds that

$$[X, Y] = \left(\xi^i \frac{\partial \eta^j}{\partial x^i} - \eta^i \frac{\partial \xi^j}{\partial x^i}\right) \frac{\partial}{\partial x^j}. \tag{2.14}$$

## 2.3 Tensor Fields

For the following, the reader is assumed to be familiar with some basic material of multilinear algebra (see, e.g., [4], Sect. 1.7).

In addition to the tangent space $T_p(M)$ at a point $p \in M$, we shall consider the dual space (cotangent space) $T_p^*(M)$.

Let $f$ be a differentiable function defined on an open set $U \subset M$. If $p \in U$ and $v \in T_p(M)$ is an arbitrary vector, we set

$$(df)_p(v) := v(f). \tag{2.15}$$

The mapping $(df)_p \colon T_p(M) \to \mathbb{R}$ is obviously linear, and hence $(df)_p \in T_p^*(M)$. The linear function $(df)_p$ is the *differential* of $f$ at the point $p$.

For a local coordinate system $(x^1, \ldots, x^n)$ in a neighborhood of $p$, we have

$$(df)_p \left(\frac{\partial}{\partial x^i}\right)_p = \frac{\partial f}{\partial x^i}(p). \tag{2.16}$$

In particular, the component function $x^i \colon q \mapsto x^i(q)$ satisfies

$$(dx^i)_p \left(\frac{\partial}{\partial x^j}\right)_p = \delta^i_j. \tag{2.17}$$

That is, the $n$-tuple $\{(dx^1)_p, \ldots, (dx^n)_p\}$ is a basis of $T_p^*(M)$ which is *dual* to the basis $\left\{\left(\frac{\partial}{\partial x^1}\right)_p, \ldots, \left(\frac{\partial}{\partial x^n}\right)_p\right\}$ of $T_p(M)$.

We can write $(df)_p$ as a linear combination of the $(dx^i)_p$:

$$(df)_p = \lambda_j (dx^j)_p.$$

Now

$$(df)_p \left(\frac{\partial}{\partial x^i}\right)_p = \lambda_j (dx^j)_p \left(\frac{\partial}{\partial x^i}\right)_p = \lambda_i.$$

From (2.16) we then have $\lambda_i = \frac{\partial f}{\partial x^i}(p)$ and thus

$$(df)_p = \left(\frac{\partial f}{\partial x^i}\right)(p)\,(dx^i)_p. \tag{2.18}$$

Let $T_p(M)^r_s$ be the set of tensors of rank $(r, s)$ defined on $T_p(M)$ (contravariant of rank $r$, covariant of rank $s$). If we assign to every $p \in M$ a tensor $t_p \in T_p(M)^r_s$, then the map $t: p \mapsto t_p$ defines a *tensor field of type* $(r, s)$.

Algebraic operations on tensor fields are defined pointwise; for example, the sum of two tensor fields is defined by

$$(t + s)_p = t_p + s_p.$$

Tensor products and contractions of tensor fields are defined analogously. A tensor field can also be multiplied by a function $f \in \mathscr{F}(M)$:

$$(ft)_p = f(p)\, t_p.$$

The set $\mathscr{T}^r_s(M)$ of tensor fields of type $(r, s)$ is thus a module over $\mathscr{F}(M)$.

In a coordinate neighborhood $U$, having coordinates $(x^1, \ldots, x^n)$, a tensor field can be expanded in the form

$$t = t^{i_1 \ldots i_r}_{j_1 \ldots j_s} \left(\frac{\partial}{\partial x^{i_1}} \otimes \ldots \otimes \frac{\partial}{\partial x^{i_r}}\right) \otimes (dx^{j_1} \otimes \ldots \otimes dx^{j_s}). \tag{2.19}$$

The $t^{i_1 \ldots i_r}_{j_1 \ldots i_s}$ are the components of $t$ relative to the coordinate system $(x^i, \ldots, x^n)$. If the coordinates are transformed to $(\bar{x}^1, \ldots, \bar{x}^n)$, the components of $t$ transform according to

$$\bar{t}^{i_1 \ldots i_r}_{j_1 \ldots j_s} = \frac{\partial \bar{x}^{i_1}}{\partial x^{k_1}} \cdots \frac{\partial \bar{x}^{i_r}}{\partial x^{k_r}} \frac{\partial x^{l_1}}{\partial \bar{x}^{j_1}} \cdots \frac{\partial x^{l_s}}{\partial \bar{x}^{j_s}}\, t^{k_1 \ldots k_r}_{l_1 \ldots l_s}. \tag{2.20}$$

A tensor field is of class $C^r$ if all its components are of class $C^r$. We see that this property is independent of the choice of coordinates. If two tensor fields $t$ and $s$ are of class $C^r$, then so are $s + t$ and $s \otimes t$. In the following, we shall only consider $C^\infty$ tensor fields.

Covariant tensors of order 1 are also called *one-forms*. The set of all one-forms will be denoted by $\mathscr{X}^*(M)$. The completely antisymmetric covariant tensors of higher order (differential forms) play an important role. We shall discuss these in detail in Sect. 4.

Let $t \in \mathscr{T}_s^r(M)$, $X_1, \ldots, X_s \in \mathscr{X}(M)$ and $\omega^1, \ldots, \omega^r \in \mathscr{X}^*(M)$. We consider, for every $p \in M$,

$$F(p) = t_p(\omega^1(p), \ldots, \omega^r(p), X_1(p), \ldots, X_s(p)).$$

The mapping defined by $p \mapsto F(p)$ is obviously a $C^\infty$ function, which we denote by $t(\omega^1, \ldots, \omega^r, X_1, \ldots, X_s)$. The assignment

$$(\omega^1, \ldots, \omega^r, X_1, \ldots, X_s) \mapsto t(\omega^1, \ldots, \omega^r, X_1, \ldots, X_s)$$

is $\mathscr{F}(M)$-multilinear. For every tensor field $t \in \mathscr{T}_s^r(M)$ there is thus an associated $\mathscr{F}(M)$-multilinear mapping:

$$\underbrace{\mathscr{X}^*(M) \otimes \ldots \otimes \mathscr{X}^*(M)}_{r\text{-times}} \otimes \underbrace{\mathscr{X}(M) \otimes \ldots \otimes \mathscr{X}(M)}_{s\text{-times}} \to \mathscr{F}(M).$$

One can show (see [6], p. 137) that every such mapping can be obtained in this manner. In particular, every one-form can be regarded as an $\mathscr{F}(M)$-linear function defined on the vector fields.

Let $\varphi: M \to N$ be a differentiable mapping. We define the *pull back* of a one-form $\omega$ on $N$ by

$$(\varphi^* \omega)_p(X_p) := \omega_{\varphi(p)}(T_p \varphi \cdot X_p), \qquad X_p \in T_p(M). \tag{2.21}$$

Analogously, we define the pull back of a covariant tensor field $t \in \mathscr{T}_s^0(N)$ by

$$(\varphi^* t)_p(v_1, \ldots, v_s) := t_{\varphi(p)}(T_p \varphi \cdot v_1, \ldots, T_p \varphi \cdot v_s), \qquad v_1, \ldots, v_s \in T_p(M). \tag{2.22}$$

For a function $f \in \mathscr{F}(N)$, we have

$$\varphi^*(df) = d(\varphi^* f). \tag{2.23}$$

Indeed, from (1.5) we have for $v \in T_p(M)$,

$$(\varphi^* df)_p(v) = df_{\varphi(p)}(T_p \varphi \cdot v) = T_p \varphi(v) f = v(f \circ \varphi)$$
$$= v(\varphi^*(f)) = d(\varphi^* f)(v).$$

**Definition 2.7:** A *pseudo-Riemannian metric* on a differentiable manifold $M$ is a tensor field $g \in \mathscr{T}_2^0(M)$ having the properties

(i) $g(X, Y) = g(Y, X)$ for all $X, Y \in \mathscr{X}(M)$;

(ii) for every $p \in M$, $g_p$ is a nondegenerate bilinear form on $T_p(M)$. This means that $g_p(X, Y) = 0$ for all $X \in T_p(M)$ if and only if $Y = 0$.

$g$ is a (proper) Riemannian metric if $g_p$ is positive definite at every point $p$.

A *(pseudo) Riemannian manifold* is a differentiable manifold $M$, together with a metric $g$.

If $(\theta^i, i=1, \ldots, \dim M)$ is a basis of one-forms on an open subset of $M$, we write

$$g = g_{ik}\,\theta^i \otimes \theta^k \tag{2.24}$$

or

$$ds^2 = g_{ik}\,\theta^i\,\theta^k\,, \qquad [\theta^i\,\theta^k := \tfrac{1}{2}(\theta^i \otimes \theta^k + \theta^k \otimes \theta^i)]\,. \tag{2.25}$$

Obviously

$$g_{ik} = g(e_i, e_k)\,, \tag{2.26}$$

if $(e_i)$ is the basis of (local) vector fields which is dual to $(\theta^i)$.

# 3. The Lie Derivative

Before defining the Lie derivative of tensor fields, we introduce some important concepts.

## 3.1 Integral Curves and Flow of a Vector Field

Let $X$ be a vector field (as usual $C^\infty$).

**Definition 3.1:** Let $J$ be an open interval in $\mathbb{R}$ which contains 0. A differentiable curve $\gamma: J \to M$ is called an *integral curve* of $X$, with starting point $x \in M$ (or through $x \in M$), provided

$$\dot{\gamma}(t) := \frac{d\gamma}{dt} = X(\gamma(t)) \quad \text{for every } t \in J \text{ and } \gamma(0) = x\,.$$

**Theorem 3.1:** For every $x \in M$, there is a unique maximal integral curve through $x$ of class $C^\infty$.

We denote this curve by $\phi_x: J_x \to M$. Let

$$\mathscr{D} := \{(t, x) \,|\, x \in M, t \in J_x\} \subset \mathbb{R} \times M\,.$$

For every $t \in \mathbb{R}$, let

$$\mathscr{D}_t := \{x \in M \,|\, (t, x) \in \mathscr{D}\} = \{x \in M \,|\, t \in J_x\}\,.$$

**Definition 3.2:** The mapping $\phi: \mathscr{D} \to M$, with $(t, x) \mapsto \phi_x(t)$ is called the (global) *flow* of $X$.

**Theorem 3.2:** (i) $\mathscr{D}$ is an open set in $\mathbb{R} \times M$ and $\{0\} \times M \subset \mathscr{D}$ (this implies that $\mathscr{D}_t$ is an open subset of $M$).
(ii) $\phi: \mathscr{D} \to M$ is a $C^\infty$-morphism.
(iii) For every $t \in \mathbb{R}$, $\phi_t: \mathscr{D}_t \to M$, $x \mapsto \phi(t, x)$ is a diffeomorphism from $\mathscr{D}_t$ to $\mathscr{D}_{-t}$, with inverse $\phi_{-t}$.

**Corollary:** For every $a \in M$, there is an open interval $J$, such that $0 \in J$ and an open neighborhood $U$ of $a$ in $M$ such that $J \times U \subset \mathscr{D}$.

**Definition 3.3:** $\psi := \phi | J \times U : J \times U \to M$ (i.e. the restriction of $\phi$ to $J \times U$) is called a *local flow* of $X$ at $a$.

**Theorem 3.3:** A local flow $\psi: J \times U \to M$ of $X$ at $a$ has the properties

(i) $\psi_x : J \to M,\ t \to \psi(t, x)$ is an integral curve of $X$ with starting point $x$. Thus

$$\dot{\psi}_x(t) = X(\psi_x(t)) \quad \text{for} \quad t \in J \quad \text{and} \quad \psi_x(0) = x .$$

(ii) $\psi_t : U \to M,\ x \mapsto \psi(t, x)$ is a diffeomorphism from $U$ onto $\psi_t(U)$.

(iii) If $s, t, s+t \in J$ and $x \in U$, then $\psi_{s+t}(x) = \psi_s(\psi_t(x))$. In other words,

$$\psi_{s+t} = \psi_s \circ \psi_t .$$

For this reason, one also calls $\psi$ a *local one parameter group of local diffeomorphisms.*

**Definition 3.4:** If $\mathscr{D} = \mathbb{R} \times M$, then $X$ is said to be *complete.*

In this case, we have

**Theorem 3.4:** If $X$ is complete and $\phi: \mathbb{R} \times M \to M$ is the global flow of $X$, then $\mathscr{D}_t = M$ for every $t \in \mathbb{R}$ and

(i) $\phi_t : M \to M$ is a diffeomorphism.

(ii) $\phi_s \circ \phi_t = \phi_{s+t}$ for $s, t \in \mathbb{R}$.

This means that the assignment $t \mapsto \phi_t$ is a group homomorphism: $\mathbb{R} \to \mathrm{Diff}(M)$ [$\mathrm{Diff}(M)$ denotes the group of all diffeomorphisms of $M$]; the set $(\phi_t)_{t \in \mathbb{R}}$ is a one parameter group of diffeomorphisms of $M$.

One can show that a vector field on a compact manifold is complete.

Proofs of these theorems can be found in Sect. 2.1 of [4].

## 3.2 Mappings and Tensor Fields

Let $\varphi: M \to N$ be a differentiable mapping. This mapping induces a mapping of covariant tensor fields (see also Sect. 2) by means of

**Definition 3.5:** The *pull back* $\varphi^* : \mathscr{T}^0_q(N) \to \mathscr{T}^0_q(M)$ is defined by

$$(\varphi^* t)_x(u_1, \ldots, u_q) = t_{\varphi(x)}(T_x \varphi \cdot u_1, \ldots, T_x \varphi \cdot u_q)$$

for $t \in \mathscr{T}^0_q(N)$ and for arbitrary $x \in M$ and $u_1, \ldots, u_q \in T_x(M)$.

**Remarks:**

(i) If $q = 0$, we define $\varphi^* : \mathscr{F}(N) \to \mathscr{F}(M)$ by $\varphi^*(f) = f \circ \varphi,\ f \in \mathscr{F}(N)$.

(ii) If $q = 1$, we have

$$(\varphi^* \omega)(x) = (T_x \varphi)^* \omega(\varphi(x))$$

for $\omega \in \mathcal{T}_1^0(N)$ and $x \in M$. Here, $(T_x \varphi)^*$ denotes the linear transformation dual to $T_x \varphi$.

**Theorem 3.5:**

$$\varphi^*: \bigoplus_{q=0}^{\infty} \mathcal{T}_q^0(N) \to \bigoplus_{q=0}^{\infty} \mathcal{T}_q^0(M)$$

is an $\mathbb{R}$-algebra homomorphism (of the covariant tensor algebras).

**Definition 3.6:** The vector fields $X \in \mathcal{X}(M)$ and $Y \in \mathcal{X}(N)$ are said to be *$\varphi$-related* if

$$T_x \varphi \cdot X_x = Y_{\varphi(x)} \quad \text{for every} \quad x \in M.$$

**Theorem 3.6:** Let $X_1, X_2 \in \mathcal{X}(M)$ and $Y_1, Y_2 \in \mathcal{X}(N)$. If $X_i$ and $Y_i$ are $\varphi$-related (for $i = 1, 2$) then so are $[X_1, X_2]$ and $[Y_1, Y_2]$.

In addition we have

**Theorem 3.7:** Let $t \in \mathcal{T}_q^0(N)$, $X_i \in \mathcal{X}(M)$, $Y_i \in \mathcal{X}(N)$, $i = 1, \ldots, q$. If the $X_i$ and $Y_i$ are $\varphi$-related then

$$(\varphi^* t)(X_1, \ldots, X_q) = \varphi^*(t(Y_1, \ldots, Y_q)).$$

Now let us consider in particular diffeomorphisms. First recall from linear algebra that if $E$ and $F$ are two vector spaces, and $E_s^r$, $F_s^r$ are the vector spaces of tensors of type $(r, s)$ on these spaces, and if the mapping $A: E \to F$ is an isomorphism, then it induces an isomorphism

$$A_s^r: E_s^r \to F_s^r,$$

which is defined as follows: Let $t \in E_s^r$; for arbitrary $y_1^*, \ldots, y_r^* \in F^*$ and $y_1, \ldots, y_s \in F$,

$$(A_s^r t)(y_1^*, \ldots, y_r^*, y_1, \ldots, y_s) := t(A^* y_1^*, \ldots, A^* y_r, A^{-1} y, \ldots, A^{-1} y_s).$$

Note that $A_0^1 = A$ and $A_1^0 = (A^{-1})^*$. In addition, we set $A_0^0 = \mathrm{Id}_{\mathbb{R}}$. Now let $\varphi: M \to N$ be a diffeomorphism. We define two induced mappings

$$\varphi_*: \mathcal{T}_s^r(M) \to \mathcal{T}_s^r(N) \quad \text{and}$$
$$\varphi^*: \mathcal{T}_s^r(N) \to \mathcal{T}_s^r(M),$$

which are each other's inverses, by $[T_s^r \varphi := (T\varphi)_s^r]$:

$$(\varphi_* t)_{\varphi(p)} = T_s^r \varphi(t_p) \qquad \text{for} \quad t \in \mathcal{T}_s^r(M)$$
$$(\varphi^* t)_p \quad = T_s^r \varphi^{-1}(t_{\varphi(p)}) \quad \text{for} \quad t \in \mathcal{T}_s^r(N).$$

It follows immediately that $X \in \mathcal{X}(M)$ and $\varphi_* X \in \mathcal{X}(N)$ are $\varphi$-related.

The linear extension of $\varphi_*$ and $\varphi^*$ to the complete tensor algebras results in two $\mathbb{R}$-*algebra isomorphisms* which are mutually inverse and which we also denote by $\varphi_*$, resp. $\varphi^*$.

If $\varphi^* t = t$ for some tensor field $t$, we say that $t$ is *invariant* unter the diffeomorphism $\varphi$. If $\phi_t^* s = s$ for every $t$ of a one parameter group, we say that $s$ is invariant under the one parameter group $(\phi_t)_{t \in \mathbb{R}}$.

---

**Exercise:** Give explicit representations for $\varphi_*$ and $\varphi^*$ in terms of coordinates.

---

## 3.3 The Lie Derivative

Let $X \in \mathscr{X}(M)$ and let $\phi_t$ be the flow of $X$.

**Definition 3.7:** We set

$$L_X T := \frac{d}{dt} \phi_t^* T \bigg|_{t=0} = \lim_{t \to 0} t^{-1} (\phi_t^* T - T),$$

where $T \in \mathscr{T}(M)$ is an arbitrary element of the algebra of tensor fields on $M$. $L_X T$ is called the *Lie derivative* of $T$ with respect to $X$.

**Theorem 3.8:** The Lie derivative has the following properties:

(i) $L_X$ is $\mathbb{R}$-linear, i.e. $L_X(T_1 + T_2) = L_X T_1 + L_X T_2$;
(ii) $L_X(T \otimes S) = (L_X T) \otimes S + T \otimes (L_X S)$ (Leibniz rule);
(iii) $L_X(\mathscr{T}_s^r(M)) \subseteq \mathscr{T}_s^r(M)$;
(iv) $L_X$ commutes with contractions;
(v) $L_X f = Xf = \langle df, X \rangle$, if $f \in \mathscr{F}(M) = \mathscr{T}_0^0(M)$;
(vi) $L_X Y = [X, Y]$, if $Y \in \mathscr{X}(M)$.

**Theorem 3.9:** If $X$, $Y \in \mathscr{X}(M)$ and $\lambda \in \mathbb{R}$, then

(i) $L_{X+Y} = L_X + L_Y$, $L_{\lambda X} = \lambda L_X$;
(ii) $L_{[X, Y]} = [L_X, L_Y] = L_X \circ L_Y - L_Y \circ L_X$.

**Theorem 3.10:** If $X$, $Y \in \mathscr{X}(M)$ and $\phi$, $\psi$ are the flows of $X$ and $Y$, respectively, then the following statements are equivalent:

(i) $[X, Y] = 0$
(ii) $L_X \circ L_Y = L_Y \circ L_X$
(iii) $\phi_s \circ \psi_t = \psi_t \circ \phi_s$ for all $s$ and $t$ (where both sides are defined).

**Theorem 3.11:** Let $\phi_t$ be a one-parameter transformation group and let $X$ be the corresponding vector field ("infinitesimal transformation"). A tensor field on $M$ is invariant under $\phi_t$ (i.e. $\phi_t^* T = T$ for all $t$) if and only if $L_X T = 0$.

**Theorem 3.12:** If $T \in \mathscr{T}_q^0(M)$ is a covariant tensor field and $X_1, \ldots, X_q \in \mathscr{X}(M)$, then

$$(L_X T)(X_1, \ldots, X_q) = X(T(X_1, \ldots, X_q)) - \sum_{k=1}^{q} T(X_1, \ldots [X, X_k], \ldots, X_q). \tag{3.1}$$

Proofs of these theorems can be found in Sect. I.3 of [5]. As an example, we prove Theorem 3.12 here: Consider

$$L_X(T \otimes X_1 \otimes \ldots \otimes X_q) = (L_X T) \otimes X_1 \otimes \ldots \otimes X_q + T \otimes L_X X_1 \otimes X_2 \otimes \ldots \otimes X_q + \ldots + T \otimes X_1 \otimes \ldots \otimes L_X X_q$$

and take the complete contraction (in all indices). Since $L_X$ commutes with contractions, one obtains

$$L_X(T(X_1, \ldots, X_q)) = (L_X T)(X_1, \ldots, X_q) + T(\underbrace{L_X X_1}_{[X, X_1]}, X_2, \ldots, X_q) + \ldots + T(X_1, \ldots, \underbrace{L_X X_q}_{[X, X_q]}),$$

which is identical to (3.1).

**Example:** Let $\omega \in \mathscr{X}^*(M)$ and $Y \in \mathscr{X}(M)$. Then

$$(L_X \omega)(Y) = X \langle \omega, Y \rangle - \omega([X, Y]). \tag{3.2}$$

As an application, we show that

$$L_X df = d L_X f \quad \text{for} \quad f \in \mathscr{F}(M). \tag{3.3}$$

*Proof*: According to (3.2)

$$\langle L_X df, Y \rangle = X \langle df, Y \rangle - df([X, Y]) = XYf - [X, Y]f = YXf = YL_X f = \langle dL_X f, Y \rangle.$$

*Local Coordinate Expressions for Lie Derivatives*

We choose the coordinates $\{x^i\}$ and the dual bases $\left\{\partial_i := \frac{\partial}{\partial x^i}\right\}$ and $\{dx^i\}$.

First of all, we note that

$$L_X dx^i = dL_X x^i = X^i_{,j} dx^j \tag{3.4}$$

and

$$L_X \partial_i = [X, \partial_i] = -X^j_{,i} \partial_j. \tag{3.5}$$

Now let $T \in \mathscr{T}_s^r(M)$. The determination of the components of $L_X T$ is analogous to the proof of Theorem 3.12. For $\omega^k = dx^{i_k}$ and $Y_l = \partial_{i_l}$, we form

$$L_X(T \otimes \omega^1 \otimes \ldots \otimes \omega^r \otimes Y_1 \otimes \ldots \otimes Y_s),$$

apply the product rule, and then contract completely. The result is

$$\begin{aligned}&L_X[T(dx^{i_1}, \ldots, dx^{i_r}, \partial_{j_1}, \ldots, \partial_{j_s})]\\&= (L_X T)(dx^{i_1}, \ldots, \partial_{j_s}) + T(L_X\, dx^{i_1}, \ldots, \partial_{j_s}) + \ldots + T(dx^{i_1}, \ldots, [X, \partial_{j_s}]).\end{aligned}$$

The left hand side is equal to $L_X T^{i_1 \ldots i_r}_{j_1 \ldots j_s}$.

Inserting (3.4) and (3.5) yields

$$\begin{aligned}(L_X T)^{i_1 \ldots i_r}_{j_1 \ldots j_s} = {} & X^i\, T^{i_1 \ldots i_r}_{j_1 \ldots j_s, i}\\&- T^{k i_2 \ldots i_r}_{j_1 \ldots j_s} \cdot X^{i_1}_{,k} - \text{all upper indices}\\&+ T^{i_1 \ldots i_r}_{k j_2 \ldots j_s} \cdot X^k_{,j_1} + \text{all lower indices}.\end{aligned} \tag{3.6}$$

In particular, if $\omega \in \mathscr{X}^*(M)$, then

$$(L_X \omega)_i = X^k\, \omega_{i,k} + \omega_k\, X^k_{,i}.$$

# 4. Differential Forms

Cartan's calculus of differential forms is particularly useful in general relativity. We begin our discussion with some algebraic preliminaries.

## 4.1 Exterior Algebra

Let $A$ be a commutative, associative, unitary $\mathbb{R}$-algebra and let $E$ be an $A$-module. In the following we have either $A=\mathbb{R}$ and $E$ a finite dimensional real vector space or $A=\mathscr{F}(M)$ and $E=\mathscr{X}(M)$ for some differentiable manifold $M$.

We consider the space $T_p(E)$ of $A$-valued, $p$-multilinear forms on $E$ and the subspace $\Lambda_p(E)$ of completely antisymmetric multilinear forms. In particular,

$$\Lambda_0(E) := T_0(E) := A$$

$$\Lambda_1(E) = T_1(E) = E^* \quad \text{(dual space of } E).$$

The elements of $\Lambda_p(E)$ are called *(exterior) forms of degree p*. We define the *alternation operator* on $T_p(E)$ by

$$(\mathscr{A}T)(v_1, \ldots, v_p) := \frac{1}{p!} \sum_{\sigma \in \mathscr{S}_p} (\operatorname{sgn} \sigma)\, T(v_{\sigma(1)}, \ldots, v_{\sigma(p)}) \quad \text{for} \quad T \in T_p(E), \tag{4.1}$$

where the sum in (4.1) extends over the permutation group $\mathscr{S}_p$ of $p$ objects; sgn $\sigma$ denotes the signature of the permutation $\sigma \in \mathscr{S}_p$. The following statements obviously hold:

(i) $\mathscr{A}$ is a linear mapping from $T_p(E)$ onto $\Lambda_p(E)$: $\mathscr{A}(T_p(E)) = \Lambda_p(E)$.
(ii) $\mathscr{A} \circ \mathscr{A} = \mathscr{A}$.

In particular, $\mathscr{A}\omega = \omega$ for any $\omega \in \Lambda_p(E)$.

Let $\omega \in \Lambda_p(E)$ and $\eta \in \Lambda_q(E)$. We define the *exterior product* by

$$\omega \wedge \eta := \frac{(p+q)!}{p!\,q!} \mathscr{A}(\omega \otimes \eta). \tag{4.2}$$

The exterior product has the following properties:

(i) $\wedge$ is $A$-bilinear.
(ii) $\omega \wedge \eta = (-1)^{pq} \eta \wedge \omega$.
(iii) $\wedge$ is associative: $(\omega_1 \wedge \omega_2) \wedge \omega_3 = \omega_1 \wedge (\omega_2 \wedge \omega_3)$. (4.3)

**Theorem 4.1:** Let $\theta^i$, $i = 1, 2, \ldots, n < \infty$ be a basis for $E^*$. Then the set

$$\theta^{i_1} \wedge \theta^{i_2} \wedge \ldots \wedge \theta^{i_p}, \qquad 1 \leqq i_1 < i_2 < \ldots < i_p \leqq n \tag{4.4}$$

is a basis for the space $\Lambda_p(E)$, $p \leqq n$, which has the dimension

$$\binom{n}{p} := \frac{n!}{p!\,(n-p)!}, \tag{4.5}$$

if $p > n$, $\Lambda_p(E) = \{0\}$.

The *Grassman algebra* (or *exterior algebra*) $\Lambda(E)$ is defined as the direct sum

$$\Lambda(E) := \bigoplus_{p=0}^{n} \Lambda_p(E),$$

where the exterior product is extended in a bilinear manner to the entire $\Lambda(E)$. Thus $\Lambda(E)$ is a graded, associative, unitary $A$-algebra.

*Interior Product*

For every $p \in \mathbb{N}_0$ we define the mapping $E \times \Lambda_p(E) \to \Lambda_{p-1}(E)$ by $(v, \omega) \mapsto i_v\,\omega$, where

$$(i_v\,\omega)(v_1, \ldots, v_{p-1}) = \omega(v, v_1, \ldots, v_{p-1}), \quad i_v\,\omega = 0 \quad \text{for} \quad \omega \in \Lambda_0(E). \tag{4.6}$$

The association $(v, \omega) \mapsto i_v\,\omega$ is called the *interior product* of $v$ and $\omega$. For every $p$, the interior product is an $A$-bilinear mapping and can thus be uniquely extended to an $A$-bilinear mapping

$$E \times \Lambda(E) \to \Lambda(E), \qquad (v, \omega) \mapsto i_v\,\omega.$$

For a proof of the following theorem and other unproved statements in this section see, for example, [6], Chap. III.

**Theorem 4.2:** For every fixed $v \in E$, the mapping $i_v\colon \Lambda(E) \to \Lambda(E)$ has the properties:

(i) $i_v$ is $A$-linear.
(ii) $i_v(\Lambda_p(E)) \subseteq \Lambda_{p-1}(E)$.
(iii) $i_v(\alpha \wedge \beta) = (i_v\,\alpha) \wedge \beta + (-1)^p\,\alpha \wedge i_v\,\beta$ for $\alpha \in \Lambda_p(E)$.

## 4.2 Exterior Differential Forms

Let $M$ be an $n$-dimensional $C^\infty$-manifold. For every $p = 0, 1, \ldots, n$ and every $x \in M$, we construct the spaces

$$\Lambda_p(T_x(M)) \subset T_x(M)_p^0 .$$

In particular,

$$\Lambda_0(T_x(M)) = \mathbb{R},$$
$$\Lambda_1(T_x(M)) = T_x^*(M) .$$

Furthermore,

$$\Lambda(T_x(M)) = \bigoplus_{p=0}^{n} \Lambda_p(T_x(M))$$

is the exterior algebra (of dimension $2^n$) over $T_x(M)$.

**Definition 4.1:** An *(exterior) differential form of degree p (differential p-form) on M* is a differentiable tensor field of rank $p$ which is an element of $\Lambda_p(T_x(M))$ for every $x \in M$. All statements made in Sect. 2 about tensor fields are of course also valid for differential forms.

We denote by $\Lambda_p(M)$ the $\mathscr{F}(M)$-module of $p$-forms. In addition, $\Lambda(M) = \bigoplus_{p=0}^{n} \Lambda_p(M)$ is the *exterior algebra of differential forms on M.*

In $\Lambda(M)$ the algebraic operations, in particular the exterior product, are defined pointwise.

As in Sect. 2.3, we can assign to $\omega \in \Lambda_p(M)$ and vector fields $X_1, \ldots, X_p$ the function

$$F(x) = \omega_x(X_1(x), \ldots, X_p(x)) .$$

This function will be denoted by $\omega(X_1, \ldots, X_p)$. The assignment $(X_1, \ldots, X_p) \mapsto \omega(X_1, \ldots, X_p)$ is $\mathscr{F}(M)$-multilinear and completely antisymmetric. That is, for a given element $\omega \in \Lambda_p(M)$, there is an element of the exterior algebra over the $\mathscr{F}(M)$-module $\mathscr{X}(M)$. We denote the latter by $\Lambda(\mathscr{X}(M))$. One can show that this correspondence is an isomorphism which preserves all algebraic structures. In addition, we have for the interior product

$$(i_X \omega)(x) = i_{X(x)} \omega(x), \qquad X \in \mathscr{X}(M), \quad \omega \in \Lambda(M), \quad x \in M.$$

If $(x^1, \ldots, x^n)$ is a local coordinate system in $U \subset M$, then $\omega \in \Lambda_p(M)$ can be expanded in the form

$$\begin{aligned}\omega &= \sum_{i_1 < \ldots < i_p} \omega_{i_1 \ldots i_p} dx^{i_1} \wedge dx^{i_2} \wedge \ldots \wedge dx^{i_p} \\ &= \frac{1}{p!} \sum_{i_1, \ldots, i_p} \omega_{i_1 \ldots i_p} dx^{i_1} \wedge \ldots \wedge dx^{i_p} \qquad (4.7)\end{aligned}$$

on $U$. Here $\{\omega_{i_1 \ldots i_p}\}$ are the *components* of $\omega$. In the second line of (4.7) they are totally antisymmetric in their indices.

*Differential Forms and Mappings*

Let $\varphi: M \to N$ be a morphism of manifolds.

The induced mapping $\varphi^*$ in the space of covariant tensors (see Sect. 3) defines a mapping

$$\varphi^*: \Lambda(N) \to \Lambda(M),$$

(we use the same symbol). Using Theorem 3.5, it is easy to see that

$$\varphi^*(\omega \wedge \eta) = \varphi^* \omega \wedge \varphi^* \eta; \qquad \omega, \eta \in \Lambda(N).$$

Thus, $\varphi^*$ is an $\mathbb{R}$-algebra homomorphism. If $\varphi$ is a diffeomorphism, then $\varphi^*$ is an $\mathbb{R}$-algebra isomorphism.

## 4.3 Derivations and Antiderivations

Let $M$ be an $n$-dimensional differentiable manifold and

$$\Lambda(M) = \bigoplus_{p=0}^{n} \Lambda_p(M)$$

be the graded $\mathbb{R}$-algebra of exterior differential forms on $M$.

**Definition 4.2:** A mapping $\theta: \Lambda(M) \to \Lambda(M)$ is called a *derivation (antiderivation) of degree* $k \in \mathbb{Z}$, provided

(i) $\theta$ is $\mathbb{R}$-linear

(ii) $\theta(\Lambda_p(M)) \subset \Lambda_{p+k}(M)$ for $p = 0, 1, \ldots, n$

(iii) $\theta(\alpha \wedge \beta) = \theta\alpha \wedge \beta + \alpha \wedge \theta\beta, \quad \alpha, \beta \in \Lambda(M)$ (Leibniz rule)

$[\theta(\alpha \wedge \beta) = \theta\alpha \wedge \beta + (-1)^p \alpha \wedge \theta\beta, \quad \alpha \in \Lambda_p(M), \quad \beta \in \Lambda(M)$ ("anti-Leibniz rule")].

**Theorem 4.3:** If $\theta$ and $\theta'$ are antiderivations of degree $k$ and $k'$ on $\Lambda(M)$, then $\theta \circ \theta' + \theta' \circ \theta$ is a derivation of degree $k + k'$.

*Proof:* Direct verification.

**Theorem 4.4:** Every derivation (antiderivation) on $\Lambda(M)$ is local. This means that for every $\alpha \in \Lambda(M)$ and every open submanifold $U$ of $M$,

$$\alpha | U = 0 \quad \text{implies} \quad \theta\alpha | U = 0.$$

*Proof:* Let $x \in U$. There exists[1] a function $h \in \mathscr{F}(M)$, such that $h(x) = 1$ and $h = 0$ on $M \backslash U$. As a consequence, $h\alpha = 0$ and hence also

[1] A proof of the existence of such a function can be found, for example, in [6], p. 69.

$\theta(h\alpha) = 0$ so that $\theta(h) \wedge \alpha + h\,\theta(\alpha) = 0$. If we evaluate this equation at the point $x$, we obtain $\theta(\alpha)(x) = 0$.

**Corollary:** Let $\alpha$ and $\alpha' \in \Lambda(M)$. If $\alpha | U = \alpha' | U$, then

$$\theta(\alpha) | U = \theta(\alpha') | U .$$

Thus we can define the restriction $\theta | U$ on $\Lambda(U)$. For this, choose for $p \in U$ and $\alpha \in \Lambda(U)$ an $\tilde{\alpha} \in \Lambda(M)$ such that $\alpha = \tilde{\alpha}$ for some neighborhood of $p$, and set

$$(\theta | U)(\alpha)(p) = (\theta\,\tilde{\alpha})(p) .$$

According to the corollary, this definition is independent of the extension $\tilde{\alpha}$. The existence of an extension is based on the following

**Lemma:** Let $U$ be an open submanifold of $M$ and let $K$ be a compact subset of $U$. For every $\beta \in \Lambda(U)$ there is an $\alpha \in \Lambda(M)$ such that $\beta | K = \alpha | K$ and $\alpha | (M \setminus U) = 0$.
*Proof:* There exists a function $h \in \mathscr{F}(M)$ such that $h(x) = 1$ for all $x \in K$ and $h = 0$ on $M \setminus U$ (i.e., supp $h \subseteq U$)[2]. With the aid of this function, define $\alpha \in \Lambda(M)$ by

$$\alpha(x) = \begin{cases} h(x)\,\beta(x), & x \in U \\ 0 & x \notin U . \end{cases}$$

This differential form has the desired properties.

We thus have the following

**Theorem 4.5:** Let $\theta$ be a derivation (antiderivation) on $\Lambda(M)$ and let $U$ be an open submanifold of $M$. There exists a unique derivation (antiderivation) $\theta_U$ on $\Lambda(U)$ such that for every $\alpha \in \Lambda(M)$,

$$\theta_U(\alpha | U) = (\theta\,\alpha) | U .$$

$\theta_U$ is called the derivation (antiderivation) on $\Lambda(U)$ *induced* by $\theta$.

In addition to this localization theorem, we need a globalizing theorem.

**Theorem 4.6:** Let $(U_i)_{i \in I}$ be an open covering of $M$. For every $i$, let $\theta_i$ be a derivation (antiderivation) on $\Lambda(U_i)$; we denote by $\theta_{ij}$ the derivation (antiderivation) on $\Lambda(U_i \cap U_j)$ induced by $\theta_i$ on $U_i \cap U_j$. For every $(i, j) \in I \times I$ let $\theta_{ij} = \theta_{ji}$. Then there exists precisely one derivation (antiderivation) $\theta$ on $\Lambda(M)$ such that $\theta | U_i = \theta_i$ for every $i \in I$.
*Proof:* For each $\alpha \in \Lambda(M)$ and $i \in I$, define

$$(\theta\,\alpha) | U_i = \theta_i(\alpha | U_i) .$$

---

[2] For a proof, see [6], p. 92.

Since

$$((\theta\,\alpha)\,|\,U_i)\,|\,U_j=\theta_{ij}(\alpha\,|\,U_i\cap U_j)=\theta_{ji}(\alpha\,|\,U_i\cap U_j)=((\theta\,\alpha)\,|\,U_j)\,|\,U_i\,,$$

for all $(i,j)\in I\times I$, $\theta\alpha\in\Lambda(M)$ is well defined and $\theta\colon\alpha\mapsto\theta\alpha$ is a derivation (antiderivation) which induces the mapping $\theta_i$ on every $U_i$ by construction.

Later on, the following theorem will also be useful:

**Theorem 4.7:** Let $\theta$ be a derivation (antiderivation) of degree $k$ on $\Lambda(M)$. If $\theta f=0$ and $\theta(df)=0$ for every $f\in\mathscr{F}(M)$, then $\theta\alpha=0$ for every $\alpha\in\Lambda(M)$.
*Proof:* Let $(h_i,U_i)_{i\in I}$ be an atlas of $M$ and let $\theta_i=\theta\,|\,U_i$ (as in Theorem 4.5). Each $\theta_i$ satisfies the premises of the theorem, i.e. $\theta_i f_i=0$ and $\theta_i\,df_i=0$ for every $f_i\in\mathscr{F}(U_i)$. Let $\alpha\in\Lambda_p(M)$ and let $i\in I$. $\alpha\,|\,U_i$ has the form

$$\alpha\,|\,U_i=\sum_{i_1<\ldots<i_p}\alpha_{i_1\ldots i_p}\,dx^{i_1}\wedge\ldots\wedge dx^{i_p}.$$

As a consequence, since $\theta$ is a derivation

$$\begin{aligned}(\theta\alpha)\,|\,U_i=\theta_i(\alpha\,|\,U_i)=\sum_{i_1<\ldots<i_p}\Big[&\theta_i\,\alpha_{i_1\ldots i_p}\wedge dx^{i_1}\wedge\ldots\wedge dx^{i_p}\\ &+\alpha_{i_1\ldots i_p}\sum_{k=1}^{p}dx^{i_1}\wedge\ldots\wedge\theta_i\,dx^{i_k}\wedge\ldots\wedge dx^{i_p}\Big]=0\end{aligned}$$

and similarly if $\theta$ is an antiderivation.

**Corollary:** A derivation (antiderivation) $\theta$ on $\Lambda(M)$ is uniquely determined by its value on $\Lambda_0(M)=\mathscr{F}(M)$ and on $\Lambda_1(M)$. In fact, its values on $\mathscr{F}(M)$ and all differentials of $\mathscr{F}(M)$ are already sufficient.

## 4.4 The Exterior Derivative

**Theorem 4.8:** There exists precisely one operator

$$d\colon\Lambda(M)\to\Lambda(M)$$

with the following properties:

(i) $d$ is an antiderivation of degree 1 on $\Lambda(M)$,
(ii) $d\circ d=0$,
(iii) $d$ is the differential of $f$ for every $f\in\mathscr{F}(M)$, i.e.

$$\langle df,X\rangle=Xf\quad\text{for all}\quad f\in\mathscr{F}(M),\ X\in\mathscr{X}(M).$$

$d$ is called the *exterior derivative.*

*Proof:* 1) The uniqueness of $d$ is a consequence of Theorem 4.7. 2) Let $h\colon U\to U'$ be a chart of $M$. According to Theorem 4.5, the exterior derivative (if it exists) induces an exterior derivative on $U$, which is also denoted by $d$. For

$$\alpha = \sum \alpha_{i_1 \ldots i_p} dx^{i_1} \wedge \ldots \wedge dx^{i_p} \in \Lambda_p(U) \quad (i_1 < \ldots < i_p)$$

one has, according to (i) and (ii),

$$d\alpha = \sum d\alpha_{i_1 \ldots i_p} \wedge dx^{i_1} \wedge \ldots \wedge dx^{i_p} \in \Lambda_{p+1}(U)$$

or, by (iii)

$$d\alpha = \sum_{i_0 < i_1 < \ldots < i_p} \sum_{k=0}^{p} (-1)^k \frac{\partial}{\partial x^{i_k}} \alpha_{i_0 \ldots \hat{i}_k \ldots i_p} dx^{i_0} \wedge \ldots \wedge dx^{i_p}. \tag{4.8}$$

In other words, we must have

$$(d\alpha)_{i_0 \ldots i_p} = \sum_{k=0}^{p} (-1)^k \frac{\partial}{\partial x^{i_k}} \alpha_{i_0 \ldots \hat{i}_k \ldots i_p} \qquad (i_0 < i_1 < \ldots < i_p). \tag{4.9}$$

3) For every $p$, $0 \leqq p \leqq n$ and every $\alpha \in \Lambda_p(U)$, define $d\alpha$ by the previous equation. By straightforward computation, one can show that $d\colon \Lambda(U) \to \Lambda(U)$ has the properties (i), (ii), (iii) of the theorem (exercise). Thus $d$ is an exterior derivative on $U$.
4) Let $(h_i, U_i)_{i \in I}$ be an atlas of $M$. For every $i \in I$, let $d_i\colon \Lambda(U_i) \to \Lambda(U_i)$ be the exterior derivative on $U_i$. For $(i, j) \in I \times I$, $d_i$ and $d_j$ each induce an exterior derivative $d_{ij}$ or $d_{ji}$, respectively, on $U_i \cap U_j$. According to (1), $d_{ij} = d_{ji}$. By Theorem 4.6, there exists a unique antiderivation $d$ on $\Lambda(M)$ which induces for every $i \in I$ the exterior derivative $d_i$ on $U_i$. Obviously $d \circ d = 0$ and $df$ is the differential of $f$ for every $f \in \mathscr{F}(M)$.

**Definition 4.3:** A differential form $\alpha$ such that $d\alpha = 0$ is called a *closed form*. A differential form $\alpha$ such that $\alpha = d\beta$ for some form $\beta$ is called an *exact form*. Since $d \circ d = 0$, every exact differential form $\alpha = d\beta$ is closed: $d\alpha = d(d\beta) = 0$. The converse is valid locally:

**Poincaré Lemma:** Let $\alpha$ be a closed form on $M$. For every $x \in M$, there is an open neighborhood $U$ of $x$ such that $\alpha \mid U$ is exact.

For a proof, see [6], p. 151.

*Morphisms and Exterior Derivatives*

Let $\varphi\colon M \to N$ be a morphism of manifolds. The following diagram is commutative:

$$\begin{array}{ccc} \Lambda(M) & \overset{\varphi^*}{\leftarrow} & \Lambda(N) \\ {\scriptstyle d}\downarrow & & \downarrow{\scriptstyle d} \\ \Lambda(M) & \overset{\varphi^*}{\leftarrow} & \Lambda(N) \end{array}$$

so that

$$d \circ \varphi^* = \varphi^* \circ d. \tag{4.10}$$

*Proof:* This has already been shown for functions in Sect. 2.3 and hence we have

$$(d \circ \varphi^*)\, df = d \circ d(\varphi^* f) = 0 = \varphi^* \circ d(df).$$

In other words, (4.10) is also true for differentials of functions. Now both sides of (4.10) are antiderivations on $\Lambda(N)$. By Theorem 4.7, they must be equal.

## 4.5 Relations Among the Operators $d$, $i_X$ and $L_X$

There are several important relations between the derivation $L_X$, the antiderivation $i_X$ and the exterior derivative $d$. The most important of these is

$$L_X = d \circ i_X + i_X \circ d. \tag{4.11}$$

*Proof:* 1) The operation $\theta = d \circ i_X + i_X \circ d \colon \Lambda(M) \to \Lambda(M)$ is a derivation of degree 0. For $f \in \mathscr{F}(M)$, one has

$$\theta f = d(i_X f) + i_X(df) = i_X\, df = \langle df, X \rangle = Xf$$

and

$$\theta\, df = d(i_X\, df) + i_X\, d(df) = d(i_X\, df) = d(Xf).$$

2) The Lie derivative with respect to $X$, $L_X \colon \Lambda(M) \to \Lambda(M)$ is a derivation of degree 0 and one has (see Sect. 3)

$$\begin{aligned} L_X f &= Xf \\ (L_X\, df)(Y) &= L_X(df(Y)) - df(L_X Y) \\ &= L_X(Yf) - df([X, Y]) \\ &= X(Yf) - [X, Y]f = YXf \\ &= \langle d(Xf), Y \rangle, \end{aligned}$$

which shows that

$$L_X\, df = d(Xf).$$

3) From (1) and (2) we obtain, using Theorem 4.7,

$$L_X = \theta.$$

**Corollary:** It follows from (4.11) that

$$dL_X = L_X\, d. \tag{4.12}$$

---

**Exercise:** In a similar manner, prove that

$$i_{[X, Y]} = [L_X, i_Y] \quad \text{for} \quad X, Y \in \mathscr{X}(M). \tag{4.13}$$

---

As an application of (4.11) we prove the following expression for the exterior derivative: if $\omega \in \Lambda_p(M)$, then

$$d\omega(X_1, \ldots, X_{p+1}) = \sum_{1 \leqq i \leqq p+1} (-1)^{i+1} X_i\, \omega(X_1, \ldots, \hat{X}_i, \ldots, X_{p+1}) + \sum_{i<j} (-1)^{i+j} \omega([X_i, X_j], X_1, \ldots, \hat{X}_i, \ldots, \hat{X}_j, \ldots, X_{p+1}). \tag{4.14}$$

*Proof:* 1) If $p = 0$, (4.14) states that

$$df(X) = Xf, \quad f \in \mathscr{F}(M),$$

which is obviously correct.
2) If $\omega \in \Lambda_1(M)$, we obtain from (4.11)

$$\begin{aligned}(L_X \omega)(Y) &= i_X\, d\omega(Y) + d(i_X \omega)(Y)\\ &= i_X\, d\omega(Y) + d(\omega(X))(Y)\\ &= d\omega(X, Y) + Y\omega(X).\end{aligned}$$

Using the results of Sect. 3 for $L_X \omega$, we find

$$d\omega(X, Y) = (L_X \omega)(Y) - Y\omega(X) = X\,\omega(Y) - Y\omega(X) - \omega([X, Y]),$$

which agrees with the right hand side of (4.14).
3) We now complete the proof by induction. Assume that (4.14) is correct for all differential forms having degree less than or equal to $p-1$. If $\omega \in \Lambda_p(M)$, use (4.11), the induction assumption for $i_X \omega$ and the explicit expression for $L_X$ in Sect. 3. After a short calculation, one shows that (4.14) is also true for differential forms of degree $p$.

*Covariant Derivative and Exterior Derivative*

Let $\omega \in \Lambda_p(M)$ and let $\nabla\omega$ be the covariant derivative of $\omega$ with respect to a symmetric affine connection (see Sect. 5). If we apply the alternation operator $\mathscr{A}$ to $\nabla\omega$, then

$$\mathscr{A}(\nabla\omega) = \frac{(-1)^p}{p+1}\, d\omega. \tag{4.15}$$

*Proof:* From Sect. 5, we have

$$(-1)^p\, \nabla\omega(X_1, \ldots, X_{p+1}) = X_1\, \omega(X_2, \ldots, X_{p+1}) - \sum \omega(X_2, \ldots, \nabla_{X_1} X_j, \ldots, X_{p+1}).$$

Hence

$$(-1)^p \mathscr{A}(\nabla\omega)(X_1, \ldots, X_{p+1}) = \frac{1}{p+1}\Big\{\sum (-1)^{i+1} X_i\, \omega(X_1, \ldots, \hat{X}_i, \ldots, X_{p+1}) + \sum_{i<j} (-1)^{i+j} \omega([\nabla_{X_i} X_j - \nabla_{X_j} X_i], X_1, \ldots, \hat{X}_i, \ldots, \hat{X}_j, \ldots, X_{p+1})\Big\}.$$

Since the torsion vanishes by assumption, we have

$$\nabla_{X_i} X_j - \nabla_{X_j} X_i = [X_i, X_j] .$$

Thus (4.14) for $d\omega$ shows that (4.15) is correct.

## 4.6 The *-Operation and the Codifferential

### 4.6.1 Oriented Manifolds

An atlas $\mathscr{A}$ of a differentiable manifold $M$ is *oriented* if for every pair of charts $(h, U)$ and $(k, V)$ in $\mathscr{A}$ the Jacobi matrix for the coordinate transformation $k \circ h^{-1}$ is positive.

There are manifolds (for example, the Möbius strip) which do not have an oriented atlas.

A manifold is *orientable* if it has an oriented atlas. Two atlases $\mathscr{A}_1$ and $\mathscr{A}_2$ have the same orientation if $\mathscr{A}_1 \cup \mathscr{A}_2$ is also an oriented atlas. This is an equivalence relation. An equivalence class of oriented atlases is called an *orientation of the manifold M*. An orientable manifold, together with a chosen orientation, is said to be an *oriented manifold.* A chart $(h, U)$ of $M$ is said to be positive (negative) if for every chart $(k, V)$ belonging to some oriented atlas which defines the orientation of $M$, the Jacobian of $k \circ h^{-1}$ is positive (negative). We state without proof

**Theorem 4.9:** Let $M$ be an $n$-dimensional, paracompact, orientable manifold. There exists on $M$ an $n$-form of class $C^\infty$ which does not vanish anywhere on $M$.
For a proof, see [6], p. 258.

An $n$-form on $M$ which does not vanish anywhere is called a *volume element* of $M$. We now consider a pseudo-Riemannian manifold $(M, g)$. Let $g_{ij}$ be the components of $g$ relative to the coordinate system $(x^1, \ldots, x^n)$. We set

$$|g| = |\det(g_{ij})| . \tag{4.16}$$

Let $\bar{g}_{ij}$ be the components of $g$ in terms of new coordinates $(y^1, \ldots, y^n)$ and let $|\bar{g}|$ be the absolute value of the corresponding determinant. From the transformation law

$$\bar{g}_{ij} = \frac{\partial x^k}{\partial y^i} \frac{\partial x^l}{\partial y^j} g_{kl} ,$$

it follws that

$$|\bar{g}| = \left[\det\left(\frac{\partial x^k}{\partial y^i}\right)\right]^2 |g| .$$

Now suppose that $M$ is oriented and that both coordinate systems are *positive*. We then have

$$\det(\partial x^k/\partial y^i) > 0$$

and

$$\sqrt{|\bar{g}|} = \sqrt{|g|}\,\det(\partial x^k/\partial y^i). \tag{4.17}$$

If, on the other hand, we consider an $n$-form $\omega$ with

$$\omega = a(x)\,dx^1 \wedge \ldots \wedge dx^n = b(y)\,dy^1 \wedge \ldots \wedge dy^n,$$

then

$$a = b\,\det(\partial y^i/\partial x^j). \tag{4.18}$$

Hence an $n$-form is defined on $M$ by

$$\eta := \sqrt{|g|}\,dx^1 \wedge dx^2 \wedge \ldots \wedge dx^n \tag{4.19}$$

for a positive coordinate system. According to (4.17) and (4.18), this definition is independent of the coordinate system. In addition, *$\eta$ is a volume element.*

### 4.6.2 The *-Operation

Let $(M, g)$ be an $n$-dimensional oriented pseudo-Riemannian manifold and let $\eta \in \Lambda_n(M)$ be the volume element (4.19) corresponding to $g$. We shall now use $\eta$ to associate with each form $\omega \in \Lambda_p(M)$ another form $*\omega \in \Lambda_{n-p}(M)$. We consider a positive local coordinate system $(x^1, \ldots, x^n)$ and write $\omega$ in the form

$$\omega = \frac{1}{p!}\,\omega_{i_1 \ldots i_p}\,dx^{i_1} \wedge \ldots \wedge dx^{i_p}. \tag{4.20}$$

($\omega_{i_1 \ldots i_p}$ is totally antisymmetric). In the same manner, we may write

$$\eta = \frac{1}{n!}\,\eta_{i_1 \ldots i_n}\,dx^{i_1} \wedge \ldots \wedge dx^{i_n}. \tag{4.21}$$

From (4.19) we have

$$\eta_{i_1 \ldots i_n} = \sqrt{|g|}\,\varepsilon_{i_1 \ldots i_n}, \tag{4.22}$$

where

$$\varepsilon_{i_1 \ldots i_n} = \begin{cases} +1 & \text{if } i_1, \ldots, i_n \text{ is an even permutation of } (1, \ldots, n) \\ -1 & \text{if } i_1, \ldots, i_n \text{ is an odd permutation of } (1, \ldots, n) \\ 0 & \text{otherwise.} \end{cases}$$

We now define

$$(*\omega)_{i_{p+1}\dots i_n} = \frac{1}{p!}\,\eta_{i_1\dots i_n}\,\omega^{i_1\dots i_p}, \tag{4.23}$$

where

$$\omega^{i_1\dots i_p} = g^{i_1 j_1}\dots g^{i_p j_p}\,\omega_{j_1\dots j_p}.$$

The definition (4.23) is obviously independent of the coordinate system.

In Exercise 1 below, we derive a coordinate-free expression for the *-operation.

Equation (4.23) is equivalent to

$$*\omega = \frac{1}{p!}\,\omega^{i_1\dots i_p}\,i(e_{i_p}) \circ \dots \circ i(e_{i_1})\,\eta, \tag{4.23$'$}$$

where $\{e_i\}$ is a local basis of vector fields and $i(e_j)$ denotes the interior products.

The correspondence $\omega \mapsto *\omega$ defines an isomorphism:

$$\Lambda_p(M) \to \Lambda_{n-p}(M).$$

A simple calculation shows that

$$*(*\omega) = (-1)^{p(n-p)}\,\mathrm{sgn}(g)\,\omega. \tag{4.24}$$

The contravariant components of $\eta$ are given by

$$\eta^{i_1\dots i_n} = \mathrm{sgn}(g)\frac{1}{\sqrt{|g|}}\,\varepsilon_{i_1\dots i_n}. \tag{4.25}$$

The following exercises contain further properties of the *-operation.

---

**Exercises:**

1) Let $\alpha, \beta \in \Lambda_p(M)$. Show that

$$\alpha \wedge *\beta = \beta \wedge *\alpha = (\alpha, \beta)\,\eta,$$

where $(\alpha, \beta) = \frac{1}{p!}\,\alpha_{i_1\dots i_p}\,\beta^{i_1\dots i_p}$ is the scalar product induced in $\Lambda_p(M)$.

*Solution:* Since $i(e_j)$ is an antiderivation, we can write

$$\begin{aligned}
\alpha \wedge *\beta &= \frac{1}{p!}\,\beta^{i_1\dots i_p}\,\alpha \wedge [i(e_{i_p}) \circ \dots \circ i(e_{i_1})\,\eta] \\
&= \frac{1}{p!}\,\beta^{i_1\dots i_p}\,[i(e_{i_1})\,\alpha] \wedge [i(e_{i_p}) \circ \dots \circ i(e_{i_2})\,\eta] \\
&= \frac{1}{p!}\,\beta^{i_1\dots i_p}\,[i(e_{i_p}) \circ \dots \circ i(e_{i_1})\,\alpha]\,\eta \\
&= \frac{1}{p!}\,\beta^{i_1\dots i_p}\,\alpha_{i_1\dots i_p}\,\eta\,.
\end{aligned}$$

2) Let $\alpha, \beta \in \Lambda_p(M)$. Show that

$$(*\alpha, *\beta) = \operatorname{sgn}(g)\,(\alpha, \beta).$$

*Solution:* From Exercise 1 and (4.24), we find

$$\begin{aligned}(*\alpha, *\beta)\,\eta &= *\alpha \wedge **\beta = (-1)^{p(n-p)} \operatorname{sgn}(g)\, *\alpha \wedge \beta \\ &= \operatorname{sgn}(g)\, \beta \wedge *\alpha = \operatorname{sgn}(g)\,(\alpha, \beta)\,\eta\,.\end{aligned}$$

3) Let $\alpha \in \Lambda_p(M)$ and $\beta \in \Lambda_{n-p}(M)$. Show that

$$(\alpha \wedge \beta, \eta) = (*\alpha, \beta)\,.$$

*Solution:* There is a unique $p$-form $\varphi$ such that $\beta = *\varphi$. The results of the previous exercises imply

$$(\alpha \wedge \beta, \eta) = (\alpha \wedge *\varphi, \eta) = \operatorname{sgn}(g)\,(\alpha, \varphi) = (*\alpha, *\varphi) = (*\alpha, \beta)\,.$$

4) If $\beta'$ is an $(n-p)$-form, and if for all $\alpha \in \Lambda_{n-p}(M)$ there is a $\beta \in \Lambda_p(M)$ such that

$$(\beta', \alpha) = (\beta \wedge \alpha, \eta)\,,$$

then $\beta' = *\beta$.

*Solution:* According to the hypothesis and Exercise 3, we have $(*\beta - \beta', \alpha) = 0$ for all $\alpha \in \Lambda_{n-p}(M)$. Hence $\beta' = *\beta$, since the scalar product is nondegenerate.

5) Let $\theta^1, \ldots, \theta^n$ be an oriented basis of one-forms on an open subset $U \subset M$. Show that $\eta$ can be represented in the form

$$\eta = \sqrt{|g|}\,\theta^1 \wedge \ldots \wedge \theta^n\,,$$

where $|g| = |\det g(e_i, e_k)|$, $\{e_i\}$ dual basis of $\{\theta^i\}$.

*Solution:* We set $\theta^i = a^i_j\, dx^j$, where $\{x^j\}$ defines a positive coordinate system. Then

$$\begin{aligned}\sqrt{|g|}\,\theta^1 \wedge \ldots \wedge \theta^n &= \sqrt{|g|}\, a^1_{j_1} \ldots a^n_{j_n}\, dx^{j_1} \wedge \ldots \wedge dx^{j_n} \\ &= \sqrt{|g|}\, \varepsilon_{j_1 \ldots j_n} a^1_{j_1} \ldots a^n_{j_n}\, dx^1 \wedge dx^2 \wedge \ldots \wedge dx^n \\ &= \sqrt{|g|}\, \det(a^i_j)\, dx^1 \wedge \ldots \wedge dx^n\,;\end{aligned}$$

note that $\det(a^i_j) > 0$. Now

$$\frac{\partial}{\partial x^j} = a^i_j\, e_i\,, \quad (\partial/\partial x^i, \partial/\partial x^j) = a^k_i\, a^l_j (e_k, e_l) = a^k_i\, a^l_j\, g_{kl}\,.$$

Therefore

$$\left|\det\left(\frac{\partial}{\partial x^i}, \frac{\partial}{\partial x^j}\right)\right|^{1/2} = \sqrt{|g|}\, \det(a^i_j)\,.$$

If we insert this into the above, we obtain the conclusion.

We observe also that $g^{ij} = (\theta^i, \theta^j)$ is the matrix inverse to $g_{ij} = (e_i, e_j)$.

6) Prove that, with the notation from Exercise 5,

$$*(\theta^{i_1} \wedge \ldots \wedge \theta^{i_p}) = \frac{\sqrt{|g|}}{(n-p)!} \varepsilon_{j_1 \ldots j_n} g^{j_1 i_1} \ldots g^{j_p i_p} \theta^{j_{p+1}} \wedge \ldots \wedge \theta^{j_n}.$$

*Solution:* According to the result of Exercise 4, it is sufficient to show that

$$\sqrt{|g|} \frac{1}{(n-p)!} \varepsilon_{j_1 \ldots j_n} g^{j_1 i_1} \ldots g^{j_p i_p} (\theta^{j_{p+1}} \wedge \ldots \wedge \theta^{j_n}, \theta^{i_{p+1}} \wedge \ldots \wedge \theta^{i_n})$$
$$= (\theta^{i_1} \wedge \ldots \wedge \theta^{i_p} \wedge \theta^{i_{p+1}} \wedge \ldots \wedge \theta^{i_n}, \eta).$$

Now, according to Exercise 5, the right hand side is equal to

$$\sqrt{|g|} (\theta^{i_1} \wedge \ldots \wedge \theta^{i_n}, \theta^1 \wedge \ldots \wedge \theta^n)$$
$$= \frac{1}{\sqrt{|g|}} \varepsilon_{i_1 \ldots i_n} (\eta, \eta) = \operatorname{sgn}(g) \frac{1}{\sqrt{|g|}} \varepsilon_{i_1 \ldots i_n}.$$

The left hand side is equal to

$$\sqrt{|g|} \varepsilon_{j_1 \ldots j_n} g^{j_1 i_1} \ldots g^{j_n i_n} = \operatorname{sgn}(g) \frac{1}{\sqrt{|g|}} \varepsilon_{i_1 \ldots i_n}.$$

so that both sides are in agreement.

---

### 4.6.3 The Codifferential

The codifferential $\delta\colon \Lambda_p(M) \to \Lambda_{p-1}(M)$ is defined by

$$\delta := \operatorname{sgn}(g)(-1)^{np+n} *d*. \tag{4.26}$$

Since $d \circ d = 0$, we also have

$$\delta \circ \delta = 0. \tag{4.27}$$

From (4.24) we see that the equation $\delta\omega = 0$ is equivalent to $d*\omega = 0$. Therefore, by Poincaré's Lemma, there exists locally a form $\varphi$ such that $*\omega = d\varphi$. Hence, $\omega = \pm *d\varphi = \delta\psi$, where $\psi = \pm *\varphi$.

Thus $\delta\omega = 0$ implies the local existence of a form $\psi$ with $\omega = \delta\psi$.

*Coordinate Expression for $\delta\omega$*

We wish to show that for $\omega \in \Lambda_p(M)$

$$(\delta\omega)^{i_1 \ldots i_{p-1}} = |g|^{-1/2} (|g|^{1/2} \omega^{k i_1 \ldots i_{p-1}})_{,k}. \tag{4.28}$$

*Proof:* We have ($q = n - p$):

$$(*d*\omega)^{k_1\ldots k_{p-1}} = \frac{1}{(q+1)!}\,\eta^{j_1\ldots j_{q+1}k_1\ldots k_{p-1}}(d*\omega)_{j_1\ldots j_{q+1}}. \tag{4.29}$$

The right hand side of (4.29) is a consequence of (4.23). For any form $\alpha \in \Lambda_s(M)$, we have

$$\alpha = \frac{1}{s!}\,\alpha_{i_1\ldots i_s}\,dx^{i_1}\wedge\ldots\wedge dx^{i_s},$$

$$d\alpha = \frac{1}{s!}\,\alpha_{i_1\ldots i_s,\,i_{s+1}}\,dx^{i_{s+1}}\wedge dx^{i_1}\wedge\ldots\wedge dx^{i_s}$$

$$= \frac{1}{(s+1)!}\,(s+1)\,(-1)^s\,\alpha_{[i_1\ldots i_s,\,i_{s+1}]}\,dx^{i_1}\wedge\ldots\wedge dx^{i_{s+1}},$$

so that

$$(d\alpha)_{i_1\ldots i_{s+1}} = (-1)^s\,(s+1)\,\alpha_{[i_1\ldots i_s,\,i_{s+1}]}. \tag{4.30}$$

Using this in (4.29) gives

$$(*d*\omega)^{k_1\ldots k_{p-1}} = \frac{1}{(q+1)!}\,\eta^{j_1\ldots j_{q+1}k_1\ldots k_{p-1}}(-1)^q\,(q+1)\,(*\omega)_{[j_1\ldots j_q,\,j_{q+1}]}. \tag{4.31}$$

We may again ignore the antisymmetrization. If we once more use (4.23), we obtain

$$\begin{aligned}(*d*\omega)^{k_1\ldots k_{p-1}} &= (-1)^q\,\frac{1}{q!}\,\frac{1}{p!}\,(\eta_{i_1\ldots i_p j_1\ldots j_q}\,\omega^{i_1\ldots i_p})_{,\,j_{q+1}}\quad \eta^{j_1\ldots j_{q+1}k_1\ldots k_{p-1}}\\ &= (-1)^q\,\frac{1}{p!\,q!}\,\mathrm{sgn}(g)\,\frac{1}{\sqrt{|g|}}\,(\sqrt{|g|}\,\omega^{i_1\ldots i_p})_{,\,j_{q+1}} \qquad (4.32)\\ &\quad\times \underbrace{\varepsilon_{j_1\ldots j_{q+1}k_1\ldots k_{p-1}}}_{(-1)^{p-1}\,\varepsilon_{j_1\ldots j_q k_1\ldots k_{p-1} j_{q+1}}} \cdot \underbrace{\varepsilon_{i_1\ldots i_p j_1\ldots j_q}}_{(-1)^{pq}\,\varepsilon_{j_1\ldots j_q i_1\ldots i_p}}.\end{aligned}$$

Now

$$\varepsilon_{j_1\ldots j_q k_1\ldots k_p}\,\varepsilon_{j_1\ldots j_q i_1\ldots i_p} = q!\,p!\,\delta^{k_1}_{[i_1}\,\delta^{k_2}_{i_2}\ldots\delta^{k_p}_{i_p]}. \tag{4.33}$$

If we use (4.33) in (4.32), we obtain

$$(*d*\omega)^{k_1\ldots k_{p-1}} = (-1)^q\,\mathrm{sgn}(g)\,(-1)^{p-1}\,(-1)^{pq}\,\frac{1}{\sqrt{|g|}}\,(\sqrt{|g|}\,\omega^{k_1\ldots k_p})_{,\,k_p} \tag{4.34}$$

By definition,

$$\delta\omega = \mathrm{sgn}(g)\,(-1)^{np+n}\,*d*\omega,$$

and hence

$$\delta\omega^{k_1\ldots k_{p-1}} = (-1)^{np+n}\,(-1)^q\,(-1)^{p-1}\,(-1)^{pq}\,\frac{1}{\sqrt{|g|}}\,(\sqrt{|g|}\,\omega^{k_1\ldots k_p})_{,\,k_p}.$$

Equation (4.28) follows from this.

**Exercises:**

1. Show that the star operation commutes with the pull-back by an orientation preserving isometry $\varphi\colon M \to M$, $\varphi^* g = g$. Conclude that the same is true for the codifferential. What happens if the orientation is reversed?
2. Discuss the invariance properties under isometries of an equation of the form

   $$\delta F = J,$$

   where $F$ is a 2-form and $J$ is a 1-form.

Specialize the discussion to Minkowski space.

## 4.7 The Integral Theorems of Stokes and Gauss

### 4.7.1 Integration of Differential Forms

Let $M$ be an oriented, $n$-dimensional differentiable manifold. We wish to define integrals of the type

$$\int_M \omega, \qquad \omega \in \Lambda_n(M).$$

We consider first the case where the support of $\omega$ is contained in the domain $U$ of a chart. Let $(x^1, \ldots, x^n)$ be *positive* coordinates in this region (in the following, we shall take all coordinates to be positive). If

$$\omega = f\, dx^1 \wedge \ldots \wedge dx^n, \qquad f \in \mathcal{F}(U),$$

we define

$$\int_M \omega = \int_U f(x^1, \ldots, x^n)\, dx^1 \ldots dx^n, \tag{4.35}$$

where the right hand side is an ordinary $n$-fold Lebesgue (Riemann) integral. This definition makes sense, for if the support $\omega$ is contained in the domain of a second chart with coordinates $(y^1, \ldots, y^n)$ and if

$$\omega = g\, dy^1 \wedge \ldots \wedge dy^n,$$

then (since both coordinate systems are positive)

$$\int g\, dy^1 \ldots dy^n = \int g \circ \varphi \det(\partial g^i/\partial x^k)\, dx^1 \ldots dx^n,$$

where $\varphi$ denotes the change of coordinates. Now

$$f(x^1, \ldots, x^n) = (g \circ \varphi)(x^1, \ldots, x^n) \det(\partial y^i/\partial x^k).$$

Hence, the integral (4.35) is independent of the choice of the coordinate system. We now consider an arbitrary $\omega \in \Lambda_n(M)$ with compact support.

Let $(U_i, g_i)_{i \in I}$ be a $C^\infty$ atlas of $M$, such that $(U_i)_{i \in I}$ is a locally finite covering (i.e. for every $x \in M$ there exists a neighborhood $U \subset M$ such that $U \cap U_i$ is nonempty only for a finite number of $i \in I$).

Further, let $(h_i)_{i \in I}$ be a corresponding partition of unity. This is a family of differentiable functions $(h_i)_{i \in I}$ on $M$ with the following properties:

(i) $h_i(x) \geqq 0$ for all $x \in M$;

(ii) $\operatorname{supp} h_i \subseteq U_i$;

(iii) $\sum_{i \in I} h_i(x) = 1$ for all $x \in M$.

Note that because of (ii) and the local finiteness of the covering, only a finite number of terms contribute to the sum in (iii).

**Remarks:** By definition (see Sect. 1), $M$ has a countable basis. One can show that $M$ then has a locally finite covering and also that for every such covering, a corresponding partition of unity exists. (For proofs, see Sects. II.14, 15 of [6]).

Since the support of $\omega$ is compact and $(U_i)_{i \in I}$ is a locally finite covering, the number of $U_i$ with nonempty $U_i \cap \operatorname{supp} \omega$ is *finite.* For $h_i \omega \neq 0$, we have $\operatorname{supp} h_i \omega \subset U_i$. Since also $\sum_i h_i = 1$, we have

$$\omega = \sum_{i \in I} h_i \omega .$$

Thus $\omega$ is a *finite sum* of $n$-forms each of which has its support in a single chart $U_i$. We then define

$$\int_M \omega = \sum_{i \in I} \int_M h_i \omega . \tag{4.36}$$

This definition is independent of the partition of unity. Indeed, let $(f_j, V_j)_{j \in J}$ be another $C^\infty$ atlas of $M$ such that $(V_j)_{j \in J}$ is a locally finite covering of $M$ and if $(k_j)_{j \in J}$ is a partition of unity corresponding to this covering, then $\{U_i \times V_j\}_{(i,j) \in I \times J}$ is also a $C^\infty$ atlas of $M$, such that $(U_i \cap V_j)$ is a locally finite covering and $(h_i \cdot k_j)$ is a corresponding partition of unity. Now

$$\sum_{i \in I} \int h_i \omega = \sum_{(i,j) \in I \times J} \int k_j h_i \omega \quad \text{and}$$

$$\sum_{j \in J} \int k_j \omega = \sum_{(i,j) \in I \times J} \int h_i k_j \omega ,$$

which shows that (4.36) is independent of the partition of unity.

The integral has the following properties: Let $\Lambda_n^c(M)$ be the set of $n$-forms having compact support. Then

(i) $\int_M (a_1 \omega_1 + a_2 \omega_2) = a_1 \int_M \omega_1 + a_2 \int_M \omega_2 , \quad a_1, a_2 \in \mathbb{R}; \quad \omega_1 \omega_2 \in \Lambda_n^c(M)$;

(ii) if $\varphi: M \to N$ is an orientation preserving diffeomorphism, then

$$\int_M \varphi^* \omega = \int_N \omega , \quad \omega \in \Lambda_n^c(N);$$

(iii) if the orientation $O$ of $M$ is changed, the integral changes sign:

$$\int_{(M,-O)} \omega = -\int_{(M,O)} \omega .$$

### 4.7.2 Stokes' Theorem

Let $D$ be a region (i.e. an open connected subset) in an $n$-dimensional differentiable manifold. One says that the region $D$ has a *smooth boundary* provided that for every $p \in \partial D$ (point on the boundary of $D$) there exist an open neighborhood $U$ of $p$ in $M$ and a local coordinate system $(x^1, \ldots, x^n)$ on $U$ such that (see Fig. 3)

$$U \cap \bar{D} = \{q \in U \mid x^n(q) \geqq x^n(p)\}. \tag{4.37}$$

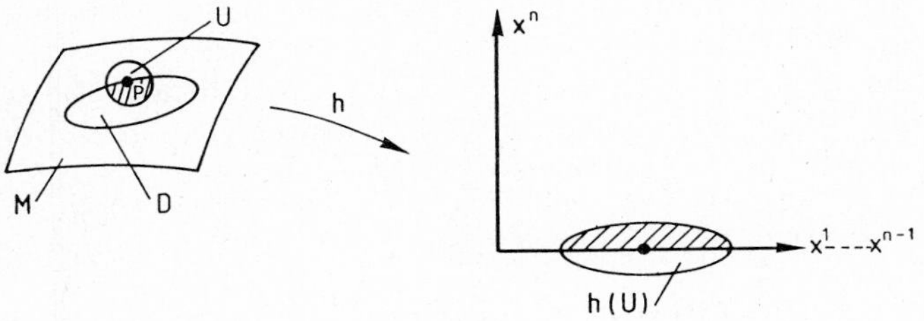


Fig. 3

One can show that $\partial D$ is a closed, $(n-1)$-dimensional submanifold of $M$ (see Sect. V.5 of [6]). If $M$ is orientable, then $\partial D$ is also orientable. If $(x^1, \ldots, x^n)$ is a positive coordinate system which satisfies (4.37), then we may choose an orientation of $\partial D$ such that $(x^1, \ldots, x^{n-1})$ is a positive coordinate system of $\partial D$. However, the following convention is more suitable: if $n$ is even, then $(x^1, \ldots, x^{n-1})$ is positive, while if $n$ is odd, $(x^1, \ldots, x^{n-1})$ is negative. Such an orientation of $\partial D$ is called the *induced* orientation of $\partial D$.

If $\bar{D}$ is compact and $\omega \in \Lambda_n(M)$, we define

$$\int_D \omega = \int_M \chi_D \, \omega ,$$

where $\chi_D$ is the characteristic function of $\bar{D}$.
Now we are ready to formulate

**Stokes' Theorem:** Let $M$ be an $n$-dimensional, oriented differentiable manifold (with denumerable basis) and let $D$ be a region of $M$ having a smooth boundary and such that $\bar{D}$ is compact. For every form $\omega \in \Lambda_{n-1}(M)$

$$\int_D d\omega = \int_{\partial D} \omega . \tag{4.38}$$

[More precisely, $\int_D d\omega = \int_{\partial D} i^*(\omega)$, where $i\colon \partial D \to M$ is the canonical injection]. In (4.38), the induced orientation of $\partial D$ has been chosen. (For a proof, see [6], Sect. V.5.)

**Application:** Let $\Omega$ be a volume element of $M$ (this implies that $M$ is an orientable manifold) and $X \in \mathscr{X}(M)$ be a vector field. We define $\operatorname{div}_\Omega X$ by

$$L_X \Omega = (\operatorname{div}_\Omega X)\, \Omega\,. \tag{4.39}$$

Now

$$L_X = d \circ i_X + i_X \circ d,$$

and hence

$$L_X \Omega = d(i_X \Omega)\,. \tag{4.40}$$

We then obtain from Stokes' theorem

$$\int_D (\operatorname{div}_\Omega X)\, \Omega = \int_{\partial D} i_X \Omega\,. \tag{4.41}$$

$i_X \Omega$ defines a measure $d\mu_X$ on $\partial D$; $d\mu_X$ depends linearly on $X$. We may then also write

$$d\mu_X = \langle X, d\sigma \rangle,$$

where $d\sigma$ is a measure-valued one-form. With these notations, we have Gauss' theorem:

$$\int_D (\operatorname{div}_\Omega X)\, \Omega = \int_{\partial D} \langle X, d\sigma \rangle\,. \tag{4.42}$$

In particular, these equations are valid for pseudo-Riemannian manifolds with

$$\Omega = \eta = \sqrt{|g|}\, dx^1 \wedge \ldots \wedge dx^n\,,$$

(in local coordinates).

*Expression for* div $X$ *in Local Coordinates*

Let

$$\Omega = a(x)\, dx^1 \wedge \ldots \wedge dx^n, \qquad X = X^i\, \partial/\partial x^i.$$

Then

$$\begin{aligned} L_X \Omega &= (X a)\, dx^1 \wedge \ldots \wedge dx^n + a \sum_{i=1}^n dx^1 \wedge \ldots \wedge d(X x^i) \wedge \ldots \wedge dx^n \\ &= \sum_{i=1}^n (X^i a_{,i} + a X^i{}_{,i})\, dx^1 \wedge \ldots \wedge dx^n \\ &= a^{-1} \sum_{i=1}^n (a X^i)_{,i}\, \Omega\,. \end{aligned}$$

Hence, we have, locally,

$$\operatorname{div}_\Omega X = a^{-1}(a X^i)_{,i}\,. \tag{4.43}$$

Let $g$ be a (positive definite) Riemannian metric and let $\eta_{\partial D}$ be the volume element on $\partial D$, which belongs to the induced Riemannian metric on $\partial D$. Then

$$i_X \eta | \partial D = (X, n)\, \eta_{\partial D}\,, \tag{4.44}$$

where $n$ is the unit outward normal on $\partial D$. Hence, the divergence theorem (4.41) can be written in the familiar form

$$\int_D (\operatorname{div} X)\, \eta = \int_{\partial D} (X, n)\, \eta_{\partial D}\,. \tag{4.45}$$

*Proof of Eq. (4.44):* We choose a positive orthonormal basis $w_1, \ldots, w_{n-1}$ of $T_p(\partial D)$. Then $(n, w_1, \ldots, w_{n-1})$ is a positive orthonormal basis of $T_p(M)$. Now $X = (X, n)\, n + Y$, $Y \in T_p(\partial D)$, and hence

$$\begin{aligned} i_X \eta(w_1, \ldots, w_{n-1}) &= \eta(X, w_1, \ldots, w_{n-1}) \\ &= (X, n)\, \eta_{\partial D}(w_1, \ldots, w_{n-1})\,. \end{aligned}$$

---

**Exercises:**

1. Show that the following identities hold for a function $f$ and a vector field $X$

$$\operatorname{div}_\Omega(fX) = X(f) + f \operatorname{div}_\Omega X\,, \tag{4.46}$$

$$\operatorname{div}_{f\Omega}(X) = \operatorname{div}_\Omega X + \frac{1}{f} X(f)\,. \tag{4.47}$$

In the second identity it is assumed that $f$ vanishes nowhere; in this case we see that

$$\operatorname{div}_{f\Omega}(X) = \frac{1}{f} \operatorname{div}_\Omega(fX)\,. \tag{4.48}$$

2. Show that

$$i_X \eta = {*}X^\flat\,, \tag{4.49}$$

where $X^\flat$ is the 1-form belonging to the vector field $X$: $X^\flat(Y) = g(X, Y)$ for all $Y \in \mathscr{X}(M)$. Conclude from this

$$\operatorname{div} X = \delta X^\flat\,. \tag{4.50}$$

---

# 5. Affine Connections

In this section we introduce an important additional structure on differentiable manifolds, thus making it possible to define a "covariant derivative" which transforms tensor fields into other tensor fields.

## 5.1 Covariant Derivative of a Vector Field

**Definition 5.1:** An *affine (linear) connection* $\nabla$ on a manifold is a mapping which assigns to every pair $X$, $Y$ of $C^\infty$ vector fields on $M$ another $C^\infty$ vector field $\nabla_X Y$ with the following properties:
(i) $\nabla_X Y$ is $\mathbb{R}$-bilinear in $X$ and $Y$;
(ii) if $f \in \mathscr{F}(M)$, then

$$\nabla_{fX} Y = f \nabla_X Y \text{ and } \nabla_X (f Y) = f \nabla_X Y + X(f)\, Y.$$

**Lemma 5.1:** Let $\nabla$ be an affine connection on $M$ and let $U$ be an open subset of $M$. If either $X$ or $Y$ vanishes on $U$, then $\nabla_X Y$ also vanishes on $U$.
*Proof:* Suppose $Y$ vanishes on $U$. Let $p \in U$. Choose a function $h \in \mathscr{F}(M)$ with $h(p) = 0$ and $h = 1$ on $M \setminus U$. From $hY = Y$ it follows that

$$\nabla_X Y = \nabla_X (h\, Y) = X(h)\, Y + h\, \nabla_X Y.$$

This vanishes at $p$. The statement about $X$ follows similarly.

This lemma shows that an affine connection $\nabla$ on $M$ induces a connection on every open submanifold $U$ of $M$. In fact, let $X, Y \in \mathscr{X}(U)$ and $p \in U$. Then there exist vector fields $\tilde{X}, \tilde{Y} \in \mathscr{X}(M)$ which coincide with $X$, resp. $Y$ on an open neighborhood of $p$ (see the continuation lemma of Sect. 4.3). We then set

$$(\nabla \,|\, U)_X Y = (\nabla_{\tilde{X}} \tilde{Y}) \,|\, U.$$

By Lemma 5.1, the right hand side is independent of the choice of $\tilde{X}$, $\tilde{Y}$.
$\nabla \,|\, U$ is obviously an affine connection on $U$.

**Lemma 5.2:** Let $X, Y \in \mathscr{X}(M)$. If $X$ vanishes at $p$, then $\nabla_X Y$ also vanishes at $p$.

*Proof:* Let $U$ be a coordinate neighborhood of $p$. On $U$, we have the representation

$$X = \xi^i \partial/\partial x^i, \qquad \xi^i \in \mathscr{F}(U),$$

where $\xi^i(p) = 0$. Then

$$(\nabla_X Y)_p = \xi^i(p)(\nabla_{\partial/\partial x^i} Y)_p = 0.$$

We define, relative to a chart $(U, x^1, \ldots, x^n)$,

$$\nabla_{\partial/\partial x^i}(\partial/\partial x^j) = \Gamma^k_{ij}\, \partial/\partial x^k. \tag{5.1}$$

The $n^3$ functions $\Gamma^k_{ij} \in \mathscr{F}(U)$ are called the *Christoffel symbols* (or *connection coefficients*) of the connection $\nabla$ in the given chart. The connection coefficients are not the components of a tensor. Their transformation properties under a coordinate transformation to the chart $(V, \bar{x}^1, \ldots, \bar{x}^n)$ are obtained from the following calculation:

On the one hand we have

$$\nabla_{\partial/\partial \bar{x}^a}\left(\frac{\partial}{\partial \bar{x}^b}\right) = \bar{\Gamma}^c_{ab} \frac{\partial}{\partial \bar{x}^c} = \bar{\Gamma}^c_{ab} \frac{\partial x^k}{\partial \bar{x}^c} \frac{\partial}{\partial x^k} \tag{5.2}$$

and on the other hand

$$\begin{aligned}\nabla_{\partial/\partial \bar{x}^a}\left(\frac{\partial}{\partial \bar{x}^b}\right) &= \nabla_{\frac{\partial x^i}{\partial \bar{x}^a}\frac{\partial}{\partial x^i}}\left(\frac{\partial x^j}{\partial \bar{x}^b}\frac{\partial}{\partial x^j}\right)\\ &= \frac{\partial x^i}{\partial \bar{x}^a}\left[\frac{\partial x^j}{\partial \bar{x}^b}\Gamma^k_{ij}\frac{\partial}{\partial x^k} + \frac{\partial}{\partial x^i}\left(\frac{\partial x^j}{\partial \bar{x}^b}\right)\frac{\partial}{\partial x^j}\right]\\ &= \frac{\partial x^i}{\partial \bar{x}^a}\frac{\partial x^j}{\partial \bar{x}^b}\Gamma^k_{ij}\frac{\partial}{\partial x^k} + \frac{\partial^2 x^j}{\partial \bar{x}^a\,\partial \bar{x}^b}\frac{\partial}{\partial x^j}.\end{aligned}$$

Comparison with (5.2) results in

$$\frac{\partial x^k}{\partial \bar{x}^c}\bar{\Gamma}^c_{ab} = \frac{\partial x^i}{\partial \bar{x}^a}\frac{\partial x^j}{\partial \bar{x}^b}\Gamma^k_{ij} + \frac{\partial^2 x^k}{\partial \bar{x}^a\,\partial \bar{x}^b} \tag{5.3}$$

or

$$\bar{\Gamma}^c_{ab} = \frac{\partial x^i}{\partial \bar{x}^a}\frac{\partial x^j}{\partial \bar{x}^b}\frac{\partial \bar{x}^c}{\partial x^k}\Gamma^k_{ij} + \frac{\partial^2 x^k}{\partial \bar{x}^a\,\partial \bar{x}^b}\frac{\partial \bar{x}^c}{\partial x^k}. \tag{5.4}$$

Conversely, if for every chart there exist $n^3$ functions $\Gamma^k_{ij}$ which transform according to (5.4) under a change of coordinates, then there exists a unique affine connection $\nabla$ on $M$ which satisfies (5.1).

For every vector field $X$ there corresponds the tensor field $\nabla X \in \mathscr{T}^1_1(M)$, defined by

$$\nabla X(Y, \omega) := \langle \omega, \nabla_Y X \rangle, \qquad \omega = \text{1-form}. \tag{5.5}$$

$\nabla X$ is the *covariant derivative* (or *absolute derivative*) of $X$. In a chart $(U, x^1, \ldots, x^n)$ let (with $\partial_i := \partial/\partial x^i$)

$$X = \xi^i \partial_i ,$$
$$\nabla X = \xi^i_{;j}\, dx^j \otimes \partial_i .$$

One has

$$\begin{aligned} \xi^i_{;j} &= \nabla X(\partial_j, dx^i) = \langle dx^i, \nabla_{\partial_j}(\xi^k \partial_k)\rangle \\ &= \langle dx^i, \xi^k_{,j} \partial_k + \xi^k \Gamma^s_{jk} \partial_s\rangle \\ &= \xi^i_{,j} + \Gamma^i_{jk} \xi^k , \end{aligned}$$

so that

$$\xi^i_{;j} = \xi^i_{,j} + \Gamma^i_{jk} \xi^k . \tag{5.6}$$

## 5.2 Parallel Transport Along a Curve

Let $\gamma: J \to M$ be a curve in $M$, and let $X$ be a vector field which is defined on an open neighborhood of $\gamma(J)$. $X$ is said to be *autoparallel* along $\gamma$ if

$$\nabla_{\dot\gamma} X = 0 \tag{5.7}$$

on $\gamma$. The vector $\nabla_{\dot\gamma} X$ is sometimes denoted $DX/dt$ (*covariant derivative along* $\gamma$).

In terms of coordinates, we have

$$X = \xi^i \partial_i, \qquad \dot\gamma = \frac{dx^i}{dt} \partial_i ,$$

$$\nabla_{\dot\gamma} X = \left(\frac{d\xi^k}{dt} + \Gamma^k_{ij} \frac{dx^i}{dt} \xi^j\right) \partial_k . \tag{5.8}$$

This shows that $\nabla_{\dot\gamma} X$ depends only on the values of $X$ along $\gamma$. In terms of coordinates, (5.7) reads

$$\frac{d\xi^k}{dt} + \Gamma^k_{ij} \frac{dx^i}{dt} \xi^j = 0 . \tag{5.9}$$

For a given curve $\gamma(t)$ with $X_0 \in T_{\gamma(0)}(M)$ there exists a unique autoparallel field $X(t)$ along $\gamma$ with $X(t) \in T_{\gamma(t)}(M)$ and $X(0) = X_0$. Since the equation for $X(t)$ is linear, there is no limitation on $t$. For every curve $\gamma$ and any two points $\gamma(s)$ and $\gamma(t)$, there is then a linear isomorphism

$$\tau_{t,s} = T_{\gamma(s)}(M) \to T_{\gamma(t)}(M) ,$$

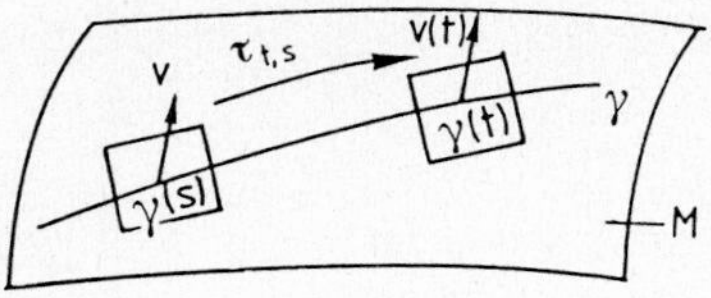


**Fig. 4**

which transforms a vector $v$ at $\gamma(s)$ into the parallel transported vector $v(t)$ at $\gamma(t)$ (see Fig. 4).

The mapping $\tau_{t,s}$ is the *parallel transport* along $\gamma$ from $\gamma(s)$ to $\gamma(t)$. We have, as a result of the uniqueness theorem for ordinary differential equations,

$$\tau_{t,s} \circ \tau_{s,r} = \tau_{t,r}, \qquad \tau_{s,s} = \text{identity}. \tag{5.10}$$

**Theorem 5.1:** Let $X$ be a vector field along $\gamma$. Then

$$\nabla_{\dot\gamma} X(\gamma(t)) = \frac{d}{ds} \tau_{t,s} X(\gamma(s))\Big|_{s=t}. \tag{5.11}$$

*Proof:* We shall work in a particular chart. By construction, $v(t) = \tau_{t,s} v_0$ with $v_0 \in T_{\gamma(s)}(M)$ satisfies the equation

$$\dot v^i + \Gamma^i_{kj} \dot x^k v^j = 0.$$

If we write $(\tau_{t,s} \circ v_0)^i = (\tau_{t,s})^i_j v_0^j$, then

$$\frac{d}{dt} (\tau_{t,s})^i_j \Big|_{t=s} = -\Gamma^i_{kj} \dot x^k.$$

Since $\tau_{t,s} = \tau_{s,t}^{-1}$ and $\tau_{s,s} = $ identity, it follows that

$$\begin{aligned} \frac{d}{ds}\Big|_{s=t} [\tau_{t,s} X(\gamma(s))]^i &= \frac{d}{ds}\Big|_{s=t} [\tau_{s,t}^{-1} X(\gamma(s))]^i \\ &= -\frac{d}{ds}\Big|_{s=t} (\tau_{s,t})^i_j X^j + \frac{d}{ds}\Big|_{s=t} X^i(\gamma(s)) \\ &= \Gamma^i_{kj} \dot x^k X^j + \frac{d}{ds}\Big|_{s=t} X^i(\gamma(s)) \\ &= \left(\frac{DX}{dt}\right)^i. \end{aligned}$$

## 5.3 Geodesics, Exponential Mapping, Normal Coordinates

A curve $\gamma$ is a *geodesic* if $\dot\gamma$ is autoparallel along $\gamma$. In local coordinates (5.9) implies that along a geodesic

$$\ddot x^k + \Gamma^k_{ij} \dot x^i \dot x^j = 0. \tag{5.12}$$

For given $\gamma(0)$, $\dot{\gamma}(0)$ there exists a unique maximal geodesic $\gamma(t)$; this is an immediate consequence of the existence and uniqueness theorems of ordinary differential equations. If $\gamma(t)$ is a geodesic, then $\gamma(at)$, $a \in \mathbb{R}$, is also a geodesic, with initial velocity $a\dot{\gamma}(0)$. Hence, for some neighborhood $V$ of $0 \in T_x(M)$, the geodesics $\gamma_v$, with $\gamma_v(0) = x$ and $\dot{\gamma}_v(0) = v \in V$, exist for all $t \in [0, 1]$. We introduce the *exponential mapping* at $x$ by $\exp_x$: $v \in V \to \gamma_v(1)$. Obviously $\gamma_{tv}(s) = \gamma_v(ts)$. If we set $s = 1$, we obtain $\exp_x(tv) = \gamma_v(t)$. The mapping $\exp_x$ is differentiable[3] at $v$ and the last equation implies $T\exp_x(0) =$ identity, since

$$T\exp_x(0)\, v = \frac{d}{dt} \exp_x(tv)\bigg|_{t=0} = \dot{\gamma}_v(0) = v$$

for all $v \in T_x(M)$. As an application of the implicit function theorem we then obtain

**Theorem 5.2:** The mapping $\exp_x$ is a diffeomorphism from a neighborhood of $0 \in T_x(M)$ to a neighborhood of $x \in M$.

This theorem permits the introduction of special coordinates. If we choose a basis $e_1, \ldots, e_n$ of $T_x(M)$, then we can represent a neighborhood of $x$ uniquely by $\exp_x(x^i e_i)$. The set $(x^1, \ldots, x^n)$ are known as *normal* or *Gaussian* coordinates. Since $\exp_x(tv) = \gamma_v(t)$, the curve $\gamma_v(t)$ has normal coordinates $x^i = v^i t$, with $v = v^i e_i$. In terms of these coordinates, (5.12) becomes

$$\Gamma^k_{ij} v^i v^j = 0;$$

hence we have $\Gamma^k_{ij}(0) + \Gamma^k_{ji}(0) = 0$. If the connection is *symmetric* $(\Gamma^k_{ij} = \Gamma^k_{ji})$, it follows that $\Gamma^k_{ij}(0) = 0$.

## 5.4 Covariant Derivative of Tensor Fields

We extend the parallel transport $\tau_{t,s}$ of Sect. 5.2 to arbitrary tensors. If $\alpha \in T^*_{\gamma(s)}(M)$, we define $\tau_{t,s}\,\alpha \in T^*_{\gamma(t)}(M)$ by

$$\langle \tau_{t,s}\,\alpha, \tau_{t,s}\, v\rangle = \langle \alpha, v\rangle \qquad \text{for all} \quad v \in T_{\gamma(s)}(M),$$

or

$$\langle \tau_{t,s}\,\alpha, w\rangle = \langle \alpha, \tau^{-1}_{t,s}\, w\rangle \qquad \text{for all} \quad w \in T_{\gamma(t)}(M). \tag{5.13}$$

If $S \in (T_{\gamma(s)}(M))^p_q$ is a tensor, let

$$(\tau_{t,s}\, S)(\alpha_1, \ldots, \alpha_p, v_1, \ldots, v_q) = S(\tau^{-1}_{t,s}\,\alpha_1, \ldots, \tau^{-1}_{t,s}\,\alpha_p, \tau^{-1}_{t,s}\, v_1, \ldots, \tau^{-1}_{t,s}\, v_q),$$
$$\alpha_i \in T^*_{\gamma(t)}(M),\; v_i \in T_{\gamma(t)}(M). \tag{5.14}$$

[3] Since $\gamma_v(t)$ depends on the initial conditions in a differentiable manner.

Now let $X$ be a vector field and $\gamma(t)$ be an integral curve of $X$, starting at $p$, so that $p=\gamma(0)$. If $S \in \mathscr{T}_s^r(M)$ is a tensor field, we define the covariant derivative in the direction of $X$ by

$$(\nabla_X S)_p = \frac{d}{ds} \tau_s^{-1} S_{\gamma(s)}\Big|_{s=0}. \tag{5.15}$$

(We have used the shorthand notation $\tau_s := \tau_{s,0}$.)
This expression is a generalization of (5.11). If $X(p)=0$ we set $(\nabla_X S)_p = 0$. For a function $f \in \mathscr{F}(M)$ we put $\nabla_X f = Xf$. Finally $\nabla_X$ is extended to a linear mapping of $\mathscr{T}(M)$.

**Proposition 5.1:** $\nabla_X$ is a derivation of the tensor algebra $\mathscr{T}(M)$.
*Proof:* Obviously, $\tau_s(S_1 \otimes S_2) = (\tau_s S_1) \otimes (\tau_s S_2)$. From this we obtain the rule for derivations as follows:

$$\begin{aligned}(\nabla_X(S_1 \otimes S_2)) &= \frac{d}{ds}\Big|_{s=0} \tau_s^{-1}(S_1(\gamma(s)) \otimes S_2(\gamma(s))) \\ &= \frac{d}{ds}\Big|_{s=0} (\tau_s^{-1} S_1(\gamma(s)) \otimes \tau_s^{-1} S_2(\gamma(s))) \\ &= \frac{d}{ds}\Big|_{s=0} \tau_s^{-1} S_1(\gamma(s)) \otimes S_2(p) + S_1(p) \otimes \frac{d}{ds}\Big|_{s=0} \tau_s^{-1} S_2(\gamma(s)) \\ &= (\nabla_X S_1)_p \otimes S_2(p) + S_2(p) \otimes (\nabla_X S_2)_p .\end{aligned}$$

**Proposition 5.2:** $\nabla_X$ commutes with contractions.
*Proof:* We give a proof for the special case $S = Y \otimes \omega$, $Y \in \mathscr{X}(M)$, $\omega \in \mathscr{X}^*(M)$. The general case is completely analogous.
If $C$ denotes the contraction operation, we have by (5.13)

$$\begin{aligned}C\,\tau_s^{-1}(Y \otimes \omega)_{\gamma(s)} &= C(\tau_s^{-1} Y_{\gamma(s)} \otimes \tau_s^{-1} \omega_{\gamma(s)}) \\ &= \langle \tau_s^{-1} \omega_{\gamma(s)}, \tau_s^{-1} Y_{\gamma(s)} \rangle \\ &= \langle \omega_{\gamma(s)}, Y_{\gamma(s)} \rangle .\end{aligned}$$

Hence

$$C\{s^{-1}[\tau_s^{-1}(Y \otimes \omega)_{\gamma(s)} - (Y \otimes \omega)_p]\} = s^{-1}[\langle \omega, Y\rangle_{\gamma(s)} - \langle \omega, Y\rangle_p].$$

In the limit $s \to 0$ we obtain, with the help of (5.15),

$$C[\nabla_X(Y \otimes \omega)] = \nabla_X[C(Y \otimes \omega)] .$$

**Application:** Since, according to Proposition 5.1,

$$\nabla_X(Y \otimes \omega) = (\nabla_X Y) \otimes \omega + Y \otimes \nabla_X \omega$$

we have, after contraction, and use of Proposition 5.2,

$$\nabla_X(\omega(Y)) = \omega(\nabla_X Y) + (\nabla_X \omega)(Y)$$

or

$$(\nabla_X \omega)(Y) = X\omega(Y) - \omega(\nabla_X Y). \tag{5.16}$$

Equation (5.16) gives an expression for the covariant derivative of a 1-form. Note that $\nabla_X \omega$ is $\mathscr{F}(M)$-linear in $X$. Because of the derivation property, $\nabla_X$ is then $\mathscr{F}(M)$-linear on the entire tensor algebra:

$$\nabla_{fX} S = f\nabla_X S \quad \text{for all} \quad f \in \mathscr{F}(M), \quad S \in \mathscr{T}(M).$$

Hence a mapping $\nabla: \mathscr{T}_p^q(M) \to \mathscr{T}_{p+1}^q(M)$ is defined by

$$(\nabla S)(X_1, \ldots, X_{p+1}, \omega_1, \ldots, \omega_q) = (\nabla_{X_{p+1}} S)(X_1, \ldots, X_p, \omega_1, \ldots, \omega_q). \tag{5.17}$$

$\nabla S$ is the *covariant derivative* of the tensor field $S$. We easily obtain a general expression for $\nabla S$, $S \in \mathscr{T}_q^p(M)$, by generalizing the previous discussion: if $\omega_i$ are 1-forms and $Y_j$ are vector fields,

$$\begin{aligned}&\nabla_X(Y_1 \otimes \ldots \otimes Y_q \otimes \omega_1 \otimes \ldots \otimes \omega_p \otimes S)\\ &\quad = \nabla_X Y_1 \otimes Y_2 \otimes \ldots \otimes S + \ldots + Y_1 \otimes \ldots \otimes Y_q \otimes \nabla_X \omega_1 \otimes \ldots \otimes S + \ldots\\ &\quad + Y_1 \otimes \ldots \otimes \omega_p \otimes \nabla_X S.\end{aligned}$$

We now take the complete contraction

$$\begin{aligned}&\nabla_X[S(Y_1, \ldots, Y_q, \omega_1, \ldots, \omega_p)]\\ &\quad = S(\nabla_X Y_1, \ldots, \omega_p) + \ldots + S(Y_1, \ldots, \nabla_X \omega_p) + (\nabla_X S)(Y_1, \ldots, \omega_p).\end{aligned}$$

From this we find the following expression for $\nabla_X S$,

$$\begin{aligned}&(\nabla_X S)(Y_1, \ldots, Y_q, \omega_1, \ldots, \omega_p)\\ &\quad = X(S(Y_1, \ldots, Y_q, \omega_1, \ldots, \omega_p)) - S(\nabla_X Y_1, Y_2, \ldots, \omega_p) - \ldots\\ &\quad - S(Y_1, \ldots, \nabla_X \omega_p).\end{aligned} \tag{5.18}$$

Finally we give also a local coordinate expression for the covariant derivative. Let $(U, x^1, \ldots, x^n)$ be a chart and let $S \in \mathscr{T}_q^p(U)$, with

$$S = S^{i_1 \ldots i_p}_{j_1 \ldots j_q} \partial_{i_1} \otimes \ldots \otimes \partial_{i_p} \otimes dx^{j_1} \otimes \ldots \otimes dx^{j_q}, \qquad X = X^i \partial_i.$$

Now

$$X S^{i_1 \ldots i_p}_{j_1 \ldots j_q} = X^k S^{i_1 \ldots i_p}_{j_1 \ldots j_q, k} \tag{5.19}$$

and

$$\nabla_X(\partial_i) = X^k \nabla_{\partial_k}(\partial_i) = X^k \Gamma^l_{ki} \partial_l. \tag{5.20}$$

In addition we have, using (5.16),

$$(\nabla_X dx^j)(\partial_i) = X\langle dx^j, \partial_i\rangle - \langle dx^j, \nabla_X \partial_i\rangle = -X^k \Gamma^j_{ki},$$

or

$$\nabla_X dx^j = -X^k \Gamma^j_{ki} dx^i. \tag{5.21}$$

If we use (5.19–21) in (5.18) for $\omega_j = dx^j$ and $Y_i = \partial_i$, we obtain an expression for the components of $\nabla S$ denoted by $S^{i_1 \dots i_p}_{j_1 \dots j_q;\,k} \equiv \nabla_k S^{i_1 \dots i_p}_{j_1 \dots j_q}$:

$$\begin{aligned} S^{i_1 \dots i_p}_{j_1 \dots j_q;\,k} = S^{i_1 \dots i_p}_{j_1 \dots j_q,\,k} &+ \Gamma^{i_1}_{kl}\, S^{l\, i_2 \dots i_p}_{j_1 \dots j_q} + \dots \\ &- \Gamma^{l}_{k j_1}\, S^{i_1 \dots i_p}_{l j_2 \dots j_q} - \dots . \end{aligned} \tag{5.22}$$

This implies that $\nabla S$ is differentiable.

In particular, we have for contravariant and covariant vector fields

$$\xi^i_{;k} = \xi^i_{,k} + \Gamma^i_{kl}\,\xi^l, \qquad \eta_{i;\,k} = \eta_{i,\,k} - \Gamma^l_{ki}\,\eta_l . \tag{5.23}$$

As is implied by Proposition 5.2, the covariant derivative of the Kronecker tensor vanishes:

$$\delta^i_{j;\,k} = 0 .$$

## 5.5 Curvature and Torsion of an Affine Connection, Bianchi Identities

Let $\nabla$ be an affine connection on $M$. The *torsion* is defined as the mapping $T\colon \mathscr{X}(M) \times \mathscr{X}(M) \to \mathscr{X}(M)$,

$$T(X, Y) = \nabla_X Y - \nabla_Y X - [X, Y] \tag{5.24}$$

and the *curvature* as the mapping $R\colon \mathscr{X}(M) \times \mathscr{X}(M) \times \mathscr{X}(M) \to \mathscr{X}(M)$,

$$R(X, Y)\, Z = \nabla_X (\nabla_Y Z) - \nabla_Y (\nabla_X Z) - \nabla_{[X, Y]}\, Z . \tag{5.25}$$

Note that

$$T(X, Y) = -\,T(Y, X), \qquad R(X, Y) = -\,R(Y, X) . \tag{5.26}$$

One may also easily verify that

$$T(fX, gY) = fg\, T(X, Y), \qquad R(fX, gY)\, hZ = fgh\, R(X, Y)\, Z$$

for all $f, g, h \in \mathscr{F}(M)$.

The mapping

$$\mathscr{X}^*(M) \times \mathscr{X}(M) \times \mathscr{X}(M) \to \mathscr{F}(M)\colon (\omega, X, Y) \mapsto \langle \omega, T(X, Y) \rangle$$

is thus a tensor field in $\mathscr{T}^1_2(M)$ and is known as the *torsion tensor.* Similarly the mapping $(\omega, Z, X, Y) \mapsto \langle \omega, R(X, Y)\, Z \rangle$ is a tensor field in $\mathscr{T}^1_3(M)$. This *curvature tensor* plays an important role in the general theory of relativity.

In local coordinates the components of the torsion tensor are given by

$$T^k_{ij} = \langle dx^k, T(\partial_i, \partial_j) \rangle = \langle dx^k, \nabla_{\partial_i} \partial_j - \nabla_{\partial_j} \partial_i \rangle$$

and hence from (5.1)

$$T^k_{ij} = \Gamma^k_{ij} - \Gamma^k_{ji}. \tag{5.27}$$

If the torsion vanishes, we have $\Gamma^k_{ij} = \Gamma^k_{ji}$ in every chart. Then $\Gamma^k_{ij}(0) = 0$ in Gaussian coordinates, as mentioned at the end of Sect. 5.3.

The components of the curvature tensor are (the ordering of the indices is important):

$$\begin{aligned} R^i_{jkl} &= \langle dx^i, R(\partial_k, \partial_l)\,\partial_j\rangle = \langle dx^i, (\nabla_{\partial_k}\nabla_{\partial_l} - \nabla_{\partial_l}\nabla_{\partial_k})\,\partial_j\rangle \\ &= \langle dx^i, \nabla_{\partial_k}\Gamma^s_{lj}\,\partial_s - \nabla_{\partial_l}\Gamma^s_{kj}\,\partial_s\rangle \\ &= \Gamma^i_{lj,k} - \Gamma^i_{kj,l} + \Gamma^s_{lj}\Gamma^i_{ks} - \Gamma^s_{kj}\Gamma^i_{ls}. \end{aligned} \tag{5.28}$$

The *Ricci tensor* is a contraction of the curvature tensor. Its components are given by

$$R_{jl} = R^i_{jil} = \Gamma^i_{lj,i} - \Gamma^i_{ij,l} + \Gamma^s_{lj}\Gamma^i_{is} - \Gamma^s_{ij}\Gamma^i_{ls}. \tag{5.29}$$

In order to formulate the next theorem, we need the following preliminaries:

An $\mathscr{F}(M)$-multilinear mapping

$$K: \underbrace{\mathscr{X}(M) \times \ldots \times \mathscr{X}(M)}_{p\text{-times}} \to \mathscr{X}(M)$$

can be regarded as a tensor field $\tilde{K} \in \mathscr{T}^1_p(M)$:

$$\tilde{K}(\omega, X_1, \ldots, X_p) = \langle \omega, K(X_1, \ldots, X_p)\rangle. \tag{5.30}$$

The covariant derivative of $K$ is naturally defined by

$$\langle \omega, (\nabla_X K)(X_1, \ldots, X_p)\rangle = (\nabla_X \tilde{K})(\omega, X_1, \ldots, X_p).$$

According to (5.18), the right hand side is

$$\begin{aligned} &(\nabla_X \tilde{K})(\omega, X_1, \ldots, X_p) \\ &\quad = X\tilde{K}(\omega, X_1, \ldots, X_p) - \tilde{K}(\nabla_X \omega, X_1, \ldots, X_p) \\ &\qquad - \tilde{K}(\omega, \nabla_X X_1, \ldots, X_p) - \ldots - \tilde{K}(\omega, X_1, \ldots, \nabla_X X_p). \end{aligned}$$

Since $\nabla_X \omega(Y) = X\omega(Y) - \omega(\nabla_X Y)$, the right hand side becomes

$$\langle \omega, \nabla_X(K(X_1, \ldots, X_p)) - \sum_i K(X_1, \ldots, \nabla_X X_i, \ldots, X_p)\rangle.$$

We may now drop $\omega$ and obtain

$$(\nabla_X K)(X_1, \ldots, X_p) = \nabla_X(K(X_1, \ldots, X_p)) - \sum_i K(X_1, \ldots, \nabla_X X_i, \ldots, X_p). \tag{5.31}$$

**Theorem 5.3:** Let $T$ and $R$ be the torsion and curvature of an affine connection $\nabla$. If $X$, $Y$ and $Z$ are vector fields, then we have
Bianchi's 1st identity:

$$\sum_{\text{cyclic}} \{R(X,Y)Z\} = \sum_{\text{cyclic}} \{T(T(X,Y),Z) + (\nabla_X T)(Y,Z)\}, \tag{5.32}$$

Bianchi's 2nd identity:

$$\sum_{\text{cyclic}} \{(\nabla_X R)(Y,Z) + R(T(X,Y),Z)\} = 0. \tag{5.33}$$

*Proof:* By (5.31) we have

$$(\nabla_X R)(Y,Z) = \nabla_X(R(Y,Z)) - R(\nabla_X Y, Z) - R(Y, \nabla_X Z) - R(Y,Z)\nabla_X.$$

The cyclic sum of the two middle terms is

$$\begin{aligned}
&-[R(\nabla_X Y, Z) + R(Y, \nabla_X Z) + R(\nabla_Y Z, X) + R(Z, \nabla_Y X) \\
&\qquad + R(\nabla_Z X, Y) + R(X, \nabla_Z Y)] \\
&= -[R(\nabla_X Y, Z) + R(Z, \nabla_Y X) + \text{cyclic permutations}] \\
&= -R(T(X,Y),Z) - R([X,Y]Z) + \text{cyclic permutations}
\end{aligned}$$

and hence the left hand side of (5.33) is equal to

$$\begin{aligned}
&\nabla_X(R(Y,Z)) - R([X,Y],Z) - R(Y,Z)\nabla_X + \text{cyclic permutations} \\
&= \nabla_X(\nabla_Y\nabla_Z - \nabla_Z\nabla_Y - \nabla_{[Y,Z]}) \\
&\quad - (\nabla_Y\nabla_Z - \nabla_Z\nabla_Y - \nabla_{[Y,Z]})\nabla_X \\
&\quad - (\nabla_{[X,Y]}\nabla_Z - \nabla_Z\nabla_{[X,Y]} - \nabla_{[[X,Y],Z]}) + \text{cyclic permutations.}
\end{aligned}$$

If one uses the Jacobi identity, the last term in the cyclic sum drops out. The remaining terms transform in pairs into each other under cyclic permutation, except for the sign, and thus cancel in pairs in the cyclic sum. This proves the second Bianchi identity.

We now prove the first Bianchi identity for $T = 0$. In that case the left hand side of (5.32) is

$$\begin{aligned}
&(\nabla_X\nabla_Y - \nabla_Y\nabla_X)Z + (\nabla_Y\nabla_Z - \nabla_Z\nabla_Y)X + (\nabla_Z\nabla_X - \nabla_X\nabla_Z)Y \\
&\qquad - \nabla_{[X,Y]}Z - \nabla_{[Y,Z]}X - \nabla_{[Z,X]}Y \\
&= \nabla_X[Y,Z] - \nabla_{[Y,Z]}X + \text{cyclic permutations} \\
&= [X,[Y,Z]] + \text{cyclic permutations} \\
&= 0.
\end{aligned}$$

## 5.6 Riemannian Connections

**Definition 5.2:** Let $(M, g)$ be a pseudo-Riemannian manifold. An affine connection is a *metric connection* if parallel transport along any smooth curve $\gamma$ in $M$ preserves the inner product: For autoparallel fields $X(t)$, $Y(t)$ along $\gamma$, $g_{\gamma(t)}(X(t), Y(t))$ is independent of $t$.

**Proposition 5.3:** An affine connection $\nabla$ is metric if and only if

$$\nabla g = 0 . \tag{5.34}$$

*Proof:* By definition an affine connection is metric if, for all $\gamma$, $Y$, $Z$

$$\frac{d}{dt} g_{\gamma(t)}(\tau_{t,s} Y_{\gamma(s)}, \tau_{t,s} Z_{\gamma(s)}) = 0 . \tag{5.35}$$

Because of the group property of $\tau_{t,s}$ this is equivalent to the same condition for $t = s$. By (5.14) and (5.15), this is equivalent to $\nabla_{\dot{\gamma}} g = 0$.

**Remark:** As a result of (5.18), $\nabla g = 0$ is equivalent to the so-called *Ricci identity*

$$X g(Y, Z) = g(\nabla_X Y, Z) + g(Y, \nabla_X Z) . \tag{5.36}$$

**Theorem 5.4:** For every pseudo-Riemannian manifold $(M, g)$, there exists a unique affine connection $\nabla$ such that
(i) torsion $= 0$ ($\nabla$ is symmetric)
(ii) $\nabla$ is metric.

*Proof:* From the Ricci identity (5.36) and (i) we obtain
a) $X g(Y, Z) = g(\nabla_Y X, Z) + g([X, Y], Z) + g(Y, \nabla_X Z)$.
From this we obtain through cyclic permutation
b) $Y g(Z, X) = g(\nabla_Z Y, X) + g([Y, Z], X) + g(Z, \nabla_Y X)$
c) $Z g(X, Y) = g(\nabla_X Z, Y) + g([Z, X], Y) + g(X, \nabla_Z Y)$.
Taking the linear combination b) + c) − a) results in

$$\begin{aligned} 2g(\nabla_Z Y, X) = &- X g(Y, Z) + Y g(Z, X) + Z g(X, Y) \\ &- g([Z, X], Y) - g([Y, Z], X) + g([X, Y], Z) . \end{aligned} \tag{5.37}$$

The right hand side is independent of $\nabla$. Since $g$ is nondegenerate, the uniqueness of $\nabla$ follows from (5.37).

*Existence:* Define the mapping $\omega: \mathscr{X}(M) \to \mathscr{F}(M)$, $\omega(X) \mathrel{\hat{=}}$ right hand side of (5.37). $\omega$ is clearly additive, and is also $\mathscr{F}(M)$-homogeneous: $\omega(fX) = f\omega(X)$, as can be demonstrated by a short computation (exercise). To the 1-form $\omega$ there corresponds a unique vector field $\nabla_Z Y$ with $g(\nabla_Z Y, X) = \omega(X)$. The mapping $\nabla: \mathscr{X}(M) \times \mathscr{X}(M) \to \mathscr{X}(M)$ defined in this manner satisfies the defining properties of an affine connection (Def. 5.1): additivity in $Y$ and $Z$ is obvious, and homogeneity in $Z$ is easily demonstrated by a short calculation. We now verify the derivation property [writing $\langle X, Y\rangle$ instead of $g(X, Y)$]:

$$\begin{aligned} 2\langle \nabla_Z fY, X\rangle &= -X\langle fY, Z\rangle + fY\langle Z, X\rangle + Z\langle X, fY\rangle \\ &\quad - \langle [Z, X], fY\rangle - \langle [fY, Z], X\rangle + \langle [X, fY], Z\rangle \\ &= 2\langle f\nabla_Z Y, X\rangle - (Xf)\langle Y, Z\rangle + (Zf)\langle X, Y\rangle \\ &\quad + (Zf)\langle Y, X\rangle + (Xf)\langle Y, Z\rangle \\ &= 2[\langle f\nabla_Z Y, X\rangle + (Zf)\langle Y, X\rangle] \\ &= 2\langle f\nabla_Z Y + (Zf)Y, X\rangle . \end{aligned}$$

This proves the derivation property

$$\nabla_Z (fY) = f\nabla_Z Y + (Zf)\,Y .$$

The affine connection constructed in this manner has vanishing torsion: (5.37) implies that

$$\langle \nabla_Z Y, X\rangle = \langle \nabla_Y Z, X\rangle + \langle [Z, Y], X\rangle .$$

Furthermore, summation of the right hand sides of (5.37) for

$$2\langle \nabla_Z Y, X\rangle + 2\langle \nabla_Z X, Y\rangle$$

shows that the Ricci identity is satisfied.

**Definition 5.3:** The unique connection on $(M, g)$ from Theorem 5.4 is called the *Riemannian connection (Levi-Cività connection).*

*Local Expressions:* We determine the Christoffel symbols of the Riemannian connection in a chart $(U, x^1, \dots, x^n)$. For his purpose take $X = \partial_k$, $Y = \partial_j$, and $Z = \partial_i$ in (5.37) and use $[\partial_i, \partial_j] = 0$, as well as $\langle \partial_i, \partial_j\rangle = g_{ij}$. The result is

$$2\langle \nabla_{\partial_i} \partial_j, \partial_k\rangle = 2\Gamma^l_{ij}\, g_{lk} = -\partial_k\langle \partial_j, \partial_i\rangle + \partial_j\langle \partial_i, \partial_k\rangle + \partial_i\langle \partial_k, \partial_j\rangle$$

or

$$g_{mk}\Gamma^m_{ij} = \tfrac{1}{2}(g_{jk,i} + g_{ik,j} - g_{ij,k}) . \tag{5.38}$$

If $(g^{ij})$ denotes the matrix inverse to $(g_{ij})$, we obtain from (5.38)

$$\Gamma^l_{ij} = \tfrac{1}{2} g^{lk}(g_{ki,j} + g_{kj,i} - g_{ij,k}) . \tag{5.39}$$

**Proposition 5.4:** The curvature tensor of a Riemannian connection has the following additional symmetry properties:

$$\langle R(X, Y)Z, U\rangle = -\langle R(X, Y)U, Z\rangle \tag{5.40}$$

$$\langle R(X, Y)Z, U\rangle = \langle R(Z, U)X, Y\rangle . \tag{5.41}$$

*Proof:* It suffices to prove these identities for vector fields with pairwise vanishing Lie brackets (e.g., for the basis fields $\partial_i$ of a chart).

Equation (5.40) is equivalent to

$$\langle R(X,Y)Z,Z\rangle = 0\,. \tag{5.40'}$$

Since $\nabla$ is a Riemannian connection,

$$\langle \nabla_X \nabla_Y Z, Z\rangle = X\langle \nabla_Y Z, Z\rangle - \langle \nabla_Y Z, \nabla_X Z\rangle$$

and

$$\langle \nabla_Y Z, Z\rangle = \tfrac{1}{2} Y\langle Z,Z\rangle\,.$$

Hence,

$$2\langle R(X,Y)Z,Z\rangle = XY\langle Z,Z\rangle - YX\langle Z,Z\rangle = 0$$

thus proving (5.40).

The first Bianchi identity (5.32) with $T=0$ implies

$$\begin{aligned}\langle R(X,Y)Z,U\rangle &= -\langle R(Y,X)Z,U\rangle\\ &= \langle R(X,Z)Y,U\rangle + \langle R(Z,Y)X,U\rangle\,.\end{aligned}$$

In addition, (5.40) and the first Bianchi identity imply

$$\begin{aligned}\langle R(X,Y)Z,U\rangle &= -\langle R(X,Y)U,Z\rangle\\ &= \langle R(Y,U)X,Z\rangle + \langle R(U,X)Y,Z\rangle\,.\end{aligned}$$

Adding these last two equations results in

$$\begin{aligned}2\langle R(X,Y)Z,U\rangle = {}&\langle R(X,Z)Y,U\rangle + \langle R(Z,Y)X,U\rangle\\ &+ \langle R(Y,U)X,Z\rangle + \langle R(U,X)Y,Z\rangle\,.\end{aligned}$$

If we interchange the pairs $X$, $Z$ and $Y$, $U$, we obtain

$$\begin{aligned}2\langle R(Z,U)X,Y\rangle = {}&\langle R(Z,X)U,Y\rangle + \langle R(X,U)Z,Y\rangle\\ &+ \langle R(U,Y)Z,X\rangle + \langle R(Y,Z)U,X\rangle\,.\end{aligned}$$

Use of (5.40) and $R(X,Y) = -R(Y,X)$ then shows that the right hand sides of both equations are in agreement.

**Remarks:** 1) The metric tensor field $g$ permits us to map $\mathscr{X}(M)$ uniquely onto $\mathscr{X}^*(M)$; $X \mapsto X^\flat$, with

$$X^\flat(Z) = g(X,Z) \quad \text{for all} \quad Z \in \mathscr{X}(M)\,.$$

This operator can be applied to tensors to produce new ones. For example, we may assign to every field $f \in \mathscr{T}_2^0(M)$ a unique field $\tilde{t} \in \mathscr{T}_1^1(M)$:

$$\tilde{t}(X, Y^\flat) = t(X,Y)\,.$$

In local coordinates, this corresponds to raising and lowering indices with the metric tensor. The inverse of the map is denoted by #.

2) Since $g_{ij}\, g^{jk} = \delta_i^k$, it follows that $g^{ij}{}_{;k} = 0$ (note that $\delta_i^j{}_{;k} = 0$).

We now give coordinate expressions for the identities satisfied by the curvature tensor of a Riemannian connection. These read, if $\sum_{(ijk)}$ denotes the cyclic sum and $R_{ijkl} := g_{is} R^s_{jkl}$,

$$\sum_{(jkl)} R^i_{jkl} = 0 \qquad \text{(1st Bianchi identity)} \tag{5.42a}$$

$$\sum_{(klm)} R^i_{jkl;m} = 0 \qquad \text{(2nd Bianchi identity)} \tag{5.42b}$$

$$R^i_{jkl} = -R^i_{jlk} \tag{5.42c}$$

$$R_{ijkl} = -R_{jikl} \tag{5.42d}$$

$$R_{ijkl} = R_{klij}\,. \tag{5.42e}$$

*Proof:* From (5.28) we have

$$R^i_{jkl} = \langle dx^i, R(\partial_k, \partial_l)\,\partial_j\rangle\,, \quad R_{ijkl} = \langle \partial_i, R(\partial_k, \partial_l)\,\partial_j\rangle \tag{5.43}$$

(5.42c) follows as a result of $R(X, Y) = -R(Y, X)$.

(5.42a) and (5.42b) follow from the first (resp. second) Bianchi identities (5.32) and (5.33) with $T = 0$.

(5.42d) and (5.42e) are a consequence of (5.40) and (5.41).

## Contracted Bianchi Identity

If, as in (5.30), we denote the Ricci tensor by $R_{ik}$ and the *scalar curvature* by

$$R = g^{ik} R_{ik}\,, \tag{5.44}$$

then the *contracted Bianchi identity*

$$(R^k_i - \tfrac{1}{2}\delta^k_i R)_{;k} = 0 \tag{5.45}$$

holds. Furthermore, the Ricci tensor is symmetric

$$R_{ik} = R_{ki}\,. \tag{5.46}$$

*Proof:* The symmetry of $R_{ik}$ follows from (5.30), i.e.

$$R_{jl} = g^{ik} R_{ijkl} \tag{5.47}$$

and (5.42e).

Now consider

$$R_j{}^m{}_{;m} = g^{ml} R_{jl;m} = g^{ml} g^{ik} R_{ijkl;m} = g^{ml} g^{ik} R_{klij;m}\,.$$

Use of the second Bianchi identity (5.42b) results in

$$R_j{}^m{}_{;m} = -g^{ml} g^{ik} (R_{kljm;i} + R_{klmi;j})\,.$$

According to (5.42c), (5.42d) and (5.47), the first term on the right hand side is equal to

$$-g^{ik} R_{kj;i} = -R^l_{j;l}.$$

From (5.42d) and (5.47), the second term is equal to $g^{ik} R_{ki;j} = R_{,j}$. Hence

$$R^k_{i;k} = \tfrac{1}{2} R_{,i} = \tfrac{1}{2} (\delta^k_i R)_{;k}.$$

The *Einstein tensor* is defined by

$$G_{ik} = R_{ik} - \tfrac{1}{2} g_{ik} R. \tag{5.48}$$

By (5.45), it satisfies the contracted Bianchi identity

$$G^k_{i;k} = 0. \tag{5.49}$$

## 5.7 The Cartan Structure Equations

Let $M$ be a differentiable manifold with an affine connection. Let $(e_1, \ldots, e_n)$ be a basis of $C^\infty$ vector fields[4] defined on an open subset $U$ (which might be the domain of a chart, for example) and let $(\theta^1, \ldots, \theta^n)$ denote the corresponding dual basis of one-forms. We define *connection forms* $\omega^i_j \in \Lambda_1(U)$ by

$$\nabla_X e_j = \omega^i_j(X)\, e_i. \tag{5.50}$$

We may generalize the definition of the Christoffel symbols (relative to the basis $\{e_i\}$) as follows:

$$\nabla_{e_k} e_j = \Gamma^i_{kj}\, e_i = \omega^i_j(e_k)\, e_i.$$

Thus

$$\omega^i_j = \Gamma^i_{kj}\, \theta^k. \tag{5.51}$$

Since $\nabla_X$ commutes with contractions, we have, according to (5.50),

$$\begin{aligned} 0 = \nabla_X \langle \theta^i, e_j \rangle &= \langle \nabla_X \theta^i, e_j \rangle + \langle \theta^i, \nabla_X e_j \rangle \\ &= \langle \nabla_X \theta^i, e_j \rangle + \underbrace{\langle \theta^i, \omega^k_j(X)\, e_k \rangle}_{\omega^i_j(X)}. \end{aligned}$$

That is,

$$\nabla_X \theta^i = -\omega^i_j(X)\, \theta^j, \tag{5.52}$$

or

$$\nabla \theta^i = -\theta^j \otimes \omega^i_j. \tag{5.52'}$$

---

[4] Such a set is called a *moving frame* (*vierbein*, or *tetrad* for $n = 4$).

If $\alpha$ is a general one-form, $\alpha = \alpha_i \theta^i$, with $\alpha_i \in \mathscr{F}(U)$, then (5.52) implies

$$\begin{aligned}\nabla_X \alpha &= (X\alpha_i)\,\theta^i + \alpha_i \nabla_X \theta^i \\ &= \langle d\alpha_i - \alpha_l\,\omega^l_i, X\rangle\,\theta^i ,\end{aligned}$$

so that

$$\nabla \alpha = \theta^i \otimes (d\alpha_i - \omega^k_i\,\alpha_k) . \tag{5.53}$$

In an analogous manner, we find for a vector field $X = X^i e_i$

$$\nabla X = e_i \otimes (dX^i + \omega^i_k X^k) . \tag{5.54}$$

The torsion and the curvature define differential forms $\Theta^i$ *(torsion forms)* and $\Omega^i_j$ *(curvature forms)* by

$$T(X, Y) = \Theta^i(X, Y)\,e_i , \tag{5.55}$$

$$R(X, Y)\,e_j = \Omega^i_j(X, Y)\,e_i . \tag{5.56}$$

**Theorem 5.5:** The torsion forms and curvature forms satisfy the *Cartan structure equations*

$$\Theta^i = d\theta^i + \omega^i_j \wedge \theta^j \tag{5.57}$$

$$\Omega^i_j = d\omega^i_j + \omega^i_k \wedge \omega^k_j . \tag{5.58}$$

*Proof:* According to the definition and (5.50) we have

$$\begin{aligned}\Theta^i(X, Y)\,e_i &= \nabla_X Y - \nabla_Y X - [X, Y] \\ &= \nabla_X(\theta^j(Y)\,e_j) - \nabla_Y(\theta^j(X)\,e_j) - \theta^j([X, Y])\,e_j \\ &= \{X\theta^j(Y) - Y\theta^j(X) - \theta^j([X, Y])\}\,e_j \\ &\quad + \{\theta^j(Y)\,\omega^i_j(X) - \theta^j(X)\,\omega^i_j(Y)\}\,e_i \\ &= d\theta^i(X, Y)\,e_i + (\omega^i_j \wedge \theta^j)(X, Y)\,e_i .\end{aligned}$$

Equation (5.57) follows from this. The derivation of (5.58) is similar: From the definition and (5.50), we have

$$\begin{aligned}\Omega^i_j(X, Y)\,e_i &= \nabla_X \nabla_Y e_j - \nabla_Y \nabla_X e_j - \nabla_{[X, Y]}\,e_j \\ &= \nabla_X(\omega^i_j(Y)\,e_i) - \nabla_Y(\omega^i_j(X))\,e_i - \omega^i_j([X, Y])\,e_i \\ &= \{X\,\omega^i_j(Y) - Y\omega^i_j(X) - \omega^i_j([X, Y])\}\,e_i \\ &\quad + \{\omega^i_j(Y)\,\omega^k_i(X) - \omega^i_j(X)\,\omega^k_i(Y)\}\,e_k \\ &= d\omega^i_j(X, Y)\,e_i + (\omega^i_k \wedge \omega^k_j)(X, Y)\,e_i .\end{aligned}$$

A comparison of the components results in (5.58).

We can expand $\Omega^i_j$ as follows

$$\Omega^i_j = R^i_{jkl}\,\theta^k \wedge \theta^l , \qquad R^i_{jkl} = -R^i_{jlk} . \tag{5.59}$$

Since

$$\langle \theta^i, R(e_k, e_l)\, e_j \rangle = \langle \theta^i, \Omega^s_j (e_k, e_l)\, e_s \rangle = \Omega^i_j (e_k, e_l)\,,$$

we have

$$R^i_{jkl} = \langle \theta^i, R(e_k, e_l)\, e_j \rangle. \tag{5.60}$$

As a result of (5.43), $R^i_{jkl}$ agree with the components of the Riemann tensor for the special case $e_i = \partial/\partial x^i$. Equation (5.60) defines these components for an arbitrary basis $\{e_i\}$. Thus the expansion coefficients of the curvature form in (5.59) are the components of the Riemann tensor.

In an analogous manner we have

$$\Theta^i = \tfrac{1}{2} T^i_{kl}\, \theta^k \wedge \theta^l, \qquad T^i_{kl} = \langle \theta^i, T(e_k, e_l) \rangle\,. \tag{5.60'}$$

**Proposition 5.5:** An affine connection $\nabla$ is metric if and only if

$$\omega_{ik} + \omega_{ki} = dg_{ik}\,, \tag{5.61}$$

where

$$\omega_{ik} = g_{ij}\, \omega^j_k \quad \text{and} \quad g_{ik} = g(e_i, e_k)\,.$$

*Proof:* As a result of Proposition 5.3, a connection is metric if and only if the Ricci identity holds:

$$\begin{aligned} dg_{ik}(X) &= X\, g_{ik} = X \langle e_i, e_k \rangle \\ &= \langle \nabla_X e_i, e_k \rangle + \langle e_i, \nabla_X e_k \rangle \\ &= \omega^j_i(X) \langle e_j, e_k \rangle + \omega^j_k(X) \langle e_i, e_j \rangle \qquad \text{(by (5.50))} \\ &= g_{jk}\, \omega^j_i(X) + g_{ij}\, \omega^j_k(X)\,. \end{aligned}$$

Thus, for a Riemannian connection the following equations hold:

$$\begin{aligned} \omega_{ij} + \omega_{ji} &= dg_{ij} \\ d\theta^i + \omega^i_j \wedge \theta^j &= 0 \\ d\omega^i_j + \omega^i_k \wedge \omega^k_j &= \Omega^i_j = \tfrac{1}{2} R^i_{jkl}\, \theta^k \wedge \theta^l\,. \end{aligned} \tag{5.63}$$

*Solution of the Structure Equations*

We expand $d\theta^i$ in terms of the basis $\theta^i$:

$$d\theta^i = -\tfrac{1}{2} C^i_{jk}\, \theta^j \wedge \theta^k, \quad C^i_{jk} = -C^i_{kj}\,. \tag{5.64}$$

Hence

$$\Gamma^i_{kl} - \Gamma^i_{lk} = C^i_{kl}. \tag{5.65}$$

(For $\theta^i = dx^i$, we have $C^i_{kl} = 0$ and the $\Gamma^i_{kl}$ are symmetric in $k$ and $l$.) In addition we set

$$dg_{ij} = g_{ij,k}\, \theta^k, \qquad g_{ij,k} = e_k(g_{ij})\,. \tag{5.66}$$

Since, according to (5.51),

$$\omega_{ij} = g_{is}\Gamma^s_{lj}\theta^l,$$

the first equation in (5.63) implies

$$g_{ij,k} = g_{is}\Gamma^s_{kj} + g_{js}\Gamma^s_{ki}. \tag{5.67}$$

We take cyclic permutations to obtain

$$g_{jk,i} = g_{js}\Gamma^s_{ik} + g_{ks}\Gamma^s_{ij} \tag{5.68}$$

$$g_{ki,j} = g_{ks}\Gamma^s_{ji} + g_{is}\Gamma^s_{jk}. \tag{5.69}$$

Forming the combination (5.67) + (5.68) − (5.69), and using (5.65) results in

$$g_{ij,k} + g_{jk,i} - g_{ki,j} = g_{is}C^s_{kj} + g_{ks}C^s_{ij} + g_{js}(\Gamma^s_{ki} + \Gamma^s_{ik}).$$

If we contract this equation with $g^{li}$, we obtain

$$\Gamma^l_{ki} + \Gamma^l_{ik} = g^{lj}(g_{ij,k} + g_{jk,i} - g_{ki,j}) - g^{lj}g_{is}C^s_{kj} - g^{lj}g_{ks}C^s_{ij}.$$

We now use (5.65) again and obtain

$$\Gamma^l_{ki} = \tfrac{1}{2}(C^l_{ki} - g_{is}g^{lj}C^s_{kj} - g_{ks}g^{lj}C^s_{ij}) + \tfrac{1}{2}g^{lj}(g_{ij,k} + g_{jk,i} - g_{ki,j}). \tag{5.70}$$

The second term vanishes for an orthonormal basis.

According to (5.51), we have

$$d\omega^i_j = d\Gamma^i_{lj} \wedge \theta^l + \Gamma^i_{lj}d\theta^l.$$

We now set $d\Gamma^i_{lj} = \Gamma^i_{lj,s}\theta^s$ $(\Gamma^i_{lj,s} = e_s\Gamma^i_{lj})$ and use (5.64). The result is

$$d\omega^i_j = \Gamma^i_{lj,s}\,\theta^s \wedge \theta^l - \tfrac{1}{2}\Gamma^i_{lj}C^l_{ab}\theta^a \wedge \theta^b,$$

or

$$d\omega^i_j = \tfrac{1}{2}(\Gamma^i_{bj,a} - \Gamma^i_{aj,b} - \Gamma^i_{lj}C^l_{ab})\,\theta^a \wedge \theta^b.$$

As a consequence, the second Cartan structure equation [the third equation in (5.63)] becomes

$$\begin{aligned} d\omega^i_j + \omega^i_l \wedge \omega^l_j &= \tfrac{1}{2}(\Gamma^i_{bj,a} - \Gamma^i_{aj,b} - \Gamma^i_{lj}C^l_{ab})\,\theta^a \wedge \theta^b \\ &\quad + \tfrac{1}{2}(\Gamma^i_{al}\Gamma^l_{bj} - \Gamma^i_{bl}\Gamma^l_{aj})\,\theta^a \wedge \theta^b = \tfrac{1}{2}R^i_{jab}\theta^a \wedge \theta^b = \Omega^i_j. \end{aligned}$$

Thus the components of the curvature tensor are given by

$$R^i_{jab} = \Gamma^i_{bj,a} - \Gamma^i_{aj,b} - \Gamma^i_{lj}C^l_{ab} + \Gamma^i_{al}\Gamma^l_{bj} - \Gamma^i_{bl}\Gamma^l_{aj}. \tag{5.71}$$

## 5.8 Bianchi Identities for the Curvature and Torsion Forms

In the following we again consider an arbitrary affine connection $\nabla$ on a manifold $M$. As before, let $\{e_i\}$, $\{\theta^i\}$ be mutually dual bases and let $\omega^i_j$ denote the connection forms.

**Proposition 5.6:** Under a change of basis

$$\bar{\theta}^i(x) = A^i_j(x)\,\theta^j(x), \tag{5.72}$$

$\omega = (\omega^i_j)$ transforms inhomogeneously; in matrix notation we have

$$\bar{\omega} = A\,\omega\,A^{-1} - dA\,A^{-1}. \tag{5.73}$$

*Proof:* In an obvious matrix notation we have, according to (5.52),

$$\begin{aligned}\nabla_X \bar{\theta} &= -\,\bar{\omega}(X)\,\bar{\theta} = \nabla_X(A\theta) = dA(X)\,\theta + A\nabla_X\theta\\ &= dA(X)\,\theta - A\,\omega(X)\,\theta = dA(X)\,A^{-1}\,\bar{\theta} - A\,\omega(X)\,A^{-1}\,\bar{\theta}.\end{aligned}$$

From this the conclusion follows immediately.

**Definition:** A tensor valued $p$-form of type $(r, s)$ is a skew symmetric $p$-multilinear mapping:

$$\phi\colon \underbrace{\mathscr{X}(M) \times \mathscr{X}(M) \times \ldots \times \mathscr{X}(M)}_{p\text{-times}} \to \mathscr{T}^r_s(M).$$

The $p$-forms

$$\phi^{i_1 \ldots i_r}_{j_1 \ldots j_s} = \phi(\theta^{i_1}, \ldots, \theta^{i_r}, e_{j_1} \ldots e_{j_s})$$

are the *components* of the tensor valued $p$-form relative to the basis $\theta^i$.

**Proposition 5.7.** For every tensor valued $p$-form $\phi$ of type $(r, s)$ there exists a unique tensor valued $(p+1)$-form $D\phi$, also of type $(r, s)$, which has the following components relative to the basis $\theta^i$:

$$(D\phi)^{i_1 \ldots i_r}_{j_1 \ldots j_s} = d\phi^{i_1 \ldots i_r}_{j_1 \ldots j_s} + \omega^{i_1}_l \wedge \phi^{l i_2 \ldots i_r}_{j_1 \ldots j_s} + \ldots - \omega^l_{j_1} \wedge \phi^{i_1 \ldots i_r}_{l j_2 \ldots j_s} - \ldots. \tag{5.74}$$

*Proof:* It is sufficient to show that the right hand side of (5.74) transforms like the components of a tensor of type $(r, s)$ under change of basis $\bar{\theta}(x) = A(x)\,\theta(x)$. Since every pair of upper and lower indices in (5.74) behaves in the same manner, we may consider a tensor valued $p$-form $\phi^i_j$ of type (1,1). The transformed components $\bar{\phi} = (\bar{\phi}^i_j)$ are given by $\bar{\phi} = A\phi A^{-1}$. In addition,

$$D\bar{\phi} = d\bar{\phi} + \bar{\omega} \wedge \bar{\phi} - (-1)^p\,\bar{\phi} \wedge \bar{\omega},$$

or, using (5.73) and $dA\,A^{-1} = -\,A\,dA^{-1}$,

$$\begin{aligned}D\bar{\phi} &= d(A\phi A^{-1}) + [A\,\omega\,A^{-1} - dA\,A^{-1}] \wedge A\phi A^{-1}\\ &\quad -(-1)^p A\phi A^{-1} \wedge [A\,\omega\,A^{-1} - dA\,A^{-1}]\\ &= dA\,\phi A^{-1} + A\,d\phi A^{-1} + (-1)^p A\,\phi\,dA^{-1} + [A\,\omega - dA] \wedge \phi\,A^{-1}\\ &\quad -(-1)^p A\phi \wedge [\omega\,A^{-1} + dA^{-1}]\\ &= A(d\phi + \omega \wedge \phi - (-1)^p\,\phi \wedge \omega]\,A^{-1}\\ &= A(D\phi)\,A^{-1}.\end{aligned}$$

The tensor valued $(p+1)$-form $D\phi$ of Proposition 5.7 is known as the *absolute exterior differential* of the tensor valued $p$-form.

*Special Cases*

1) A "usual" $p$-form is a tensor valued form of type $(0, 0)$ and for this case we have $D = d$.
2) A tensor field $t \in \mathcal{T}^r_s(M)$ is a tensor valued 0-form of type $(r, s)$ and

$$Dt = \nabla t. \tag{5.75}$$

Indeed, for a coordinate basis $\theta^i = dx^i$, (5.74) reduces with the help of (5.51) to (5.22).

It follows trivially from (5.74) that $D$ satisfies the antiderivation rule for two tensor valued forms:

$$D(\phi \wedge \psi) = D\phi \wedge \psi + (-1)^p \phi \wedge D\psi, \quad p = \text{degree of } \phi.$$

Here, $\wedge$ denotes exterior multiplication of the components[5].

**Remarks**

1) A connection is metric if and only if $Dg = 0$.
2) The basis $\theta^i$ is a tensor valued 1-form of type $(1, 0)$. The first structure equation can be written in the form

$$D\theta^i = \Theta^i$$

as one easily infers from (5.74).

We may now write the Bianchi identities in a very compact form. Obviously, $\Theta^i$ and $\Omega^i_j$ are tensor valued 2-forms of type $(1, 0)$ and $(1, 1)$, respectively [see the defining equations (5.55), (5.56)].

**Proposition 5.8:** The torsion and curvature forms satisfy the following identities:

$$D\Theta^i = \Omega^i_j \wedge \theta^j \quad \text{(1st Bianchi identity)} \tag{5.76}$$

$$D\Omega^i_j = 0. \quad \text{(2nd Bianchi identity)} \tag{5.77}$$

*Proof:* Using (5.74) and the Cartan structure equations we have, in matrix notation,

$$\begin{aligned} D\Theta &= d\Theta + \omega \wedge \Theta \\ &= d(d\theta + \omega \wedge \theta) + \omega \wedge (d\theta + \omega \wedge \theta) \\ &= d\omega \wedge \theta - \omega \wedge d\theta + \omega \wedge d\theta + \omega \wedge \omega \wedge \theta = \Omega \wedge \theta. \end{aligned}$$

In an analogous manner,

$$D\Omega = d\Omega + \omega \wedge \Omega - \Omega \wedge \omega = d\Omega + \omega \wedge d\omega - d\omega \wedge \omega$$

$$d\Omega = d(d\omega + \omega \wedge \omega) = d\omega \wedge \omega - \omega \wedge d\omega$$

---

[5] If $\phi$ is a tensor valued $p$-form and $\psi$ a tensor valued $q$-form, then $(\phi \wedge \psi)(X_1, \ldots, X_{p+q})$ $= \frac{1}{p!\,q!} \sum_{\sigma \in \mathcal{S}_{p+q}} \operatorname{sgn}(\sigma)\, \phi(X_{\sigma(1)} \ldots X_{\sigma(p)}) \otimes \psi(X_{\sigma(p+1)} \ldots X_{\sigma(p+q)})$.

and hence

$$D\Omega = 0.$$

We now show that (5.76) and (5.77) are equivalent to the previous formulations of the Bianchi identities (5.32), (5.33).

In a natural basis $\theta^i = dx^i$, (5.76) has the form (see also (5.60′))

$$\begin{aligned} D\Theta^i &= d\Theta^i + \omega^i_j \wedge \Theta^j \\ &= d(\tfrac{1}{2}\, T^i_{kl}\, dx^k \wedge dx^l) + \omega^i_j \wedge \tfrac{1}{2}\, T^j_{kl}\, dx^k \wedge dx^l \\ &= \Omega^i_j \wedge \theta^j = \tfrac{1}{2}\, R^i_{jkl}\, dx^k \wedge dx^l \wedge dx^j \end{aligned}$$

or

$$(T^i_{kl,j} + \Gamma^i_{js} T^s_{kl})\, dx^j \wedge dx^k \wedge dx^l = R^i_{jkl}\, dx^j \wedge dx^k \wedge dx^l.$$

Thus

$$\sum_{(jkl)} T^i_{kl,j} = \sum_{(jkl)} (R^i_{jkl} - \Gamma^i_{js} T^s_{kl}).$$

We now use (5.27) and obtain

$$\begin{aligned} \sum_{(jkl)} T^i_{kl;j} &= \sum_{(jkl)} (T^i_{kl,j} + \Gamma^i_{js} T^s_{kl} - \Gamma^s_{jk} T^i_{sl} - \Gamma^s_{jl} T^i_{ks}) \\ &= \sum_{(jkl)} (T^i_{kl,j} + \Gamma^i_{js} T^s_{kl} + T^s_{jl} T^i_{ks}). \end{aligned}$$

Hence

$$\sum_{(jkl)} R^i_{jkl} = \sum_{(jkl)} (T^i_{kl;j} - T^s_{jl} T^i_{ks}). \tag{5.78}$$

This equation is valid in any basis $(\theta^i)$ and is simply (5.32) written in terms of components.

Similarly one finds from (5.77)

$$\sum_{(klm)} R^i_{jkl;\, m} = \sum_{(klm)} T^s_{km} R^i_{jsl}, \tag{5.79}$$

which is (5.33) in terms of components.

Let $\phi$ be a tensor valued $p$-form. If we apply the absolute exterior differential twice and use (5.74) and the second structure equation, we obtain immediately

$$(D^2\phi)^{i_1 \dots i_r}_{j_1 \dots j_s} = \Omega^{i_1}_l \wedge \phi^{l i_2 \dots i_r}_{j_1 \dots j_s} + \dots - \Omega^l_{j_1} \wedge \phi^{i_1 \dots i_r}_{l j_2 \dots j_s} - \dots. \tag{5.80}$$

**Examples**

1) In particular, we have for a function $f$

$$D^2 f = 0.$$

Now $Df = df = f_{,i}\, dx^i = f_{;i}\, dx^i$; according to the antiderivation rule,

$$D^2 f = Df_{;i} \wedge dx^i + f_{;i} D\, dx^i.$$

According to the first structure equation $D\,dx^i = \Theta^i = \frac{1}{2} T^i_{kl}\,dx^k \wedge dx^l$. Hence, with (5.75), we have

$$D^2 f = f_{;i;k}\,dx^k \wedge dx^i + \tfrac{1}{2} f_{;s} T^s_{kl}\,dx^k \wedge dx^l$$

so that

$$f_{;i;k} - f_{;k;i} = T^s_{ik} f_{;s}. \tag{5.81}$$

2) For a vector field $\xi^i$ (5.80) becomes

$$D^2 \xi^i = \Omega^i_l \xi^l = \tfrac{1}{2} R^i_{jkl} \xi^j\,dx^k \wedge dx^l.$$

On the other hand,

$$\begin{aligned} D^2 \xi^i &= D(\xi^i_{;l}\,dx^l) = D\xi^i_{;l} \wedge dx^l + \xi^i_{;s}\,\Theta^s \\ &= \xi^i_{;l;k}\,dx^k \wedge dx^l + \xi^i_{;s} \tfrac{1}{2} T^s_{kl}\,dx^k \wedge dx^l. \end{aligned}$$

Direct comparison results in

$$\xi^i_{;l;k} - \xi^i_{;k;l} = R^i_{jkl} \xi^j + T^s_{lk} \xi^i_{;s}. \tag{5.82}$$

In particular, if $\Theta = 0$, we have

$$f_{;i;k} = f_{;k;i} \tag{5.81'}$$

$$\xi^i_{;l;k} - \xi^i_{;k;l} = R^i_{jkl} \xi^j. \tag{5.82'}$$

## 5.9 Locally Flat Manifolds

**Definition:** Let $(M, g)$ and $(N, h)$ be two pseudo-Riemannian manifolds. A diffeomorphism $\varphi\colon M \to N$ is an *isometry* if $\varphi^* h = g$. Every pseudo-Riemannian manifold which is isometric to $(\mathbb{R}^n, \mathring{g})$, with

$$\mathring{g} = \sum_{i=1}^{n} \varepsilon_i\,dx^i\,dx^i, \quad \varepsilon_i = \pm 1 \tag{5.83}$$

is said to be *flat*. If a space is locally isometric to a flat space, then it is said to be *locally flat*.

**Theorem 5.6:** A pseudo-Riemannian space is locally flat if and only if the curvature of the Riemannian connection vanishes.

*Proof:* If the space is locally flat, the curvature vanishes. In order to see this, choose a chart in the neighborhood of a point for which the metric has the form (5.83). In this neighborhood the Christoffel symbols vanish and hence also the curvature.

In order to prove the converse, we use Theorem 5.7. According to this theorem, parallel transport is locally independent of the path when $\Omega^i_j = 0$. Hence it is possible to parallel displace a basis of vectors in $T_p(M)$, $p \in M$, to all points in some neighborhood of $p$. In this manner we obtain local basis fields $e_i$ which have vanishing covariant derivative. For the dual basis $\{\theta^i\}$ corresponding to $\{e_i\}$ we have

$$d\theta^i(e_j, e_k) = -\theta^i([e_j, e_k]).$$

However, if the connection is symmetric, we have

$$[e_j, e_k] = \nabla_{e_j} e_k - \nabla_{e_k} e_j = 0.$$

Consequently $d\theta^i = 0$. According to Poincaré's Lemma there exist local functions $x^i\colon U \to \mathbb{R}$, $p \in U$, for which $\theta^i = dx^i$. If we choose these $x^i$ as coordinates, then $e_i = \partial/\partial x^i$ and

$$0 = \nabla_{e_i} e_j = \Gamma^k_{ij} e_k.$$

Thus the $\Gamma^i_{ij}$ vanish identically in the coordinate system $x^i$ in a neighborhood of $p$. If the connection is pseudo-Riemannian, then the metric coefficients $g_{ik}$ are constant in the neighborhood $U$ [see, e.g., (5.61)]. In this case, they can be transformed to the normal form (5.83) by a suitable choice of coordinates.

**Theorem 5.7:** For an affine connection, parallel transport is independent of the path if and only if the curvature tensor vanishes.

We prove this theorem after the following heuristic discussion. Let $\gamma\colon [0, 1] \to M$ be a closed path, with $\gamma(0) = p$. We displace an arbitrary vector $v_0 \in T_p(M)$ parallel along $\gamma$ and obtain the field $v(t) = \tau_t v_0$, $v(t) \in T_{\gamma(t)}(M)$. We assume that the closed path is sufficiently small that we can work in a particular chart. Then $\dot v^i = -\Gamma^i_{jk} \dot x^j v^k$. We are interested in

$$\begin{aligned}\Delta v^i &= v^i(1) - v^i(0) = \int_0^1 \dot v^i \, dt = -\int_0^1 \Gamma^i_{jk}(\gamma(t)) \, v^k \dot x^j \, dt \\ &= -\int_\gamma \Gamma^i_{kj} v^j dx^k = -\int_\gamma \omega^i_j v^j.\end{aligned} \tag{5.84}$$

We shall assume that the coordinate system is geodetic at the point $p = \gamma(0)$. We now evaluate the path integral (5.84) for an infinitesimal loop. For this purpose we use

$$\omega^i_j v^j = \omega^i_j (v^j - v^j_0) + \omega^i_j v^j_0$$

and replace $\omega^i_j$ by $\Gamma^i_{kj}(p) \, dx^k = 0$ in the first term of the right hand side. Using Stokes' theorem and the second structure equation, we obtain, when $f$ denotes a surface enclosed by $\gamma$,

$$\begin{aligned}\Delta v^i &\simeq -\int_\gamma \omega^i_j v^j_0 = -\int_f d\omega^i_j v^j_0 \\ &\simeq -\tfrac{1}{2} R^i_{jkl}(p) \, v^j_0 \int_f dx^k \wedge dx^l.\end{aligned} \tag{5.85}$$

Equation (5.85) implies heuristically Theorem 5.7. We now give a more rigorous proof of this theorem.

*Proof of Theorem 5.7:* Consider the 1-parameter family of curves $H(t, s)$, $(\alpha \leqq t \leqq \beta)$, $s \in J$, with $H(\alpha, s) = p$ and $H(\beta, s) = q$, for all $s \in J$. Let $v \in T_p(M)$ and let $Y(t, s)$ be the set of vectors which one obtains by parallel displacement of $v$ along $t \mapsto H(t, s)$ for every fixed $s$. Let $D_1$ be the field tangent to the curves $t \mapsto H(t, s)$ and $D_2$ be the field tangent to the curves $s \mapsto H(t, s)$ (for fixed $t$). With this notation, we have $Y(\alpha, s) = v$, $s \in J$, and

$$(\nabla_{D_1} Y)_{t,s} = 0, \quad (\nabla_{D_2} Y)_{\alpha, s} = 0. \tag{5.86}$$

Naturally, $[D_1, D_2] = 0$ since differentiation with respect to $s$ and $t$ is commutative. Now suppose that the curvature vanishes. It follows from (5.86) that

$$\nabla_{D_1} \nabla_{D_2} Y\,|_{t,s} = \nabla_{D_2} \nabla_{D_1} Y\,|_{t,s} = 0,$$

which means that the family $t \mapsto \nabla_{D_2} Y$ is autoparallel along $t \mapsto H(t, s)$. However, (5.86) then implies

$$\nabla_{D_2} Y\,|_{t,s} = 0. \tag{5.87}$$

One thus obtains the same result whether $v$ is transported parallel along the path segment $\gamma_1$ or $\gamma_2$ (see Fig. 5).

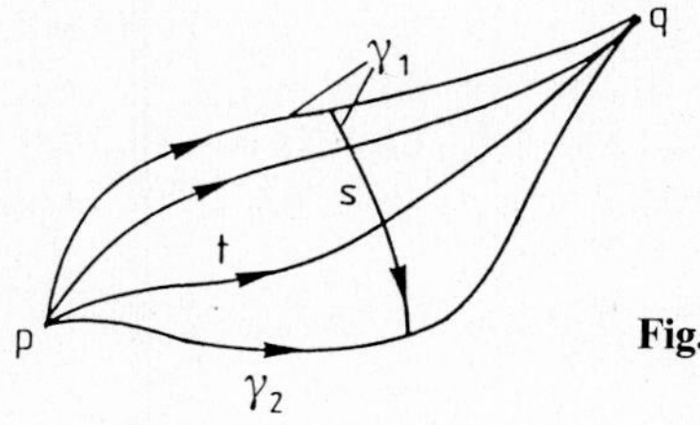


**Fig. 5**

By going to the limit, one sees that parallel displacement to $q$ is independent of the path.

Conversely, assume that parallel displacement is independent of the path. Then (5.87) is valid, and from this we may conclude that

$$(\nabla_{D_1} \nabla_{D_2} - \nabla_{D_2} \nabla_{D_1})\, Y = R(D_1, D_2)\, Y = 0$$

for all $t$, $s$, and hence that $\Omega = 0$.

## 5.10 Table of Important Formulae

The following table summarizes some of the important formulae which have been obtained in the foregoing sections and will constantly be used throughout the book.

1. Vector fields on $M$ with the bracket $[X, Y]$ form a Lie algebra over $\mathbb{R}$ and a module over the associative algebra $\mathscr{F}(M)$ of $C^\infty$ functions. In local coordinates:

$$X = \xi^i \partial_i, \quad Y = \eta^j \partial_j, \quad [X, Y] = (\xi^i \eta^j_{,i} - \eta^i \xi^j_{,i}) \partial_j .$$

2. For a diffeomorphism $\varphi$, $\varphi_* [X, Y] = [\varphi_* X, \varphi_* Y]$.
3. The differential forms on a manifold form a real associative algebra with $\wedge$ as multiplication. Furthermore, $\alpha \wedge \beta = (-1)^{kl} \beta \wedge \alpha$ for $k$ and $l$ forms $\alpha$ and $\beta$, respectively.
4. If $\varphi$ is a map, $\varphi^* (\alpha \wedge \beta) = \varphi^* \alpha \wedge \varphi^* \beta$.
5. The exterior differential $d$ is an antiderivation of degree $+1$, in particular

$$d(\alpha \wedge \beta) = d\alpha \wedge \beta + (-1)^k \alpha \wedge d\beta \qquad \text{for} \quad \alpha \text{ a } k\text{-form.}$$

Furthermore,

$$d \circ d = 0 .$$

6. For $\alpha$ a $k$-form and $X_0, \ldots, X_k$ vector fields:

$$d\alpha(X_0, \ldots, X_k) = \sum_{i=0}^{k} (-1)^i X_i (\alpha(X_0, \ldots, \hat{X}_i, \ldots, X_k)) + \sum_{i<j} (-1)^{i+j} \alpha([X_i, X_j], X_0, \ldots, \hat{X}_i, \ldots, \hat{X}_j, \ldots, X_k) .$$

7. For a map $\varphi$, $\varphi^* d\alpha = d\varphi^* \alpha$.
8. (Poincaré Lemma) If $d\alpha = 0$, then $\alpha$ is locally exact; that is, there is a neighborhood $U$ of each point on which $\alpha = d\beta$.
9. The interior derivative $i_X$ is an antiderivation of degree $-1$ and for $f \in \mathscr{F}(M)$, $i_{fX} \alpha = f i_X \alpha = i_X f \alpha$. Also $i_X \circ i_X = 0$.
10. For a diffeomorphism $\varphi$, $\varphi^* i_X \alpha = i_{\varphi^* X} \varphi^* \alpha$.
11. The Lie derivative $L_X$ is a derivation of the tensor algebra $\mathscr{T}(M)$ of degree 0, is $\mathbb{R}$-linear in $X$ and for a diffeomorphism $\varphi$, $\varphi^* L_X T = L_{\varphi^* X} \varphi^* T$, $T \in \mathscr{T}(M)$.
12. For a covariant tensor field $T \in \mathscr{T}^0_k(M)$,

$$(L_X T)(X_1, \ldots, X_k) = X(T(X_1, \ldots, X_k)) - \sum_{i=1}^{k} T(X_1, \ldots, [X, X_i], \ldots, X_k) .$$

13. In local coordinates we have for a tensor field $T \in \mathscr{T}^r_s(M)$

$$(L_X T)^{i_1 \ldots i_r}_{j_1 \ldots j_s} = X^i T^{i_1 \ldots i_r}_{j_1 \ldots j_s, i} - T^{k i_2 \ldots i_r}_{j_1 \ldots j_s} X^{i_1}_{,k} - (\text{all upper indices}) + T^{i_1 \ldots i_r}_{k j_2 \ldots j_s} X^k_{,j_1} + (\text{all lower indices}).$$

14. The following identities hold for differential forms

$$L_X \alpha = d i_X \alpha + i_X d\alpha$$

$$L_{fX}\,\alpha = f\,L_X\alpha + df\wedge i_X\alpha$$
$$L_{[X,Y]} = L_X\circ L_Y - L_Y\circ L_X$$
$$i_{[X,Y]} = L_X\circ i_Y - i_Y\circ L_X$$
$$L_X\circ d = d\circ L_X$$
$$L_X\circ i_X = i_X\circ L_X\,.$$

15. The Hodge star operator for an oriented $n$-dimensional pseudo-Riemannian manifold $(M, g)$ with volume element $\eta$ determined by $g$ is a linear isomorphism $*\colon \Lambda_k(M)\to\Lambda_{n-k}(M)$ and satisfies for $\alpha$, $\beta\in\Lambda_k(M)$:

$$\alpha\wedge *\beta = \beta\wedge *\alpha = (\alpha,\beta)\,\eta,\qquad *(*\alpha) = (-1)^{k(n-k)}\,\mathrm{sgn}\,(g)\,\alpha$$
$$(*\alpha, *\beta) = \mathrm{sgn}\,(g)\,(\alpha,\beta)\,.$$

If $\theta^1,\dots,\theta^n$ is an oriented basis of 1-forms, then

$$\eta = \sqrt{|g|}\,\theta^1\wedge\dots\wedge\theta^n,\qquad g = g_{ik}\,\theta^i\otimes\theta^k\,,$$
$$|g| = |\det(g_{ik})|$$

and

$$*(\theta^{i_1}\wedge\dots\wedge\theta^{i_k}) = \sqrt{|g|}\,\frac{1}{(n-k)!}\,\varepsilon_{j_1\dots j_n}\,g^{j_1 i_1}\dots g^{j_k i_k}\,\theta^{j_{k+1}}\wedge\dots\wedge\theta^{j_n}.$$

16. The codifferential $\delta\colon \Lambda_k(M)\to\Lambda_{k-1}(M)$ is defined by

$$\delta = \mathrm{sgn}\,(g)\,(-1)^{n(k+1)}*d*$$

and satisfies

$$\delta\circ\delta = 0\,.$$

In local coordinates we have for $\beta\in\Lambda_k(M)$

$$(\delta\beta)^{i_1\dots i_{k-1}} = |g|^{-1/2}\,(|g|^{1/2}\,\beta^{i\,i_1\dots i_{k-1}})_{,i}\,.$$

17. The covariant derivative $\nabla_X$ of an affine connection on $M$ is a derivation of the tensor algebra $\mathcal{T}(M)$, which commutes with all contractions. For $S\in\mathcal{T}^p_q(M)$,

$$(\nabla_X S)(Y_1,\dots,Y_q,\omega_1,\dots,\omega_p) = X(S(Y_1,\dots,Y_q,\omega_1,\dots,\omega_p))$$
$$- S(\nabla_X Y_1, Y_2,\dots,\omega_p) - \dots - S(Y_1,\dots,\nabla_X\omega_p)\,.$$

In local coordinates

$$S^{i_1\dots i_p}_{j_1,\dots,j_q;k} = S^{i_1\dots i_p}_{j_1\dots j_q,k} + \Gamma^{i_1}_{kl}\,S^{l i_2\dots i_p}_{j_1\dots j_q} + \dots - \Gamma^{l}_{k j_1}\,S^{i_1\dots i_p}_{l j_2\dots j_q} - \dots\,.$$

18. Let $(e_1,\dots,e_n)$ be a moving frame with the dual basis $(\theta^1,\dots,\theta^n)$ of 1-forms. The connection forms $\omega^i_j$ are given by

$$\nabla_X e_j = \omega^i_j(X)\,e_i$$

or

$$\nabla_X \theta^i = -\,\omega^i_j(X)\,\theta^j .$$

The torsion and curvature forms satisfy the structure equations

$$d\theta^i + \omega^i_j \wedge \theta^j = \Theta^i$$
$$d\omega^i_j + \omega^i_k \wedge \omega^k_j = \Omega^i_j .$$

The expansion coefficients of $\Omega^i_j$,

$$\Omega^i_j = \tfrac{1}{2} R^i_{jkl}\,\theta^k \wedge \theta^l$$

are the components of the Riemann tensor.

For the Riemannian connection, with metric $g = g_{ik}\,\theta^i \otimes \theta^k$, we have in addition

$$\omega_{ik} + \omega_{ki} = dg_{ik} \quad \text{with} \quad \omega_{ik} := g_{ij}\,\omega^j_k .$$

19. The absolute exterior differential of a tensor valued $p$-form of type $(r, s)$ is given by

$$(D\phi)^{i_1\ldots i_r}_{j_1\ldots j_s} = d\phi^{i_1\ldots i_r}_{j_1\ldots j_s} + \omega^{i_1}_l \wedge \phi^{l i_2\ldots i_r}_{j_1\ldots j_s} + \ldots - \omega^l_{j_1} \wedge \phi^{i_1\ldots i_r}_{l j_2\ldots j_s} - \ldots .$$

We have

$$(D^2\phi)^{i_1\ldots i_r}_{j_1\ldots j_s} = \Omega^{i_1}_l \wedge \phi^{l i_2\ldots i_r}_{j_1\ldots j_s} + \ldots - \Omega^l_{j_1} \wedge \phi^{i_1\ldots i_r}_{l j_2\ldots j_s} - \ldots .$$

The Bianchi-identities are

$$D\Theta^i = \Omega^i_j \wedge \theta^j$$
$$D\Omega^i_j = 0 .$$

20. The coordinate expression of the Riemann tensor is

$$R^i_{jkl} = \Gamma^i_{lj,k} - \Gamma^i_{kj,l} + \Gamma^s_{lj}\Gamma^i_{ks} - \Gamma^s_{kj}\Gamma^i_{ls} .$$

For the Levi-Civita connection

$$\Gamma^l_{ij} = \tfrac{1}{2} g^{lk}(g_{ki,j} + g_{kj,i} - g_{ij,k}) .$$

The Ricci-tensor is

$$R_{jl} = R^i_{jil} .$$

---

**Exercises:** The *Laplacian* on differential forms is defined by

$$\Delta = d \circ \delta + \delta \circ d$$

1. Show that

$$*\Delta = \Delta *$$
$$\delta\Delta = \Delta\delta$$
$$d\Delta = \Delta d .$$

2. Prove that $\Delta$ is self-adjoint, in the sense that

$$\langle \Delta\alpha, \beta\rangle = \langle \alpha, \Delta\beta\rangle ,$$

where

$$\langle \alpha, \beta\rangle = \int_M (\alpha, \beta)\, \eta = \int_M \alpha \wedge *\beta$$

by showing that $d$ and $\delta$ are adjoint to each other, up to a sign:

$$\langle d\alpha, \beta\rangle = -\langle \alpha, \delta\beta\rangle, \quad \alpha \in \Lambda_{p-1}(M), \quad \beta \in \Lambda_p(M) .$$

3. Show that on a proper Riemannian manifold the form $\omega$ is *harmonic,* i.e. $\Delta\,\omega = 0$, if and only if $d\omega = \delta\omega = 0$.
4. Derive a coordinate expression for $\Delta\,\omega$.
5. Prove that a harmonic function with compact support on a Riemannian manifold is constant.

---

# Part II

# General Theory of Relativity

# Introduction

The discovery of the general theory of relativity (GR) has often been justly praised as one of the greatest intellectual achievements of a single human being. At the ceremonial presentation of Hubacher's bust of Einstein in Zürich, W. Pauli said:

"The general theory of relativity then completed and – in contrast to the special theory – worked out by Einstein alone without simultaneous contributions by other researchers, will forever remain the classic example of a theory of perfect beauty in its mathematical structure."

Let us also quote M. Born:

"(The general theory of relativity) seemed and still seems to me at present to be the greatest accomplishment of human thought about nature; it is a most remarkable combination of philosophical depth, physical intuition and mathematical ingenuity. I admire it as a work of art."

The origin of the GR is all the more remarkable when one considers that, aside from a minute rotation of the perihelion of Mercury, which remained unexplained even after all perturbations in the Newtonian theory were taken into account, no experimental necessity for going beyond the Newtonian theory of gravitation existed. Purely theoretical considerations led to the genesis of GR. The Newtonian law of gravitation as an action-at-a-distance law is not compatible with the special theory of relatively. Einstein and other workers were thus forced to try to develop a relativistic theory of gravitation. It is remarkable that Einstein was soon convinced that gravitation has no place in the framework of special relativity. In his lecture *On the Origins of the General Theory of Relativity* Einstein enlarged on this subject as follows:

"I came a first step closer to a solution of the problem when I tried to treat the law of gravitation within the framework of special relativity theory. Like most authors at that time, I tried to formulate a field law for gravity, since the introduction of action at a distance was no longer possible, at least in any natural way, due to the elimination of the concept of absolute simultaneity.

The simplest and most natural procedure was to retain the scalar Laplacian gravitational potential and to add a time derivative to the Poisson equation in such a way that the requirements of the special theory would be satisfied. In addition, the law of motion for a point mass in a gravitational field had to be adjusted to the requirements of special relativity. Just how to do this was not so clear, since the inertial mass of a body might depend on the gravitational potential. In fact, this was to be expected in view of the principle of mass-energy equivalence.

However, such considerations led to a result which made me extremely suspicious. According to classical mechanics, the vertical motion of a body in a vertical gravitational field is independent of the horizontal motion. This is connected with the fact that in such a gravitational field the vertical acceleration of a mechanical system, or of its center of mass, is independent of its kinetic energy. Yet, according to the theory which I was investigating, the gravitational acceleration was not independent of the horizontal velocity or of the internal energy of the system.

This in turn was not consistent with the well known experimental fact that all bodies experience the same acceleration in a gravitational field. This law, which can also be formulated as the law of equality of inertial and gravitational mass, now struck me in its deep significance. I wondered to the highest degree about its validity and supposed it to be the key to a deeper understanding of inertia and gravitation. I did not seriously doubt its strict validity even without knowing the result of the beautiful experiment of Eötvös, which – if I remember correctly – I only heard of later. I now gave up my previously described attempt to treat gravitation in the framework of the special theory as inadequate. It obviously did not do justice to precisely the most fundamental property of gravitation."

In the first chapter we shall discuss several arguments to show that a satisfactory theory of gravitation cannot be formulated within the framework of the special theory of relativity.

After nearly ten years of hard work, Einstein finally conceived the general theory of relativity. One can see how hard he struggled with this theory in the following passage in a letter from Einstein to A. Sommerfeld:

"At present I occupy myself exclusively with the problem of gravitation and now believe that I shall master all difficulties with the help of a friendly mathematician here (Marcel Grossmann). But one thing is certain, in all my life I have never labored nearly as hard, and I have become imbued with great respect for mathematics, the subtler part of which I had in my simple-mindedness regarded as pure luxury until now. Compared with this problem, the original relativity is child's play."

In GR the space-time structure of the special theory of relativity (SR) is generalized. Einstein arrived at this generalization as a result of his *principle of equivalence*, according to which gravitation can be *"locally" transformed away* in a freely falling, nonrotating system. (In the space age, this has become obvious to everybody.) This means that on an infinitesimal scale, relative to a *local* inertial system, such as we have just described, special relativity remains valid. The metric field of the SR varies, however, over finite regions of space-time. Expressed mathematically, space-time is described by a (differentiable) pseudo-Riemannian manifold. The metric field $g$ has the signature[1] $(+---)$ and describes not only the metric properties of space and time as well as its causality properties, but also the gravitational field. It thus becomes a *dynamical element*. General relativity thus *unifies geometry and gravitation*. It is a mathematical fact that one can introduce coordinates in a Lorentz manifold so that, at a given point $x$, one has

$$\text{(i)} \qquad (g_{\mu\nu}(x)) = \begin{pmatrix} 1 & & & 0 \\ & -1 & & \\ & & -1 & \\ 0 & & & -1 \end{pmatrix}$$

$$\text{(ii)} \qquad g_{\mu\nu,\lambda}(x) = 0 \,.$$

In such a coordinate system, the gravitational field is locally transformed away (up to higher order inhomogeneities). The kinematical structure of GR thus permits local inertial systems. According to the equivalence principle, the well-known laws of SR (for example, Maxwell's equations) should hold in this system. This requirement permits one to formulate the non-gravitational laws in the presence of a gravitational field. The equivalence principle thus prescribes the coupling of mechanical and electromagnetic systems to external gravitational fields. Using the equivalence principle, Einstein derived important results (such as the gravitational red shift of light) long before the theory was complete.

The metric field depends on the gravitating masses and energies which are present. This dependence is expressed quantitatively in *Einstein's field equations*, which are the core of the theory. The field equations couple the metric field to the energy-momentum tensor of matter by means of nonlinear partial differential equations. Einstein discovered these after a long and difficult search, and showed that they are almost uniquely determined by only a few requirements. At the same time, he was able to show that his theory reduces to the Newtonian one, in lowest order, for the case of weak fields and slow motions of the masses which serve as sources of the gravitational field.

---

[1] In this case, the pseudo-Riemannian manifold is called a *Lorentz manifold*.

However, the new theory also described higher order effects, such as the precession of the perihelion of Mercury. Einstein once commented on these magnificant achievements:

"In the light of present knowledge, these achievements seem to be almost obvious, and every intelligent student grasps them without much trouble. Yet the prolonged years of intuitive searching in the dark with its tense yearning, its alternation of confidence and fatigue and the final break-through are known only to him who has experienced it."

Contrary to original expectations, GR had little immediate influence on further developments in physics. With the development of quantum mechanics, the theory of matter and electromagnetism took a direction which had little connection with the ideas of GR. (Einstein himself, and some other important physicists, did not participate in these developments and clung to classical field concepts.) Nothing would indicate that such phenomena would not be compatible with special relativity.

As a consequence, GR was only studied by a rather small group of mathematically oriented specialists.

GR is a very special non-Abelian gauge theory. Thus, Einstein's theory of gravitation is closely related to modern developments in theoretical physics. At the present time, many physicists hope that all the fundamental interactions can be described within the framework of (possibly unified) nonabelian gauge theories.

Important astronomical discoveries in the 1960s and 1970s have brought GR and relativistic astrophysics to the forefront of present day research in physics. We know today that objects having extremely strong gravitational field exist in the Universe. Catastrophic events, such as stellar collapse or explosions in the centers of galaxies give rise to not only strong, but also rapidly varying, gravitational fields. This is where GR finds its proper applications. For sufficiently massive objects, gravitation dominates at some point over all other interactions, due to its universally attractive and long range character. Not even the most repulsive nuclear forces can always prevent the final collapse to a black hole. So-called horizons appear in the space-time geometry, behind which matter disappears and thus, for all practical purposes of physics and astrophysics, ceases to exist. For such dramatic events GR must be used in its full ramifications, and is no longer merely a small correction to the Newtonian theory. The possible discovery of a black hole in 1972 would thus be, provided all doubts as to its existence can be removed, one of the most important events in the history of astronomy.

As a result of these astronomical observations, theorists have performed relevant investigations of the stability of gravitating systems, gravitational collapse and the physics of black holes. Some of these issues will be discussed in Part III of this book.

# Chapter 1. The Principle of Equivalence

As we have already mentioned in the introduction, the principle of equivalence is one of the foundation pillars of the general theory of relativity. It leads naturally to the kinematical framework of general relativity and determines to a large extent the coupling of physical systems to external gravitational fields. This will be discussed in detail in this chapter.

## 1.1 Characteristic Properties of Gravitation

Four basic interactions are known to modern physics: the strong and weak interactions, electromagnetism and gravity. Only the latter two are long range, thus permitting a classical description in the macroscopic limit. Today a highly successful quantum electrodynamics exists; however, we do not yet have a quantum theory of gravity.

### 1.1.1 Strength of the Gravitational Interaction

Gravity is by far the weakest of the four fundamental interactions. If we compare the gravitational and electrostatic force between two protons, we find

$$\frac{Gm_{\mathrm{p}}^2}{r^2} = 0.8 \times 10^{-36} \frac{e^2}{r^2} .$$

From this, we are let to introduce, by analogy with the dimensionless fine structure constant $\alpha = e^2/\hbar c \simeq 1/137$, a "gravitational fine structure constant"

$$\alpha_{\mathrm{G}} = \frac{Gm_{\mathrm{p}}^2}{\hbar c} = 5.9 \times 10^{-39}.$$

A comparison of the Bohr radius $a_{\mathrm{B}}$ with the corresponding gravitational Bohr radius $(a_{\mathrm{B}})_{\mathrm{G}}$ enables us to envisage the meaning of this miniscule number. We have

$$a_{\mathrm{B}} = \frac{\hbar^2}{m_{\mathrm{e}}\, e^2} = 0.53 \times 10^{-8}\ \mathrm{cm}\,,$$

$$(a_{\mathrm{B}})_{\mathrm{G}} = \frac{\hbar^2}{m_{\mathrm{e}}\, G m_{\mathrm{e}}\, m_{\mathrm{p}}} = a_{\mathrm{B}} \frac{e^2}{G m_{\mathrm{p}}\, m_{\mathrm{e}}} = \frac{\alpha}{\alpha_{\mathrm{G}}} \frac{m_{\mathrm{p}}}{m_{\mathrm{e}}}\, a_{\mathrm{B}}$$

$$\approx 10^{31}\ \mathrm{cm} \approx 10^{13}\ \text{light years},$$

which is much larger than the "radius of the Universe". Gravity does not become important until one is dealing with rather large masses. For sufficiently large masses, it will, however, sooner or later predominate over all other interactions and will lead to the catastrophic collapse to a black hole. One can show (see Chap. 6) that this is always the case for stars having a mass greater than about

$$\alpha_{\mathrm{G}}^{-3/2}\, m_{\mathrm{p}} \simeq 2\, M_{\odot}\,.$$

Gravity wins because it is not only long range, but also universally attractive. (By comparison, the electromagnetic forces cancel to a large extent due to the alternating signs of the charges, and the Fermi statistics of the electrons.) In addition, not only matter, but also antimatter, and every other form of energy acts as a source for gravitational fields. At the same time, gravity also acts on every form of energy.

### 1.1.2 Universality of the Gravitational Interaction

Since the time of Galilei, we know that all bodies fall at the same rate. This means that for an appropriate choice of units, the inertial mass is equal to the gravitational mass. As we have seen in the introduction, Einstein was profoundly astonished by this fact. The equality of the inertial and gravitational masses has been experimentally verified to an accuracy of one part in $10^{12}$ (Bessel, Eötvös, Dicke, Braginski and Panov). This remarkable fact suggests the validity of the following *universality property:*

The motion of a test body in a gravitational field is independent of its mass and composition (at least when one neglects interactions of spin or of a quadrupole moment with field gradients).

For the Newtonian theory, universality is of course a consequence of the equality of inertial and gravitational masses. We postulate that it holds generally, in particular also for large velocities and strong fields.

### 1.1.3 Precise Formulation of the Principle of Equivalence

The equality of inertial and gravitational masses provides experimental support for the principle of equivalence. By this we mean the following:

**Einstein's Principle of Equivalence:** In an arbitrary gravitational field no local experiment can distinguish a freely falling nonrotating system (local inertial system) from a uniformly moving system in the absence of a gravitational field.

Briefly, we may say that gravity can be locally transformed away. Today of course, this is a well known fact to anyone who has watched space flight on television.

**Remarks:**

1) The principle of equivalence implies (among other things) that inertia and gravity cannot be uniquely separated.
2) The given form of the principle of equivalence is somewhat vague, since it is not entirely clear what a local experiment is. At this point, the principle is thus of a heuristic nature. Further on (Sect. 1.3) we shall replace the principle of equivalence by a mathematical requirement which can be regarded as an idealized form of this heuristic principle.
3) One might at first sight think that the universality property of gravity implies the principle of equivalence. However, this is not the case, as can be seen from the following counterexample. Consider a fictitious world in which, by a suitable choice of units, the electric charge is equal to the mass of the particles and in which there are no negative charges. In a classical framework, there are no objections to such a theory, and by definition, the universality property is satisfied. However, the principle of equivalence is not satisfied. Consider a homogeneous magnetic field. Since the radii and axes of the spiral motion are arbitrary, there is no transformation to an accelerated frame of reference which can remove the effect of the magnetic field on all particles at the same time.

---

**Exercise:** Consider a homogeneous electric field in the $z$-direction and a charged particle with $e=m$. Show that a particle which was originally at rest moves faster in the vertical direction than a particle which was originally moving horizontally.

---

### 1.1.4 Gravitational Red Shift as Evidence for the Validity of the Principle of Equivalence

According to the principle of equivalence, all effects of a homogeneous gravitational field are identical to those existing in a uniformly accelerated frame in the absence of gravity.

We consider two experimenters in a space ship, which experiences the uniform accelerating $g$. Let $h$ be the distance between the observers in the direction of $g$ (see Fig. 1.1). At the time $t=0$ the lower observer emits a photon in the direction of the upper observer. We shall assume that at $t=0$ the space ship is at rest relative to an inertial system. At

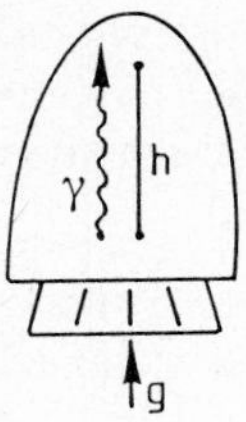


Fig. 1.1

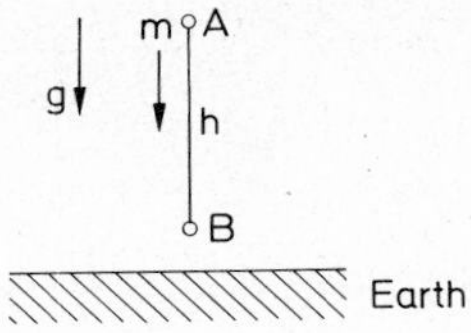


Fig. 1.2

the time $t \simeq h/c$ the photon reaches the upper observer (we neglect corrections of relative order $v/c$ or higher). At this time the observer has the velocity $v = g\,t = g\,h/c$ and he will therefore observe the photon with a Doppler shift $z = \Delta\lambda/\lambda \simeq v/c = g\,h/c^2$. According to the principle of equivalence, the same red shift is also to be expected in the presence of a uniform gravitational field. In this case, we may write, in place of $g\,h/c^2$,

$$z = \Delta\phi/c^2, \tag{1.1.1}$$

where $\phi$ is the Newtonian gravitational potential. Equation (1.1.1) has been verified to an accuracy of 1% in a terrestrial experiment, by making use of the Mössbauer effect (*Pound* and *Snider*, 1965) and to an accuracy $2 \times 10^{-4}$ in a rocket experiment (*Versot* and *Levine*, 1976).

At the time when Einstein formulated his principle of equivalence (around 1911), the prediction (1.1.1) could not be directly verified. Einstein was able to convince himself of its validity indirectly, since (1.1.1) is also a consequence of the conservation of energy.

To see this, consider two points A and B, with separation $h$ in a homogeneous gravitational field (Fig. 1.2).

Let a mass $m$ fall with initial velocity 0 from A to B. According to the Newtonian theory, it has the kinetic energy $m\,g\,h$ at point B. Now let us assume that at B the entire energy of the falling body (rest energy plus kinetic energy) is annihilated to a photon, which subsequently returns to the point A. If the photon did not interact with the gravitational field, we could convert it back to the mass $m$ and gain the energy $mgh$ in each cycle of such a process. In order to preserve the conservation of energy, the photon must experience a red shift. Its energy must satisfy

$$E_{\text{lower}} = E_{\text{upper}} + m\,g\,h = m\,c^2 + m\,g\,h = E_{\text{upper}}(1 + g\,h/c^2)\,.$$

For the wavelength we then have

$$1 + z = \lambda_{\text{upper}}/\lambda_{\text{lower}} = h\,\nu_{\text{lower}}/h\,\nu_{\text{upper}} = E_{\text{lower}}/E_{\text{upper}} = 1 + g\,h/c^2\,,$$

in perfect agreement with (1.1.1).

## 1.2 Special Relativity and Gravitation

As we have mentioned in the introduction, Einstein recognized very early that gravity does not fit naturally into the framework of special relativity. In this section, we shall discuss some arguments which demonstrate this.

### 1.2.1 The Gravitational Red Shift is not Consistent with Special Relativity

According to SR, a clock moving along the timelike world line $x^\mu(\lambda)$ measures the time interval

$$\Delta\tau = \int_{\lambda_1}^{\lambda_2} \sqrt{\eta_{\mu\nu}\frac{dx^\mu}{d\lambda}\frac{dx^\nu}{d\lambda}}\, d\lambda\,, \tag{1.2.1}$$

where

$$(\eta_{\mu\nu}) = \begin{pmatrix} 1 & & & 0 \\ & -1 & & \\ & & -1 & \\ 0 & & & -1 \end{pmatrix}.$$

In the presence of a gravitational field, (1.2.1) can no longer be valid, as is shown by the following argument.

Consider the red shift experiment in the Earth's gravitational field and assume that a special relativistic theory of gravity exists, which need not be further specified here. For this experiment we may neglect all masses other than that of the Earth and regard the Earth as being at rest relative to some inertial system. In a space-time diagram (height $z$ above the Earth's surface vs. time), the Earth's surface, the emitter and the absorber all move along world lines of constant $z$ (see Fig. 1.3).

The sender emits at a fixed frequency from the time $S_1$ to $S_2$. The photons move along world lines $\gamma_1$ and $\gamma_2$, which are not necessarily

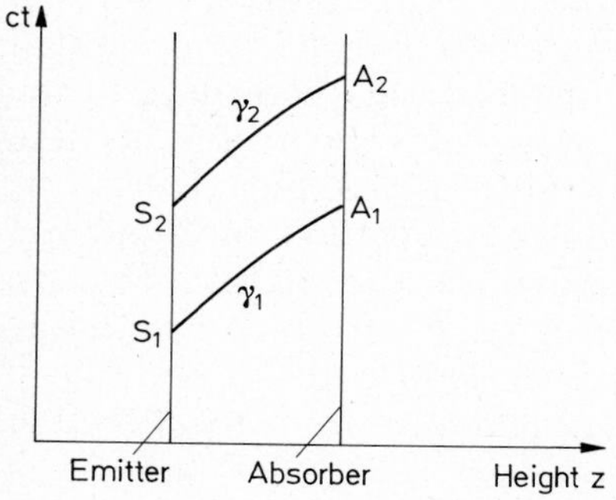


Fig. 1.3

straight lines at an angle of 45° due to a possible interaction with the gravitational field, but which must be *parallel*, since we are dealing with a static situation. Thus, if the flat Minkowski geometry holds and the time measurement is made using (1.2.1), it follows that the time difference between $S_1$ and $S_2$ is equal to the time difference between $A_1$ and $A_2$. In this case, there would be no red shift. This indicates that at the very least (1.2.1) is no longer valid. The argument does not exclude the possibility that the metric $g_{\mu\nu}$ is conformally flat (see, however, Sect. 1.2.3).

### 1.2.2 Global Inertial Systems Cannot be Realized in the Presence of Gravitational Fields

In Newtonian-Galilean mechanics and in special relativity, the law of inertia distinguishes a special class of equivalent frames of reference (inertial systems). Due to the universality of gravitation, only the free fall of electrically neutral test bodies can be regarded as particularly distinguished motion in the presence of gravitational fields. Such bodies experience, however, relative accelerations. Hence the law of inertia is no longer valid and the concept of an inertial system cannot be defined operationally. We have thus been deprived of an essential foundation of the special theory of relativity.

We no longer have any reason to describe space-time as a linear space. The absolute, integrable affine structure of the space-time manifold in Newtonian-Galilean mechanics and in special relativity was, after all, suggested by the law of inertia. A more satisfactory theory should account for inertia and gravity in terms of a single, indecomposable structure.

### 1.2.3 The Deflection of Light Rays

A superficial consideration would indicate that the deflection of light in a gravitational field would follow from the principle of equivalence. The argument goes as follows: We consider the famous Einstein elevator in an elevator shaft attached to the Earth. A light ray is emitted perpendicular to the direction of motion of the cabin. According to the principle of equivalence, the light ray propagates in a straight line inside the cabin (and relative to the cabin). Since the elevator is accelerated relative to the Earth, one would expect that the light ray propagates along a parabolic path relative to the Earth. This is, however, not necessarily so. Indeed, we shall see later that it is possible to construct a theory which satisfies the principle of equivalence, but in which there is no deflection of light. Where is the error, or what further hidden assumptions have we made?

In any case, the deflection of light is an experimental fact (the precise magnitude of the effect does not concern us here). This implies that the metric (if one exists) *cannot be conformally flat* in the presence of a gravitational field. This means that it does not differ from the flat metric simply by a space-dependent scale function. A metric of the form $g_{\mu\nu}(x) = \lambda(x)\,\eta_{\mu\nu}$ (in a suitable coordinate system) defines the same light cones as $\eta_{\mu\nu}$ and is thus not compatible with the empirical deflection of light.

### 1.2.4 Theories of Gravity in Flat Space-Time

In spite of these arguments one may ask, as many have done, how far one gets with a theory of gravity in Minkowski space, along the lines of electrodynamics, admitting the nonobservability of a flat metric. Such attempts have shown that a consequent development finally allows the elimination of the flat metric leading to a description in terms of a "curved" metric which has a direct physical interpretation. The originally postulated Poincaré invariance turns out to be physically meaningless and plays no useful role (see [21]). We may summarize as follows:

In the presence of gravitational fields, Minkowski space can no longer be physically realized. If one requires from the theory that the defining concepts have an empirically verifiable meaning, then it is more sensible to relate its assertions to the orbits of point masses or light rays, rather than to an unobservable Minkowski space.

These considerations should not make us forget that a theory cannot be deduced from empirical facts alone. In this connection, the following words of Einstein should be kept in mind:

"(the scientist) appears as *realist* insofar as he seeks to describe a world independent of the acts of perception; as *idealist* insofar as he looks upon the concepts and theories as the free inventions of the human spirit (not logically derivable from what is empirically given); as *positivist* insofar as he considers his concepts and theories justified *only* to the extent to which they furnish a logical representation of relations among sensory experiences. He may even appear as *Platonist* or *Pythagorean* insofar as he considers the viewpoint of logical simplicity as an indispensable and effective tool of his research."

In the previous sections we have, for the most part, taken a positivistic attitude. The other aspects also have a place in this book.

## 1.3 Space and Time as a Lorentzian Manifold, Mathematical Formulation of the Principle of Equivalence

"Either, therefore, the reality which underlies space must form a discrete manifold, or we must seek the ground of its metric relations (measure-conditions) outside it, in binding forces which act upon it." B. Riemann.

The discussion of Sect. 1.2 has shown that space and time cannot be represented as a Minkowski space when a gravitational field is present. According to the principle of equivalence, special relativity remains, however, valid in "infinitesimal" regions. This suggests that the space-time manifold has a distinguished metric tensor. This tensor field $g_{\mu\nu}(x)$ must vary from point to point in such a way that it is not possible to find a coordinate system in which $g_{\mu\nu}(x) = \eta_{\mu\nu}$ in a finite region of space-time. This should be possible only when no true gravitational fields are present. We postulate therefore:

The mathematical model for space and time (i.e. the set of all events) in the presence of gravitational fields is a pseudo-Riemannian manifold $M$ whose metric $g$ has the same signature as the Minkowski metric.

The pair $(M, g)$ is called a *Lorentz manifold* and $g$ is called a *Lorentzian metric* (for a precise definition, see Sect. 2 of Part I).

As in Minkowski space, the metric $g$ determines the causal relationships. Light signals emitted at the point $x \in M$ follow the surface of the forward light cone (to $g$) (see Sect. 1.4.1). We assume that it is possible to distinguish the forward light cone from the backward light cone in a continuous manner. If this is the case, we shall say that space-time is *time-orientable*. Correspondingly, the tangent vector to the world line $\gamma(s)$ of a point mass is timelike and lies in the forward light cone for every point on $\gamma(s)$. (See Fig. 1.4.)

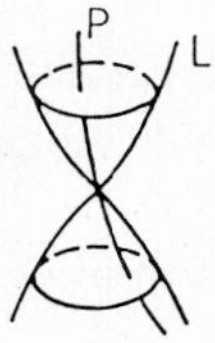


**Fig. 1.4.** Light cone $L$. $P$ is the world line of a particle

At the same time, we also interpret the metric $g$ as the *gravitational-inertial potential.*

In the general theory of relativity, the gravitational field, the metric properties, and causal structure of space-time are all described by one and the same quantity $g$.

In a Lorentz manifold, there are in general no distinguished coordinate systems (except in particularly symmetric cases). Therefore, the requirement that physical laws be *covariant* with respect to the group $K(M)$ of differentiable (i.e. $C^\infty$) coordinate transformations is obvious. By this we mean

**Definition:** A system of equations is *covariant with respect to* $K(M)$ (we also say *generally covariant*) provided that for every element of $K(M)$, the quantities appearing in the equations can be transformed to new quantities in such a way that

(i) the assignment preserves the group structure of $K(M)$;
(ii) both the original and the transformed quantities satisfy the same system of equations.

Only generally covariant laws have an intrinsic meaning in the Lorentz manifold. If a suitable calculus is used, these can be formulated in a coordinate-free manner.

From Sect. 5.3 of Part I, we know that in a neighborhood of every point $x_0$ a coordinate system exists, such that

$$\begin{aligned} g_{\mu\nu}(x_0) &= \eta_{\mu\nu} \\ g_{\mu\nu,\lambda}(x_0) &= 0 . \end{aligned} \tag{1.3.1}$$

Such coordinates are said to be inertial at $x_0$. This is interpreted as a local inertial system. *The metric g describes the behavior of clocks and measuring sticks in this local inertial system, exactly as in special relativity.* This fact will be important in Sect. 1.6. In such a system, the usual laws of electrodynamics, mechanics, etc. in the special relativistic form are locally valid. The form of these laws in an arbitrary system is to a large extend determined by the following two requirements (we shall discuss possible ambiguities in Sect. 1.4):

1. Aside from the metric and its derivatives, the laws should contain only quantities which are also present in the special theory of relativity[1].

2. The laws must be generally covariant and reduce to the special relativistic form at the origin of a locally inertial coordinate system.

These requirements provide a mathematical formulation of the principle of equivalence.

## 1.4 Physical Laws in the Presence of External Gravitational Fields

We now apply the mathematical formulation of the principle of equivalence, as it was presented in the previous section, and shall discuss possible ambiguities at the end of this section. Familiarity with the concept of covariant differentiation (Sects. 5.1–5.6 of Part I) will be assumed.

[1] It is not permitted to introduce in addition to $g_{\mu\nu}$ other "external" (absolute) elements such as a flat metric which is independent of $g$.

### 1.4.1 Motion of a Test Body in a Gravitational Field and Paths of Light Rays

In a local inertial system with origin $p \in M$, the orbit $x^\mu(s)$ through the point $p$ of a test body which is not subjected to any (nongravitational) external forces, is, according to the principle of equivalence, given by

$$\frac{d^2x^\mu}{ds^2}=0 \quad (\text{at } p), \tag{1.4.1}$$

if $s$ is the arc length, i.e. if

$$g_{\mu\nu}\frac{dx^\mu}{ds}\frac{dx^\nu}{ds}=1. \tag{1.4.2}$$

If we introduce the Christoffel symbols

$$\Gamma^\mu_{\alpha\beta}=\tfrac{1}{2}g^{\mu\nu}(g_{\alpha\nu,\beta}+g_{\beta\nu,\alpha}-g_{\alpha\beta,\nu}) \tag{1.4.3}$$

then at the point $p$, we may write instead of (1.4.1), since $\Gamma^\mu_{\alpha\beta}(p)=0$,

$$\frac{d^2x^\mu}{ds^2}+\Gamma^\mu_{\alpha\beta}\frac{dx^\alpha}{ds}\frac{dx^\beta}{ds}=0. \tag{1.4.4}$$

This equation of a geodesic is generally covariant and hence is valid in every frame of reference and at every point of the path, since $p$ is arbitrary. Note that (1.4.2) and (1.4.4) are compatible. This follows from the result of the following

---

**Exercise:** Use (1.4.4) to show that

$$\frac{d}{ds}\left(g_{\mu\nu}\frac{dx^\mu}{ds}\frac{dx^\nu}{ds}\right)=0.$$

---

Using the same arguments, the following equations for the path of a light ray are obtained ($\lambda$ is an affine parameter):

$$\frac{d^2x^\mu}{d\lambda^2}+\Gamma^\mu_{\alpha\beta}\frac{dx^\alpha}{d\lambda}\frac{dx^\beta}{d\lambda}=0 \tag{1.4.5}$$

$$g_{\mu\nu}\frac{dx^\mu}{d\lambda}\frac{dx^\nu}{d\lambda}=0. \tag{1.4.6}$$

As before, (1.4.5) and (1.4.6) are compatible.

Equation (1.4.4) can be regarded as the generalization of the Galilean law of inertia in the presence of a gravitational field. In this form, it is unnecessary to introduce separately the inertial and gravitational masses, merely in order to equate them later. In this manner a certain "magic" element of the Newtonian theory is eliminated. Because of (1.4.4) one may call the connection coefficients $\Gamma^\mu_{\alpha\beta}$ the gravitational-inertial *field strength* relative to $(x^\mu)$.

### 1.4.2 Energy and Momentum Conservation in the Presence of an External Gravitational Field

According to the special theory of relativity, the energy-momentum tensor of a closed system satisfies the conservation law

$$T^{\mu\nu}_{\ ,\nu} = 0.$$

In the presence of a gravitational field, we define a corresponding tensor field on $(M, g)$ such that it reduces to the special relativistic form in a local inertial system.

**Example:** For an ideal fluid we have (with $c = 1$)

$$T^{\mu\nu} = (p + \varrho)\, u^\mu u^\nu - p\, g^{\mu\nu}, \tag{1.4.7}$$

where $p$ is the pressure, $\varrho$ the energy density and $u^\mu$ the velocity field, with

$$g_{\mu\nu} u^\mu u^\nu = 1. \tag{1.4.8}$$

We shall discuss a general method of constructing the energy-momentum tensor in the framework of the Lagrangian formalism later.

In a local inertial system with origin $p \in M$ we have, by the principle of equivalence,

$$T^{\mu\nu}_{\ ,\nu} = 0 \quad \text{at } p.$$

We may just as well write

$$T^{\mu\nu}_{\ ;\nu} = 0 \quad \text{at } p,$$

where the semicolon denotes the covariant derivative of the tensor field. This equation is generally covariant, and hence is valid in any coordinate system. We thus arrive at

$$T^{\mu\nu}_{\ ;\nu} = 0. \tag{1.4.9}$$

From this example, we conclude that the physical laws of special relativity are changed in the presence of a gravitational field simply by the substitution of covariant derivatives for ordinary derivatives (comma → semicolon). This is an expression of the principle of equivalence. In this manner the coupling of the gravitational field to physical systems is determined in an extremely simple manner.

We may write (1.4.9) as follows:
General calculational rules give

$$T^{\mu\nu}_{\ ;\sigma} = T^{\mu\nu}_{\ ,\sigma} + \Gamma^\mu_{\sigma\lambda} T^{\lambda\nu} + \Gamma^\nu_{\sigma\lambda} T^{\mu\lambda}$$

and hence

$$T^{\mu\nu}_{\ ;\nu} = T^{\mu\nu}_{\ ,\nu} + \Gamma^\mu_{\nu\lambda} T^{\lambda\nu} + \Gamma^\nu_{\nu\lambda} T^{\mu\lambda}.$$

Now we have[2]

$$\Gamma^{\nu}_{\nu\lambda} = \frac{1}{\sqrt{-g}}\,\partial_\lambda \sqrt{-g}\ ,$$

where $g$ is the determinant of $(g_{\mu\nu})$. Hence (1.4.9) is equivalent to

$$\frac{1}{\sqrt{-g}}\,\partial_\nu(\sqrt{-g}\ T^{\mu\nu}) + \Gamma^{\mu}_{\nu\lambda} T^{\lambda\nu} = 0\,. \tag{1.4.10}$$

Because of the second term in (1.4.10), this is no longer a conservation law. We cannot form any constants of the motion from (1.4.10). This should also not be expected, since the system under consideration can exchange energy and momentum with the gravitational field.

Equations (1.4.9) [or (1.4.10)] and (1.4.7) provide the basic hydrodynamic equations for an ideal fluid in the presence of a gravitational field.

In the derivation of the field equations for the gravitational field, (1.4.9) will play an important role.

---

**Exercises:**

1. Contract Eq. (1.4.9) with $u^\mu$ and show that the stress-energy tensor (1.4.7) for a perfect fluid leads to

   $$\nabla_u \varrho = -(\varrho + p)\,\nabla\cdot u\,.$$

2. Contract Eq. (1.4.9) with the "projection tensor"

   $$h_{\mu\nu} = g_{\mu\nu} - u_\mu u_\nu$$

   and derive the following relativistic *Euler equation* for a perfect fluid:

   $$(\varrho + p)\,\nabla_u u = \nabla p - (\nabla_u p)\,u\,.$$

3. Show that ($\eta$: volume form corresponding to $g$):

   $$L_u\,\eta = (\nabla\cdot u)\,\eta\,,$$

   i.e., $\mathrm{div}_\eta\, u = \nabla\cdot u$ (see Sect. 4.7.2 of Part I).

---

[2] Since $g\,g^{\mu\nu}$ is the minor of $g_{\mu\nu}$, we have

$$\partial_\alpha g = g\,g^{\mu\nu}\,\partial_\alpha g_{\mu\nu}.$$

Hence

$$\Gamma^{\nu}_{\nu\alpha} = g^{\mu\nu}\tfrac{1}{2}(\partial_\alpha g_{\mu\nu} + \partial_\nu g_{\mu\alpha} - \partial_\mu g_{\nu\alpha}) = \tfrac{1}{2} g^{\mu\nu}\,\partial_\alpha g_{\mu\nu}$$
$$= \frac{1}{2g}\,\partial_\alpha g = \frac{1}{\sqrt{-g}}\,\partial_\alpha(\sqrt{-g}\,).$$

### 1.4.3 Electrodynamics

In the absence of gravitational fields, Maxwell's equations are

$$F^{\mu\nu}{}_{,\nu} = -4\pi j^{\mu}, \tag{1.4.11}$$

$$F_{\mu\nu,\lambda} + F_{\nu\lambda,\mu} + F_{\lambda\mu,\nu} = 0, \tag{1.4.12}$$

where

$$(F^{\mu\nu}) = \begin{pmatrix} 0 & -E_1 & -E_2 & -E_3 \\ E_1 & 0 & -B_3 & B_2 \\ E_2 & B_3 & 0 & -B_1 \\ E_3 & -B_2 & B_1 & 0 \end{pmatrix} \tag{1.4.13}$$

and $j^{\mu}$ is the charge-current 4-vector

$$j^{\mu} = (\varrho, \boldsymbol{J}). \tag{1.4.14}$$

In the presence of a gravitational field, we define $F^{\mu\nu}$ and $j^{\mu}$ such that (1) they transform as tensor fields and (2) they reduce to (1.4.13, 14) in local inertial systems.

With the same arguments as in Sect. 1.4.2 the Maxwell equations in the presence of a gravitational field must have the form

$$F^{\mu\nu}{}_{;\nu} = -4\pi j^{\mu} \tag{1.4.15}$$

$$F_{\mu\nu;\lambda} + F_{\nu\lambda;\mu} + F_{\lambda\mu;\nu} = 0, \tag{1.4.16}$$

where now

$$F_{\mu\nu} = g_{\mu\alpha} g_{\nu\beta} F^{\alpha\beta}. \tag{1.4.17}$$

Since $F^{\mu\nu}$ and $F_{\mu\nu}$ are antisymmetric, we may also write

$$\frac{1}{\sqrt{-g}} \partial_{\nu}(\sqrt{-g}\, F^{\mu\nu}) = -4\pi j^{\mu} \tag{1.4.18}$$

$$F_{\mu\nu,\lambda} + F_{\nu\lambda,\mu} + F_{\lambda\mu,\nu} = 0. \tag{1.4.19}$$

---

**Exercise:** Verify (1.4.18) and (1.4.19). [See also (4.15) in Part I.]

---

From (1.4.15) or (1.4.18) we obtain the conservation law

$$j^{\mu}{}_{;\mu} = 0, \tag{1.4.20}$$

or

$$\partial_{\mu}(\sqrt{-g}\, j^{\mu}) = 0. \tag{1.4.21}$$

Equation (1.4.19) is the integrability condition for the local existence of an electromagnetic potential $A_{\mu}$. This follows from Poincaré's lemma (see Sect. 4 of Part I):

$$F_{\mu\nu} = A_{\nu,\mu} - A_{\mu,\nu}. \tag{1.4.22}$$

In terms of the potential, (1.4.15) reads

$$A^{\nu;\mu}{}_{;\nu} - A^{\mu;\nu}{}_{;\nu} = -4\pi j^{\mu}. \tag{1.4.23}$$

The energy-momentum tensor of the electromagnetic field is given by

$$T^{\mu\nu} = \frac{1}{4\pi}\left[F^{\mu}{}_{\lambda} F^{\lambda\nu} + \tfrac{1}{4} g^{\mu\nu} F^{\sigma\lambda} F_{\sigma\lambda}\right]. \tag{1.4.24}$$

The Lorentz equation of motion for a charged point mass is generalized to

$$m\left(\frac{d^2x^{\mu}}{ds^2} + \Gamma^{\mu}_{\alpha\beta}\frac{dx^{\alpha}}{ds}\frac{dx^{\beta}}{ds}\right) = e F^{\mu}{}_{\nu}\frac{dx^{\nu}}{ds}. \tag{1.4.25}$$

*Formulation of Electrodynamics with the Exterior Calculus*

Let $F$ be the 2-form

$$F = \tfrac{1}{2} F_{\mu\nu}\, dx^{\mu} \wedge dx^{\nu}. \tag{1.4.26}$$

The homogeneous Maxwell equations (1.4.19) can then be written as

$$dF = 0. \tag{1.4.27}$$

The current form $J$ is defined as

$$J = j_{\mu}\, dx^{\mu}. \tag{1.4.28}$$

Since the codifferential has the coordinate representation

$$(\delta F)^{\mu} = -\frac{1}{\sqrt{-g}}\left(\sqrt{-g}\; F^{\mu\nu}\right)_{,\nu}$$

(Part I, Sect. 4.6.3), we may write the inhomogeneous Maxwell equations in the concise form

$$\delta F = 4\pi J, \tag{1.4.29}$$

exactly as in special relativity. Only the meaning of the *-operation has been changed.

Since $\delta \circ \delta = 0$ we obtain current conservation from (1.4.29)

$$\delta J = 0. \tag{1.4.30}$$

Equation (1.4.27) implies, by Poincaré's lemma, the (local) existence of a potential $A = A_{\mu}\, dx^{\mu}$ with

$$F = dA. \tag{1.4.31}$$

---

**Exercises:**

1. Show that the 2-form (1.4.26) has the same Hodge dual $*F$ for two conformally related metrics $g$ and $\bar{g} = e^{2\phi} g$. (More generally, this is true for any $p$-form on a $2p$-dimensional pseudo-Riemannian manifold.) Conclude from this that the source-free Maxwell equa-

tions are *conformally invariant*. This means that they are invariant under any diffeomorphism $\sigma$ of $(M, g)$ with the property

$$\sigma^* g = e^{2\phi} g, \quad \phi \in \mathscr{F}(M).$$

2. Consider again two conformally related metrics $g$ and $\bar{g} = e^{2\phi} g$.
   a) Compute the connection $\bar{\Gamma}$ of $\bar{g}$ in terms of the connection $\Gamma$ of $g$. The result is:

   $$\bar{\Gamma}^{\alpha}_{\mu\nu} = \Gamma^{\alpha}_{\mu\nu} + \delta^{\alpha}_{\mu} \phi_{,\nu} + \delta^{\alpha}_{\nu} \phi_{,\mu} - g_{\mu\nu} g^{\alpha\beta} \phi_{,\beta}.$$

   b) If $\nabla$ and $\bar{\nabla}$ denote the covariant derivative belonging to $g$ and $\bar{g}$, respectively, show that

   $$\bar{\nabla}_\alpha (\bar{g}^{\alpha\mu} \bar{g}^{\beta\nu} F_{\mu\nu}) = e^{-4\phi} \nabla_\alpha F^{\alpha\beta}.$$

3. Let $\Box := d \circ \delta + \delta \circ d$ be the *d'Alembertian* (Laplacian for a Lorentz manifold).
   a) Derive the wave equation $\Box F = 0$ from the source-free Maxwell equations.
   b) Show that the equation

   $$\Box \psi + \tfrac{1}{6} R \psi = 0$$

   for a scalar field $\psi$ is invariant under conformal transformations for a suitable transformation law of $\psi$.

---

### 1.4.4 Ambiguities

In connection with (1.4.23), we now point out certain ambiguities which can arise when applying the principle of equivalence.

In flat space-time, (1.4.23) can be written as

$$A^{\nu,\mu}{}_{,\nu} - A^{\mu,\nu}{}_{,\nu} = -4\pi j^\mu . \tag{1.4.32}$$

Since partial differentiation is commutative, we may also write

$$A^{\nu}{}_{,\nu}{}^{,\mu} - A^{\mu,\nu}{}_{,\nu} = -4\pi j^\mu . \tag{1.4.33}$$

If we now replace in (1.4.32) and (1.4.33) the partial derivatives by covariant derivatives, we obtain different equations, namely,

$$A^{\nu;\mu}{}_{;\nu} - A^{\mu;\nu}{}_{;\nu} = -4\pi j^\mu \tag{1.4.32'}$$

$$A^{\nu}{}_{;\nu}{}^{;\mu} - A^{\mu;\nu}{}_{;\nu} = -4\pi j^\mu . \tag{1.4.33'}$$

The difference arises because covariant differentiation is not commutative. We have (from Sect. 5.8 of Part I)

$$A^{\nu;\mu}{}_{;\nu} = A^{\nu}{}_{;\nu}{}^{;\mu} + R^{\mu}_{\nu} A^{\nu},$$

where $R^\mu_\nu$ is the Ricci tensor. Thus, in place of (1.4.33′), we could also write

$$A^{\nu;\mu}{}_{;\nu} - A^{\mu;\nu}{}_{;\nu} + R^\mu_\nu A^\nu = -4\pi j^\mu . \tag{1.4.33″}$$

A comparison with (1.4.33′) exhibits the additional term $R^\mu_\nu A^\nu$, which contains second derivatives of $g_{\mu\nu}$.

This example shows that the transition from special relativity to general relativity involves an unavoidable ambiguity in all equations which contain higher derivatives, due to the fact that covariant differentiation is not commutative. Note the analogy to the problem of the ordering of operators in the transition from classical mechanics to quantum mechanics.

In practice, one hardly ever encounters uncertainties. For example, one should not regard (1.4.32) as a fundamental equation, but rather go back to the original Maxwell equations (1.4.11) and (1.4.12). However, it is not possible to give a general prescription.

## 1.5 The Newtonian Limit

In order to make contact with the Newtonian theory, let us consider a particle which is moving slowly in a weak gravitational field.

For a weak field, we may introduce a coordinate system which is nearly Lorentzian. In this system we have

$$g_{\mu\nu} = \eta_{\mu\nu} + h_{\mu\nu}, \qquad |h_{\mu\nu}| \ll 1 . \tag{1.5.1}$$

Furthermore, a slowly moving particle has $dx^0/ds \cong 1$ and we may neglect $dx^i/ds$, $i = 1, 2, 3$ in comparison to $dx^0/ds$ in (1.4.4). We then obtain

$$\frac{d^2x^i}{dt^2} \cong \frac{d^2x^i}{ds^2} = -\Gamma^i_{\alpha\beta}\frac{dx^\alpha}{ds}\frac{dx^\beta}{ds} \cong -\Gamma^i_{00} . \tag{1.5.2}$$

Thus, only the components $\Gamma^i_{00}$ appear in the equation of motion:

$$\Gamma^i_{00} \cong \tfrac{1}{2} h_{00,i} - h_{0i,0} . \tag{1.5.3}$$

We now make the additional assumption that the gravitational field is stationary or quasistationary. We may then neglect the second term on the right hand side of (1.5.3) and obtain, with $\boldsymbol{x} = (x^1, x^2, x^3)$

$$\frac{d^2\boldsymbol{x}}{dt^2} = -\tfrac{1}{2}\nabla h_{00} . \tag{1.5.4}$$

This coincides with the Newtonian equation of motion

$$\ddot{\boldsymbol{x}} = -\nabla\phi ,$$

if we set $h_{00} = 2\phi + \text{const.}$

Since $\phi$ and $h_{00}$ vanish at large distances from all masses, we have (reintroducing $c$)

$$g_{00} = 1 + 2\phi/c^2 . \tag{1.5.5}$$

Space-time is strongly curved only when $\phi/c^2$ is not much smaller than unity. As is seen from (1.5.5), only the component $g_{00}$ of the metric plays a role in the Newtonian limit. However, this does not mean that the other components of $h_{\mu\nu}$ must be small in comparison to $h_{00}$.

We give a few numerical examples:

| $\phi/c^2$ | on the surface of |
|---|---|
| $10^{-9}$ | the Earth |
| $10^{-6}$ | the Sun |
| $10^{-4}$ | a white dwarf |
| $10^{-1}$ | a neutron star |
| $10^{-39}$ | a proton |

In the Newtonian limit, the Laplace equation for $\phi$ will follow from Einstein's field equations.

## 1.6 The Red Shift in a Static Gravitational Field

We consider a clock in an arbitrary gravitational field which moves along an arbitrary timelike world line (not necessarily in free fall). According to the principle of equivalence, the clock rate is unaffected by the gravitational field when one observes it from a local inertial system. Let $\Delta t$ be the time between "ticks" of a clock *at rest* in some inertial system in the absence of a gravitational field. In the local inertial system $\{\xi^\mu\}$ under consideration, we then have for the coordinate differential $d\xi^\mu$ between two ticks

$$\Delta t = (\eta_{\mu\nu}\, d\xi^\mu\, d\xi^\nu)^{1/2} .$$

In an arbitrary coordinate system we obviously have

$$\Delta t = (g_{\mu\nu}\, d\xi^\mu\, d\xi^\nu)^{1/2} .$$

Hence

$$\frac{dt}{\Delta t} = \left( g_{\mu\nu} \frac{d\xi^\mu}{dt} \frac{d\xi^\nu}{dt} \right)^{-1/2} \tag{1.6.1}$$

where $dt = dx^0$ denotes the time interval between two ticks relative to the system $\{x^\mu\}$. If the clock is at rest relative to this system $(dx^i/dt = 0)$, we have in particular

$$\frac{dt}{\Delta t} = (g_{00})^{-1/2} . \tag{1.6.2}$$

This is true for *any* clock. For this reason, we cannot verify (1.6.1) or (1.6.2) locally. However, we can compare the time dilations at two different points with each other. For this purpose, we specialize the discussion to the case of a *stationary* field. We choose the coordinates $x^\mu$ such that the $g_{\mu\nu}$ are independent of $t$. Now consider two clocks at rest at the points 1 and 2. (One can convince oneself that the clocks are at rest in any other coordinate system in which the $g_{\mu\nu}$ are independent of time. The concept "at rest" has an intrinsic meaning for stationary fields. See Sect. 1.9 for a geometrical discussion.)

Let a periodic wave be emitted at point 2. Since the field is stationary, the time (relative to our chosen coordinate system) which a wave crest needs to move from point 2 to point 1 is constant[3]. The time between the arrival of successive crests (or troughs) at point 1 is thus equal to the time $dt_2$ between their emission at point 2, which is, according to (1.6.2)

$$dt_2 = \Delta t \, [g_{00}(x_2)]^{-1/2} .$$

If on the other hand, we consider the same atomic transition at point 1, then, according to (1.6.2) the time $dt_1$ between two wave crests, as observed at point 1 is

$$dt_1 = \Delta t \, [g_{00}(x_1)]^{-1/2} .$$

For a given atomic transition, the ratio of frequencies observed at point 1 for light emitted at the points 2 and 1, respectively, is equal to

$$\frac{\nu_2}{\nu_1} = \sqrt{\frac{g_{00}(x_2)}{g_{00}(x_1)}} . \tag{1.6.3}$$

For weak fields, $g_{00} \cong 1 + 2\phi$, with $|\phi| \ll 1$, and we have

$$\frac{\Delta\nu}{\nu} = \frac{\nu_2}{\nu_1} - 1 \cong \phi(x_2) - \phi(x_1) , \tag{1.6.4}$$

in agreement with our previous result.

As an example, we apply (1.6.4) to sunlight observed at the surface of the Earth. At the surface of the Sun, $\phi = -2.12 \times 10^{-6}$. The

[3] From $g_{\mu\nu}\, dx^\mu dx^\nu = 0$, we have

$$dt = (g_{00})^{-1} \cdot [-g_{i0}\, dx^i - \sqrt{(g_{i0}\, g_{j0} - g_{ij}\, g_{00})\, dx^i dx^j}\,] .$$

The time interval being discussed is equal to the integral of the right hand side from 2 to 1 and is thus constant.

gravitational potential at the surface of the Earth is negligible by comparison. We thus obtain a red shift of about 2 ppm. This would be extremely difficult to observe, since thermal effects (which give rise to Doppler shifts) completely mask the gravitational red shift. As we have mentioned before, the red shift can be measured quite precisely on the Earth. In Sect. 1.9 we shall derive the red shift in a more elegant manner using methods of differential geometry.

## 1.7 Fermat's Principle for Static Gravitational Fields

In the following we shall study in more detail light rays in a static gravitational field. A characteristic property of a static field is that in suitable coordinates the metric splits as follows:

$$ds^2 = g_{00}(\mathbf{x})\, dt^2 + g_{ik}(\mathbf{x})\, dx^i\, dx^k . \tag{1.7.1}$$

Thus there are no off-diagonal elements $g_{0i}$ and the $g_{\mu\nu}$ are independent of time. We shall give an intrinsic definition of a static field in Sect. 1.9.

If $\lambda$ is an affine parameter, the paths $x^\mu(\lambda)$ of light rays can be characterised by the variational principle (using classical notation)

$$\delta \int_1^2 g_{\mu\nu} \frac{dx^\mu}{d\lambda} \frac{dx^\nu}{d\lambda}\, d\lambda = 0 , \tag{1.7.2}$$

where the endpoints of the path are held fixed. In addition, we have

$$g_{\mu\nu} \frac{dx^\mu}{d\lambda} \frac{dx^\nu}{d\lambda} = 0 . \tag{1.7.3}$$

---

**Exercise:** Verify (1.7.2).

---

Consider now a static space-time with a metric of the form (1.7.1). If we vary only $t(\lambda)$, we have

$$\begin{aligned}
\delta \int_1^2 g_{\mu\nu} \frac{dx^\mu}{d\lambda} \frac{dx^\nu}{d\lambda}\, d\lambda &= \int_1^2 2 g_{00} \frac{dt}{d\lambda}\, \delta \frac{dt}{d\lambda}\, d\lambda \\
&= \int_1^2 2 g_{00} \frac{dt}{d\lambda} \frac{d}{d\lambda} (\delta t)\, d\lambda \\
&= 2 g_{00} \frac{dt}{d\lambda}\, \delta t \,\Big|_1^2 - 2 \int_1^2 \frac{d}{d\lambda} \left( g_{00} \frac{dt}{d\lambda} \right) \delta t\, d\lambda .
\end{aligned} \tag{1.7.4}$$

The variational principle (1.7.2) thus implies

$$g_{00} \frac{dt}{d\lambda} = \text{const.}$$

We normalize $\lambda$ such that

$$g_{00}\frac{dt}{d\lambda}=1\,. \tag{1.7.5}$$

Now consider a general variation of the path $x^\mu(\lambda)$, for which only the *spatial* endpoints $x^i(\lambda)$ are held fixed, while the condition $\delta t=0$ at the endpoints is dropped. For such variations, we find from (1.7.4) and (1.7.5), making use of the variational principle (1.7.2)

$$\delta\int_1^2 g_{\mu\nu}\frac{dx^\mu}{d\lambda}\frac{dx^\nu}{d\lambda}\,d\lambda=2\,\delta t\,\Big|_1^2=2\,\delta\int_1^2 dt\,. \tag{1.7.6}$$

If the orbit which was varied is also traversed at the speed of light (just as the original path), the left hand side of (1.7.6) is equal to zero and for the varied light-like curves we have also

$$g_{00}^{1/2}\,dt=d\sigma\,, \tag{1.7.7}$$

where

$$d\sigma^2=\gamma_{ik}\,dx^i\,dx^k=-\,g_{ik}\,dx^i\,dx^k\,. \tag{1.7.8}$$

($d\sigma^2$ = 3-dimensional Riemannian metric on the spatial sections.)

We thus have

$$\delta\int_1^2 dt=0=\delta\int_1^2 (g_{00})^{-1/2}\,d\sigma\,. \tag{1.7.9}$$

This is *Fermat's principle of least time.* The second equality in (1.7.9) determines the spatial path of the light ray. The time has been completely eliminated in this formulation: (1.7.9) is valid for an arbitrary portion of the path of the light ray, for any variation such that the ends are held fixed.

A comparison with Fermat's principle in optics shows that the role of the index of refraction has been taken over by $(g_{00})^{-1/2}$.

Equation (1.7.9) states that the path of a light ray is a geodesic in the spatial sections for the metric with coefficients $-(g_{00})^{-1/2}\,g_{ik}$. This result is useful for calculating the bending of light rays in a gravitational field.

## 1.8 Geometric Optics in a Gravitational Field

In most instances gravitational fields vary even over macroscopic distances so little that the propagation of light and radio waves can be described in the geometric optics limit. We shall derive in this section the laws of geometric optics in the presence of gravitational fields from Maxwell's equations (see also the corresponding discussion in books on optics). In addition to the geodesic equation for light rays, we shall find a simple propagation law for the polarization vector.

The following characteristic lengths are important:

1) $\lambda$ = wavelength

2) a typical length $L$ over which the amplitude, polarization and wavelength of the wave vary significantly (example: the radius of curvature of a wave front)

3) A typical "radius of curvature" for the geometry. More precisely, take

$$R = \left| \begin{array}{l} \text{typical component of the Riemannian tensor} \\ \text{in a typical local inertial system} \end{array} \right|^{-1/2}$$

The region of validity for geometric optics is

$$\lambda \ll L \quad \textit{and} \quad \lambda \ll R\,. \tag{1.8.1}$$

Consider a wave which is highly monochromatic in regions having a size $< L$ (more general cases can be treated via Fourier analysis). Now separate the four-vector potential $A_\mu$ into a rapidly varying real phase $\psi$ and a slowly varying complex amplitude

$$A_\mu = \mathrm{Re}\,\{\text{amplitude} \cdot e^{i\psi}\}\,.$$

Let

$$\varepsilon = \lambda/\min\,(L, R)\,.$$

We may expand

$$\text{Amplitude} = a_\mu + \varepsilon\, b_\mu + \ldots,$$

where $a_\mu$, $b_\mu$, ... are independent of $\lambda$. Since $\psi \propto \lambda^{-1}$, we replace $\psi$ by $\psi/\varepsilon$. We thus seek a solution of the form

$$A_\mu = \mathrm{Re}\,\{(a_\mu + \varepsilon\, b_\mu + \ldots)\, e^{i\psi/\varepsilon}\}\,. \tag{1.8.2}$$

In the following let

$$k_\mu = \partial_\mu \psi \qquad \text{(wave number)} \tag{1.8.3}$$

$$a = (a_\mu \bar{a}^\mu)^{1/2} \qquad \text{(scalar amplitude)} \tag{1.8.4}$$

$$f_\mu = a_\mu/a \qquad \text{(polarization vector)}; \tag{1.8.5}$$

$f_\mu$ is a complex unit vector. By definition, light rays are integral curves of the vector field $k^\mu$ and are hence perpendicular to surfaces of constant phase $\psi$ (wave fronts).

Now insert the geometric-optics ansatz (1.8.2) into Maxwell's equations. In vacuum, these are given by (1.4.23),

$$A^{\nu;\mu}{}_{;\nu} - A^{\mu;\nu}{}_{;\nu} = 0\,. \tag{1.8.6}$$

We use the identity

$$A^{\nu;\mu}{}_{;\nu} = A^{\mu;\nu}{}_{;\nu} + R^{\mu}_{\nu} A^{\nu} \tag{1.8.7}$$

and require the Lorentz gauge condition

$$A^{\nu}{}_{;\nu} = 0 \,. \tag{1.8.8}$$

Equation (1.8.6) then takes the form

$$A^{\mu;\nu}{}_{;\nu} - R^{\mu}_{\nu} A^{\nu} = 0 \,. \tag{1.8.9}$$

If we now insert (1.8.2) into the Lorentz condition, we obtain

$$0 = A^{\nu}_{;\nu} = \mathrm{Re}\left\{\left[\mathrm{i}\,\frac{k_\mu}{\varepsilon}\,(a^\mu + \varepsilon\, b^\mu + \ldots) + (a^\mu + \varepsilon\, b^\mu + \ldots)_{;\mu}\right] \mathrm{e}^{\mathrm{i}\,\psi/\varepsilon}\right\} . \tag{1.8.10}$$

From the leading term, it follows that

$$k_\mu\, a^\mu = 0 \,;$$

(the amplitude is perpendicular to the wave vector), or equivalently

$$k_\mu\, f^\mu = 0 \,. \tag{1.8.11}$$

The next order in (1.8.10) leads to

$$k_\mu\, b^\mu = \mathrm{i}\, a^\mu_{;\mu} \,.$$

Now substitute (1.8.2) in (1.8.9) to obtain

$$\begin{aligned} 0 &= -A^{\mu;\nu}{}_{;\nu} + R^{\mu}_{\nu} A^{\nu} \\ &= \mathrm{Re}\left\{\left[\frac{1}{\varepsilon^2}\, k^\nu k_\nu\, (a^\mu + \varepsilon\, b^\mu + \ldots) - 2\,\frac{\mathrm{i}}{\varepsilon}\, k^\nu\, (a^\mu + \varepsilon\, b^\mu + \ldots)_{;\nu} \right.\right. \\ &\qquad - \frac{\mathrm{i}}{\varepsilon}\, k^\nu_{;\nu}\, (a^\mu + \varepsilon\, b^\mu + \ldots) \\ &\qquad \left.\left. - (a^\mu + \ldots)^{;\nu}_{;\nu} + R^\mu_\nu\, (a^\nu + \ldots)\right] \mathrm{e}^{\mathrm{i}\,\psi/\varepsilon}\right\} . \end{aligned} \tag{1.8.12}$$

This gives, in order $\varepsilon^{-2}$,

$$k^\nu k_\nu\, a^\mu = 0 \,,$$

which is equivalent to

$$k^\nu k_\nu = 0 \,, \tag{1.8.13}$$

(wave vector is null). The terms of order $\varepsilon^{-1}$ give

$$k^\nu k_\nu\, b^\mu - 2\,\mathrm{i}\,(k^\nu\, a^\mu_{;\nu} + \tfrac{1}{2}\, k^\nu_{;\nu}\, a^\mu) = 0 \,.$$

Thus, as a result of (1.8.13), we have

$$k^\nu\, a^\mu_{;\nu} = -\tfrac{1}{2}\, k^\nu_{;\nu}\, a^\mu \,. \tag{1.8.14}$$

As a consequence of these equations, we obtain the geodesic law for the propagation of light rays as follows. Equation (1.8.13) implies

$$0=(k^\nu k_\nu)_{;\mu}=2k^\nu k_{\nu;\mu}\,.$$

Now $k_\nu=\psi_{,\nu}$ and since $\psi_{;\nu;\mu}=\psi_{;\mu;\nu}$, we obtain, after interchanging indices,

$$k^\nu k_{\mu;\nu}=0 \qquad (\nabla_k k=0)\,. \tag{1.8.15}$$

We have thus demonstrated that, as a consequence of Maxwell's equations, the paths of *light rays are null geodesics.*

Now consider the amplitude $a^\mu=a\,f^\mu$. From (1.8.14) we have

$$2a\,k^\nu a_{,\nu}=2a\,k^\nu a_{;\nu}=k^\nu(a^2)_{;\nu}=k^\nu(a_\mu\bar a^\mu)_{;\nu}$$

$$=\bar a^\mu k^\nu a_{\mu;\nu}+a_\mu k^\nu \bar a^\mu_{;\nu}\overset{(1.8.14)}{=}-\tfrac{1}{2}k^\nu_{;\nu}(\bar a^\mu a_\mu+a_\mu\bar a^\mu)\,,$$

so that

$$k^\nu a_{,\nu}=-\tfrac{1}{2}k^\nu_{;\nu}a\,. \tag{1.8.16}$$

This can be regarded as a propagation law for the scalar amplitude. If we now insert $a^\mu=a\,f^\mu$ into (1.8.14) we obtain

$$0=k^\nu(a\,f^\mu)_{;\nu}+\tfrac{1}{2}k^\nu_{;\nu}a\,f^\mu$$

$$=a\,k^\nu f^\mu_{;\nu}+f^\mu[k^\nu a_{;\nu}+\tfrac{1}{2}k^\nu_{;\nu}a]\overset{(1.8.16)}{=}a\,k^\nu f^\mu_{;\nu}$$

or

$$k^\nu f^\mu_{;\nu}=0\,, \qquad (\nabla_k f=0)\,. \tag{1.8.17}$$

We thus see that *the polarization vector $f^\mu$ is perpendicular to the light rays and is parallel-propagated along them.*

**Remark:** The gauge condition (1.8.11) is consistent with the other equations: Since the vectors $k^\mu$ and $f^\mu$ are parallel transported along the rays, one must specify the condition $k_\mu f^\mu=0$ at only one point on the ray. For the same reason, the equations $f_\mu\bar f^\mu=1$ and $k_\mu k^\mu=0$ are preserved as the wave propagates.

Equation (1.8.16) can be rewritten as follows: after multiplying by $a$, we have

$$(k^\nu\nabla_\nu)\,a^2+a^2\nabla_\nu k^\nu=0$$

or

$$(a^2k^\mu)_{;\mu}=0\,; \tag{1.8.18}$$

thus $a^2k^\mu$ is a conserved "current".

Quantum mechanically this has the meaning of a conservation law for the number of photons. Of course, the photon number is not in general conserved; it is an adiabatic invariant, in other words, a quantity which varies very slowly for $R\gg\lambda$, in comparison to the photon frequency.

**Exercise:** Show that the energy-momentum tensor, averaged over a wavelength, is

$$\langle T^{\mu\nu}\rangle = \frac{1}{8\pi} a^2 k^\mu k^\nu .$$

In particular, the energy flux is

$$\langle T^{0j}\rangle = \langle T^{00}\rangle n^j, \; n^j = k^j/k^0 .$$

The Eqs. (1.8.15) and (1.8.18) imply

$$8\pi \nabla_\nu \langle T^{\mu\nu}\rangle = \nabla_\nu (a^2 k^\mu k^\nu) = \nabla_\nu (a^2 k^\nu) k^\mu + a^2 k^\nu \nabla_\nu k^\mu = 0 .$$

## 1.9 Static and Stationary Fields

In this section the results of Sects. 3 and 4 of Part I will be used.

A stationary field was defined in Sect. 1.5 by the existence of a coordinate system $\{x^\mu\}$ in which the $g_{\mu\nu}$ are independent of $t = x^0$. We shall now reformulate this "naive" definition in an invariant form.

Let $K = \partial/\partial x^0$, so that $K^\mu = (1, \mathbf{0})$. The Lie derivative of the metric tensor is then

$$(L_K g)_{\mu\nu} = K^\lambda g_{\mu\nu,\lambda} + g_{\lambda\nu} K^\lambda_{,\mu} + g_{\mu\lambda} K^\lambda_{,\nu} = 0 + 0 + 0 , \tag{1.9.1}$$

so that

$$L_K g = 0 . \tag{1.9.2}$$

**Definition 1.9.1:** A vector field $K$ which satisfies (1.9.2) is called a *Killing field* (or an *infinitesimal isometry*).

**Definition 1.9.2:** The metric tensor $g$ of a Lorentz manifold $(M, g)$ is *stationary* if there exists a timelike Killing field.

From this definition we can derive the existence of local coordinates in which the $g_{\mu\nu}$ are independent of time. Consider a three-dimensional spacelike submanifold $S$ and the integral curves of $K$ passing through $S$ (Fig. 1.5).

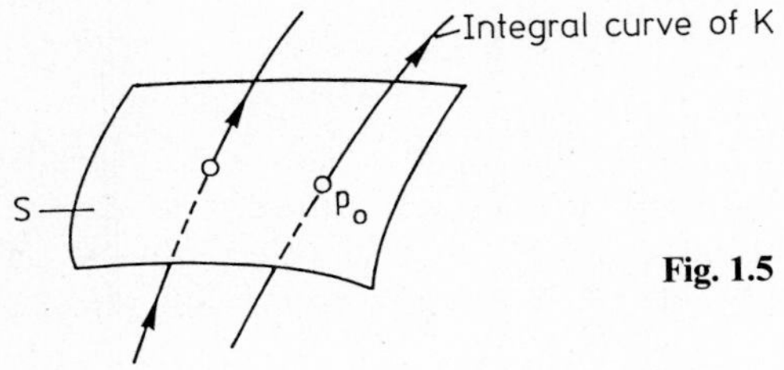


**Fig. 1.5**

Choose arbitrary coordinates $(x^1, x^2, x^3)$ in $S$ and let these be carried along by the flow: If $\phi_t$ is the flow of $K$ and $p = \phi_t(p_0)$ with $p_0 \in S$, then the (Lagrange)-coordinates of $p$ are given by

$$(t, x^1(p_0), x^2(p_0), x^3(p_0)) .$$

In terms of these coordinates, we have

$$K = \partial/\partial x^0 \qquad (x^0 = t) .$$

From $L_K g = 0$ and (1.9.1) we obtain

$$g_{\mu\nu,0} + 0 + 0 = 0 .$$

**Remark:** A Lorentz manifold may have a Killing field which is timelike only in some open domain, and is either lightlike or spacelike outside this region. (We shall become acquainted with some important examples later on.) We then say that the metric field is stationary in the region under consideration.

Static fields are special cases of stationary fields.

**Definition 1.9.3:** A stationary field $g$ of a Lorentz manifold $(M, g)$ with a timelike Killing field $K$ is *static* if the 1-form $\boldsymbol{K}$ corresponding to $K$ satisfies

$$\boldsymbol{K} \wedge d\boldsymbol{K} = 0 . \tag{1.9.3}$$

**Remark:** One knows that (1.9.3) is necessary and sufficient for the existence of a local three dimensional spacelike submanifold perpendicular to $K$ through every point of $M$, due to a theorem of Frobenius (see Sect. 8 of [6]). We do not need this theorem in what follows, since we shall derive the existence of such a foliation directly from (1.9.2) and (1.9.3).

More specifically, we demonstrate the local existence of a function $f$ such that

$$\boldsymbol{K} = (K, K)\, df . \tag{1.9.4}$$

The surfaces $f = \text{const}$ are then the above-mentioned spacelike (local) hypersurfaces orthogonal to $K$.

*Proof of Eq. (1.9.4)* (in local coordinates): First we need Killing's equation

$$K_{\mu;\nu} + K_{\nu;\mu} = 0 , \tag{1.9.5}$$

which can be deduced as follows. In normal coordinates around a point $p$ Eq. (1.9.5) reduces to

$$K_{\mu,\nu} + K_{\nu,\mu} = 0 \tag{1.9.6}$$

at that point. The validity of this equation follows immediately from (1.9.1) and (1.9.2).

In components (1.9.3) reads

$$K_\mu K_{\nu,\lambda} + K_\nu K_{\lambda,\mu} + K_\lambda K_{\mu,\nu} = 0\,. \tag{1.9.7}$$

The left hand side does not change if ordinary derivatives are replaced by covariant derivatives. If we then multiply (1.9.7) by $K^\lambda$ and use (1.9.5), we obtain

$$-K_\mu (K^\lambda K_\lambda)_{;\nu} + K_\nu (K^\lambda K_\lambda)_{;\mu} + K^\lambda K_\lambda (K_{\mu;\nu} - K_{\nu;\mu}) = 0\,.$$

This implies

$$[K_\nu/(K,K)]_{,\mu} - [K_\mu/(K,K)]_{,\nu} = 0\,.$$

According to Poincaré's Lemma, there exists a $C^\infty$ function $f$ such that

$$K_\mu = (K,K)\, f_{,\mu}\,.$$

Coordinate-free Derivation: From the general formulae we have (see Sect. 5.10 of Part I):

$$(L_K g)(X,Y) = K(X,Y) - ([K,X],Y) - (X,[K,Y]) \tag{1.9.8}$$

$$d\boldsymbol{K}(X,Y) = X\boldsymbol{K}(Y) - Y\boldsymbol{K}(X) - \boldsymbol{K}([X,Y]) \tag{1.9.9}$$

$$\begin{aligned}(\boldsymbol{K}\wedge d\boldsymbol{K})(K,X,Y) &= \tfrac{1}{2}[\boldsymbol{K}(K)\, d\boldsymbol{K}(X,Y) + \boldsymbol{K}(X)\, d\boldsymbol{K}(Y,K) \\ &\quad + \boldsymbol{K}(Y)\, d\boldsymbol{K}(K,X)] = 0\,.\end{aligned} \tag{1.9.10}$$

If we now use (1.9.9) in (1.9.10), it follows that

$$\begin{aligned}2\boldsymbol{K}\wedge d\boldsymbol{K}(K,X,Y) &= (K,K)[X\boldsymbol{K}(Y) - Y\boldsymbol{K}(X) - \boldsymbol{K}([X,Y])] \\ &\quad + (K,X)[Y\boldsymbol{K}(K) - K\boldsymbol{K}(Y) - \boldsymbol{K}([Y,K])] \\ &\quad + (K,Y)[K\boldsymbol{K}(K) - X\boldsymbol{K}(K) - \boldsymbol{K}([K,X])] = 0\,.\end{aligned} \tag{1.9.11}$$

From

$$L_K g\,(K,X) = K(K,X) - (K,[K,X]) = 0\,,$$

one concludes that

$$K\boldsymbol{K}(X) = \boldsymbol{K}([K,X]) \tag{1.9.12a}$$

and similarly

$$K\boldsymbol{K}(Y) = \boldsymbol{K}([K,Y])\,. \tag{1.9.12b}$$

If we use these equations in (1.9.11) we obtain

$$(K,K)\, d\boldsymbol{K}(X,Y) + \boldsymbol{K}(X)\, Y\boldsymbol{K}(K) - \boldsymbol{K}(Y)\, X\boldsymbol{K}(K) = 0\,. \tag{1.9.13}$$

Let us denote $(K,K)$ by $h$; then (1.9.13) becomes

$$h\, d\boldsymbol{K}(X,Y) + \boldsymbol{K}(X)\, dh((Y) - \boldsymbol{K}(Y)\, dh(X) = 0$$

or

$$h\, d\boldsymbol{K} + \boldsymbol{K}\wedge dh = 0\,.$$

Hence

$$d(\boldsymbol{K}/h) = 0\,.$$

The statement follows again as a consequence of Poincaré's Lemma.

The two derivations illustrate that computations with components are sometimes quicker.

We choose the function $f$ in (1.9.4) as the *time t*. $K$ is perpendicular to the surfaces of constant time. Now consider an integral curve $\gamma(\lambda)$ of $K$:

$$\frac{d\gamma(\lambda)}{d\lambda} = K(\gamma(\lambda))\,.$$

Let $t(\lambda)$ be the time along $\gamma$ (with arbitrary zero point). Then

$$\frac{d\gamma}{dt}\frac{dt}{d\lambda} = K(\gamma(t))\,.$$

Thus

$$\langle \boldsymbol{K}, K\rangle_{\gamma(t)} = \frac{dt}{d\lambda}\left\langle \boldsymbol{K}, \frac{d\gamma}{dt}\right\rangle = (K, K)_{\gamma(t)}\,.$$

From (1.9.4), i.e., from

$$\boldsymbol{K} = (K, K)\,dt\,, \tag{1.9.14}$$

we find, on the other hand,

$$\left\langle \boldsymbol{K}, \frac{d\gamma}{dt}\right\rangle = (K, K)\left\langle dt, \frac{d\gamma}{dt}\right\rangle = (K, K)\,,$$

and hence $dt/d\lambda = 1$, which implies that $t = \lambda + \text{const}$. This shows that an orthogonal trajectory to the surfaces $t = \text{const}$, with $t$ as parameter, is an integral curve of $K$.

**Remarks:**

1) The flow of $K$ maps the hypersurfaces $t = \text{const}$ isometrically onto each other.
2) An *observer at rest* moves along an integral curve of $K$.
3) If there is only one timelike Killing field, then the time defined by (1.9.14) is distinguished.

If we now introduce Lagrangian coordinates, with the role of the hypersurface $S$ above now taken by a surface $t = \text{const}$, the metric is given by

$$ds^2 = g_{00}(\boldsymbol{x})\,dt^2 + g_{ik}(\boldsymbol{x})\,dx^i\,dx^k\,. \tag{1.9.15}$$

In Sect. 1.7, we used the possibility of introducing coordinates such that $ds^2$ has the form (1.9.15) as a "naive" definition of a static field.

Conversely (1.9.15) implies the geometrical Definition 1.9.3 (at least locally). Indeed, we already know that $K=\partial/\partial t$ is a Killing field. In addition, $\boldsymbol{K}=g_{00}\,dt$, so that $d\boldsymbol{K}=g_{00,i}\,dx^i\wedge dt$ and hence

$$\boldsymbol{K}\wedge d\boldsymbol{K}=g_{00,i}\,g_{00}\,dt\wedge dx^i\wedge dt=0\,.$$

*The Red Shift Revisited*

*First Derivation:* In the limit of geometrical optics we had

$$F_{\mu\nu}=\mathrm{Re}\,\{f_{\mu\nu}\,\mathrm{e}^{\mathrm{i}\psi}\}\,.$$

Light rays are the integral curves of $k^\mu=\psi^{,\mu}$ which are also null geodesics. Since also $k^\mu\,\psi_{,\mu}=0$, the light rays propagate along surfaces of constant phase.

Now consider the world lines of a transmitter and an observer, and two light rays which connect the two, as in Fig. 1.6. The corresponding phases are $\psi=\psi_0$ and $\psi=\psi_0+\Delta\psi$. Let $\Delta s_i$ be the interval between the points at which the two light rays intersect the world line $i$ $(i=1,2)$, and let $u_1$ and $u_2$ be tangent vectors to the world lines, with $(u_1,u_1)=(u_2,u_2)=1$. Obviously

$$u_1^\mu\,(\partial_\mu\psi)_1\,\Delta s_1=\Delta\psi=u_2^\mu\,(\partial_\mu\psi)_2\,\Delta s_2\,. \tag{1.9.16}$$

If $\nu_1$ and $\nu_2$ are the frequencies assigned to the light by observers 1 and 2, respectively, then (1.9.16) gives

$$\nu_1/\nu_2=\Delta s_2/\Delta s_1=(k_\mu\,u^\mu)_1/(k_\mu\,u^\mu)_2\,. \tag{1.9.17}$$

This equation represents the combined effects of the Doppler shift and the gravitational red shift, and is also valid in SR.

Now consider a stationary field with Killing vector $K$. Let both observer and transmitter be at rest relative to the field, so that

$$K=(K,K)^{1/2}\,u\,. \tag{1.9.18}$$

As a result of Killing's equation (1.9.5), we have

$$k^\beta\,\nabla_\beta\,(k^\alpha\,K_\alpha)=(k^\beta\,\nabla_\beta k^\alpha)\,K_\alpha+k^\alpha\,k^\beta\,\nabla_\beta\,K_\alpha=0\,.$$

Hence $k^\alpha K_\alpha$ is *constant* along the light ray.

It then follows from (1.9.17) and (1.9.18) that

$$\frac{\nu_1}{\nu_2}=\frac{(K,K)_2^{1/2}}{(K,K)_1^{1/2}}\,. \tag{1.9.19}$$

In an adapted coordinate system, $K=\partial/\partial t$ and $(K,K)=g_{00}$. We then obtain, as before,

$$\frac{\nu_1}{\nu_2}=\sqrt{\frac{g_{00}|_2}{g_{00}|_1}}\,. \tag{1.9.20}$$

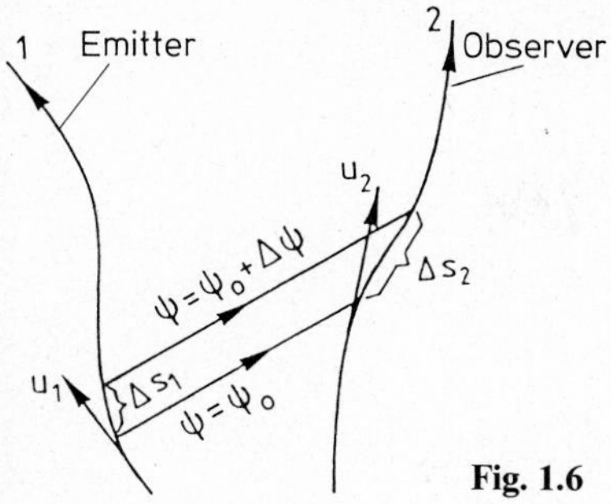

Fig. 1.6

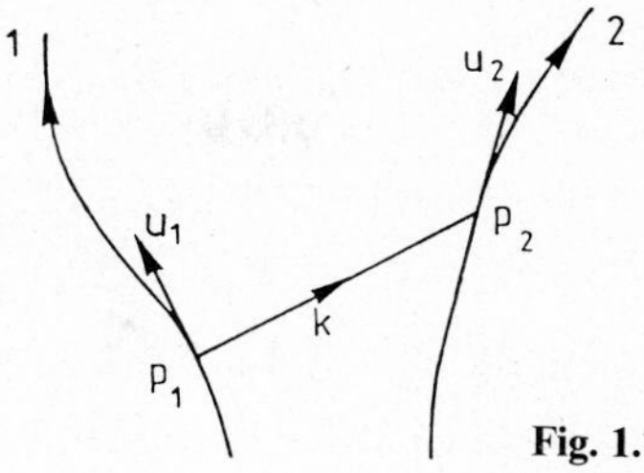

Fig. 1.7

*Second Derivation:* Our starting point is geometric optics. The two observers, having four-velocities $u_1$ and $u_2$, can be connected to each other by a null geodesic, whose tangent vector is denoted by $k$ (see Fig. 1.7).

Let $t_1$ and $t_2$ be the proper times of the two observers. We assume that for a finite $t_1$-interval, null geodesics exist which can be received by the second observer. These geodesics can be parametrized by the observation times at 1 or at 2. This defines a function $t_2(t_1)$. The frequency ratio $r$ is clearly the derivative of this function,

$$r = dt_2/dt_1 . \tag{1.9.21}$$

The null geodesics can be parametrized such that their affine parameter $s$ is zero along the world line of the first observer and unity along that of the second observer. The points of the null geodesics can thus be parametrized by the pair $(t_1, s)$. We then have[4], when $(t_1, s) \mapsto \lambda(t_1, s)$,

$$k = (\partial/\partial s)_{\lambda(t_1,s)} . \tag{1.9.22}$$

We consider also the field of tangent vectors[4]

$$V = (\partial/\partial t_1)_{\lambda(t_1,s)} . \tag{1.9.23}$$

Obviously

$$V|_{s=0} = u_1 , \tag{1.9.24}$$

while

$$V|_{s=1} = \frac{dt_2}{dt_1}\,\frac{\partial}{\partial t_2} = r\,u_2 . \tag{1.9.25}$$

We show below that $(V, k)$ is constant along a null geodesic. Then

$$(V, k)_1 = (u_1, k) = (V, k)_2 = r\,(u_2, k) ,$$

---

[4] More precisely $k = \lambda_*(\partial/\partial s)$ and $V = \lambda_*(\partial/\partial t_1)$.

so that

$$r=(u_1,k)/(u_2,k)\,, \tag{1.9.26}$$

in agreement with (1.9.17). The rest is as before.
*Proof that* $(V,k)=$ const:
A null geodesic satisfies

$$(k,k)=0\,, \qquad \nabla_k k=0\,. \tag{1.9.27}$$

For the Levi-Civita connection we have

$$T(X,Y)=\nabla_X Y-\nabla_Y X-[X,Y]=0 \tag{1.9.28}$$

$$X(Y,Z)=(\nabla_X Y,Z)+(Y,\nabla_X Z)\,. \tag{1.9.29}$$

From (1.9.27) and (1.9.29) it follows that

$$\frac{\partial}{\partial s}(V,k)=k\,(V,k)=(\nabla_k V,k)+(V,\nabla_k k)=(\nabla_k V,k)\,. \tag{1.9.30}$$

According to (1.9.22) and (1.9.23), we have[5]

$$[V,k]=0\,.$$

Hence, using (1.9.28) and (1.9.30),

$$\frac{\partial}{\partial s}(V,k)=(\nabla_k V,k)=(\nabla_V k,k)=\tfrac{1}{2}V\,(k,k)=0\,.$$

---

**Exercise:** Consider a stationary source-free electromagnetic field in a static Lorentz manifold $(M,g)$ in adapted coordinates (1.9.15),

$$g=g_{00}\,dt^2-h_{ik}\,dx^i\,dx^k\,.$$

Show that the time independent scalar potential $\varphi$ satisfies the equation

$$\Delta\,(h_F)\,\varphi=0\,,$$

where $\Delta\,(h_F)$ is the Laplacian for the "Fermat metric"

$$h_{ik}^F=\frac{h_{ik}}{\sqrt{g_{00}}}\,,$$

which already played a role in Sect. 1.7.

---

[5] Since the vector fields $k$ and $V$ are defined on a submanifold, the meaning of the commutator of $V$ and $k$ is not immediately clear. If $Q$ is a compact square in $\mathbb{R}^2$ and if the image $\lambda(Q)$ is contained in a coordinate neighborhood $U$, one can extend the fields $k$ and $V$ to $C^\infty$ vector fields $\tilde{k}$ and $\tilde{V}$ on $U$. We then define

$$[V,k]_{\lambda(s,t_1)}=[\tilde{V},\tilde{k}]_{\lambda(s,t_1)}, \qquad (s,t_1)\in Q\,.$$

From the coordinate representation of the Lie product (see Sect. 2.2 of Part I), one easily sees that this definition is independent of the extensions $\tilde{k}$, $\tilde{V}$. This is also a consequence of Theorem 3.6 of Part I.

*Hint:* It is convenient to use the conformal invariance of the source-free Maxwell equations.

---

## 1.10 Local Reference Frames and Fermi Transport

In this section we shall answer the following questions:
1) Suppose that an observer moves along a timelike worldline in a gravitational field (not necessarily in free fall). One might, for example, consider an astronaut in a space capsule. For practical reasons he will choose a coordinate system in which all apparatus attached to his capsule is at rest. What is the equation of motion of a freely falling test body in this coordinate system?
2) How should the observer orient his space ship so that "Coriolis forces" do not appear?
3) How does one describe the motion of a gyroscope? One might expect that it will not rotate relative to the frame of reference, provided the latter is chosen such that Coriolis forces are absent.
4) Is it possible to find a special frame of reference for an observer at rest in a stationary field, which one might call Copernican? What is the equation of motion of a spinning top in such a frame of reference? Under what conditions will it not rotate relative to the Copernican frame?

### 1.10.1 Precession of the Spin in a Gravitational Field

By spin we mean either the polarization vector of a particle (i.e. the expectation value of the spin operator for a particle in a particular quantum mechanical state) or the intrinsic angular momentum of a rigid body, such as a gyroscope.

In both cases this is initially defined only relative to a local inertial system in which the body is at rest (its *local rest system*). In this system the spin is described by a three vector $\boldsymbol{S}$. For a gyroscope or for an elementary particle, the equivalence principle implies that in the local rest system, in the absence of external forces,

$$\frac{d}{dt}\boldsymbol{S}(t) = 0. \tag{1.10.1}$$

(We assume that the interaction of the gyroscope's quadrupole moment with inhomogeneities of the gravitational field can be neglected; this effect is studied in an exercise at the end of this section). We now define a four vector $S$ which reduces to $(0, \boldsymbol{S})$ in the local rest system. This last requirement can be expressed in the covariant form

$$(S, u) = 0, \tag{1.10.2}$$

where $u$ is the four-velocity.

We shall now rewrite (1.10.1) in a covariant form. For this we consider $\nabla_u S$. In the local rest system (denoted by R) we have

$$(\nabla_u S)_{\mathrm{R}} = \left(\frac{dS^0}{dt}, \frac{d}{dt}\boldsymbol{S}\right) = \left(\frac{dS^0}{dt}, \boldsymbol{0}\right). \tag{1.10.3}$$

It follows from (1.10.2) that

$$(\nabla_u S, u) = -(S, \nabla_u u) = -(S, a), \tag{1.10.4}$$

where

$$a = \nabla_u u. \tag{1.10.5}$$

Hence

$$(\nabla_u S, u) = \left.\frac{dS^0}{dt}\right|_{\mathrm{R}} = -(S, a). \tag{1.10.6}$$

From (1.10.3) and (1.10.6) we then have

$$(\nabla_u S)_{\mathrm{R}} = (-(S, a), \boldsymbol{0}) = -((S, a)\, u)_{\mathrm{R}}. \tag{1.10.7}$$

The desired covariant equation is thus

$$\nabla_u S = -(S, a)\, u. \tag{1.10.8}$$

Equation (1.10.2) is consistent with (1.10.8). Indeed from (1.10.8) we find

$$(u, \nabla_u S) = -(S, a)(u, u) = -(S, a) = -(S, \nabla_u u),$$

so that $\nabla_u (u, S) = 0$.

**Application: Thomas Precession.** For Minkowski space (1.10.8) reduces to

$$\dot{S} = -(S, \dot{u})\, u, \tag{1.10.9}$$

where the dot means differentiation with respect to the proper time. One can easily derive the Thomas precession from this equation.

Let $x(\tau)$ denote the path of a particle. The instantaneous rest system (at time $\tau$) is obtained from the laboratory system via the special Lorentz transformation $\Lambda(\boldsymbol{\beta})$. With respect to this family of instantaneous rest systems $S$ hast the form: $S = (0, \boldsymbol{S}(t))$, where $t$ is the time in the laboratory frame. We obtain the equation of motion for $S(t)$ easily from (1.10.9). Since $S$ is a four vector, we have, in the laboratory frame,

$$S = \left(\gamma\,\boldsymbol{\beta}\cdot\boldsymbol{S}, \boldsymbol{S} + \boldsymbol{\beta}\frac{\gamma^2}{\gamma+1}\boldsymbol{\beta}\cdot\boldsymbol{S}\right). \tag{1.10.10}$$

In addition

$$u = (\gamma, \gamma\,\boldsymbol{\beta}), \quad \dot{u} = (\dot{\gamma}, \dot{\gamma}\,\boldsymbol{\beta} + \gamma\,\dot{\boldsymbol{\beta}}). \tag{1.10.11}$$

Hence

$$(S,\, \dot{u}) = \dot{\gamma}\, \gamma\, \boldsymbol{\beta} \cdot \boldsymbol{S} - (\dot{\gamma}\, \boldsymbol{\beta} + \gamma\, \dot{\boldsymbol{\beta}}) \left( \boldsymbol{S} + \boldsymbol{\beta} \frac{\gamma^2}{\gamma+1} \boldsymbol{\beta} \cdot \boldsymbol{S} \right)$$

$$= -\gamma \left( \dot{\boldsymbol{\beta}} \cdot \boldsymbol{S} + \frac{\gamma^2}{\gamma+1} \dot{\boldsymbol{\beta}} \cdot \boldsymbol{\beta}\ \boldsymbol{\beta} \cdot \boldsymbol{S} \right). \tag{1.10.12}$$

From (1.10.9) and (1.10.12) we then obtain

$$(\gamma\, \boldsymbol{\beta} \cdot \boldsymbol{S})^{\cdot} = \gamma^2 \left( \dot{\boldsymbol{\beta}} \cdot \boldsymbol{S} + \frac{\gamma^2}{\gamma+1} \dot{\boldsymbol{\beta}} \cdot \boldsymbol{\beta}\ \boldsymbol{\beta} \cdot \boldsymbol{S} \right)$$

$$\left( \boldsymbol{S} + \boldsymbol{\beta} \frac{\gamma^2}{\gamma+1} \boldsymbol{\beta} \cdot \boldsymbol{S} \right)^{\cdot} = \boldsymbol{\beta}\, \gamma^2 \left( \dot{\boldsymbol{\beta}} \cdot \boldsymbol{S} + \frac{\gamma^2}{\gamma+1} \dot{\boldsymbol{\beta}} \cdot \boldsymbol{\beta}\ \boldsymbol{\beta} \cdot \boldsymbol{S} \right).$$

After some rearrangement, one finds

$$\dot{\boldsymbol{S}} = \boldsymbol{S} \times \boldsymbol{\omega}_T,$$

where $\boldsymbol{\omega}_T = \dfrac{\gamma - 1}{\beta^2}\, \boldsymbol{\beta} \times \dot{\boldsymbol{\beta}}$.

This is the well known expression for the Thomas precession.

---

**Exercise:** Carry out the rearrangements leading to the last two equations.

---

### 1.10.2 Fermi Transport

Let $\gamma(s)$ be a timelike curve with tangent vector $u = \dot{\gamma}$ satisfying $(u,\, u) = 1$.

The *Fermi derivative* $\mathbb{F}_u$ of a vector field $X$ along $\gamma$ is defined by

$$\mathbb{F}_u X = \nabla_u X + (X,\, a)\, u - (X,\, u)\, a, \tag{1.10.13}$$

where $a = \nabla_u u$. Since $(S,\, u) = 0$ we may write (1.10.8) in the form

$$\mathbb{F}_u S = 0. \tag{1.10.14}$$

It is easy to show that the Fermi derivative (1.10.13) has the following important properties:

(i) $\mathbb{F}_u = \nabla_u$, if $\gamma$ is a geodesic;
(ii) $\mathbb{F}_u u = 0$;
(iii) If $\mathbb{F}_u X = \mathbb{F}_u Y = 0$ along $\gamma$, then $(X,\, Y)$ is constant along $\gamma$;
(iv) If $(X,\, u) = 0$ along $\gamma$, then

$$\mathbb{F}_u X = (\nabla_u X)_\perp. \tag{1.10.15}$$

Here $\perp$ denotes the projection perpendicular to $u$. These properties show that the Fermi derivative is a natural generalization of $\nabla_u$.

We say that a vector field $X$ is *Fermi transported* along $\gamma$ if $\mathbb{F}_{\dot\gamma} X = 0$. Since this equation is linear in $X$, Fermi transport defines (analogously to parallel transport) a two parameter family of isomorphisms

$$\tau^{\mathbb{F}}_{t,s}: T_{\gamma(s)}(M) \to T_{\gamma(t)}(M).$$

One can show that

$$\mathbb{F}_{\dot\gamma} X(\gamma(t)) = \frac{d}{ds} \tau^{\mathbb{F}}_{t,s} X(\gamma(s))\,|_{s=t}.$$

The proof is similar to that of (5.11) in Part I.

As in the case of the covariant derivative (Sect. 5.4 of Part I), the Fermi derivative can be extended to arbitrary tensor fields such that the following properties hold:

(i) $\mathbb{F}_u$ transforms a tensor of type $(r, s)$ into another tensor of the same type;
(ii) $\mathbb{F}_u$ commutes with contractions;
(iii) $\mathbb{F}_u(S \otimes T) = (\mathbb{F}_u S) \otimes T + S \otimes (\mathbb{F}_u T)$;
(iv) $\mathbb{F}_u f = df/ds$, when $f$ is a function;
(v) $\tau^{\mathbb{F}}_{t,s}$ induces linear isomorphisms:

$$T_{\gamma(s)}(M)^r_s \to T_{\gamma(t)}(M)^r_s.$$

We now consider the world line $\gamma(\tau)$ of an accelerated observer ($\tau$ is the proper time).

Let $u = \dot\gamma$ and let $\{e_i\}$, $i = 1, 2, 3$ be an arbitrary orthonormal frame along $\gamma$ perpendicular to $e_0 := \dot\gamma = u$. We then have

$$(e_\mu, e_\nu) = \eta_{\mu\nu}.$$

If $a := \nabla_u u$, it follows from $(u, u) = 1$ that $(a, u) = 0$. We now set

$$\omega_{ij} = -(\nabla_u e_i, e_j) = -\omega_{ji}. \tag{1.10.16}$$

If $e^\mu = \eta^{\mu\nu} e_\nu$, we have

$$\begin{aligned}\nabla_u e_i &= (\nabla_u e_i, e^\alpha)\, e_\alpha \\ &= (\nabla_u e_i, u)\, u - (\nabla_u e_i, e_j)\, e_j \\ &= -(e_i, \nabla_u u)\, u + \omega_{ij}\, e_j,\end{aligned}$$

so that

$$\nabla_u e_i = -(e_i, a)\, u + \omega_{ij}\, e_j. \tag{1.10.17}$$

Adding a vanishing term, this can be rewritten as

$$\nabla_u e_i = (e_i, u)\, a - (e_i, a)\, u + \omega_{ij}\, e_j.$$

Let

$$(\omega_{\alpha\beta}) = \begin{pmatrix} 0 & 0 \\ 0 & \omega_{ij} \end{pmatrix}. \tag{1.10.18}$$

Then

$$\nabla_u e_\alpha = (e_\alpha, u)\, a - (e_\alpha, a)\, u - \omega_{\alpha\beta}\, e^\beta,$$

since for $\alpha = 0$ the right hand side is equal to $(u, u)\, a - (u, a)\, u = a = \nabla_u u$. This can be written in the form [using (1.10.13)]

$$\mathbb{F}_u e_\alpha = -\omega_{\alpha\beta}\, e^\beta. \tag{1.10.19}$$

$\omega_{\alpha\beta}$ thus describes the deviation from Fermi transport. For a spinning top we have $\mathbb{F}_u S = 0$, $(S, u) = 0$. If we write $S = S^i e_i$ then

$$0 = \mathbb{F}_u S = \frac{dS^i}{d\tau} e_i + S^j\, \mathbb{F}_u e_j = \frac{dS^i}{d\tau} e_i - S^j\, \omega_{j\,i}\, e^i,$$

and thus

$$dS^i/d\tau = \omega_{ij}\, S^j. \tag{1.10.20}$$

Thus the top precesses relative to the frame $\{e_i\}$ with angular velocity $\boldsymbol{\Omega}$ where

$$\omega_{ij} = \varepsilon_{ijk}\, \Omega^k. \tag{1.10.21}$$

We may write (1.10.20) in three-dimensional vector notation

$$\frac{d\boldsymbol{S}}{d\tau} = \boldsymbol{S} \times \boldsymbol{\Omega}. \tag{1.10.22}$$

If the frame $\{e_i\}$ is Fermi transported along $\gamma$, then clearly $\boldsymbol{\Omega} = 0$. We shall evaluate (1.10.16) for the angular velocity in a number of instances. A first example is given in Sect. 1.10.4.

### 1.10.3 The Physical Difference Between Static and Stationary Fields

We consider now an observer at rest in a stationary space-time with the Killing field $K$. The observer thus moves along an integral curve $\gamma(\tau)$ of $K$. His four velocity $u$ is

$$u = (K, K)^{-1/2}\, K. \tag{1.10.23}$$

We choose now an orthonormal tetrad $\{e_i\}$ along $\gamma$ which is *Lie-transported:*

$$L_K e_i = 0, \quad i = 1, 2, 3. \tag{1.10.24}$$

Note that under Lie transport (1.10.24) the $e_i$ remain perpendicular to $K$ and hence to $u$. Indeed, it follows from

$$0 = L_K g(X, Y) = K(X, Y) - (L_K X, Y) - (X, L_K Y)$$

that the orthogonality of $X$ and $Y$ is preserved when $L_K X = L_K Y = 0$. Note that $K$ itself is Lie-transported: $L_K K = [K, K] = 0$.

The $\{e_i\}$ can then be interpreted as "axes at rest" and define a "Copernican system". We are interested in the change of the spin relative to this system. Our starting point is (1.10.16) or, making use of (1.10.23)

$$\omega_{ij} = -(K, K)^{-1/2}(e_j, \nabla_K e_i). \tag{1.10.25}$$

Now

$$0 = T(K, e_i) = \nabla_K e_i - \nabla_{e_i} K - [K, e_i]$$

and

$$[K, e_i] = L_K e_i = 0.$$

Hence (1.10.25) implies

$$\omega_{ij} = -(K, K)^{-1/2}(e_j, \nabla_{e_i} K) = -(K, K)^{-1/2} \nabla \boldsymbol{K}(e_j, e_i).$$

In order to convince oneself of the correctness of the last step, one may write down component expressions for all the terms. Since $\omega_{ij}$ is antisymmetric

$$\omega_{ij} = (K, K)^{-1/2} \tfrac{1}{2} [\nabla \boldsymbol{K}(e_i, e_j) - \nabla \boldsymbol{K}(e_j, e_i)]$$

or, since any one-form $\varphi$ satisfies

$$\nabla\varphi(X, Y) - \nabla\varphi(Y, X) = -d\varphi(X, Y),$$

we have also

$$\omega_{ij} = -\tfrac{1}{2}(K, K)^{-1/2} d\boldsymbol{K}(e_i, e_j). \tag{1.10.26}$$

We shall subsequently show that

$$\omega_{ij} = 0, \quad \text{if and only if} \quad \boldsymbol{K} \wedge d\boldsymbol{K} = 0. \tag{1.10.27}$$

From this it follows that a *Copernican system does not rotate if and only if the stationary field is static.*

The one-form $*(\boldsymbol{K} \wedge d\boldsymbol{K})$ can be regarded as a measure of the "absolute" rotation, because the vector $\Omega = \Omega^k e_k$ can be expressed in the form

$$\boldsymbol{\Omega} = \tfrac{1}{2}(K, K)^{-1/2} * (\boldsymbol{K} \wedge d\boldsymbol{K}) \tag{1.10.28}$$

where $\boldsymbol{\Omega}$ denotes the one-form corresponding to $\Omega$.

*Proof of (1.10.28):* Let $\{\theta^\mu\}$ denote the dual basis of $\{e_\mu\}$. From well-known properties of the *-operation (see Sect. 4.6.2 of Part I), we have

$$\theta^\mu \wedge (\boldsymbol{K} \wedge d\boldsymbol{K}) = \eta(\theta^\mu, *(\boldsymbol{K} \wedge d\boldsymbol{K})). \tag{1.10.29}$$

Since $\boldsymbol{K} = (K, K)^{1/2}\theta^0$, the left hand side of (1.10.29) is equal to $(K, K)^{1/2}\theta^\mu \wedge \theta^0 \wedge d\boldsymbol{K}$ and vanishes for $\mu = 0$. From (1.10.26) we conclude that

$$d\boldsymbol{K} = -(K, K)^{1/2}\omega_{ij}\theta^i \wedge \theta^j + \text{terms containing } \theta^0.$$

Hence we have

$$\theta^\mu \wedge (\boldsymbol{K} \wedge d\boldsymbol{K})$$
$$= \begin{cases} 0 & \text{for} \quad \mu = 0 \\ -(K,K)\,\varepsilon_{ijl}\,\Omega^l \theta^k \wedge \theta^0 \wedge \theta^i \wedge \theta^j = 2(K,K)\,\eta\,\Omega^k & \text{for} \quad \mu = k, \end{cases}$$

where we used $\theta^k \wedge \theta^0 \wedge \theta^i \wedge \theta^j = -\varepsilon_{ijk}\,\eta$.

From this and (1.10.29) we get

$$(\theta^\mu, *(\boldsymbol{K} \wedge d\boldsymbol{K})) = \begin{cases} 0, & \text{if} \quad \mu = 0 \\ 2(K,K)\,\Omega^k, & \text{if} \quad \mu = k. \end{cases}$$

The left hand side of this expression is equal to the contravariant components of $*(\boldsymbol{K} \wedge d\boldsymbol{K})$ and hence (1.10.28) follows.

Obviously (1.10.28) implies (1.10.27).

### 1.10.4 Spin Rotation in a Stationary Field

The spin rotation relative to the Copernican system is given by (1.10.28). We now write this in terms of suitable coordinates. Let $K = \partial/\partial t$ and $g_{\mu\nu}$ be independent of $t = x^0$. Then

$$\boldsymbol{K} = g_{00}\,dt + g_{0i}\,dx^i,$$
$$d\boldsymbol{K} = g_{00,k}\,dx^k \wedge dt + g_{0i,k}\,dx^k \wedge dx^i$$
$$\boldsymbol{K} \wedge d\boldsymbol{K} = (g_{00}\,g_{0i,j} - g_{0i}\,g_{00,j})\,dt \wedge dx^j \wedge dx^i + g_{0k}\,g_{0i,j}\,dx^k \wedge dx^j \wedge dx^i.$$

From Exercise 6 of Sect. 4.6.2 of Part I, we then have

$$\begin{aligned} *(\boldsymbol{K} \wedge d\boldsymbol{K}) &= g_{00}^2 \left(\frac{g_{0i}}{g_{00}}\right)_{,j} *(dt \wedge dx^j \wedge dx^i) + g_{0k}\,g_{0i,j} *(dx^k \wedge dx^j \wedge dx^i) \\ &= \frac{g_{00}^2}{\sqrt{-g}}\,\varepsilon_{ijl}\left(\frac{g_{0i}}{g_{00}}\right)_{,j} \Bigg[(g_{lk}\,dx^k + g_{l0}\,dx^0) \\ &\qquad - \left(g_{l0}\,dx^0 + \frac{g_{l0}\,g_{k0}}{g_{00}}\,dx^k\right)\Bigg], \end{aligned}$$

where we used $*(dt \wedge dx^j \wedge dx^i) = \eta^{0jil} g_{l\mu}\,dx^\mu = -(-g)^{-1/2}\,\varepsilon_{ijl}\,g_{l\mu}\,dx^\mu$ in the last step.

We thus obtain

$$\boldsymbol{\Omega} = \frac{g_{00}}{2\sqrt{-g}}\,\varepsilon_{ijl}\left(\frac{g_{0i}}{g_{00}}\right)_{,j}\left(g_{lk} - \frac{g_{l0}\,g_{k0}}{g_{00}}\right) dx^k \tag{1.10.30}$$

and it follows immediately that

$$\Omega = \frac{g_{00}}{2\sqrt{-g}}\,\varepsilon_{ijk}\left(\frac{g_{0i}}{g_{00}}\right)_{,j}\left(\partial_k - \frac{g_{k0}}{g_{00}}\,\partial_0\right). \tag{1.10.31}$$

We shall apply this equation to the field outside of a rotating star or black hole later. At sufficiently large distances we have

$$g_{00} \cong 1, \quad g_{ij} \cong -\delta_{ij}, \quad g_{0k}/g_{00} \ll 1$$

and a good approximation to (1.10.31) is given by

$$\Omega \cong -\tfrac{1}{2}\, \varepsilon_{ijk}\, g_{0i,j}\, \partial_k . \tag{1.10.32}$$

Since $e_k \cong \partial_k$, the gyroscope rotates relative to the Copernican frame with angular velocity

$$\boldsymbol{\Omega} \cong \tfrac{1}{2} \nabla \times \boldsymbol{g}, \quad \boldsymbol{g} := (g_{01}, g_{02}, g_{03}) . \tag{1.10.33}$$

We have shown that in a stationary (but not static) field, a gyroscope rotates relative to the Copernican system (relative to the "fixed stars") with angular velocity (1.10.31). In a weak field this can be approximated by (1.10.33). This means that *the rotation of a star drags along the local inertial system.* We shall discuss possible experimental tests of this effect later.

### 1.10.5 Local Coordinate Systems

We again consider the world line $\gamma(\tau)$ of an (accelerated) observer. Let $u = \dot{\gamma}$ and let $\{e_i\}$ be an arbitrary orthonormal frame along $\gamma$ which is perpendicular to $e_0 = u$. As before, let $a = \nabla_u u$. Now construct a local coordinate system as follows: at every point on $\gamma(\tau)$ consider spacelike geodesics $\alpha(s)$ perpendicular to $u$, with proper length $s$ as affine parameter. Thus $\alpha(0) = \gamma(\tau)$; let $\dot{\alpha}(0) = n \perp u$. In order to distinguish the various geodesics, we denote the geodesic through $\gamma(\tau)$ in the direction $n$ with affine parameter $s$ by $\alpha(s, n, \tau)$. We have

$$n = \left(\frac{\partial}{\partial s}\right)_{\alpha(0, n, \tau)}, \quad (n, n) = -1 . \tag{1.10.34}$$

Every point $p \in M$ in the vicinity of the observer's world line lies on precisely one of these geodesics. If $p = \alpha(s, n, \tau)$ and $n = n^j e_j$ we assign the following coordinates to $p$:

$$(x^0(p), \ldots, x^3(p)) = (\tau, s\, n^1, s\, n^2, s\, n^3) . \tag{1.10.35}$$

This means

$$\begin{aligned} x^0(\alpha(s, n, \tau)) &= \tau \\ x^j(\alpha(s, n, \tau)) &= s\, n^j = -s\, n_j = -s\, (n, e_j) . \end{aligned} \tag{1.10.36}$$

*Calculation of the Christoffel Symbols Along $\gamma(\tau)$*

Along the observer's world line, we have by construction

$$\frac{\partial}{\partial x^\alpha} = e_\alpha, \quad \text{along } \gamma(\tau) , \tag{1.10.37}$$

and hence

$$g_{\alpha\beta} = (\partial_\alpha, \partial_\beta) = \eta_{\alpha\beta}, \quad \text{along } \gamma(\tau). \tag{1.10.38}$$

In general,

$$\nabla_u e_\alpha = \nabla_{e_0} e_\alpha = \Gamma^\beta_{0\alpha} e_\beta$$

and thus

$$(e_\beta, \nabla_u e_\alpha) = \eta_{\beta\gamma} \Gamma^\gamma_{0\alpha}. \tag{1.10.39}$$

In particular,

$$\Gamma^0_{00} = (u, \nabla_u u) = 0 \tag{1.10.40}$$

$$\Gamma^j_{00} = -(e_j, \nabla_u u) = -(e_j, a) = a^j \tag{1.10.41}$$

$$\Gamma^0_{0j} = (u, \nabla_u e_j) \overset{(1.10.17)}{=} -(e_j, a) = a^j. \tag{1.10.42}$$

Let

$$\omega_{jk} = \varepsilon_{ijk}\, \omega^i. \tag{1.10.43}$$

We thus have

$$\Gamma^j_{0k} = \varepsilon_{ikj}\, \omega^i. \tag{1.10.44}$$

The remaining Christoffel symbols can be read off from the equation for the geodesics $\alpha(s, n, \tau)$. According to (1.10.36) the coordinate expression for such a geodesic is

$$x^0(s) = \text{const}$$
$$x^j(s) = s\, n^j.$$

Consequently $d^2x^\alpha/ds^2 = 0$ along a geodesic. On the other hand, a geodesic satifies the equation

$$0 = \frac{d^2x^\alpha}{ds^2} + \Gamma^\alpha_{\beta\gamma} \frac{dx^\beta}{ds} \frac{dx^\gamma}{ds} = \Gamma^\alpha_{jk}\, n^j n^k.$$

Hence, along the observer's world line $\gamma$ we have

$$\Gamma^\alpha_{jk} = 0. \tag{1.10.45}$$

The partial derivatives of the metric coefficients can be determined from the Christoffel symbols. The general relation is

$$0 = g_{\alpha\beta;\gamma} = g_{\alpha\beta,\gamma} - \Gamma^\mu_{\alpha\gamma} g_{\mu\beta} - \Gamma^\mu_{\beta\gamma} g_{\alpha\mu}. \tag{1.10.46}$$

If we now substitute (1.10.38) and our previously derived results for the Christoffel symbols, we find

$$\left.\begin{array}{ll} g_{\alpha\beta,0} = 0, & g_{ik,l} = 0 \\ g_{00,j} = 2a^j, & g_{0j,k} = \varepsilon_{jkl}\, \omega^l \end{array}\right\} \text{ along } \gamma. \tag{1.10.47}$$

These relations and $g_{\alpha\beta}=\eta_{\alpha\beta}$ along $\gamma$ imply that the line element near $\gamma$ is given by

$$\begin{aligned} ds^2 = {} & (1+2\boldsymbol{a}\cdot\boldsymbol{x})(dx^0)^2 + 2\,\varepsilon_{jkl}\,x^k\,\omega^l\,dx^0\,dx^j \\ & - \delta_{jk}\,dx^j\,dx^k + O(|\boldsymbol{x}|^2)\,dx^\alpha\,dx^\beta . \end{aligned} \tag{1.10.48}$$

From this we see that:

(i) Acceleration leads to the additional term

$$\delta g_{00} = 2\,\boldsymbol{a}\cdot\boldsymbol{x} . \tag{1.10.49}$$

(ii) Since the observer's coordinate axes rotate ($\omega^i \neq 0$), the metric has a "nondiagonal" term

$$g_{0j} = \varepsilon_{jkl}\,x^k\,\omega^l = (\boldsymbol{x}\times\boldsymbol{\omega})^j . \tag{1.10.50}$$

(iii) The lowest order corrections are not affected by the curvature. The curvature shows itself in second order.

(iv) If $a=\nabla_u u=0$ and $\boldsymbol{\omega}=0$ (no acceleration and no rotation) we have a local inertial system ($g_{\alpha\beta}=\eta_{\alpha\beta}$, $\Gamma^{\alpha}_{\beta\gamma}=0$) *along* $\gamma(\tau)$.

*Motion of a Test Body*

Suppose that the observer whose world line is $\gamma(\tau)$ (an astronaut in a space capsule, for example) observes a nearby freely falling body. This obeys the equation of motion

$$\frac{d^2x^\alpha}{d\lambda^2} + \Gamma^{\alpha}_{\beta\gamma}\frac{dx^\beta}{d\lambda}\frac{dx^\gamma}{d\lambda} = 0 .$$

We now replace the proper time $\lambda$ of the test body by the coordinate time $t$:

$$d/d\lambda = (dt/d\lambda)\,d/dt =: \gamma\,d/dt. \text{ Since } dx^\alpha/d\lambda = \gamma\,(1, dx^k/dt) ,$$

we have

$$\frac{d^2x^\alpha}{dt^2} + \frac{1}{\gamma}\frac{d\gamma}{dt}\frac{dx^\alpha}{dt} + \Gamma^{\alpha}_{\beta\gamma}\frac{dx^\beta}{dt}\frac{dx^\gamma}{dt} = 0 . \tag{1.10.51}$$

For $\alpha=0$, this becomes

$$\frac{1}{\gamma}\frac{d\gamma}{dt} + \Gamma^{0}_{\beta\gamma}\frac{dx^\beta}{dt}\frac{dx^\gamma}{dt} = 0 .$$

Substitution of this into (1.10.51) for $\alpha=j$ gives

$$\frac{d^2x^j}{dt^2} + \left(-\frac{dx^j}{dt}\Gamma^{0}_{\beta\gamma} + \Gamma^{j}_{\beta\gamma}\right)\frac{dx^\beta}{dt}\frac{dx^\gamma}{dt} = 0 . \tag{1.10.52}$$

Now let the velocity of the test body be small. To first order in $v^j = dx^j/dt$ the equation of motion is

$$\frac{dv^j}{dt} - v^j \Gamma^0_{00} + \Gamma^j_{00} + 2\Gamma^j_{k0} v^k = 0 . \tag{1.10.53}$$

Since the particle is falling in the vicinity of the observer, the spatial coordinates (1.10.36) are small. To first order in $x^k$ and $v^k$ we have

$$\frac{dv^j}{dt} = v^j \Gamma^0_{00}|_{x=0} - \Gamma^j_{00}|_{x=0} - x^k \Gamma^j_{00,k}|_{x=0} - 2v^k \Gamma^j_{k0}|_{x=0} .$$

The Christoffel symbols for $\boldsymbol{x} = \mathbf{0}$ have already been determined. If we insert these, we find

$$\frac{dv^j}{dt} = -a^j - 2\varepsilon_{jik}\,\omega^i v^k - x^k \Gamma^j_{00,k}|_{x=0} .$$

The quantity $\Gamma^j_{00,k}$ is obtained from the Riemann tensor:

$$R^\alpha_{\beta\gamma\delta} = \Gamma^\alpha_{\beta\delta,\gamma} - \Gamma^\alpha_{\beta\gamma,\delta} + \Gamma^\alpha_{\gamma\mu}\Gamma^\mu_{\beta\delta} - \Gamma^\alpha_{\delta\mu}\Gamma^\mu_{\beta\gamma} ,$$

so that

$$\Gamma^j_{00,k} = R^j_{0k0} + \Gamma^j_{0k,0} - \Gamma^j_{k\mu}\Gamma^\mu_{00} + \Gamma^j_{0\mu}\Gamma^\mu_{0k} .$$

For $\boldsymbol{x} = 0$, we have

$$\Gamma^j_{0k,0} = -\varepsilon_{jkm}\,\omega^m_{,0}$$

$$\Gamma^j_{k\mu}\Gamma^\mu_{00} = 0$$

$$\Gamma^j_{0\mu}\Gamma^\mu_{0k} = \Gamma^j_{00}\Gamma^0_{0k} + \Gamma^j_{0m}\Gamma^m_{0k} = a^j a^k + \varepsilon_{mjn}\,\omega^n\,\varepsilon_{kml}\,\omega^l .$$

Hence

$$x^k \Gamma^j_{00,k} = x^k (R^j_{0k0} - \varepsilon_{jkm}\,\omega^m_{,0} + a^j a^k + \varepsilon_{mjn}\,\omega^n\,\varepsilon_{kml}\,\omega^l) .$$

Returning to three-dimensional vector notation we end up with

$$\dot{\boldsymbol{v}} = -\boldsymbol{a}(1 + \boldsymbol{a}\cdot\boldsymbol{x}) - 2\boldsymbol{\omega}\times\boldsymbol{v} - \boldsymbol{\omega}\times(\boldsymbol{\omega}\times\boldsymbol{x}) - \dot{\boldsymbol{\omega}}\times\boldsymbol{x} + \boldsymbol{f} , \tag{1.10.54}$$

where

$$f^j := R^j_{00k}\,x^k . \tag{1.10.55}$$

The first term in (1.10.54) is the usual "inertial acceleration", including the relativistic correction $(1 + \boldsymbol{a}\cdot\boldsymbol{x})$, which is a result of (1.10.49). The terms containing $\boldsymbol{\omega}$ are well-known from classical mechanics. The force $R^j_{00k}\,x^k$ is a consequence of the inhomogeneity of the gravitational field.

If the frame is Fermi transported ($\boldsymbol{\omega} = 0$), the Coriolis force vanishes. If the observer is freely falling, then also $\boldsymbol{a} = 0$ and only the "tidal force" $f^j$ remains. This cannot be transformed away. We shall discuss this in more detail at the beginning of the next chapter.

**Exercises:**

1. A nonspherical body in an inhomogeneous gravitational field experiences a torque which results in a time dependence of the spin four vector. Suppose that the center of mass is freely falling along a geodesic with four velocity $u$. Show that $S$ satisfies the equation of motion

$$\nabla_u S^\varrho = \eta^{\varrho\beta\alpha\mu} u_\mu u^\sigma u^\lambda t_{\beta\nu} R^\nu_{\sigma\alpha\lambda},$$

where $t_{\beta\nu}$ is the "reduced quadrupole moment tensor"

$$t_{ij} = \int \varrho \left(x^i x^j - \tfrac{1}{3}\delta_{ij}\right) d^3x$$

in the rest frame of the center of mass and $t^{\alpha\beta} u_\beta = 0$. It is assumed that the Riemann tensor is determined by an external field which is nearly constant over distances comparable to the size of the test body.

2. Show that

$$H = \tfrac{1}{2} g^{\mu\nu} (\Pi_\mu - e A_\mu)(\Pi_\nu - e A_\nu)$$

is the Hamiltonian which describes the motion of a charged particle with charge $e$ in a gravitational field ($\Pi_\mu$ is the *canonical* momentum).

# Chapter 2. Einstein's Field Equations

"There is something else that I have learned from the theory of gravitation: No collection of empirical facts, however comprehensive, can ever lead to the setting up of such complicated equations. A theory can be tested by experience, but there is no way from experience to the formulation of a theory. Equations of such complexity as are the equations of the gravitational field can be found only through the discovery of a logically simple mathematical condition which determines the equations completely or [at least] almost completely. Once one has those sufficiently strong formal conditions, one requires only little knowledge of facts for the setting up of a theory; in the case of the equations of gravitation it is the four-dimensionality and the symmetric tensor as expression for the structure of space which, together with the invariance concerning the continuous transformation-group, determine the equations almost completely." A. Einstein

In the previous chapter we examined the kinematical framework of the general theory of relativity and the effect of gravitational fields on physical systems. The core of the theory, however, consists of Einstein's field equations, which relate the metric field to matter. After a discussion of the physical meaning of the curvature tensor, we shall first give a simple physical motivation of the field equations and will show then that they are determined by only a few natural requirements [1].

## 2.1 Physical Meaning of the Curvature Tensor

In a local inertial system the "field strengths" of the gravitational field (the Christoffel symbols) can be transformed away. Due to its tensor character, this is not possible for the curvature. Physically, the curvature describes the "tidal forces" as we shall see in the following. [This was indicated already in the discussion of (1.10.54).]

Consider a family of timelike geodesics having the property that in a sufficiently small open subspace of the Lorentz manifold $(M, g)$ pre-

[1] For an excellent historical account of Einstein's struggle which culminated in the final form of his gravitational field equations, presented on November 25 (1915), we refer to: A. Pais, 'Subtle is the Lord, The Science and the Life of Albert Einstein', Oxford University Press, 1982 (Chap. 14).

cisely one geodesic passes through every point. Such a collection is called a *congruence of timelike geodesics.* It might represent a swarm of freely falling bodies. The tangent field to this set of curves, with proper time as curve parameter, is denoted by $u$. Note that $(u, u) = 1$.

Now let $\gamma(t)$ be a curve transversal to the congruence being investigated. This means that the tangent vector $\dot{\gamma}$ is never parallel to $u$ in the region under consideration, as indicated in Fig. 2.1.

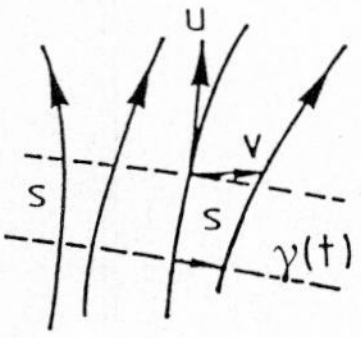


**Fig. 2.1**

Imagine that every point on the curve $\gamma(t)$ moves a distance $s$ along the geodesic which passes through that point. Let the resulting point be denoted by $\gamma(t, s)$. For every $t$, the mapping $s \mapsto \gamma(t, s)$ is a geodesic and $u = \gamma_* \frac{\partial}{\partial s}$. Let $v = \gamma_* \frac{\partial}{\partial t}$. Obviously[2] $[u, v] = 0$, which means that

$$L_u v = 0 . \tag{2.1.1}$$

$v$ represents the separation of points which are moved the same distance along neighboring curves of the congruence, beginning at arbitrary starting points. According to (2.1.1) this displacement vector is Lie transported: If $\phi_s$ is the flow of $u$, the field $v$ is invariant under $\phi_s$, i.e. $\phi_s^* v = v$ (see Sect. 3.3 of Part I).

If one adds a multiple of $u$ to $v$, the resulting vector represents the separation of points on the same two neighboring curves, but at different distances along the curve. However, we are interested only in the distance between the curves and not in the distance between particular points on these curves. This is represented by the projection of $v$ on the subspace of the tangent space which is perpendicular to $u$. Thus the (infinitesimal) separation vector is

$$n = v - (v, u)\, u . \tag{2.1.2}$$

Since $(u, u) = 1$, $n$ is indeed perpendicular to $u$. We now show that $n$ is also Lie transported. We have

$$L_u n = [u, n] = [u, v] - [u, (v, u)\, u] = -[u\,(v, u)]\, u .$$

The Ricci identity and $(u, u) = 1$ imply

$$0 = \nabla_v (u, u) = (\nabla_v u, u) .$$

[2] Recall that it is shown in the footnote on p. 110 that $[u, v]$ is defined, even when $v$ is only determined on a submanifold.

Since also $\nabla_v u = \nabla_u v$ (use (2.1.1) and the fact that the torsion vanishes), it follows that

$$u\,(u, v) = (\nabla_u u, v) + (u, \nabla_u v) = (u, \nabla_v u) = 0\,. \tag{2.1.3}$$

Hence

$$L_u n = 0\,. \tag{2.1.4}$$

Now

$$\nabla_u^2 v = \nabla_u \nabla_u v = \nabla_u (\nabla_v u) = (\nabla_u \nabla_v - \nabla_v \nabla_u)\, u$$

or, due to (2.1.1),

$$\nabla_u^2 v = R\,(u, v)\, u\,. \tag{2.1.5}$$

(In the mathematical literature this is called the *Jacobi equation* for the Jacobi field $v$; see e.g. [5], Sect. 1 of Chap. VIII.)

We now show that $n$ also satisfies such an equation. From (2.1.3) it follows that

$$\begin{aligned} \nabla_u n &= \nabla_u v - (u\,(v, u))\, u - (v, u)\, \nabla_u u \\ &= \nabla_u v\,. \end{aligned} \tag{2.1.6}$$

Furthermore

$$R\,(u, n)\, u = R\,(u, v)\, u - (v, u)\, R\,(u, u)\, u = R\,(u, v)\, u\,. \tag{2.1.7}$$

From (2.1.5, 6, 7) we thus obtain the *equation of geodesic deviation*

$$\nabla_u^2 n = R\,(u, n)\, u\,; \tag{2.1.8}$$

$n$ is a Jacobi field which is everywhere perpendicular to $u$.

Now let $\{e_j\}$, $i = 1, 2, 3$ be an orthonormal frame perpendicular to $u$, which is parallel transported along an arbitrary geodesic of the congruence (according to the discussion in Sect. 1.10, this is a nonrotating frame of reference). Since $n$ is orthogonal to $u$, it has the expansion

$$n = n^i e_i\,. \tag{2.1.9}$$

Note that

$$\nabla_u n = (u\, n^i)\, e_i + n^i \nabla_u e_i = \frac{dn^i}{ds}\, e_i\,.$$

Now take $e_0 = u$; from (2.1.8) we obtain

$$\frac{d^2 n^i}{ds^2}\, e_i = n^j R\,(e_0, e_j)\, e_0 = n^j R^i_{00j}\, e_i$$

or

$$\frac{d^2 n^i}{ds^2} = R^i_{00j}\, n^j\,. \tag{2.1.10}$$

If we set $K_{ij} = R^i_{0 0 j}$, (2.1.10) can be written in matrix form:

$$\frac{d^2}{ds^2}\boldsymbol{n} = K\boldsymbol{n}\,. \tag{2.1.11}$$

We have $K = K^T$ and

$$\operatorname{tr} K = R^i_{00i} = R^\mu_{00\mu} = -R_{00}\,. \tag{2.1.12}$$

Equation (2.1.10) [or (2.1.11)] describes the relative acceleration of neighboring freely falling test bodies.

*Comparison with Newtonian Theory*

The paths of two neighboring test bodies, denoted by $x^i(t)$ and $x^i(t) + n^i(t)$, satisfy the equations of motion

$$\ddot{x}^i(t) = -\left(\frac{\partial\phi}{\partial x^i}\right)_{\boldsymbol{x}(t)}$$

$$\ddot{x}^i(t) + \ddot{n}^i(t) = -\left(\frac{\partial\phi}{\partial x^i}\right)_{\boldsymbol{x}(t)+\boldsymbol{n}(t)}.$$

If we take the difference of these two equations and retain only terms linear in $\boldsymbol{n}$, we obtain

$$\ddot{n}^i(t) = -(\partial_i\partial_j\phi)\, n^j(t)\,.$$

Thus we again find (2.1.11), but with

$$K = -(\partial_i\partial_j\phi)\,.$$

Note that $\operatorname{tr} K = -\Delta\phi$. In matter-free space we thus have $\operatorname{tr} K = 0$. As we shall see from the field equations, this is also true in GR.

We thus have the correspondence

$$R^i_{0j0} \leftrightarrow \partial_i\partial_j\phi\,. \tag{2.1.13}$$

In particular,

$$R_{\mu\nu}u^\mu u^\nu \leftrightarrow \Delta\phi\,. \tag{2.1.14}$$

**Summary:** Variations of the gravitational field are described by the Riemann tensor. The tensor character of this quantity implies that such inhomogeneities, unlike the "field strengths" $\Gamma^\mu_{\alpha\beta}$, cannot be transformed away. Relative accelerations ("tidal forces") of freely falling test bodies are described by the Riemann tensor, via the equation of geodesic deviation (2.1.8).

These remarks are also important for understanding the effect of a gravitational wave on a mechanical detector.

**Exercises:**

1. Use the results of Sect. 1.5 to show that in the Newtonian limit the components $R^i_{0j0}$ of the curvature tensor are given approximately by $\partial_i \partial_j \phi$.
2. Write the Newtonian equations of motion for a potential as a geodesic equation in a four dimensional affine space. Compute the Christoffel symbols and the Riemann tensor, and show that the affine connection is not metric.
3. Generalize the considerations of this section to an arbitrary congruence of timelike curves (replace parallel transport everywhere by Fermi transport). Derive, in particular, the following generalization of Eq. (2.1.8)

$$\mathbb{F}_u^2 n = R(u, n)\, u + (\nabla_n a)_\perp - (a, n)\, a\,,$$

where $a = \nabla_u u$ and $\perp$ denotes the projection orthogonal to $u$. (Important consequences of this equation, in particular the Landau-Raychaudhuri equation, are studied, for instance, in [15], Sect. 4.1.)
4. Consider a static field (1.9.15) in adapted coordinates. Show that the function $\varphi = \sqrt{g_{00}}$ satisfies the equation

$$\frac{1}{\varphi} \overset{(3)}{\Delta} \varphi = R_{\mu\nu} u^\mu u^\nu\,,$$

where $\overset{(3)}{\Delta}$ denotes the Laplacian of the spatial sections $x^0 = \text{const}$, and where $(u^\mu) = \frac{1}{\varphi} (\partial/\partial t)$.

*Hint:* Use the Killing equation for the Killing field $K = \partial/\partial t$ and $d(K^\flat/\varphi^2) = 0$ to show first that $\varphi\, u_{\alpha;\beta} = -u_\beta\, \varphi_{,\alpha}$. This implies that $a = \nabla_u u$ is given by $a_\alpha = -(1/\varphi)\, \varphi_{,\alpha}$ and that $u_{\alpha;\beta} = a_\alpha u_\beta$. Then express the divergence of $a_\alpha$ in two different ways.

The result can also be obtained in a straightforward manner with the help of the structure equations.

## 2.2 The Gravitational Field Equations

"You will be convinced of the general theory of relativity once you have studied it. Therefore I am not going to defend it with a single word."

(A. Einstein, on a postcard to A. Sommerfeld, Feb. 8, 1916.)

In the absence of true gravitational fields one can always find a coordinate system in which $g_{\mu\nu} = \eta_{\mu\nu}$ everywhere. This is the case when the Riemann tensor vanishes (modulo global questions, as discussed in Sect. 5.9 of Part I). (One might call this fact the "zeroth law of gravitation".) In the presence of gravitational fields, space will become curved.

The field equations must describe the dependence of the curvature of the metric field on the mass and energy distributions by means of partial differential equations.

*Heuristic "Derivation"*

As a starting point for the arguments which will lead to the field equations we recall from Sect. 1.5 that for a weak, stationary field generated by a nonrelativistic mass distribution, the component $g_{00}$ of the metric tensor (relative to a suitable coordinate system which is nearly Lorentzian on a global scale) is related to the Newtonian potential by

$$g_{00} \cong 1 + 2\phi . \tag{2.2.1}$$

The potential $\phi$ satisfies the Poisson equation

$$\Delta\phi = 4\pi G \varrho . \tag{2.2.2}$$

For nonrelativistic matter, $\varrho$ is approximately equal to the energy density $T_{00}$ ($T_{\mu\nu}$ is the energy-momentum tensor),

$$T_{00} \cong \varrho .$$

Hence in this limit

$$\Delta g_{00} \cong 2\Delta\phi \cong 8\pi G T_{00} . \tag{2.2.3}$$

From Sect. 2.1 we have the correspondence

$$R_{00} \leftrightarrow \Delta\phi . \tag{2.2.4}$$

Thus instead of (2.2.3) we could write

$$R_{00} \cong 4\pi G T_{00} . \tag{2.2.5}$$

This would suggest the generally covariant equation

$$R_{\mu\nu} = 4\pi G T_{\mu\nu} . \tag{2.2.6}$$

This equation was also considered (and published) by Einstein. However, it cannot be the correct field law. If we take the trace of both sides, we obtain

$$R = 4\pi G T , \qquad T = T^{\mu}_{\mu} .$$

Hence we must also have

$$R^{\mu\nu} - \tfrac{1}{2} g^{\mu\nu} R = 4\pi G (T^{\mu\nu} - \tfrac{1}{2} g^{\mu\nu} T) . \tag{2.2.7}$$

If we now take the divergence and use the reduced Bianchi identity (Sect. 5.6 of Part I), we find

$$T^{\mu\nu}_{\ ;\nu} - \tfrac{1}{2} g^{\mu\nu} T_{,\nu} = 0 .$$

From Sect. 1.4 we know that the energy-momentum tensor must satisfy

$$T^{\mu\nu}_{;\nu}=0\,.$$

This would then imply $T=\text{const}$, which is generally not the case. This difficulty is avoided if we modify the right hand side of (2.2.7) as follows

$$R^{\mu\nu}-\tfrac{1}{2}g^{\mu\nu}R=8\pi G T^{\mu\nu}\,. \tag{2.2.8}$$

Taking the trace of (2.2.8) results in

$$R_{\mu\nu}=8\pi G\,(T_{\mu\nu}-\tfrac{1}{2}g_{\mu\nu}T) \tag{2.2.9}$$

instead of (2.2.6). For weak fields and nonrelativistic matter, there is no inconsistency between (2.2.8, 9) and (2.2.5), since

$$\begin{aligned}T_{00}-\tfrac{1}{2}g_{00}T &\cong \tfrac{1}{2}\ (\text{energy density} + 3\times \text{pressure})\\ &\cong \tfrac{1}{2}\ (\text{energy density}).\end{aligned}$$

Equations (2.2.8) [or (2.2.9)] are the *Einstein field equations.* After a number of earlier incomplete attempts, they were announced in November 1915.

---

**Exercise:** Use the result of the last exercise in Sect. 2.1 to show that a static field of a perfect fluid satisfies the (exact) equation

$$\frac{1}{\varphi}\overset{(3)}{\Delta}\varphi=4\pi G\,(\varrho+3p)\,.$$

In particular, a static vacuum manifold satisfies

$$\overset{(3)}{\Delta}\varphi=0\,,$$

i.e., $\varphi$ is harmonic on spatial sections. What can be concluded from this if the spatial sections are diffeomorphic to $\mathbb{R}^3$ and $\varphi$ vanishes at infinity (see the exercises at the end of Part I)?

---

*The Question of Uniqueness*

Now that we have motivated the field equations with simple physical arguments, we must ask how uniquely they are determined. The following investigation will show that one has very little freedom.

Equation (2.2.3) suggests that the ten potentials $g_{\mu\nu}$ satisfy equations of the form

$$\mathscr{D}_{\mu\nu}[g]=T_{\mu\nu}\,. \tag{2.2.10}$$

Here $\mathscr{D}_{\mu\nu}[g]$ is a tensor constructed from $g_{\mu\nu}$ and its first and second derivatives. Furthermore the identity

$$\mathscr{D}^{\mu\nu}_{;\nu}=0 \tag{2.2.11}$$

should hold.

**Remarks:** 1) The requirement that the field equations contain derivatives of $g_{\mu\nu}$ only up to second order is certainly reasonable. If this were not the case, one would have to specify the initial values (for the Cauchy problem) not only of the metric and its first derivative, but also higher derivatives on a spacelike surface, in order to determine the development of the metric field.
2) Equation (2.2.11) is required in order that $T^{\mu\nu}_{;\nu}=0$ be a consequence of the field equations. Since, on the other hand, this is also a consequence of the non-gravitational laws (see Sect. 1.4), the coupled system of equations contains *four identities.* As we shall subsequently discuss in more detail, general covariance of the theory indeed requires the existence of four identities.

It is very satisfying that one can prove

**Theorem 2.1:** A tensor $\mathscr{D}_{\mu\nu}[g]$ with the required properties has in four dimensions the form

$$\mathscr{D}_{\mu\nu}[g]=a\,G_{\mu\nu}+b\,g_{\mu\nu}\,. \tag{2.2.12}$$

Here $G_{\mu\nu}$ is the Einstein tensor [defined in Eq. (5.48) of Part I]:

$$G_{\mu\nu}=R_{\mu\nu}-\tfrac{1}{2}\,g_{\mu\nu}R\,. \tag{2.2.13}$$

**Remark:** This theorem has been proved only recently (see [24] and work cited therein). Previously one had made the additional assumption that the tensor $\mathscr{D}_{\mu\nu}[g]$ was *linear in the second derivatives.* A justification of this is obtained by examining (2.2.3). It is remarkable that it is unnecessary to postulate linearity in the second derivatives only in four dimensions. It is also by no means trivial that it is not necessary to require the symmetry of $\mathscr{D}_{\mu\nu}$. The theorem thus shows that a gravitational field of the form (2.2.10) cannot be coupled to a nonsymmetric energy-momentum tensor.

We prove first the following

**Theorem 2.2:** Let $K[g]$ be a functional which assigns to every smooth pseudo-Riemannian metric field $g$ a tensor field (of arbitrary type) such that the components of $K[g]$ at the point $p$ depend smoothly on $g_{ij}$ and its derivatives $g_{ij,k}$, $g_{ij,kl}$ at $p$. Then $K$ is determined pointwise by $g$ and the curvature tensor by elementary operations of tensor algebra.
*Proof:* In normal coordinates (Sect. 5.3 of Part I) the geodesics are straight lines $x^i(s)=s\,a^i$. Hence

$$\Gamma^k_{ij}(s\,a^m)\,a^i a^j=0\,. \tag{2.2.14}$$

Since $\Gamma$ is symmetric, we have $\Gamma^k_{ij}(0)=0$ and hence $g_{ij,k}=0$. Furthermore, at $x=0$ we can write $g_{ij}$ in normal form: $g_{ij}(0)=\mathring{g}_{ij}=\varepsilon_i\,\delta_{ij}$, with $\varepsilon_i = \pm 1$. In a neighborhood of $x = 0$ we may then expand

$$g_{ik} = \mathring{g}_{ik} + \tfrac{1}{2}\beta_{ik,rs}\, x^r x^s + \dots .$$

Thus, if $\Gamma_{ikr} := g_{ij}\Gamma^j_{kr}$,

$$\begin{aligned}\Gamma_{ikr} &= \tfrac{1}{2}(g_{ik,r} + g_{ir,k} - g_{kr,i}) \\ &= \tfrac{1}{2}(\beta_{ik,rs}\, x^s + \beta_{ir,ks}\, x^s - \beta_{kr,is}\, x^s) + \dots .\end{aligned} \tag{2.2.15}$$

From (2.2.14) we have

$$\Gamma^k_{ij}(x)\, x^i x^j = 0$$

and hence

$$\Gamma_{ikr}(x)\, x^k x^r = 0 . \tag{2.2.16}$$

If we insert (2.2.15) into (2.2.16) we obtain

$$(\beta_{ik,rs} + \beta_{ir,ks} - \beta_{kr,is})\, x^k x^r x^s = 0 . \tag{2.2.17}$$

We shall use the abbreviation

$$\beta_{ik,rs} = \left(\frac{i\,k}{r\,s}\right) = \left.\frac{\partial^2 g_{ik}}{\partial x^r\, \partial x^s}\right|_{x=0} . \tag{2.2.18}$$

The $x^k$ in (2.2.17) are arbitrary; hence the contribution to the coefficients which is symmetric in $(k, r, s)$ must vanish. In other words, the following cylic sum must vanish

$$\sum_{(k,r,s)} \left[\left(\frac{i\,k}{r\,s}\right) + \left(\frac{i\,r}{k\,s}\right) - \left(\frac{k\,r}{i\,s}\right)\right] = 0 \tag{2.2.19}$$

or

$$2\left[\left(\frac{i\,k}{r\,s}\right) + \left(\frac{i\,r}{s\,k}\right) + \left(\frac{i\,s}{k\,r}\right)\right] - \left(\frac{r\,s}{i\,k}\right) - \left(\frac{k\,s}{i\,r}\right) - \left(\frac{k\,r}{i\,s}\right) = 0 . \tag{2.2.20}$$

The negative terms are obtained from the positive ones by interchanging numerators and denominators. Instead of $i$, one can distinguish $k$ or $r$:

$$2\left[\left(\frac{i\,k}{r\,s}\right) + \left(\frac{k\,r}{s\,i}\right) + \left(\frac{k\,s}{i\,r}\right)\right] - \left(\frac{r\,s}{i\,k}\right) - \left(\frac{i\,s}{k\,r}\right) - \left(\frac{i\,r}{k\,s}\right) = 0$$

$$2\left[\left(\frac{r\,k}{i\,s}\right) + \left(\frac{i\,r}{s\,k}\right) + \left(\frac{r\,s}{k\,i}\right)\right] - \left(\frac{i\,s}{r\,k}\right) - \left(\frac{k\,s}{i\,r}\right) - \left(\frac{k\,i}{r\,s}\right) = 0 .$$

If we add these three equations, we obtain

$$\left(\frac{i\,k}{r\,s}\right) + \left(\frac{r\,i}{k\,s}\right) + \left(\frac{k\,r}{i\,s}\right) = 0 . \tag{2.2.21}$$

Finally, interchange $i$ and $s$ in (2.2.20); the result is

$$2\left[\left(\frac{s\,k}{r\,i}\right)+\left(\frac{s\,r}{i\,k}\right)+\left(\frac{s\,i}{k\,r}\right)\right]-\underbrace{\left(\frac{r\,i}{s\,k}\right)-\left(\frac{k\,i}{s\,r}\right)-\left(\frac{k\,r}{i\,s}\right)}_{0}=0\,.$$

If we now again interchange $i$ and $s$, we obtain

$$\left(\frac{i\,k}{r\,s}\right)+\left(\frac{i\,r}{s\,k}\right)+\left(\frac{i\,s}{k\,r}\right)=0\,. \tag{2.2.22}$$

A comparison with (2.2.21) yields

$$\left(\frac{i\,s}{k\,r}\right)=\left(\frac{k\,r}{i\,s}\right). \tag{2.2.23}$$

From the general expression for the Riemann tensor, one obtains in normal coordinates

$$R_{iklm}=\tfrac{1}{2}\left[\left(\frac{i\,m}{k\,l}\right)+\left(\frac{k\,l}{m\,i}\right)-\left(\frac{i\,l}{k\,m}\right)-\left(\frac{k\,m}{i\,l}\right)\right]\overset{(2.2.23)}{=}\left(\frac{i\,m}{k\,l}\right)-\left(\frac{i\,l}{k\,m}\right).$$

Using this and (2.2.22) one obtains

$$\begin{aligned}R_{iklm}+R_{lkim}&=\left(\frac{i\,m}{k\,l}\right)-\left(\frac{i\,l}{k\,m}\right)+\left(\frac{l\,m}{k\,i}\right)-\left(\frac{i\,l}{k\,m}\right)\\&\overset{(2.2.23)}{=}\left(\frac{i\,m}{k\,l}\right)+\left(\frac{k\,i}{l\,m}\right)-2\left(\frac{i\,l}{k\,m}\right)\overset{(2.2.22)}{=}-3\left(\frac{i\,l}{k\,m}\right).\end{aligned}$$

Hence, in normal coordinates

$$\begin{aligned}g_{ik}&=\mathring{g}_{ik}-\tfrac{1}{6}\left(R_{irks}+R_{kris}\right)x^r x^s+\ldots,\\g_{ik}&=\mathring{g}_{ik}-\tfrac{1}{3}R_{irks}x^r x^s+\ldots\,.\end{aligned} \tag{2.2.24}$$

Thus $R_{irks}$ expressed in terms of normal coordinates determines all the second derivatives of $g_{ik}$ at $p$. The first derivatives vanish. Since $R_{irks}$ is a tensor, the conclusion of the theorem follows immediately.

We now prove Theorem 2.1 under the additional assumption that $\mathscr{D}_{\mu\nu}[g]$ is linear in the second derivatives. (For the generalization we refer once more to [24].)

Since the $R^{\mu}_{\alpha\beta\gamma}$ are linear in the second derivatives of the $g_{\mu\nu}$, Theorem 2.2 implies that $\mathscr{D}_{\mu\nu}[g]$ must have the following form

$$\mathscr{D}_{\mu\nu}[g]=c_1 R_{\mu\nu}+c_2 R\,g_{\mu\nu}+c_3\,g_{\mu\nu}\,.$$

As a result of the contracted Bianchi identity $G^{\mu\nu}_{;\nu}=0$ [$G_{\mu\nu}$ given by (2.2.13)], we must have

$$\mathscr{D}^{\mu\nu}_{;\nu}=\left(c_2+\tfrac{1}{2}c_1\right)\left(g^{\mu\nu}R\right)_{;\nu}=\left(c_2+\tfrac{1}{2}c_1\right)g^{\mu\nu}R_{,\nu}=0\,.$$

Hence $c_2+\frac{1}{2}c_1=0$.

Theorem 2.1 requires that field equations having the structure (2.2.10) must be of the form

$$G_{\mu\nu}+\Lambda g_{\mu\nu}=\varkappa T_{\mu\nu}. \tag{2.2.25}$$

**Remarks:** 1) The field equations (2.2.25) are *nonlinear* partial differential equations, even in vacuum. This must be the case, since every form of energy, including the "energy of the gravitational field", is a source of gravitational fields. This nonlinearity makes a detailed analysis of the field equations extremely difficult. Fortunately, one has found a number of physically relevant exact solutions.
2) The field equations (2.2.25) contain two constants $\Lambda$ and $\varkappa$. In order to determine them, we consider the Newtonian limit. We need

$$R_{\mu\nu}=\partial_\lambda\Gamma^\lambda_{\nu\mu}-\partial_\nu\Gamma^\lambda_{\lambda\mu}+\Gamma^\sigma_{\nu\mu}\Gamma^\lambda_{\lambda\sigma}-\Gamma^\sigma_{\lambda\mu}\Gamma^\lambda_{\nu\sigma}.$$

As in Sect. 1.5, we can introduce a nearly Lorentzian system for weak, stationary fields, in which

$$g_{\mu\nu}=\eta_{\mu\nu}+h_{\mu\nu}, \qquad |h_{\mu\nu}|\ll 1 \tag{2.2.26}$$

and $h_{\mu\nu}$ is independent of time.

We may neglect the quadratic terms in the Ricci tensor,

$$R_{\mu\nu}\cong\partial_\lambda\Gamma^\lambda_{\mu\nu}-\partial_\nu\Gamma^\lambda_{\lambda\mu}.$$

Since the field is stationary, we have

$$R_{00}\cong\partial_l\Gamma^l_{00}$$

and

$$\Gamma^l_{00}\cong\tfrac{1}{2}\partial_l g_{00}.$$

In the Newtonian limit

$$g_{00}\cong 1+2\phi.$$

We thus find, as already implied by (2.1.14),

$$R_{00}\cong\Delta\phi. \tag{2.2.27}$$

If $\Lambda=0$, (2.2.25) is equivalent to (2.2.9), with $\varkappa=8\pi G$. For nonrelativistic matter $|T_{ij}|\ll|T_{00}|$ and hence we obtain the Poisson equation (2.2.2) for the (0, 0) component of (2.2.9), using (2.2.27). A small nonzero $\Lambda$ in (2.2.25) leads to

$$\Delta\phi=4\pi G\varrho+\Lambda. \tag{2.2.28}$$

Note that instead of the result of the last Exercise (p. 129), we obtain now as an exact equation for $\varphi=\sqrt{g_{00}}\cong 1+\phi$

$$\frac{1}{\varphi}\overset{(3)}{\Delta}\varphi=4\pi G(\varrho+3p)+\Lambda.$$

Thus the term in $\Lambda$ corresponds to an effective mass density of magnitude

$$\varrho_{\mathrm{eff}} = \frac{\Lambda}{4\pi G}.$$

In the following we shall set $\Lambda = 0$. This is reasonable, since only then is the flat metric a solution of the vacuum field equations ($T_{\mu\nu} = 0$). In any case, there is no astronomical evidence for $\Lambda \neq 0$. From cosmological arguments one knows that $\varrho_{\mathrm{eff}}$ is certainly very small ($\varrho_{\mathrm{eff}} < 10^{-29}\,\mathrm{g/cm^3}$). As long as no empirical reasons for the introduction of the so-called "cosmological constant" exist, one should not introduce any arbitrary constants which are not present in the Newtonian theory. However, from the point of view of present day quantum field theory, the smallness of the cosmological constant is a deep mystery. The vacuum expectation value $\langle T_{\mu\nu} \rangle_{\mathrm{vac}}$ of the energy-momentum operator must be proportional to $g_{\mu\nu}$,

$$\langle T_{\mu\nu} \rangle_{\mathrm{vac}} = g_{\mu\nu}\, \varrho_{\mathrm{vac}}$$

and thus acts like a cosmological term with

$$\Lambda = 8\pi G\, \varrho_{\mathrm{vac}}.$$

In the evolution of the very early Universe, we expect various phase transitions, which are associated with a hierarchical spontaneous breakdown of unifying gauge symmetries and other symmetries. The structure of the vacuum thereby changes, and the corresponding changes of $\varrho_{\mathrm{vac}}$ should be gravitationally relevant. Since these changes are huge compared to the observational limits for the present cosmological constant, we see no reason why the latter should be so small.

3) The energy-momentum tensor $T^{\mu\nu}$ must be obtained from a theory of matter (just as is the case for the current density $j^\mu$ in Maxwell's equations). We shall discuss the construction of $T^{\mu\nu}$ (and $j^\mu$) in the framework of a Lagrangian field theory in Sect. 2.3. In practical applications one often uses simple phenomenological models (such as ideal or viscous fluids).

4) We shall derive the field equations again in Sect. 2.3 from a variational principle, as was first done by Hilbert.

---

**Exercises:** *Introductory Remarks:* One can obtain field equations different from those of Einstein if one limits a priori the degrees of freedom of the metric tensor (beyond the requirement that the signature be Lorentzian). For example, one could require that the metric be *conformally flat*, i.e. that $g$ has the form

$$g = \phi^2 \eta, \tag{2.2.29}$$

where $\eta$ is a flat metric (so that the corresponding Riemann tensor vanishes) and $\phi$ is a function. (It can be shown that this is the case locally if and only if the so-called Weyl tensor vanishes.) Since the metric (2.2.29) contains only one scalar degree of freedom $\phi(x)$, we need a scalar equation. The only possibility which satisfies the requirement that $\phi$ appears linearly in the second derivatives is

$$R = 24\pi GT, \tag{2.2.30}$$

where $T = T^{\mu}_{\mu}$ and $R$ is the scalar Riemann curvature. The constants have been chosen such that in the Newtonian limit (2.2.30) reduces to the Poisson equation. This scalar theory was considered by Einstein and Fokker; in the linearized approximation it agrees with the original Nordstrøm theory. As we have emphasized previously, there is no bending of light rays in such a theory. In addition, the precession of the perihelion of Mercury is not correctly predicted.

However, this scalar theory *does* satisfy the principle of equivalence. Thus we have an example to demonstrate that the equivalence principle alone (without additional assumptions) does not predict the bending of light rays.

1) Choose a coordinate system such that (2.2.29) has the form

$$g_{\mu\nu} = \phi^2 \eta_{\mu\nu}, \quad (\eta_{\mu\nu}) = \mathrm{diag}\,(1, -1, -1, -1) \tag{2.2.31}$$

and show that in this case

$$R = -6\,\phi^{-3}\,\eta^{\mu\nu}\,\partial_\mu\,\partial_\nu\,\phi. \tag{2.2.32}$$

Equation (2.2.30) then reads

$$\Box\,\phi = -4\pi\,GT\,\phi^3, \quad \Box := \eta^{\mu\nu}\,\partial_\mu\,\partial_\nu \tag{2.2.33}$$

and this reduces indeed to the Poisson equation in the Newtonian limit.

2) Consider a static, spherically symmetric solution for a point source subject to the boundary condition

$$g_{\mu\nu} \to \eta_{\mu\nu} \quad \text{for} \quad r \to \infty.$$

3) Write down the differential equations for planetary orbits (geodesics for $g_{\mu\nu}$). Show that the precession of the perihelion is given by

$$\Delta\varphi = -\pi\left(\frac{4\pi\,GM}{L}\right)^2,$$

where $M$ is the central mass (solar mass) and $L = r^2\,\dot{\varphi}$ is the constant in the law of equal surfaces ($L$ is proportional to the orbital angular momentum). We shall see that this result is equal to $-\frac{1}{6}\,\Delta\varphi_{\mathrm{Einstein}}$!

---

## 2.3 Lagrangian Formalism

In this section we shall derive the field equations from a Hamiltonian variational principle. We shall also give a general formula for the energy-momentum tensor in the framework of a Lagrangian field theory, and show that its divergence vanishes as a consequence of the equations for the matter fields. In this connection we shall discuss the necessity of having four identities in the coupled system of gravitational field equations and matter equations, and point out the analogy with electrodynamics.

### 2.3.1 Hamilton's Principle for the Vacuum Field Equations

Under a change of coordinates, $\varphi$, the metric tensor transforms according to

$$g_{\mu\nu}=\frac{\partial\bar{x}^{\sigma}}{\partial x^{\mu}}\frac{\partial\bar{x}^{\varrho}}{\partial x^{\nu}}\bar{g}_{\sigma\varrho}\circ\varphi .$$

Hence

$$\sqrt{|g|}=|\det(\partial\bar{x}^{\sigma}/\partial x^{\mu})|\sqrt{|\bar{g}|} .$$

It follows that if $f$ is a continuous function on a Lorentz manifold $(M, g)$ having support in a coordinate patch $U\subset M$, with coordinates $\{x^{\mu}\}$, then the integral

$$\int_M f\,dv := \int f(x)\sqrt{|g|}\,d^4x \tag{2.3.1}$$

*is independent of the coordinate system.*

For a continuous function having compact support which is not contained in a coordinate patch, we define the integral by making use of a partition of unity, exactly as for the integration of differential forms in Sect. 4.7.1 of Part I. If the manifold is orientable, then

$$\int_M f\,dv=\int_M f\eta,$$

where $\eta$ is the 4-form corresponding to the metric (see Sect. 4.6.1 of Part I). In positive local coordinates,

$$\eta=\sqrt{-g}\;dx^0\wedge dx^1\wedge dx^2\wedge dx^3 .$$

The linear functional thus defined on the set of continuous functions having compact support defines in a well-known manner a measure $dv$ on the $\sigma$-algebra of Borel sets of $M$. This is called the measure determined by the Lorentz metric.

The Hamiltonian variational principle for the field equations in vacuum is

$$\delta\int_D R\,dv=0 . \tag{2.3.2}$$

Here $D$ is a compact region having a smooth boundary $\partial D$. The variations of the metric are assumed to vanish on $\partial D$.

*Proof:* We have [3]

$$\delta \int_D R\, dv = \int_D \delta(g^{\mu\nu} R_{\mu\nu} \sqrt{-g}\,)\, d^4x$$
$$= \int_D \delta R_{\mu\nu} g^{\mu\nu} \sqrt{-g}\; d^4x + \int_D R_{\mu\nu} \delta(g^{\mu\nu} \sqrt{-g}\,)\, d^4x\,. \tag{2.3.3}$$

We first consider variations of $R_{\mu\nu}$. We had

$$R_{\mu\nu} = \partial_\alpha \Gamma^\alpha_{\mu\nu} - \partial_\nu \Gamma^\alpha_{\mu\alpha} + \Gamma^\varrho_{\mu\nu} \Gamma^\alpha_{\varrho\alpha} - \Gamma^\varrho_{\mu\alpha} \Gamma^\alpha_{\varrho\nu}.$$

For a geodesic system at a point $p \in D$ we have in $p$:

$$\delta R_{\mu\nu} = \partial_\alpha (\delta \Gamma^\alpha_{\mu\nu}) - \partial_\nu (\delta \Gamma^\alpha_{\mu\alpha})$$

and hence

$$g^{\mu\nu} \delta R_{\mu\nu} = g^{\mu\nu} \partial_\alpha (\delta \Gamma^\alpha_{\mu\nu}) - g^{\nu\alpha} \partial_\alpha (\delta \Gamma^\mu_{\mu\nu})\,. \tag{2.3.4}$$

The variations of the Christoffel symbols transform as tensors, since their transformation properties are given by (Sect. 5.1. of Part I)

$$\bar{\Gamma}^\gamma_{\alpha\beta} = \frac{\partial x^\mu}{\partial \bar{x}^\alpha} \frac{\partial x^\nu}{\partial \bar{x}^\beta} \frac{\partial \bar{x}^\gamma}{\partial x^\lambda} \Gamma^\lambda_{\mu\nu} + \frac{\partial^2 x^\lambda}{\partial \bar{x}^\alpha \partial \bar{x}^\beta} \cdot \frac{\partial \bar{x}^\gamma}{\partial x^\lambda}\,.$$

Hence

$$w^\alpha := g^{\mu\nu} \delta \Gamma^\alpha_{\mu\nu} - g^{\alpha\nu} \delta \Gamma^\mu_{\mu\nu} \tag{2.3.5}$$

is a vector field. In a geodesic system (2.3.4) and (2.3.5) imply

$$g^{\mu\nu} \delta R_{\mu\nu} = w^\alpha_{;\alpha}. \tag{2.3.6}$$

This equation is covariant and hence it holds in every coordinate system. Since the variations vanish on the boundary of $D$, we may ignore the first integral in (2.3.3) by Gauss' theorem (Sect. 4.7.2 of Part I). For the second term in (2.3.3) we need $\delta g^{\mu\nu}$ and $\delta \sqrt{-g}$. It follows from

$$g^{\mu\alpha} g_{\alpha\beta} = \delta^\mu_\beta$$

that

$$\delta g^{\mu\alpha} g_{\alpha\beta} + g^{\mu\alpha} \delta g_{\alpha\beta} = 0\,.$$

If we contract this with $g^{\nu\beta}$, we obtain

$$\delta g^{\mu\nu} = -g^{\mu\alpha} g^{\nu\beta} \delta g_{\alpha\beta}.$$

In addition (see Sect. 1.4.2)

$$\delta(\sqrt{-g}\,) = -(2\sqrt{-g}\,)^{-1} \delta g = -(2\sqrt{-g}\,)^{-1} g\, g^{\alpha\beta} \delta g_{\alpha\beta}$$

---

[3] Choose for $D$ a region which is contained in a coordinate patch. It is possible to generalize the procedure to arbitrary regions $D$ (provided they have the assumed properties) by using a partition of unity.

so that

$$\delta(\sqrt{-g}) = \tfrac{1}{2}\sqrt{-g}\; g^{\alpha\beta}\,\delta g_{\alpha\beta}.$$

Using this we have

$$\delta(g^{\mu\nu}\sqrt{-g}) = \sqrt{-g}\;(\tfrac{1}{2} g^{\mu\nu} g^{\alpha\beta} - g^{\mu\alpha} g^{\nu\beta})\,\delta g_{\alpha\beta}. \tag{2.3.7}$$

If we insert this into (2.3.3), we obtain

$$\delta \int_D R\, dv = -\int_D (R^{\mu\nu} - \tfrac{1}{2} g^{\mu\nu} R)\,\delta g_{\mu\nu}\, dv = -\int_D G^{\mu\nu}\,\delta g_{\mu\nu}\, dv\,. \tag{2.3.8}$$

The validity of Hamilton's principle (2.3.2) follows from (2.3.8).

### 2.3.2 Another Derivation of the Bianchi Identity and its Meaning

We can use the variational equation (2.3.8) to obtain another derivation of the Bianchi identity (this presentation assumes familiarity with Sect. 4 of Part I). For the moment we shall keep the discussion general, so that it can be used later in another context.

For this purpose we consider a 4-form $\Omega[\psi]$, which is a functional of certain tensor fields $\psi$. This functional is assumed to be invariant in the following sense:

$$\varphi^*(\Omega[\psi]) = \Omega[\varphi^*(\psi)] \quad \text{for all} \quad \varphi \in \mathrm{Diff}(M)\,. \tag{2.3.9}$$

In particular, (2.3.9) is valid for the flow $\phi_t$ of a vector field $X$. Hence it follows by differentiating with respect to $t$ for $t = 0$ that

$$L_X(\Omega[\psi]) = \frac{d}{dt}\bigg|_{t=0} \Omega[\phi_t^*(\psi)].$$

We integrate this equation over a region $D$ which has a smooth boundary and compact closure. It follows from the Cartan formula (Sect. 4.5 of Part I)

$$L_X = d\circ i_X + i_X \circ d$$

and Stokes' theorem that when $X$ vanishes on the boundary $\partial D$,

$$\int_D L_X \Omega = \int_D d i_X \Omega = \int_{\partial D} i_X \Omega = 0$$

$$= \int_D \frac{d}{dt}\bigg|_{t=0} \Omega[\phi_t^*(\psi)].$$

Thus

$$\int_D \frac{d}{dt}\bigg|_{t=0} \Omega[\phi_t^*(\psi)] = 0\,. \tag{2.3.10}$$

In particular, we obtain from (2.3.8) and (2.3.10), for $\delta g_{\mu\nu} = (L_X g)_{\mu\nu}$, and $\Omega = (R\,\eta)[g]$,

$$\int_D G^{\mu\nu}(L_X g)_{\mu\nu}\,\eta = 0\,. \tag{2.3.11}$$

The components of the Lie derivative $L_X g$ are (see Sect. 3 of Part I)

$$(L_X g)_{\mu\nu} = X^\lambda g_{\mu\nu,\lambda} + g_{\lambda\nu} X^\lambda_{,\mu} + g_{\mu\lambda} X^\lambda_{,\nu}. \tag{2.3.12}$$

In a geodesic system this is equal to

$$(L_X g)_{\mu\nu} = X_{\mu,\nu} + X_{\nu,\mu}.$$

Thus, in an arbitrary system we have

$$(L_X g)_{\mu\nu} = X_{\mu;\nu} + X_{\nu;\mu}. \tag{2.3.13}$$

Hence

$$\int_D G^{\mu\nu}(X_{\mu;\nu} + X_{\nu;\mu})\, \eta = 0$$

or, due to the symmetry of $G^{\mu\nu}$,

$$\int_D G^{\mu\nu} X_{\mu;\nu}\, \eta = 0.$$

In place of this we write

$$\int_D (G^{\mu\nu} X_\mu)_{;\nu}\, \eta - \int_D G^{\mu\nu}_{;\nu} X_\mu\, \eta = 0.$$

According to Gauss' theorem the first term vanishes. Since $D$ and the values of $X_\mu$ in $D$ are arbitrary, the contracted Bianchi identity

$$G^{\mu\nu}_{;\nu} = 0$$

must hold.

**Discussion:** The Bianchi identity shows that the vacuum field equations $G_{\mu\nu} = 0$ are not mutually independent. That this must be the case becomes apparant from the following consideration. Suppose that $g$ is a vacuum solution. For every diffeomorphism $\phi$ the diffeomorphically transformed field $\phi^*(g)$ must also be a vacuum solution, since $g$ and $\phi^*(g)$ are physically equivalent. The field equations can thus only determine equivalence classes of diffeomorphic metric fields. Since Diff$(M)$ has four "degrees of freedom", we expect four identities.

Equivalently, one may argue as follows:
For every solution of the $g_{\mu\nu}$ as functions of particular coordinates, one can find a coordinate transformation such that four of the $g_{\mu\nu}$ can be made to take on arbitrary values. Hence only six of the ten field equations for the $g_{\mu\nu}$ can be independent.

A similar situation exists in electrodynamics. The fields $A_\mu$ are physically equivalent within a particular "gauge class". The function $\Lambda$ in the gauge transformation

$$A_\mu \to A_\mu + \partial_\mu \Lambda$$

is arbitrary and for this reason an identity must hold in the field equations for the $A_\mu$. In vacuum, these become

$$\Box A_\mu - \partial_\mu \partial_\nu A^\nu = 0.$$

The left hand side identically satisfies

$$\partial^\mu(\Box A_\mu - \partial_\mu \partial_\nu A^\nu) = 0,$$

and this is the identity we are looking for.

We shall generalize this discussion to the coupled matter-field system further on.

### 2.3.3 Energy-Momentum Tensor in a Lagrangian Field Theory

In a Lagrangian field theory it is possible to give a general formula for the energy-momentum tensor.

Let $\mathscr{L}$ be the Lagrangian density for a set of "matter fields" $\psi_A, A=1,\ldots,N$ (we include the electromagnetic field among the $\psi$'s). For simplicity, we consider only tensor fields, since we have not discussed the description of spinor fields in Lorentz manifolds. If we assume that we know $\mathscr{L}$ from local physics in flat space, the principle of equivalence prescribes the form of $\mathscr{L}$ in the presence of a gravitational field. We must simply write $g_{\mu\nu}$ in place of $\eta_{\mu\nu}$ and replace ordinary derivatives by covariant ones, as was discussed in Sect. 1.4. Thus (dropping the index $A$) we have

$$\mathscr{L} = \mathscr{L}(\psi, \nabla\psi, g). \tag{2.3.14}$$

**Example:** The Lagrangian density for the electromagnetic field is

$$\mathscr{L} = -(16\pi)^{-1} F_{\mu\nu} F_{\sigma\varrho} g^{\mu\sigma} g^{\nu\varrho}.$$

The matter field equations are obtained by means of a variational principle

$$\delta \int_D \mathscr{L}\eta = 0,$$

where the fields $\psi$ are varied inside of $D$ such that the variations vanish on the boundary. As usual, $\delta$ means differentiation with respect to the parameter of a 1-parameter family of field variations at the point 0.

Now

$$\delta \int_D \mathscr{L}\eta = \int_D (\delta\mathscr{L})\,\eta = \int_D \left(\frac{\partial\mathscr{L}}{\partial\psi}\delta\psi + \frac{\partial\mathscr{L}}{\partial\nabla\psi}\delta\nabla\psi\right)\eta. \tag{2.3.15}$$

The covariant derivative $\nabla$ commutes with the variational derivative $\delta$. (This is obvious when one considers coordinate expressions for the various quantities, since $\delta$ commutes with ordinary differentiation.)

Hence

$$\frac{\partial\mathscr{L}}{\partial(\nabla\psi)}\delta(\nabla\psi) = \nabla\cdot\left(\frac{\partial\mathscr{L}}{\partial(\nabla\psi)}\delta\psi\right) - \left(\nabla\cdot\frac{\partial\mathscr{L}}{\partial(\nabla\psi)}\right)\delta\psi. \tag{2.3.16}$$

The first term on the right hand side is the divergence of vector field (this becomes clear when one inserts indices). Hence, by Gauss' theorem it does not contribute to (2.3.15). From (2.3.15) we thus obtain

$$\int_D \left[\frac{\partial \mathscr{L}}{\partial \psi} - \nabla \cdot \frac{\partial \mathscr{L}}{\partial (\nabla \psi)}\right] \delta\psi\, \eta = 0,$$

which implies with the usual argument the Euler-Lagrange equations for the matter fields

$$\frac{\partial \mathscr{L}}{\partial \psi_A} - \nabla_\mu \frac{\partial \mathscr{L}}{\partial (\nabla_\mu \psi_A)} = 0. \tag{2.3.17}$$

In order to obtain an expression for the energy-momentum tensor, we consider variations of the action integral, which are induced by variations in the metric. $\mathscr{L}$ depends on the metric both explicitly and implicitly, through the covariant derivatives of the matter fields. In addition, the 4-form $\eta$ is an invariant functional of $g$. Hence

$$\delta \int_D \mathscr{L}\,\eta = \int_D \left\{\left[\frac{\partial \mathscr{L}}{\partial (\nabla_\lambda \psi)}\, \delta(\nabla_\lambda \psi) + \frac{\partial \mathscr{L}}{\partial g_{\mu\nu}}\, \delta g_{\mu\nu}\right] \eta + \mathscr{L}\,\delta\eta\right\}. \tag{2.3.18}$$

We first consider $\delta\eta$. In local coordinates,

$$\eta = \sqrt{-g}\; dx^0 \wedge \ldots \wedge dx^3,$$

and thus

$$\delta\eta = -(2\sqrt{-g}\,)^{-1}\, \delta g\; dx^0 \wedge \ldots \wedge dx^3.$$

Since

$$\delta g = g\, g^{\mu\nu}\, \delta g_{\mu\nu},$$

it follows that

$$\delta\eta = \tfrac{1}{2}\sqrt{-g}\;\; g^{\mu\nu} \cdot \delta g_{\mu\nu}\, dx^0 \wedge \ldots \wedge dx^3$$

or

$$\delta\eta = \tfrac{1}{2}\, g^{\mu\nu}\, \delta g_{\mu\nu}\, \eta. \tag{2.3.19}$$

Even if $\delta\psi = 0$, we have $\delta(\nabla_\lambda \psi) \neq 0$, since the Christoffel symbols vary. Now $\delta\Gamma^\mu_{\alpha\beta}$ is a tensor. In normal coordinates about a point $p$ we have at that point

$$\delta\Gamma^\mu_{\alpha\beta} = \tfrac{1}{2}\, g^{\mu\nu}\left[(\delta g_{\nu\alpha})_{;\beta} + (\delta g_{\nu\beta})_{;\alpha} - (\delta g_{\alpha\beta})_{;\nu}\right]. \tag{2.3.20}$$

Since both sides of (2.3.20) are tensors, this is true in any coordinate system. Using (2.3.20) it is possible to express $\delta(\nabla_\lambda \psi)$ in terms of $\delta g_{\alpha\beta;\gamma}$. After integration by parts, one can write the variation of the action integral in the form

$$\delta \int_D \mathscr{L}\,\eta = -\tfrac{1}{2} \int_D T^{\mu\nu}\, \delta g_{\mu\nu}\, \eta, \tag{2.3.21}$$

whereby only $g$ is varied. We identify the tensor $T^{\mu\nu}$ in (2.3.21) with the energy-momentum tensor. Note that $T^{\mu\nu}$ is automatically symmetric. As a first justification for this interpretation, we note that its covariant divergence vanishes.

*Proof of* $\nabla \cdot T = 0$

The proof of $T^{\mu\nu}_{;\nu} = 0$ is very similar to the derivation of the Bianchi identity in Sect. 2.2. We start with (2.3.10), which is valid for every invariant functional $\mathscr{L}[\psi, g]$ (i.e., $\phi^*(\mathscr{L}[\psi, g]) = \mathscr{L}[\phi^*\psi, \phi^* g]$ for all $\phi \in \mathrm{Diff}(M)$):

$$\int_D \left.\frac{d}{dt}\right|_{t=0} (\mathscr{L}\eta[\phi_t^*\psi, \phi_t^* g]) = 0. \tag{2.3.22}$$

As usual, $\phi_t$ denotes the flow of a vector field $X$, which vanishes on the boundary of $D$. The variations of the $\psi$'s do not contribute, due to the matter equations (2.3.17). Hence, using (2.3.21), we have

$$\int_D T^{\mu\nu}(L_X g)_{\mu\nu}\,\eta = 0. \tag{2.3.23}$$

Exactly as in the proof of the contracted Bianchi identity, we obtain

$$T^{\mu\nu}_{;\nu} = 0. \tag{2.3.24}$$

For familiar systems, the definition (2.3.21) leads to the usual expressions for the energy-momentum tensor. We show this for the electromagnetic field. In this case,

$$\mathscr{L} = -\frac{1}{16\pi} F_{\mu\nu} F_{\sigma\varrho} g^{\mu\sigma} g^{\nu\varrho}.$$

If we vary only $g$ and use (2.3.19), we obtain

$$\delta \int_D \mathscr{L}\eta = \int_D [\delta\mathscr{L}\eta + \mathscr{L}\delta\eta] = \int_D [\delta\mathscr{L} + \tfrac{1}{2}\mathscr{L} g^{\mu\nu}\delta g_{\mu\nu}]\,\eta$$

$$= \int_D \left[-\frac{1}{8\pi} F_{\mu\nu} F_{\sigma\varrho} g^{\mu\sigma}\delta g^{\nu\varrho} + \tfrac{1}{2}\mathscr{L} g^{\mu\nu}\delta g_{\mu\nu}\right]\eta.$$

However,

$$g^{\nu\beta} g_{\alpha\beta} = \delta^{\nu}_{\alpha} \Rightarrow \delta g^{\nu\beta} g_{\alpha\beta} + g^{\nu\beta}\delta g_{\alpha\beta} = 0,$$

so that

$$\delta g^{\nu\varrho} = -g^{\nu\beta} g^{\alpha\varrho}\delta g_{\alpha\beta}.$$

Hence

$$\delta \int_D \mathscr{L}\eta = \frac{1}{8\pi}\int_D [F_{\mu\nu} F_{\sigma\varrho} g^{\mu\sigma} g^{\nu\beta} g^{\alpha\varrho}\delta g_{\alpha\beta} - \tfrac{1}{4} F_{\mu\nu} F^{\mu\nu} g^{\alpha\beta}\delta g_{\alpha\beta}]\,\eta.$$

Thus we obtain from (2.3.21)

$$T^{\alpha\beta} = -\frac{1}{4\pi}\left[F^{\sigma\beta} F_{\sigma}^{\ \alpha} - \tfrac{1}{4} F_{\mu\nu} F^{\mu\nu} g^{\alpha\beta}\right]$$

and

$$T_{\alpha\beta} = -\frac{1}{4\pi}\left[F_{\alpha\mu} F_{\beta\nu} g^{\mu\nu} - \tfrac{1}{4} g_{\alpha\beta} F_{\mu\nu} F^{\mu\nu}\right]. \tag{2.3.25}$$

This agrees with the well known expression (see Sect. 1.4).

**Remark:** In SR the energy-momentum tensor is usually derived from translation invariance. This results in the conserved canonical energy-momentum tensor, which is not in general symmetric, but which can always be symmetrized. One could also use the above procedure in SR: replace *formally* $\eta_{\mu\nu}$ by a curved metric $g_{\mu\nu}$, use (2.3.21) and then set $g_{\mu\nu}$ equal to the flat metric at the end. The resulting tensor is then divergenceless [by (2.3.24)] and symmetric. The relationship of the two procedures is discussed in [25].

### 2.3.4 Analogy with Electrodynamics

The definition of $T^{\mu\nu}$ is similar to the definition of the current density $j^{\mu}$ in electrodynamics. We decompose the full Lagrangian $\mathscr{L}$ for matter into a part $\mathscr{L}_{\mathrm{F}}$ for the electromagnetic field and a term $\mathscr{L}_{M}$, which contains the interaction between the charged fields and the electromagnetic field. As before,

$$\mathscr{L}_{\mathrm{F}} = -\frac{1}{16\pi} F_{\mu\nu} F^{\mu\nu}, \quad F_{\mu\nu} = A_{\nu,\mu} - A_{\mu,\nu}.$$

By varying only the $A_{\mu}$, such that the variation vanishes on the boundary of $D$, we obtain

$$\begin{aligned}
\delta \int_D \mathscr{L}_{\mathrm{F}}\, \eta &= +\frac{1}{8\pi} \int_D F^{\mu\nu}\, \delta(A_{\mu,\nu} - A_{\nu,\mu})\, \eta \\
&= +\frac{1}{8\pi} \int_D F^{\mu\nu}\, \delta(A_{\mu;\nu} - A_{\nu;\mu})\, \eta \\
&= +\frac{1}{4\pi} \int_D F^{\mu\nu}\, \delta A_{\mu;\nu}\, \eta \\
&= -\frac{1}{4\pi} \int_D F^{\mu\nu}_{\ \ ;\nu}\, \delta A_{\mu}\, \eta .
\end{aligned} \tag{2.3.26}$$

We define the current density $j^{\mu}$ by

$$\delta \int_D \mathscr{L}_{\mathrm{M}} = -\int_D j^{\mu}\, \delta A_{\mu}\, \eta, \tag{2.3.27}$$

(only the $A_{\mu}$ are varied).

Maxwell's equations follow from Hamilton's variational principle

$$\delta \int_D \mathscr{L}\eta = 0, \tag{2.3.28}$$

together with (2.3.26) and (2.3.27). Equation (2.3.27) is analogous to (2.3.21). The divergence of the electromagnetic current vanishes as a consequence of the gauge invariance of the action $\int \mathscr{L}_{\mathrm{M}}\, \eta$. Let us show this in detail.

If we subject all the fields to a gauge transformation with gauge function $s\Lambda(x)$ and denote the derivative with respect to $s$ at the point $s=0$ by $\delta_\Lambda$, then the $\delta_\Lambda \psi$ of the matter fields (with the exception of the $A_\mu$) do not contribute to the variational equation

$$\delta_\Lambda \int_D \mathscr{L}_{\mathrm{M}}\, \eta = 0,$$

as a result of the field equations for matter. Using (2.3.27) and $\delta_\Lambda A_\mu = \Lambda_{,\mu}$, we obtain

$$0 = \delta_\Lambda \int_D \mathscr{L}_{\mathrm{M}}\, \eta = -\int_D j^\mu \Lambda_{,\mu}\, \eta.$$

If $\Lambda$ vanishes on the boundary of $D$, we obtain, after an integration by parts,

$$\begin{aligned} 0 &= -\int_D j^\mu \Lambda_{,\mu}\, \eta = -\int_D j^\mu \Lambda_{;\mu}\, \eta \\ &= -\int_D (j^\mu \Lambda)_{;\mu}\, \eta + \int_D j^\mu_{;\mu} \Lambda\, \eta \\ &= \int_D j^\mu_{;\mu} \Lambda\, \eta. \end{aligned}$$

Hence

$$j^\mu_{;\mu} = 0. \tag{2.3.29}$$

This equation is thus a consequence of Maxwell's equations on the one hand, since $F^{\mu\nu}_{;\nu;\mu} = 0$, and of the matter equations on the other hand. This means that the coupled system of Maxwell's equations and matter equations (the Euler-Lagrange equations for $\mathscr{L}_{\mathrm{M}}$) is not independent. This must be the case, since the transformed fields $\{A^\Lambda_\mu, \psi^\Lambda\}$, obtained from a solution $\{A_\mu, \psi\}$ of the coupled equations for matter and the electromagnetic field must again be a physically equivalent solution. (These remarks are easily generalized to Yang-Mills theories.)

### 2.3.5 Meaning of the Equation $\nabla \cdot T = 0$

In general relativity the equation $T^{\mu\nu}_{;\nu} = 0$ plays a completely analogous role. For every solution $(g, \psi)$ of the coupled system of field equations and matter equations (we now include the electromagnetic field among the matter fields), and every $\phi \in \mathrm{Diff}(M)$, $\{\phi^*(g), \phi^*(\psi)\}$ must also be a physically equivalent solution. We expect thus four identities, since Diff$(M)$ has four degrees of freedom. (Instead of "active" diffeomorphisms, we could just as well consider "passive" coordinate transformations.) These four identities are a result of the fact that (as was first emphasized by Hilbert) $\nabla \cdot T = 0$ is a consequence of both the matter equations and the gravitational field equations. The role of the gauge group of electrodynamics is now taken over by the group Diff$(M)$, which of course is non-Abelian.

*The Equations of Motion and* $\nabla \cdot T = 0$

If we take $T^{\mu\nu}$ to be the energy-momentum tensor of an ideal fluid with pressure $p = 0$ (incoherent dust), so that

$$T^{\mu\nu} = \varrho\, u^\mu u^\nu, \tag{2.3.30}$$

and assume that the quantity of fluid is conserved, so that

$$L_u(\varrho\, \eta) = 0,$$

which is equivalent to (see the exercises at the end of Sect. 4.7 of Part I)

$$(\varrho\, u^\mu)_{;\mu} = 0,$$

then it follows from $T^{\mu\nu}_{;\nu} = 0$ that

$$(\varrho\, u^\nu)_{;\nu}\, u^\mu + \varrho\, u^\nu u^\mu_{;\nu} = 0.$$

This means that

$$\nabla_u u = 0, \tag{2.3.31}$$

which implies that the integral curves of $u$ (streamlines) are geodesics. This law of motion is thus a consequence of the field equations.

### 2.3.6 Variational Principle for the Coupled System

Einstein's field equations and the matter equations follow from the variational principle

$$\delta \int_D [-(16\pi\, G)^{-1} R + \mathscr{L}]\, \eta = 0, \tag{2.3.32}$$

since according to (2.3.8) and (2.3.21), variation of $g$ only results in

$$\delta \int_D R\, \eta = -\int_D G^{\mu\nu}\, \delta g_{\mu\nu}\, \eta,$$

$$\delta \int_D \mathscr{L}\, \eta = -\tfrac{1}{2} \int_D T^{\mu\nu}\, \delta g_{\mu\nu}\, \eta.$$

Hence

$$G_{\mu\nu} = 8\pi\, G\, T_{\mu\nu}.$$

## 2.4 Nonlocalizability of the Gravitational Energy

In SR the conservation laws for energy and momentum of a closed system are a consequence of the invariance with respect to translations in time and space. In general, translations are not symmetry transformations of a Lorentz manifold and for this reason a general conservation law for energy and momentum does not exist in GR. This has been disturbing to many people, but one will simply have to get used to this fact. If one tries to find an "energy-momentum tensor for the gravitational field", one is on the wrong track. This is also clear since the gravitational field ($\Gamma^{\mu}_{\alpha\beta}$) can be transformed away at any point. If there is no field, there is no energy and no momentum.

However, it is still possible to define the *total* energy and total momentum of an isolated system with an asymptotically flat geometry. We shall discuss this in detail in Sect. 2.6.

At this point, we shall merely clarify the following: if a Killing field $K$ exists (i.e. a field $K$ such that $L_K g = 0$), then a conservation law can be derived from $\nabla \cdot T = 0$. We construct

$$P^{\mu} = T^{\mu\nu} K_{\nu}. \tag{2.4.1}$$

We have

$$P^{\mu}_{;\mu} = T^{\mu\nu}_{;\mu} K_{\nu} + T^{\mu\nu} K_{\nu;\mu} = \tfrac{1}{2} T^{\mu\nu}(K_{\mu;\nu} + K_{\nu;\mu}) = 0\,, \tag{2.4.2}$$

since (2.3.13) shows that

$$L_K g = 0\,, \quad \text{if and only if} \quad K_{\mu;\nu} + K_{\nu;\mu} = 0\,. \tag{2.4.3}$$

Equation (2.4.3) is the so-called *Killing equation.*

If $D$ is a region having a smooth boundary $\partial D$ and compact closure $\bar{D}$, then Gauss' theorem (see Sect. 4.7.2 of Part I) implies

$$\int_{\partial D} P^{\mu}\, d\sigma_{\mu} = \int_{D} P^{\mu}_{;\mu}\, dv = 0\,, \tag{2.4.4}$$

where $dv$ is the measure corresponding to $g$.

In SR one has ten Killing fields, corresponding to the ten dimensional Lie algebra of the inhomogeneous Lorentz group. The corresponding quantities (2.4.1) are precisely the ten classically conserved variables. In a Lorentz system, the ten independent Killing fields are

$$^{(\alpha)}T = \partial/\partial x^{\alpha}\,, \tag{2.4.5}$$

which generate the translations, and

$$^{(\alpha,\beta)}M = \eta_{\alpha\gamma} x^{\gamma} \frac{\partial}{\partial x^{\beta}} - \eta_{\beta\gamma} x^{\gamma} \frac{\partial}{\partial x^{\alpha}}\,, \tag{2.4.6}$$

which generate the homogeneous Lorentz transformations. One easily verifies that the Killing equation is satisfied for (2.4.5) and (2.4.6). The conserved quantities (2.4.1) are $T^{\mu\alpha}$ for (2.4.5) and

$$T^{\mu\beta}x^{\alpha} - T^{\mu\alpha}x^{\beta}$$

for (2.4.6), in other words the well-known angular momentum density.

## 2.5 The Tetrad Formalism

In this section we shall formulate Einstein's field equations in terms of differential forms. This has certain advantages, also for practical calculations. We shall use Sects. 5.7 and 5.8 of Part I.

In the following, we let $(\theta^{\alpha})$ be a (local) basis of 1-forms and $(e_{\alpha})$ be the corresponding dual basis of (local) vector fields. We shall often take these bases to be orthonormal, so that

$$g = \mathring{g}_{\mu\nu}\,\theta^{\mu} \otimes \theta^{\nu}, \quad (\mathring{g}_{\mu\nu}) = \begin{pmatrix} 1 & & & 0 \\ & -1 & & \\ & & -1 & \\ 0 & & & -1 \end{pmatrix}. \tag{2.5.1}$$

Under a change of basis

$$\bar{\theta}^{\alpha}(x) = A^{\alpha}{}_{\beta}(x)\,\theta^{\beta}(x) \tag{2.5.2}$$

the connection forms $\omega = (\omega^{\alpha}{}_{\beta})$ transform inhomogeneously

$$\bar{\omega} = A\,\omega\,A^{-1} - dA\,A^{-1}, \tag{2.5.3}$$

while the curvature $\Omega = (\Omega^{\alpha}{}_{\beta})$ transforms as a tensor

$$\bar{\Omega} = A\,\Omega\,A^{-1}. \tag{2.5.4}$$

With respect to an orthonormal basis we have (see Sct. 5.7 of Part I)

$$\omega_{\alpha\beta} + \omega_{\beta\alpha} = 0\,, \tag{2.5.5}$$

so that $\omega \in \mathrm{so}(1, 3)$ (the Lie algebra of the Lorentz group). For orthonormal tetrad fields $(\theta^{\alpha})$, (2.5.2) is a position dependent Lorentz transformation

$$\bar{\theta}(x) = \Lambda(x)\,\theta(x), \quad \Lambda(x) \in L_{+}^{\uparrow} \tag{2.5.6}$$

$$\bar{\omega}(x) = \Lambda(x)\,\omega(x)\,\Lambda^{-1}(x) - d\Lambda(x)\,\Lambda^{-1}(x). \tag{2.5.7}$$

For any given point $x_0$ there is always a $\Lambda(x)$ such that $\bar{\omega}(x_0) = 0$; according to (2.5.7), $\Lambda(x)$ must satisfy the equation $\Lambda^{-1}(x_0)\,d\Lambda(x_0) = \omega(x_0)$. One can easily convince oneself that this is always possible, since $\omega(x_0) \in \mathrm{so}(1, 3)$. If $\omega(x_0) = 0$, then $\theta^{\alpha}(x)$ describes a local inertial frame at $x_0$. At the point $x_0$, the exterior covariant derivative $D$ coincides with the exterior derivative $d$.

Obviously the metric (2.5.1) is invariant under the transformation (2.5.6). By analogy to the gauge transformations of electrodynamics (or of Yang-Mills theories) the local Lorentz transformations (2.5.6) are also called gauge transformations. General relativity is invariant with respect to such transformations and is thus a (special) non-Abelian gauge theory. The tetrad fields $\theta^\alpha$ can be regarded as the potentials of the gravitational field, since by (2.5.1), variations of the $\theta^\alpha$ induce variations in the metric.

The Einstein action $*R = R\,\eta$ can be represented in terms of a tetrad field as follows:

$$*R = \eta^{\mu\nu} \wedge \Omega_{\mu\nu}, \tag{2.5.8}$$

where

$$\eta^{\mu\nu} = *(\theta^\mu \wedge \theta^\nu). \tag{2.5.9}$$

Indices are raised or lowered with $g_{\mu\nu}$, appearing in $g = g_{\mu\nu}\,\theta^\mu \otimes \theta^\nu$.

*Proof of (2.5.8):* Clearly

$$\eta_{\alpha\beta} \wedge \Omega^{\alpha\beta} = \tfrac{1}{2}\,\eta_{\alpha\beta}\,R^{\alpha\beta}{}_{\mu\nu} \wedge \theta^\mu \wedge \theta^\nu .$$

From Exercise 6 of Sect. 4.6.2 of Part I, we have

$$*(\theta^\mu \wedge \theta^\nu) = \tfrac{1}{2}\,\eta_{\alpha\beta\sigma\varrho}\,g^{\alpha\mu}\,g^{\beta\nu}\,\theta^\sigma \wedge \theta^\varrho,$$

which implies that

$$\eta_{\alpha\beta} = \tfrac{1}{2}\,\eta_{\alpha\beta\sigma\varrho}\,\theta^\sigma \wedge \theta^\varrho. \tag{2.5.10}$$

Hence

$$\eta_{\alpha\beta} \wedge \theta^\mu \wedge \theta^\nu = \tfrac{1}{2}\,\eta_{\alpha\beta\sigma\varrho}\,\theta^\sigma \wedge \theta^\varrho \wedge \theta^\mu \wedge \theta^\nu = (\delta^\mu_\alpha\,\delta^\nu_\beta - \delta^\mu_\beta\,\delta^\nu_\alpha)\,\eta,$$

and thus

$$\eta_{\alpha\beta} \wedge \Omega^{\alpha\beta} = \tfrac{1}{2}\,(\delta^\mu_\alpha\,\delta^\nu_\beta - \delta^\mu_\beta\,\delta^\nu_\alpha)\,R^{\alpha\beta}_{\mu\nu}\,\eta = R\,\eta = *\,R\,.$$

*Variation of Tetrad Fields*

Under a variation $\delta\theta^\alpha$ of the orthonormal tetrad fields we have

$$\delta(\eta_{\alpha\beta} \wedge \Omega^{\alpha\beta}) = \delta\theta^\alpha \wedge (\eta_{\alpha\beta\gamma} \wedge \Omega^{\beta\gamma}) + \text{exact differential}, \tag{2.5.11}$$

where

$$\eta^{\alpha\beta\gamma} = *(\theta^\alpha \wedge \theta^\beta \wedge \theta^\gamma). \tag{2.5.12}$$

*Proof of (2.5.11):* As for (2.5.10), one obtains the expression

$$\eta_{\alpha\beta\gamma} = \eta_{\alpha\beta\gamma\delta}\,\theta^\delta. \tag{2.5.13}$$

Now

$$\delta(\eta_{\alpha\beta} \wedge \Omega^{\alpha\beta}) = \delta\eta_{\alpha\beta} \wedge \Omega^{\alpha\beta} + \eta_{\alpha\beta} \wedge \delta\Omega^{\alpha\beta}.$$

Using (2.5.10) and (2.5.13) results in

$$\delta\eta_{\alpha\beta} = \tfrac{1}{2}\,\delta\,(\eta_{\alpha\beta\gamma\delta}\,\theta^\gamma \wedge \theta^\delta) = \delta\theta^\gamma \wedge \eta_{\alpha\beta\gamma}\,.$$

If we now use the second structure equation, we obtain

$$\delta\Omega^{\alpha\beta} = d\delta\omega^{\alpha\beta} + \delta\omega^\alpha{}_\lambda \wedge \omega^{\lambda\beta} + \omega^\alpha{}_\lambda \wedge \delta\omega^{\lambda\beta}.$$

Hence

$$\begin{aligned}\delta\,(\eta_{\alpha\beta} \wedge \Omega^{\alpha\beta}) &= \delta\theta^\lambda \wedge (\eta_{\alpha\beta\gamma} \wedge \Omega^{\alpha\beta}) + d\,(\eta_{\alpha\beta} \wedge \delta\omega^{\alpha\beta}) \\ &\quad - d\eta_{\alpha\beta} \wedge \delta\omega^{\alpha\beta} + \eta_{\alpha\beta} \wedge (\delta\omega^\alpha{}_\lambda \wedge \omega^{\lambda\beta} + \omega^\alpha{}_\lambda \wedge \delta\omega^{\lambda\beta})\,.\end{aligned}$$

The last line is equal to $\delta\omega^{\alpha\beta} \wedge D\,\eta_{\alpha\beta}$. However,

$$D\,\eta_{\alpha\beta} = 0 \tag{2.5.14}$$

for the Levi-Cività connection, as we shall show below. We thus obtain

$$\delta\,(\eta_{\alpha\beta} \wedge \Omega^{\alpha\beta}) = \delta\theta^\lambda \wedge (\eta_{\alpha\beta\lambda} \wedge \Omega^{\alpha\beta}) + d\,(\eta_{\alpha\beta} \wedge \delta\omega^{\alpha\beta})\,, \tag{2.5.15}$$

which proves (2.5.11).

*Proof of (2.5.14):* Relative to an orthonormal system $\eta_{\alpha\beta\gamma\delta}$ is constant. Since orthonormality is preserved under parallel transport (for a metric connection) we conclude that

$$D\,\eta_{\alpha\beta\gamma\delta} = 0\,. \tag{2.5.16}$$

Together with (2.5.10) and the first structure equation, this implies

$$\begin{aligned}D\,\eta_{\alpha\beta} &= \tfrac{1}{2}\,D\,(\eta_{\alpha\beta\mu\nu}\,\theta^\mu \wedge \theta^\nu) = \eta_{\alpha\beta\mu\nu}\,D\theta^\mu \wedge \theta^\nu \\ &= \eta_{\alpha\beta\mu\nu}\,\Theta^\mu \wedge \theta^\nu = \Theta^\mu \wedge \eta_{\alpha\beta\mu}\end{aligned}$$

or, for the Levi-Cività connection, (2.5.14).

In the same manner, one can show that

$$D\,\eta_{\alpha\beta\gamma} = 0\,. \tag{2.5.14'}$$

The total Lagrangian density is given by

$$\mathscr{L} = -\,(16\,\pi\,G)^{-1}\,{*R} + \mathscr{L}_{\text{mat}}\,. \tag{2.5.17}$$

Under variation of the $\theta^\alpha$ let

$$\delta\mathscr{L}_{\text{mat}} = -\,\delta\theta^\alpha \wedge {*T_\alpha}\,, \tag{2.5.18}$$

where $*T_\alpha$ are the 3-forms of energy and momentum for matter. Thus, together with (2.5.15) we have

$$\delta\mathscr{L} = -\,\delta\theta^\alpha \wedge ((16\,\pi\,G)^{-1}\,\eta_{\alpha\beta\gamma} \wedge \Omega^{\beta\gamma} + {*T_\alpha}) + \text{exact differential},$$

and the field equations can be written as

$$-\tfrac{1}{2}\,\eta_{\alpha\beta\gamma} \wedge \Omega^{\beta\gamma} = 8\,\pi\,G\,{*T_\alpha}\,. \tag{2.5.19}$$

Since the $\eta_{\alpha\beta\gamma}$ and $\Omega^{\beta\gamma}$ transform as tensors, (2.5.19) holds not only for orthonormal tetrads. $T_\alpha$ is related to the energy-momentum tensor $T_{\alpha\beta}$ by

$$T_\alpha = T_{\alpha\beta}\,\theta^\beta\,. \tag{2.5.20}$$

With this identification, the Eqs. (2.5.19) are in fact equivalent to the field equations in the classical form

$$R_{\mu\nu} - \tfrac{1}{2}\,g_{\mu\nu}\,R = 8\,\pi\,G\,T_{\mu\nu}\,. \tag{2.5.21}$$

We show this by explicit calculation. Using (2.5.20) and

$$\eta^\alpha := {*\theta^\alpha}\,, \tag{2.5.22}$$

$$\eta_\alpha = \frac{1}{3!}\,\eta_{\alpha\beta\gamma\delta}\,\theta^\beta \wedge \theta^\gamma \wedge \theta^\delta = \tfrac{1}{3}\,\theta^\beta \wedge \eta_{\alpha\beta}\,, \tag{2.5.23}$$

(2.5.19) can be written in the form

$$-\tfrac{1}{4}\,\eta_{\alpha\mu\nu} \wedge \theta^\sigma \wedge \theta^\varrho\,R^{\mu\nu}{}_{\sigma\varrho} = 8\,\pi\,G\,T_{\alpha\beta}\,\eta^\beta\,.$$

In this we use the middle two of the following easily proved identities:

$$\begin{aligned}
\theta^\beta \wedge \eta_\alpha &= \delta^\beta_\alpha\,\eta \\
\theta^\gamma \wedge \eta_{\alpha\beta} &= \delta^\gamma_\beta\,\eta_\alpha - \delta^\gamma_\alpha\,\eta_\beta \\
\theta^\delta \wedge \eta_{\alpha\beta\gamma} &= \delta^\delta_\gamma\,\eta_{\alpha\beta} + \delta^\delta_\beta\,\eta_{\gamma\alpha} + \delta^\delta_\alpha\,\eta_{\beta\gamma} \\
\theta^\varepsilon \wedge \eta_{\alpha\beta\gamma\delta} &= \delta^\varepsilon_\delta\,\eta_{\alpha\beta\gamma} - \delta^\varepsilon_\gamma\,\eta_{\delta\alpha\beta} + \delta^\varepsilon_\beta\,\eta_{\gamma\delta\alpha} - \delta^\varepsilon_\alpha\,\eta_{\beta\gamma\delta}
\end{aligned} \tag{2.5.24}$$

and obtain for the left hand side

$$\begin{aligned}
&-\tfrac{1}{4}\,R^{\mu\nu}{}_{\sigma\varrho}\,[\delta^\varrho_\nu\,(\delta^\sigma_\mu\,\eta_\alpha - \delta^\sigma_\alpha\,\eta_\mu) + \delta^\varrho_\mu\,(\delta^\sigma_\alpha\eta_\nu - \delta^\sigma_\nu\,\eta_\alpha) + \delta^\varrho_\alpha\,(\delta^\sigma_\nu\,\eta_\mu - \delta^\sigma_\mu\,\eta_\nu)] \\
&\quad = -\tfrac{1}{2}\,R^{\mu\nu}{}_{\mu\nu}\,\eta_\alpha + R^{\mu\nu}{}_{\alpha\nu}\,\eta_\mu = R^{\beta\nu}{}_{\alpha\nu}\,\eta_\beta - \tfrac{1}{2}\,R^{\mu\nu}{}_{\mu\nu}\,\delta^\beta_\alpha\,\eta_\beta = (R^\beta_\alpha - \tfrac{1}{2}\,\delta^\beta_\alpha R)\,\eta_\beta\,.
\end{aligned}$$

This proves (2.5.21).

It follows from (2.5.14′) and the second Bianchi identity $D\,\Omega = 0$ that the absolute exterior differential of the left hand side of (2.5.19) vanishes. Hence Einstein's field equations imply

$$D\,{*T_\alpha} = 0\,. \tag{2.5.25}$$

This equation is equivalent to $\nabla \cdot T = 0$. (Show this!)

*Consequences of the Invariance Properties of $\mathscr{L}$*

Let us write (2.5.11) in the form

$$-\tfrac{1}{2}\,\delta{*R} = \delta\theta^\alpha \wedge {*G_\alpha} + \text{exact differential} \tag{2.5.26}$$

with

$${*G_\alpha} = -\tfrac{1}{2}\,\eta_{\alpha\beta\gamma} \wedge \Omega^{\beta\gamma}\,. \tag{2.5.27}$$

According to the derivation of (2.5.21),

$$G_\alpha = G_{\alpha\beta}\,\theta^\beta\,, \tag{2.5.28}$$

where, as usual $G_{\alpha\beta} = R_{\alpha\beta} - \frac{1}{2} g_{\alpha\beta} R$. We now consider two special types of variations $\delta\theta^\alpha$.

*A. Local Lorentz Invariance:* For an infinitesimal Lorentz transformation in (2.5.6) we have

$$\delta\theta^\alpha(x) = \lambda^\alpha{}_\beta(x)\,\theta^\beta(x)\,, \qquad (\lambda^\alpha{}_\beta(x)) \in \mathrm{so}(1,3)\,.$$

Since $*R$ is invariant under such transformations, it follows from (2.5.26) that

$$0 = \lambda_{\alpha\beta}(\theta^\beta \wedge *G^\alpha - \theta^\alpha \wedge *G^\beta) + \text{exact differential.}$$

If we integrate this over a region $D$ such that the support of $\lambda$ is compact and contained in $D$, it follows, with $\lambda_{\alpha\beta} + \lambda_{\beta\alpha} = 0$, that

$$\theta^\alpha \wedge *G^\beta = \theta^\beta \wedge *G^\alpha \tag{2.5.29}$$

or, using (2.5.28)

$$\theta^\alpha \wedge \eta_\gamma G^{\beta\gamma} = \theta^\beta \wedge \eta_\gamma G^{\alpha\gamma}\,,$$

so that, using $\theta^\alpha \wedge \eta_\gamma = \delta^\alpha_\gamma \eta$, we have

$$G^{\alpha\beta} = G^{\beta\alpha}\,. \tag{2.5.30}$$

In the same manner, the local Lorentz invariance of $\mathscr{L}_{\text{mat}}$, together with the equations for matter, imply

$$T_{\alpha\beta} = T_{\beta\alpha}\,. \tag{2.5.31}$$

*B. Invariance under Diff(M):* We now produce the variations by means of Lie derivatives $L_X$ such that the support of $X$ is compact and contained in $D$. This time we have, as a result of the invariance of $*R$ under Diff$(M)$,

$$L_X\theta^\alpha \wedge *G_\alpha + \text{exact differential} = 0 \tag{2.5.32}$$

or

$$\int_D L_X\theta^\alpha \wedge *G_\alpha = 0\,. \tag{2.5.33}$$

Since

$$L_X\theta^\alpha = d i_X\theta^\alpha + i_X \underbrace{d\theta^\alpha}_{-\,\omega^\alpha{}_\beta \wedge \theta^\beta}\,,$$

we have

$$\begin{aligned} L_X\theta^\alpha \wedge *G_\alpha &= d\,(i_X\theta^\alpha \wedge *G_\alpha) - i_X\theta^\alpha \wedge d{*G_\alpha} \\ &\quad - \underbrace{i_X(\omega^\alpha{}_\beta \wedge \theta^\beta)}_{(i_X\omega^\alpha{}_\beta) \wedge \theta^\beta - \omega^\alpha{}_\beta \wedge i_X\theta^\beta} \wedge *G_\alpha \\ &= - i_X\theta^\alpha \wedge [d{*G_\alpha} - \omega^\beta{}_\alpha \wedge *G_\beta] - (i_X\omega^\alpha{}_\beta) \wedge \theta^\beta \wedge *G_\alpha \\ &\quad + \text{exact differential.} \end{aligned}$$

The second term vanishes due to (2.5.29) and the antisymmetry of $\omega_{\alpha\beta}$. According to this,

$$L_X\theta^\alpha \wedge *G_\alpha = -\underbrace{(i_X\theta^\alpha)}_{X^\alpha} D*G_\alpha + \text{exact differential}, \tag{2.5.34}$$

and from (2.5.33)

$$\int_D X^\alpha D*G_\alpha = 0. \tag{2.5.35}$$

We again obtain the contracted Bianchi identity

$$D*G_\alpha = 0. \tag{2.5.36}$$

In an identical manner, the invariance of $\mathscr{L}_{\text{mat}}$, together with the equations for matter, imply

$$D*T_\alpha = 0. \tag{2.5.37}$$

These are well-known results in a new form.

**Example:** The Lagrangian of the electromagnetic field is

$$\mathscr{L}_{\text{em}} = -(8\pi)^{-1} F \wedge *F. \tag{2.5.38}$$

We compute its variation with respect to simultaneous variations of $F$ and $\theta^\alpha$. We have

$$\delta(F \wedge *F) = \delta F \wedge *F + F \wedge \delta*F. \tag{2.5.39}$$

The second term of (2.5.39) can be computed as follows: From Exercise 1 of Part I, Sect. 4.6.2, we have

$$\theta^\alpha \wedge \theta^\beta \wedge *F = F \wedge *(\theta^\alpha \wedge \theta^\beta) = F \wedge \eta^{\alpha\beta}$$

and hence

$$\delta(\theta^\alpha \wedge \theta^\beta) \wedge *F + (\theta^\alpha \wedge \theta^\beta) \wedge \delta*F = \delta F \wedge \eta^{\alpha\beta} + F \wedge \delta\eta^{\alpha\beta}.$$

Multiplication by $\frac{1}{2} F_{\alpha\beta}$ results in

$$\underbrace{\tfrac{1}{2} F_{\alpha\beta}\, \delta(\theta^\alpha \wedge \theta^\beta)}_{\delta\theta^\alpha \wedge F_{\alpha\beta}\theta^\beta} \wedge *F + F \wedge \delta*F = \delta F \wedge *F + \tfrac{1}{2} F \wedge \underbrace{\delta\eta^{\alpha\beta}}_{\delta\theta^\gamma \wedge \eta^{\alpha\beta}{}_\gamma} \cdot F_{\alpha\beta}.$$

If we insert this into (2.5.39), we obtain

$$\delta(-\tfrac{1}{2} F \wedge *F) = -\delta F \wedge *F + \tfrac{1}{2}\delta\theta^\alpha \wedge [F_{\alpha\beta}\theta^\beta \wedge *F - \tfrac{1}{2} F \wedge F^{\mu\nu}\eta_{\alpha\mu\nu}]. \tag{2.5.40}$$

In the square bracket, use $F_{\alpha\beta}\theta^\beta = i_\alpha F$, with $i_\alpha := i_{e_\alpha}$, as well as $F^{\mu\nu}\eta_{\alpha\mu\nu} = 2 i_\alpha *F$, which is obtained from

$$i_\alpha *F = i_\alpha(\tfrac{1}{2} F^{\mu\nu}\eta_{\mu\nu}) = \tfrac{1}{4} F^{\mu\nu}\eta_{\mu\nu\varrho\sigma}\, i_\alpha(\theta^\varrho \wedge \theta^\sigma) = \tfrac{1}{2} F^{\mu\nu}\eta_{\mu\nu\alpha}.$$

This gives

$$\delta(-(8\pi)^{-1} F \wedge *F) = -\frac{1}{4\pi}\delta F \wedge *F - \delta\theta^\alpha \wedge *T^{\text{elm}}_\alpha, \tag{2.5.41}$$

where

$$*T^{\mathrm{elm}}_{\alpha} = -\frac{1}{8\pi}\left[i_\alpha F \wedge *F - F \wedge i_\alpha {*F}\right]. \tag{2.5.42}$$

We now use (2.5.24) to rewrite the square bracket in (2.5.40) in the form

$$\begin{aligned}
[\ldots] &= F_{\alpha\beta}\,\theta^\beta \wedge *F - \tfrac{1}{2} F \wedge F^{\mu\nu}\,\eta_{\mu\nu\alpha} \\
&= \tfrac{1}{2} F_{\alpha\beta} F^{\sigma\varrho} \underbrace{\theta^\beta \wedge \eta_{\sigma\varrho}}_{\delta^\beta_\varrho \eta_\sigma - \delta^\beta_\sigma \eta_\varrho} \\
&\quad - \tfrac{1}{4} F_{\sigma\varrho} F^{\mu\nu} \underbrace{\theta^\sigma \wedge \theta^\varrho \wedge \eta_{\alpha\mu\nu}}_{\delta^\varrho_\nu(\delta^\sigma_\mu \eta_\alpha - \delta^\sigma_\alpha \eta_\mu) + \delta^\varrho_\mu(\delta^\sigma_\alpha \eta_\nu - \delta^\sigma_\nu \eta_\alpha) + \delta^\varrho_\alpha(\delta^\sigma_\nu \eta_\mu - \delta^\sigma_\mu \eta_\nu)} \\
&= -\tfrac{1}{2} F_{\mu\nu} F^{\mu\nu}\,\eta_\alpha + 2 F_{\alpha\nu} F^{\mu\nu}\,\eta_\mu .
\end{aligned}$$

We thus also have

$$T^{\mathrm{elm}}_{\alpha} = \frac{1}{4\pi}\left(F_{\alpha\nu} F_\mu^{\ \nu}\,\theta^\mu + \tfrac{1}{4} F_{\mu\nu} F^{\mu\nu}\,\theta_\alpha\right) = T^{\mathrm{elm}}_{\alpha\beta}\,\theta^\beta . \tag{2.5.43}$$

---

**Exercises**

1) Show that the Einstein Lagrangian density can be written relative to an orthonormal basis as follows

$$\begin{aligned}
\tfrac{1}{2}\,\eta_{\alpha\beta} \wedge \Omega^{\alpha\beta} = &-\tfrac{1}{2}\,(d\theta^\alpha \wedge \theta^\beta) \wedge *(d\theta_\beta \wedge \theta_\alpha) \\
&+ \tfrac{1}{4}\,(d\theta^\alpha \wedge \theta_\alpha) \wedge *(d\theta^\beta \wedge \theta_\beta) + \text{exact differential.}
\end{aligned} \tag{2.5.44}$$

*Hint:* Set $d\theta^\alpha = \frac{1}{2} F^\alpha_{\beta\gamma}\,\theta^\beta \wedge \theta^\gamma$ and express both sides in terms of $F^\alpha_{\beta\gamma}$.

2) Rewrite the vacuum field equations in terms of the "dual" curvature forms

$$\tilde{\Omega}_{\alpha\beta} = \tfrac{1}{2}\,\eta_{\alpha\beta\gamma\delta}\,\Omega^{\gamma\delta}$$

and show that they are automatically satisfied if the curvature is (anti-)selfdual[4]:

$$\tilde{\Omega}_{\alpha\beta} = \pm\,\Omega_{\alpha\beta} .$$

What follows from $*\tilde{\Omega}_{\alpha\beta} = \pm\,\Omega_{\alpha\beta}$?

3) Equation (2.3.27) can be written in the form

$$\delta \int_D \mathscr{L}_{\mathrm{M}}\,\eta = ) \int_D *J \wedge \delta A .$$

---

[4] For Lorentz manifolds this remark is uninteresting, because the (anti-)selfduality implies that the curvature vanishes. (Prove this!) In the Euclidean path integral approach to quantum gravity, one is interested in solutions of the (classical) *Euclidean* Einstein equations. (Of particular interest are finite action solutions, called *gravitational instantons* because of the close analogy to the Yang-Mills instantons.) A variety of solutions with selfdual curvature are known in the Euclidean case.

Use this and (2.5.41) to derive Maxwell's equation from the variational principle

$$\delta \int_D \left( -\frac{1}{8\pi} F \wedge {}^*F + \mathscr{L}_M \eta \right) = 0 .$$

---

## 2.6 Energy, Momentum, and Angular Momentum of Gravity for Isolated Systems

In this section we shall derive conservation laws for energy, momentum and angular momentum of gravitating systems having asymptotically flat geometry (see [64] for a precise definition of this concept). The results of the previous sections will turn out to be extremely useful.

We first rewrite the field equations in the form of a continuity equation, from which differential "conservation laws" follow immediately.

Our starting point is (2.5.19) in the form

$$-\tfrac{1}{2} \Omega_{\beta\gamma} \wedge \eta^{\beta\gamma}{}_{\alpha} = 8\pi G {}^* T_{\alpha}. \tag{2.6.1}$$

We now use the second structure equation [5]

$$\Omega_{\beta\gamma} = d\omega_{\beta\gamma} - \omega_{\sigma\beta} \wedge \omega^{\sigma}{}_{\gamma} \tag{2.6.2}$$

and rearrange the contribution of the first term on the left hand side of (2.6.1):

$$d\omega_{\beta\gamma} \wedge \eta^{\beta\gamma}{}_{\alpha} = d(\omega_{\beta\gamma} \wedge \eta^{\beta\gamma}{}_{\alpha}) + \omega_{\beta\gamma} \wedge d\eta^{\beta\gamma}{}_{\alpha}. \tag{2.6.3}$$

From (5.14′) we have $D\eta^{\beta\gamma}{}_{\alpha} = 0$, so that

$$d\eta^{\beta\gamma}{}_{\alpha} + \omega^{\beta}{}_{\sigma} \wedge \eta^{\sigma\gamma}{}_{\alpha} + \omega^{\gamma}{}_{\sigma} \wedge \eta^{\beta\sigma}{}_{\alpha} - \omega^{\sigma}{}_{\alpha} \wedge \eta^{\beta\gamma}{}_{\sigma} = 0 .$$

If this is used in (2.6.3), the result is

$$\begin{aligned} d\omega_{\beta\gamma} \wedge \eta^{\beta\gamma}{}_{\alpha} &= d(\omega_{\beta\gamma} \wedge \eta^{\beta\gamma}{}_{\alpha}) \\ &\quad + \omega_{\beta\gamma} \wedge (-\omega^{\beta}{}_{\sigma} \wedge \eta^{\sigma\gamma}{}_{\alpha} - \omega^{\gamma}{}_{\sigma} \wedge \eta^{\beta\sigma}{}_{\alpha} + \omega^{\sigma}{}_{\alpha} \wedge \eta^{\beta\gamma}{}_{\sigma}) . \end{aligned}$$

---

[5] Using $dg_{\alpha\beta} = \omega_{\alpha\beta} + \omega_{\beta\alpha}$ one easily derives from

$$\Omega^{\alpha}{}_{\beta} = d\omega^{\alpha}{}_{\beta} + \omega^{\alpha}{}_{\lambda} \wedge \omega^{\lambda}{}_{\beta}$$

that

$$\Omega_{\beta\gamma} = d\omega_{\beta\gamma} - \omega_{\sigma\beta} \wedge \omega^{\sigma}{}_{\gamma} .$$

In a similar manner, one obtains from the first structure equation

$$d\theta_{\beta} - \omega^{\sigma}{}_{\beta} \wedge \theta_{\sigma} = 0 .$$

If we add the contribution of the second term of (2.6.2) to this, we obtain from (2.6.1) the Einstein field equations in the following form:

$$-\tfrac{1}{2}\, d(\omega_{\beta\gamma}\wedge\eta^{\beta\gamma}{}_{\alpha}) = 8\pi G\, {}^*(T_\alpha + t_\alpha)\,, \tag{2.6.4}$$

where

$$ {}^*t_\alpha = \frac{1}{16\pi G}\,\omega_{\beta\gamma}\wedge(\omega_{\sigma\alpha}\wedge\eta^{\beta\gamma\sigma} - \omega^{\gamma}{}_{\sigma}\wedge\eta^{\beta\sigma}{}_{\alpha})\,. \tag{2.6.5}$$

Equation (2.6.4) implies the conservation laws

$$d({}^*T_\alpha + {}^*t_\alpha) = 0\,. \tag{2.6.6}$$

This indicates that ${}^*t_\alpha$ should be interpreted as energy and momentum 3-forms of the gravitational field. However, we know from the discussion of Sect. 2.4 that these quantities cannot be localized. This is reflected here by the fact that ${}^*t_\alpha$ does not transform as a tensor with respect to gauge transformations. When $\omega_{\beta\gamma}(x) = 0$ (which can always be made to hold at some given point $x$), then ${}^*t_\alpha = 0$. Conversely, ${}^*t_\alpha$ vanishes even in flat space only with respect to global Lorentz systems. Hence, a physical meaning can at best be assigned to integrals of ${}^*t_\alpha$ over spacelike hypersurfaces. For isolated systems with asymptotically flat geometry, this is in fact the case, when one chooses the reference frame $(\theta^\alpha)$ to be asymptotically Lorentzian.

In order to be able to define the total angular momentum in such cases, we need a conservation law of the form (2.6.6) so that $t_{\alpha\beta}$ (in $t_\alpha = t_{\alpha\beta}\,\theta^\beta$) is *symmetric* with respect to a natural basis. Unfortunately, this is not the case for (2.6.5). For this reason, we rewrite the Einstein equations in still another form[6].

As a starting point we again take (2.5.19), and insert

$$\eta^{\alpha\beta\gamma} = \eta^{\alpha\beta\gamma\delta}\,\theta_\delta\,. \tag{2.6.7}$$

Together with (2.6.2) this gives (the basis is not necessarily orthonormal)

$$-\tfrac{1}{2}\,\eta^{\alpha\beta\gamma\delta}\,\theta_\delta\wedge(d\omega_{\beta\gamma} - \omega_{\sigma\beta}\wedge\omega^{\sigma}{}_{\gamma}) = 8\pi G\, {}^*T^\alpha.$$

In the first term, we perform a "partial integration" and use the first structure equation (see the footnote on p. 154)

$$d(\omega_{\beta\gamma}\wedge\theta_\delta) = \theta_\delta\wedge d\omega_{\beta\gamma} - \omega_{\beta\gamma}\wedge\underbrace{d\theta_\delta}_{\omega_{\sigma\delta}\wedge\theta^\sigma}\,.$$

Hence

$$-\tfrac{1}{2}\,\eta^{\alpha\beta\gamma\delta}\, d(\omega_{\beta\gamma}\wedge\theta_\delta) = 8\pi G\, {}^*(T^\alpha + t^\alpha_{L-L})\,, \tag{2.6.8}$$

where the right hand side now contains the so-called Landau-Lifshitz 3-form. It is given by

$$ {}^*t^\alpha_{L-L} = -\frac{1}{16\pi G}\,\eta^{\alpha\beta\gamma\delta}(\omega_{\sigma\beta}\wedge\omega^{\sigma}{}_{\gamma}\wedge\theta_\delta - \omega_{\beta\gamma}\wedge\omega_{\sigma\delta}\wedge\theta^\sigma)\,. \tag{2.6.9}$$

---

[6] This discussion follows partly Sect. 4.2.11 of [19].

We now multiply (2.6.8) by $\sqrt{-g}$; it follows from

$$\eta^{\alpha\beta\gamma\delta} = -(-g)^{-1/2}\,\varepsilon_{\alpha\beta\gamma\delta}\,,$$

that

$$-d\,(\sqrt{-g}\;\eta^{\alpha\beta\gamma\delta}\,\omega_{\beta\gamma} \wedge \theta_\delta) = 16\,\pi\,G\sqrt{-g}\;\; (*T^\alpha + *t^\alpha_{L-L})$$

or

$$-d(\sqrt{-g}\;\omega^{\beta\gamma} \wedge \eta^\alpha{}_{\beta\gamma}) = 16\,\pi\,G\sqrt{-g}\;\;(*T^\alpha + *t^\alpha_{L-L})\,. \tag{2.6.10}$$

From this we obtain the differential conservation law

$$d\,(\sqrt{-g}\;*(T^\alpha + t^\alpha_{L-L})) = 0\,. \tag{2.6.11}$$

In a natural basis $\theta^\alpha = dx^\alpha$, $t^{\alpha\beta}$ corresponding to (2.6.9) is now symmetric

$$dx^\varrho \wedge *t^\alpha_{L-L} = dx^\alpha \wedge *t^\varrho_{L-L}\,. \tag{2.6.12}$$

---

**Exercise:** Use the fact that $\Gamma^\mu{}_{\alpha\beta} = \Gamma^\mu{}_{\beta\alpha}$ in a natural basis to verify (2.6.12).

---

As before, the Landau-Lifshitz 3-forms (2.6.9) do not transform as a tensor under gauge transformations.

In the following we use the notation

$$\tau^\alpha = T^\alpha + t^\alpha_{L-L}\,. \tag{2.6.13}$$

From (2.6.11), i.e.

$$d\,(\sqrt{-g}\;*\tau^\alpha) = 0 \tag{2.6.14}$$

and the symmetry of $\tau^\alpha$,

$$dx^\alpha \wedge *\tau^\beta = dx^\beta \wedge *\tau^\alpha \tag{2.6.15}$$

we obtain

$$d\,(\sqrt{-g}\;*M^{\alpha\beta}) = 0\,, \tag{2.6.16}$$

where

$$*M^{\alpha\beta} = x^\alpha *\tau^\beta - x^\beta *\tau^\alpha\,. \tag{2.6.16$'$}$$

In fact

$$d\,(\sqrt{-g}\;*M^{\alpha\beta}) = dx^\alpha \wedge *\tau^\beta - dx^\beta \wedge *\tau^\alpha = 0\,.$$

*Interpretation*

We consider an isolated system with asymptotically flat geometry. All coordinate systems will be assumed to be asymptotically Lorentzian.

If $\Sigma$ is a spacelike surface, we interpret

$$P^\alpha = \int_\Sigma \sqrt{-g}\;*\tau^\alpha \tag{2.6.17}$$

as the total four momentum and

$$J^{\alpha\beta} = \int_{\Sigma} \sqrt{-g}\; {}^*M^{\alpha\beta} \tag{2.6.18}$$

as the total angular momentum of the isolated system. These quantities can be decomposed into contributions from matter and from the gravitational field. $P^\alpha$ and $J^{\alpha\beta}$ are constant in time if the gravitational fields fall off sufficiently fast at spacelike infinity. This behavior is expected for a stationary mass distribution. For non-stationary situations, these quantities vary with time due to the emission of gravitational radiation.

We can use the field equations to express $P^\alpha$ and $J^{\alpha\beta}$ in terms of two-dimensional flux integrals.

If we integrate (2.6.10) over a three-dimensional spacelike region $D_3$, we obtain

$$16\pi G \int_{D_3} \sqrt{-g}\; {}^*\tau^\alpha = -\int_{\partial D_3} \sqrt{-g}\; \omega^{\beta\gamma} \wedge \eta^\alpha{}_{\beta\gamma}\,.$$

Hence

$$P^\alpha = -\frac{1}{16\pi G} \oint \sqrt{-g}\; \omega^{\beta\gamma} \wedge \eta^\alpha{}_{\beta\gamma}\,. \tag{2.6.19}$$

The region of integration must be extended over a "surface at infinity". One obtains the same expression for the four-momentum from (2.6.4).

We now write the total angular momentum also as a flux integral. If we use the field equations [in the form (2.6.10)] in (2.6.16′), we obtain

$$\begin{aligned} 16\pi G \sqrt{-g}\; {}^*M^{\varrho\alpha} &= x^\varrho\, dh^\alpha - x^\alpha\, dh^\varrho \\ &= d\,(x^\varrho h^\alpha - x^\alpha h^\varrho) - (dx^\varrho \wedge h^\alpha - dx^\alpha \wedge h^\varrho)\,. \end{aligned} \tag{2.6.20}$$

where

$$h^\alpha := -\sqrt{-g}\; \omega^{\beta\gamma} \wedge \eta^\alpha{}_{\beta\gamma}\,. \tag{2.6.21}$$

The second term on the right hand side can also be written as an exact differential. We have

$$\begin{aligned} dx^\varrho \wedge h^\alpha - dx^\alpha \wedge h^\varrho &= \sqrt{-g}\; \omega^{\beta\gamma} \wedge \underbrace{dx^\varrho \wedge \eta^\alpha{}_{\beta\gamma}}_{\delta^\varrho_\gamma \eta^\alpha{}_\beta + \delta^\varrho_\beta \eta_\gamma{}^\alpha + g^{\varrho\alpha}\, \eta_{\beta\gamma}} - (\alpha \leftrightarrow \varrho) \\ &= \sqrt{-g}\; [\omega^{\beta\varrho} \wedge \eta^\alpha{}_\beta + \omega^{\varrho\beta} \wedge \eta_\beta{}^\alpha - (\alpha \leftrightarrow \varrho)] \\ &= \sqrt{-g}\; (\omega_\beta{}^\varrho \wedge \eta^{\alpha\beta} + \omega^\varrho{}_\beta \wedge \eta^{\beta\alpha} - \omega_\beta{}^\alpha \wedge \eta^{\varrho\beta} - \omega^\alpha{}_\beta \wedge \eta^{\beta\varrho})\,. \end{aligned}$$

We now use $D\eta^{\varrho\alpha} = 0$, which means that

$$d\eta^{\varrho\alpha} + \omega^\varrho{}_\beta \wedge \eta^{\beta\alpha} + \omega^\alpha{}_\beta \wedge \eta^{\varrho\beta} = 0\,,$$

and obtain

$$dx^\varrho \wedge h^\alpha - dx^\alpha \wedge h^\varrho = \sqrt{-g}\; [\omega_\beta{}^\varrho \wedge \eta^{\alpha\beta} - (\varrho \leftrightarrow \alpha) - d\eta^{\varrho\alpha}]\,. \tag{2.6.22}$$

However, if we insert (using $\omega^\mu{}_\nu = \Gamma^\mu{}_{\alpha\nu} dx^\alpha$)

$$\omega_\beta{}^\varrho \wedge \eta^{\alpha\beta} = \Gamma_{\beta\mu}{}^\varrho\, dx^\mu \wedge \eta^{\alpha\beta} = \Gamma_\beta{}^{\beta\varrho}\, \eta^\alpha - \Gamma_\beta{}^{\alpha\varrho}\, \eta^\beta$$

in (2.6.22), we obtain

$$dx^\varrho \wedge h^\alpha - dx^\alpha \wedge h^\varrho = \sqrt{-g}\,(\Gamma^{\beta}{}_{\beta}{}^{\varrho}\,\eta^\alpha - \Gamma^{\beta}{}_{\beta}{}^{\alpha}\,\eta^\varrho - d\eta^{\varrho\alpha}).$$

Since

$$\Gamma^{\beta}{}_{\beta}{}^{\varrho} = \frac{1}{\sqrt{-g}}\, g^{\mu\varrho}\,\partial_\mu \sqrt{-g}\,,$$

we finally obtain

$$\begin{aligned} dx^\varrho \wedge h^\alpha - dx^\alpha \wedge h^\varrho &= -\sqrt{-g}\,d\eta^{\varrho\alpha} + \partial_\mu\sqrt{-g}\,\underbrace{(g^{\mu\varrho}\,\eta^\alpha - g^{\mu\alpha}\,\eta^\varrho)}_{dx^\mu \wedge \eta^{\alpha\varrho}} \\ &= -\sqrt{-g}\,d\eta^{\varrho\alpha} - d\sqrt{-g}\wedge\eta^{\varrho\alpha}\,, \end{aligned}$$

so that

$$dx^\varrho \wedge h^\alpha - dx^\alpha \wedge h^\varrho = -d\,(\sqrt{-g}\,\eta^{\varrho\alpha})\,. \tag{2.6.23}$$

If we insert this result in (2.6.20), we obtain for the total angular momentum (2.6.18)

$$J^{\varrho\alpha} = \frac{1}{16\pi G}\oint[(x^\varrho\,h^\alpha - x^\alpha\,h^\varrho) + \sqrt{-g}\,\eta^{\varrho\alpha}]$$

or

$$J^{\varrho\alpha} = \frac{1}{16\pi G}\oint\sqrt{-g}\,[(x^\varrho\eta^{\alpha}{}_{\beta\gamma} - x^\alpha\,\eta^{\varrho}{}_{\beta\gamma})\wedge\omega^{\beta\gamma} + \eta^{\varrho\alpha}]\,. \tag{2.6.24}$$

$P^\alpha$ and $J^{\alpha\beta}$ are gauge invariant in the following sense: Under a transformation

$$\begin{aligned} \theta(x) &\to A(x)\,\theta(x) \\ \omega(x) &\to A(x)\,\omega(x)\,A^{-1}(x) - dA(x)\,A^{-1}(x)\,, \end{aligned} \tag{2.6.25}$$

which reduces asymptotically to the identity, the flux integrals (2.6.19) and (2.6.24) remain invariant. *Proof:* The homogeneous contributions to (2.6.25) obviously do not change the flux integrals. The inhomogeneous term gives an additional surface integral of an exact differential, which vanishes by Stokes' theorem. Thus $P^\alpha$ and $J^{\alpha\beta}$ transform as a four-vector (resp. a tensor) under every transformation which leaves the flat metric $\mathring{g}_{\mu\nu}$ asymptotically invariant, since every such transformation can be represented as the product of a Lorentz transformation (with respect to which $P^\alpha$ and $J^{\alpha\beta}$ transform as tensors) and a transformation which reduces to the identity asymptotically.

In order to establish the connection with presentations found in other texts (for example, Sect. 101 of [12]), we use the result of the following exercise. With this result, we can write the field equation (2.6.10) in the form

$$H^{\mu\alpha\nu\beta}{}_{,\alpha\beta} = 16\pi G(-g)(T^{\mu\nu} + t^{\mu\nu}_{L-L})\,. \tag{2.6.26}$$

The expression (2.6.9) for $t^{\mu\nu}_{L-L}$ can be computed explicitly. The result is (see Eq. (101.7) of [12])

$$(-g)\,t^{\alpha\beta}_{L-L} = \frac{1}{16\pi G}\{\tilde{g}^{\alpha\beta}{}_{,\lambda}\,\tilde{g}^{\lambda\mu}{}_{,\mu} - \tilde{g}^{\alpha\lambda}{}_{,\lambda}\,\tilde{g}^{\beta\mu}{}_{,\mu}$$
$$+\tfrac{1}{2}g^{\alpha\beta}g_{\lambda\mu}\tilde{g}^{\lambda\nu}{}_{,\varrho}\tilde{g}^{\varrho\mu}{}_{,\nu} - (g^{\alpha\lambda}g_{\mu\nu}\tilde{g}^{\beta\nu}{}_{,\varrho}\tilde{g}^{\mu\varrho}{}_{,\lambda}$$
$$+g^{\beta\lambda}g_{\mu\nu}\tilde{g}^{\alpha\nu}{}_{,\varrho}\tilde{g}^{\mu\varrho}{}_{,\lambda}) + g_{\mu\lambda}g^{\nu\varrho}\tilde{g}^{\alpha\lambda}{}_{,\nu}\tilde{g}^{\beta\mu}{}_{,\varrho} \tag{2.6.27}$$
$$+\tfrac{1}{8}(2g^{\alpha\lambda}g^{\beta\mu} - g^{\alpha\beta}g^{\lambda\mu})(2g_{\nu\varrho}g_{\sigma\tau} - g_{\varrho\sigma}g_{\nu\tau})\,\tilde{g}^{\nu\tau}{}_{,\lambda}\tilde{g}^{\varrho\sigma}{}_{,\mu}\}\,.$$

This expression is quadratic in the $\tilde{g}^{\alpha\beta}{}_{,\mu}$, where $\tilde{g}^{\alpha\beta} := \sqrt{-g}\;g^{\alpha\beta}$.

---

**Exercise:** Show that the left hand side of the field equations (2.6.10) can be written in the form

$$-d(\sqrt{-g}\;\omega^{\alpha\beta}\wedge\eta^{\mu}{}_{\alpha\beta}) = \frac{1}{\sqrt{-g}}\,H^{\mu\alpha\nu\beta}{}_{,\alpha\beta}\,\eta_{\nu}\,, \tag{2.6.28}$$

where

$$H^{\mu\alpha\nu\beta} := \tilde{g}^{\mu\nu}\tilde{g}^{\alpha\beta} - \tilde{g}^{\alpha\nu}\tilde{g}^{\beta\mu}\,,\qquad \tilde{g}^{\mu\nu} := \sqrt{-g}\;g^{\mu\nu}\,, \tag{2.6.29}$$

is the so-called Landau-Lifshitz "superpotential".

*Solution:* First note that

$$-\sqrt{-g}\;\omega^{\alpha\beta}\wedge\eta^{\mu}{}_{\alpha\beta} = -(-g)\,(\omega^{\alpha\beta}\wedge dx^{\lambda})\,g^{\mu\gamma}\,\varepsilon_{\gamma\alpha\beta\lambda}$$
$$= -(-g)\,g^{\mu\gamma}g^{\alpha\tau}g^{\beta\varrho}\tfrac{1}{2}(g_{\sigma\tau,\varrho} + g_{\tau\varrho,\sigma} - g_{\varrho\sigma,\tau})\,dx^{\sigma}\wedge dx^{\lambda}\,\varepsilon_{\gamma\alpha\beta\lambda}\,.$$

Hence

$$d(\sqrt{-g}\;\omega^{\alpha\beta}\wedge\eta^{\mu}{}_{\alpha\beta})$$
$$= -\tfrac{1}{2}\varepsilon_{\gamma\alpha\beta\lambda}\{(-g)\,g^{\mu\gamma}g^{\alpha\tau}g^{\beta\varrho}(g_{\sigma\tau,\varrho} + g_{\tau\varrho,\sigma} - g_{\varrho\sigma,\tau})\}_{,\varkappa}\cdot dx^{\varkappa}\wedge dx^{\sigma}\wedge dx^{\lambda}\,.$$

Due to symmetry, the second term does not contribute. If we denote the left hand side of the last equation by $(\sqrt{-g})^{-1}H^{\mu\nu}\eta_{\nu}$, we have

$$(\sqrt{-g})^{-1}H^{\mu\nu}\eta = -\tfrac{1}{2}\varepsilon_{\gamma\alpha\beta\lambda}\{\ldots\}_{,\varkappa}\underbrace{dx^{\nu}\wedge dx^{\varkappa}\wedge dx^{\sigma}\wedge dx^{\lambda}}_{\frac{1}{\sqrt{-g}}\varepsilon_{\nu\varkappa\sigma\lambda}\eta}$$

or [see Eq. (4.33) of Part I]

$$H^{\mu\nu} = -\tfrac{1}{2}\,3!\,\delta^{\nu}_{[\gamma}\delta^{\varkappa}_{\alpha}\delta^{\sigma}_{\beta]}\{(-g)\,g^{\mu\gamma}g^{\alpha\tau}g^{\beta\varrho}(g_{\sigma\tau,\varrho} - g_{\varrho\sigma,\tau})\}_{,\varkappa}\,.$$

Since the last factor is antisymmetric in $\varrho$ and $\tau$, it is no longer necessary to antisymmetrize in $\alpha$ and $\beta$. Hence only a cyclic sum over $(\gamma, \alpha, \beta)$ remains in the expression for $H^{\mu\nu}$:

$$H^{\mu\nu} = -\sum_{(\gamma,\alpha,\beta)}\delta^{\nu}_{\gamma}\delta^{\varkappa}_{\alpha}\delta^{\sigma}_{\beta}\{\ldots\}_{,\varkappa}\,.$$

The first term in this sum is

$$-\{(-g)\,g^{\mu\nu}g^{\alpha\tau}g^{\beta\varrho}(g_{\beta\tau,\varrho} - g_{\varrho\beta,\tau})\}_{,\alpha}\,. \tag{*}$$

We now use

$$g^{\beta\varrho} g_{\beta\varrho,\tau} = g^{-1} g_{,\tau}$$

and

$$g^{\alpha\tau} g^{\beta\varrho} g_{\beta\tau,\varrho} = - g^{\alpha\varrho}{}_{,\varrho} .$$

Hence (*) is equal to $[g^{\mu\nu}(-g\, g^{\alpha\beta})_{,\beta}]_{,\alpha}$. In an analogous manner one can simplify the other terms in the cyclic sum and easily finds

$$H^{\mu\nu} = H^{\mu\alpha\nu\beta}{}_{,\alpha\beta}$$

which is what we wanted to show.

---

**Remark:** More than twenty years ago Arnowitt, Deser and Misner conjectured that an isolated gravitational system with nonnegative local matter density must have nonnegative total energy (measured at infinity). This has finally been shown to be true by *Schoen* and *Yau* [65]; see also [66], and references therein.

## 2.7 Remarks on the Cauchy Problem

A study of the Cauchy problem provides a deeper understanding of the structure of Einstein's field equations. A detailed investigation of this difficult problem can be found in Chap. 7 of [21]. We restrict ourselves here to a few simple remarks.

*Nature of the Problem*

The Cauchy problem in GR can be stated as follows (for simplicity we consider only the vacuum case): Let $\mathscr{S}$ be a given three dimensional manifold and $\alpha$ initial data on it. We seek a four dimensional Lorentz manifold $(M, g)$ and an embedding $\sigma: \mathscr{S} \to M$ such that $g$ satisfies the field equations, agrees with the initial conditions on $\sigma(\mathscr{S})$, and that $\sigma(\mathscr{S})$ is a *Cauchy surface* for $(M, g)$. This means the following: $M = D(\mathscr{S})$, with $D(\mathscr{S}) = D^+(\mathscr{S}) \cup D^-(\mathscr{S})$, where $D^+(\mathscr{S})$ denotes the set of points $p$ which have the property that every non-spacelike curve through $p$ which cannot be extended to the past, intersects $\mathscr{S}$. $D^+(\mathscr{S})$ is also called the *domain of dependence* of $\mathscr{S}$. $D^-(\mathscr{S})$ is defined correspondingly.

The set $(M, \sigma, g)$ is said to be a *development* of $(\mathscr{S}, \alpha)$. A different development $(M', \sigma', g')$ is an *extension* of $(M, \sigma, g)$ provided an injective differentiable mapping $\varphi: M \to M'$ exists, which leaves the image of $\mathscr{S}$ pointwise invariant and which transforms $g$ into $g'$.

It is important to keep in mind that every $(M, \sigma, g')$ is an extension of $(M, \sigma, g)$ if $g' = \varphi_* g$ and $\varphi$ is a diffeomorphism which leaves $\sigma(\mathscr{S})$

pointwise invariant. In this sense, a development of $(\mathscr{S}, \alpha)$ is *not unique*. In order to obtain a unique development, one must introduce four *gauge conditions*. One can show (a heuristic justification follows) that if the initial data $\alpha$ satisfy certain subsidiary conditions on $\mathscr{S}$, then a development of $(\mathscr{S}, \alpha)$ exists. Furthermore, a maximal development (i.e. one which is an extension of all other developments) exists. Of course, this is unique only if four gauge conditions are imposed. Furthermore, on $U \subset D^+(\mathscr{S})$, $g$ depends only on $J^-(U) \cap \mathscr{S}$, where $J^-(U)$ is the causal past of $U$. This dependence is continuous, provided $U$ has a compact closure in $D^+(\mathscr{S})$.

*Heuristic Consideration of the Local Problem*

We consider the local problem of a Cauchy development. Let $g_{\mu\nu}$ and $g_{\mu\nu,0}$ be given on the hypersurface $x^0 = t$. The field equations do not permit us to determine all the $g_{\mu\nu,00}$ at the time $t$. In other words, the computer cannot calculate the development of the $g_{\mu\nu}$. One sees this from the contracted Bianchi identity which implies that

$$G^{\mu 0}_{,0} \equiv - G^{\mu i}_{,i} - \Gamma^{\mu}_{\nu\lambda} G^{\lambda\nu} - \Gamma^{\nu}_{\nu\lambda} G^{\mu\lambda} . \tag{2.7.1}$$

The right hand side contains at most two differentiations with respect to time and this must then also be true for the left hand side. This shows that $G^{\mu 0}$ contains only first derivatives with respect to time. Hence we learn nothing about the time evolution from the field equations

$$G^{\mu 0} = 0 . \tag{2.7.2}$$

These must be regarded as *constraint equations* for the initial data, i.e. for $g_{\mu\nu}$ and $g_{\mu\nu,0}$ at the time $t$. This leaves as "dynamical" equations only the remaining six field equations

$$G^{ij} = 0 . \tag{2.7.3}$$

We thus have a fourfold ambiguity for the ten second derivatives $g_{\mu\nu,00}$, whose origin should be clear by now. To remove these ambiguities, we must impose four gauge conditions. As an example, we may choose the "harmonic coordinate conditions" $\delta\, dx^\lambda = 0$, where $\delta$ denotes the codifferential. Explicitly, these conditions are

$$(\sqrt{-g}\; g^{\mu\nu})_{,\nu} = 0 . \tag{2.7.4}$$

From this we obtain

$$(\sqrt{-g}\; g^{\mu 0})_{,00} = - (\sqrt{-g}\; g^{\mu i})_{,0i} . \tag{2.7.5}$$

Now the ten equations (2.7.3) and (2.7.5) are sufficient to determine the second derivatives of the $g_{\mu\nu}$.

It is important that the solutions of the initial value problem satisfy the constraint equations (2.7.2) also at every later time. This happens automatically for the following reason. Independent of the field equations, we have

$$G^{\mu\nu}_{;\nu} = 0 .$$

For $x^0 = t$, we impose the constraints (2.7.2). Together with (2.7.3), $G^{\mu\nu}$ vanishes everywhere on $x^0 = t$, and hence the constraint equations (2.7.2) propagate.

Thus, with given initial data for $g_{\mu\nu}$ and $g_{\mu\nu,0}$ which satisfy the subsidiary condition (2.7.2) at the initial time, there is a solution of the initial value problem. (The method can be programmed on a computer.)

It is possible to generalize the above discussion to the case $T_{\mu\nu} \neq 0$. For this, it is necessary to keep in mind that $T^{\mu\nu}_{;\nu} = 0$, independent of the field equations (see Sect. 3.5).

---

**Exercises:** Let $\mathscr{S}$ be a spacelike three dimensional submanifold of $(M, g)$ with induced metric $\bar{g}$. We introduce an adapted orthonormal frame on an open subset of $M$ such that $e_i$ $(i = 1, 2, 3)$ are tangent to $\mathscr{S}$ at points of $\mathscr{S}$, and consequently $e_0$ is normal to $\mathscr{S}$ at points of $\mathscr{S}$. The dual basis of $e_\mu$ is denoted by $\theta^\mu$ and $\omega^\mu_\nu, \Omega^\mu_\nu$ are as usual the connection and curvature forms relative to this basis. The corresponding quantities of the submanifold $\mathscr{S}$ relative to $\theta^i$ are denoted by $\bar{\omega}^i_j$, $\bar{\Omega}^i_j$.

1. Use the first structure equation and show that

$$\omega^i_j = \bar{\omega}^i_j \quad \text{on} \quad T\mathscr{S}, \tag{2.7.6}$$

$$\omega^0_j = K_{ij}\,\theta^j \quad \text{on} \quad T\mathscr{S}, \tag{2.7.7}$$

$$K_{ij} = K_{ji}. \tag{2.7.8}$$

   $K_{ij}$ are the components of the *second fundamental form (extrinsic curvature)* of $\mathscr{S}$ and the first two equations are equivalent to the *Gauss formulas* for submanifolds.

2. Use the second structure equation and derive *Gauss's Equation,*

$$\Omega^i_j = \bar{\Omega}^i_j + \omega^i_0 \wedge \omega^0_j \quad \text{on} \quad T\mathscr{S} \tag{2.7.9}$$

   and the *Codazzi-Mainardi Equation,*

$$\Omega^0_j = d\omega^0_j + \omega^0_i \wedge \omega^i_j \quad \text{on} \quad T\mathscr{S} \tag{2.7.10}$$

3. Derive the following formulas for the components $G_{0\mu}$ of the Einstein tensor:

$$G_{0i} = \Omega^j_0\,(e_j, e_i) \tag{2.7.11}$$

$$G_{00} = \tfrac{1}{2}\,\Omega^j_i\,(e_j, e_i)\,, \tag{2.7.12}$$

   which show that their values on $\mathscr{S}$ require only the curvature forms $\Omega^\mu_\nu$ restricted to $T\mathscr{S}$.

Use the previous results to prove the following formulae:

$$G_{0i} = \bar{\nabla}_j K_{ij} - \bar{\nabla}_i (\mathrm{Tr}\, K) \quad \text{on} \quad \mathscr{S} \tag{2.7.13}$$

$$G_{00} = \tfrac{1}{2} \bar{R} - \tfrac{1}{2} [\mathrm{Tr}(K \cdot K) - (\mathrm{Tr}\, K)^2] \quad \text{on} \quad \mathscr{S}, \tag{2.7.14}$$

where $\bar{\nabla}$ is the covariant derivative on $\mathscr{S}$ and $\bar{R}$ the curvature scalar of $\mathscr{S}$.

---

Equations (2.7.13) and (2.7.14) show again that the field equations (2.7.2) are *constraint equations* for the initial data of the Cauchy problem.

**Remark:** The submanifold $\mathscr{S}$ is called *totally geodesic* if every geodesic $\gamma$ with $\gamma(0) \in \mathscr{S}$ and $\dot{\gamma}(0) \in T_{\gamma(0)}\mathscr{S}$ remains in $\mathscr{S}$ on some interval $(-\varepsilon, \varepsilon)$. This is equivalent to the property that $\nabla_X Y$ is tangent to $\mathscr{S}$ whenever $X$ and $Y$ are (prove this!). Now for $Y = Y^j e_j$

$$\nabla_X Y = \nabla_X (Y^j e_j) = X(Y^j)\, e_j + Y^j \omega_j^\alpha(X)\, e_\alpha .$$

We conclude that $\mathscr{S} \subset M$ is totally geodesic if and only if the second fundamental form $K$ of $\mathscr{S}$ is zero.

If $K$ vanishes the (0, 0)-component of the field equation takes a remarkably simple form:

$$\bar{R} = 16\pi G T(n, n), \tag{2.7.15}$$

where $n = e_0$ is the unit normal field to $\mathscr{S}$.

## 2.8 Characteristics of the Einstein Field Equations

In this section we investigate the propagation of gravitational wave fronts. It will turn out that these are always null hypersurfaces.

It is instructive to consider first the characteristics of the generalized wave equation. The d'Alembertian operator $\Box_g : \Lambda_p(M) \to \Lambda_p(M)$ of a Lorentz manifold $(M, g)$ is:

$$\Box_g = d\delta + \delta d. \tag{2.8.1}$$

In particular, for a function $\psi$ we have

$$\Box_g \psi = \delta\, d\psi. \tag{2.8.2}$$

The coordinate representation of (2.8.2) is [see Eq. (4.6.28) of Part I]:

$$\Box_g \psi = (-g)^{-1/2} (\sqrt{-g}\, g^{\mu\nu} \psi_{,\nu})_{,\mu} . \tag{2.8.3}$$

The Cauchy problem for the generalized wave equation

$$\Box_g \psi = 0 \tag{2.8.4}$$

consists of finding a solution of (2.8.4) when $\psi$ and its first derivative are given on a hypersurface

$$u(x) = 0. \tag{2.8.5}$$

This problem does not have a unique solution if the hypersurface is chosen such that the wave equation (2.8.4) does not determine the second derivatives of $\psi$ on the hypersurface. In this case, the hypersurface is said to be a *characteristic surface*, or a *characteristic* of the differential equation (2.8.4). The second derivatives can be discontinuous on a characteristic hypersurface. For this reason, a (moving) wavefront *must* be a characteristic.

Let us assume that $\psi'' = \partial^2\psi/\partial u^2$ has a discontinuity on the surface (2.8.5). Then the coefficient of $\psi''$ in $\Box_g \psi$ must vanish. Since (2.8.4) is equivalent to $d^* d\psi = 0$, it follows from

$$\begin{aligned} d\psi &= \psi' \, du + \ldots \\ {}^*d\psi &= \psi' \, {}^*du + \ldots \\ d^*d\psi &= \psi'' \, du \wedge {}^*du + \text{continuous terms} \end{aligned}$$

that $du \wedge {}^*du = 0$. However, from Exercise 1 of Sect. 4.6.2 in Part I, we have

$$du \wedge {}^*du = (du, du)\, \eta .$$

We have thus shown that the normal vectors to a characteristic surface must be lightlike.

We now prove the corresponding statement for the Einstein field equations, following Sect. 4.2.13 of [19].

Let $(\theta^\alpha)$ be an orthonormal tetrad field which satisfies the Einstein field equations

$$-\eta_{\alpha\beta\gamma} \wedge \Omega^{\beta\gamma} = 16\pi G \, {}^*T_\alpha , \tag{2.8.6}$$

but wich has a discontinuity in the second derivative with respect to a local coordinate $u$ on the hypersurface $u(x) = 0$. The part of $d\theta^\alpha$ which has discontinuous first derivatives must be proportional to $du$:

$$d\theta^\alpha = C^\alpha{}_\beta \, du \wedge \theta^\beta + \text{continuous terms}. \tag{2.8.7}$$

We decompose $C_{\alpha\beta} = g_{\alpha\gamma} C^\gamma{}_\beta$ into symmetric and antisymmetric parts:

$$C_{\alpha\beta} = S_{\alpha\beta} + A_{\alpha\beta}, \quad S_{\alpha\beta} = S_{\beta\alpha}, \quad A_{\alpha\beta} = -A_{\beta\alpha}. \tag{2.8.8}$$

In the following, all equations are understood modulo continuous terms. The connection forms are

$$\omega_{\alpha\beta} = -A_{\alpha\beta}\, du + S_\alpha n_\beta - S_\beta n_\alpha , \tag{2.8.9}$$

with $S_\alpha = S_{\alpha\beta}\theta^\beta$ and $du = n_\alpha \theta^\alpha$; $\omega_{\alpha\beta}$ is obviously antisymmetric and

$$\begin{aligned} \omega^\alpha{}_\beta \wedge \theta^\beta &= -A^\alpha{}_\beta \, du \wedge \theta^\beta + S^\alpha{}_\gamma n_\beta \theta^\gamma \wedge \theta^\beta - n^\alpha S_{\beta\gamma} \theta^\gamma \wedge \theta^\beta \\ &= -(A^\alpha{}_\beta + S^\alpha{}_\beta)\, du \wedge \theta^\beta , \end{aligned}$$

so that, by (2.8.7) the second structure equation is satisfied. By assumption, only the term $d\omega_{\alpha\beta}$ can contribute a discontinuity to the curvature. From (2.8.9) we find

$$\Omega_{\alpha\beta} = dS_\alpha n_\beta - dS_\beta n_\alpha = du \wedge (S'_\alpha n_\beta - S'_\beta n_\alpha), \tag{2.8.10}$$

with $S'_\alpha = S'_{\alpha\beta}\theta^\beta$. From this it follows (with $i_\alpha = i_{e_\alpha}$ and $e_\alpha$ the basis dual to $\theta^\alpha$)

$$\begin{aligned} i_\alpha \Omega^\alpha{}_\beta \wedge du &= (i_\alpha du)(S'^\alpha n_\beta - S'_\beta n^\alpha) \wedge du \\ &= (n_\alpha S'^\alpha n_\beta - S'_\beta n^2) \wedge du; \end{aligned}$$

we have used $i_\alpha du = \langle du, e_\alpha \rangle = n_\alpha$. This in turn implies

$$(i_\alpha \Omega^\alpha{}_\beta n_\gamma - i_\alpha \Omega^\alpha{}_\gamma n_\beta) \wedge du = n^2 (S'_\gamma n_\beta - S'_\beta n_\gamma) \wedge du = n^2 \Omega_{\gamma\beta}.$$

However, $i_\alpha \Omega^\alpha{}_\beta = R_{\beta\sigma}\theta^\sigma$. According to Einstein's field equations (2.8.6), the left hand side is continuous. Hence, either $\Omega_{\beta\gamma}$ is continuous or $n^2 = 0$.

"True" discontinuities will show up in the curvature. Even in flat space, it is possible to construct tetrad fields with discontinuous second derivatives, which obviously have nothing to do with discontinuities in the metric field.

We have thus shown that true discontinuities can only propagate along null hypersurfaces. This means that variations of the gravitational field propagate at the speed of light.

---

**Exercise:** Determine the characteristics of Maxwell's vacuum equations on a Lorentz manifold.

*Solution:* The discontinuities in the first derivatives of $F$ must be such that they do not appear in $dF$. Hence $F$ has the form (using previous notation):

$$F = f_\alpha du \wedge dx^\alpha + \text{continuous terms}.$$

Modulo continuous terms, we then have

$$d{*F} = f'_\alpha du \wedge {*(du \wedge dx^\alpha)}.$$

Since the current in $\delta F = 4\pi J$ is continuous, the form $du \wedge {*(du \wedge dx^\alpha)}$ must vanish.

Now for any 1-form $\theta$ and a $p$-form $\omega$ we have

$$\theta \wedge {*\omega} = (-1)^{p+1} {*(i_{\theta^\#}\omega)},$$

where $\theta^\#$ denotes the vector field which is dual to $\theta$. (Prove this identity!). Hence we find for $n = \operatorname{grad} u\,(=(du)^\#)$:

$$i_n(du \wedge dx^\alpha) = 0$$

or $n^2 dx^\alpha - du\, n^\alpha = 0$. Applying $i_\beta$ to this equation leads to

$$n^2 \delta^\alpha_\beta - n^\alpha n_\beta = 0.$$

Taking the trace gives, as expected, $n^2 = 0$. The characteristics are thus again null surfaces.

---

# Chapter 3. The Schwarzschild Solution and Classical Tests of General Relativity

"... Imagine my joy at the feasibility of general covariance and the result that the equations give the perihelian motion of Mercury correctly. For a few days I was beside myself with joyous exitement."

(A. Einstein, to P. Ehrenfest, January 17, 1916)

The solution of the field equations, which describes the field outside of a spherically symmetric mass distribution, was found by Karl Schwarzschild only two months after Einstein published his field equations. Schwarzschild performed this work under rather unusual conditions. In the spring and summer of 1915 he was assigned to the eastern front. There he came down with an infectious disease and in the fall of 1915 he returned seriously ill to Germany. He died only a few months later, on May 11, 1916. In this short time, he wrote two significant papers, in spite of his illness. One of these dealt with the Stark effect in the Bohr-Sommerfeld theory, and the other solved the Einstein field equations for a static, spherically symmetric field. From this solution he derived the precession of the perihelion of Mercury and the bending of light rays at the surface of the sun. Einstein had calculated these effects previously, by solving the field equations in the post-Newtonian approximation.

## 3.1 Derivation of the Schwarzschild Solution

We choose the manifold to be $M = \mathbb{R} \times \mathbb{R}_+ \times S^2$. In polar coordinates and a natural basis, the metric has the form

$$g = e^{2a(r)}\, dt^2 - [e^{2b(r)}\, dr^2 + r^2 (d\vartheta^2 + \sin^2 \vartheta\, d\varphi^2)]. \tag{3.1.1}$$

We use the shorthand notation $dt^2$ for $dt \otimes dt$, etc.[1] The coordinate $r$ ("radius") is suitably normalized, so that a circle of radius $r$ has the circumference $2\pi r$.

The functions $a(r)$ and $b(r)$ approach zero asymptotically; thus $g$ is asymptotically flat. We must now insert the ansatz (3.1.1) into the field equations. For this, it is necessary to compute the Ricci tensor (or the Einstein tensor) corresponding to the metric (3.1.1). This is accomplished most quickly with the help of the Cartan calculus (the traditional computation using the Christoffel symbols is given in detail in Sect. 6.1 of [17]).

We choose the following basis of 1-forms:

$$\theta^0 = e^a\, dt\,, \ \theta^1 = e^b\, dr\,, \ \theta^2 = r\, d\vartheta\,, \ \theta^3 = r \sin\vartheta\, d\varphi\,. \tag{3.1.2}$$

The metric (3.1.1) then reads

$$g = g_{\mu\nu}\,\theta^\mu \otimes \theta^\nu, \quad (g_{\mu\nu}) = \mathrm{diag}\,(1, -1, -1, -1)\,. \tag{3.1.3}$$

Thus, the basis $(\theta^\alpha)$ is orthonormal. Hence the connection forms $\omega^\mu_{\ \nu}$ satisfy

$$\omega_{\mu\nu} + \omega_{\nu\mu} = 0\,, \quad \omega_{\mu\nu} := g_{\mu\lambda}\,\omega^\lambda_{\ \nu}\,. \tag{3.1.4}$$

In order to determine these from the first structure equation, we compute the exterior derivative (with $a' = da/dr$, etc.)

$$\begin{aligned}
d\theta^0 &= a'\, e^a\, dr \wedge dt \\
d\theta^1 &= 0 \\
d\theta^2 &= dr \wedge d\vartheta \\
d\theta^3 &= \sin\vartheta\, dr \wedge d\varphi + r\cos\vartheta\, d\vartheta \wedge d\varphi\,.
\end{aligned}$$

We express the right hand sides in terms of the basis $\theta^\sigma \wedge \theta^\varrho$, obtaining

$$\begin{aligned}
&d\theta^0 = a'\, e^{-b}\,\theta^1 \wedge \theta^0, \quad d\theta^1 = 0 \\
&\theta^2 = r^{-1}\, e^{-b}\,\theta^1 \wedge \theta^2 \quad d\theta^3 = r^{-1}\,[e^{-b}\,\theta^1 \wedge \theta^3 + \cot\vartheta\,\theta^2 \wedge \theta^3]\,.
\end{aligned} \tag{3.1.5}$$

When this is compared with the first structure equation $d\theta^\alpha = -\omega^\alpha_{\ \beta} \wedge \theta^\beta$, one expects the following expressions for the connection forms:

$$\begin{aligned}
&\omega^0_{\ 1} = \omega^1_{\ 0} = a'\, e^{-b}\,\theta^0, \quad \omega^0_{\ 2} = \omega^2_{\ 0} = \omega^0_{\ 3} = \omega^3_{\ 0} = 0 \\
&\omega^2_{\ 1} = -\,\omega^1_{\ 2} = r^{-1}\, e^{-b}\,\theta^2 \\
&\omega^3_{\ 1} = -\,\omega^1_{\ 3} = r^{-1}\, e^{-b}\,\theta^3 \\
&\omega^3_{\ 2} = -\,\omega^2_{\ 3} = r^{-1}\cot\vartheta\,\theta^3\,.
\end{aligned} \tag{3.1.6}$$

This ansatz satisfies indeed (3.1.4) and the first structure equation. On the other hand, we know that the solution is unique (see Sect. 5.7 of Part I).

---

[1] The Lorentz manifold $(M, g)$, with $g$ given by (3.1.1), is *spherically symmetric* in the sense of the following

**Definition:** A Lorentz manifold is *spherically symmetric* if it admits the group SO(3) as an isometry group, such that the group orbits are two-dimensional spacelike surfaces.

Conversely, it is possible to show that in a static, spherically symmetric space-time, one can always introduce coordinates such that $g$ has the form (3.1.1). We shall prove this in the Appendix to Chap. 3.

The determination of the curvature forms $\Omega^\mu_{\ \nu}$ from the second structure equation is now straightforward:

$$\begin{aligned}\Omega^0_{\ 1} &= d\omega^0_{\ 1} + \omega^0_{\ k} \wedge \omega^k_{\ 1} = d\omega^0_{\ 1} = d(a'\,\mathrm{e}^{-b}\,\theta^0)\\ &= (a'\,\mathrm{e}^{-b})'\,dr \wedge \theta^0 + a'\,\mathrm{e}^{-b}\,d\theta^0\\ &= (a'\,\mathrm{e}^{-b})'\,\mathrm{e}^{-b}\,\theta^1 \wedge \theta^0 + (a'\,\mathrm{e}^{-b})^2\,\theta^1 \wedge \theta^0\\ \Omega^0_{\ 1} &= -\mathrm{e}^{-2b}(a'^2 - a'\,b' + a'')\,\theta^0 \wedge \theta^1\\ \Omega^0_{\ 2} &= d\omega^0_{\ 2} + \omega^0_{\ k} \wedge \omega^k_{\ 2} = \omega^0_{\ 1} \wedge \omega^1_{\ 2} = -\mathrm{e}^{-2b}\frac{a'}{r}\,\theta^0 \wedge \theta^2.\end{aligned}$$

One obtains the other components in a similar manner.

We summarize the results for later use ($\Omega^\mu_{\ \nu}$ is proportional to $\theta^\mu \wedge \theta^\nu$; the indices 2 and 3 are equivalent):

$$\begin{aligned}\Omega^0_{\ 1} &= e^{-2b}(a'\,b' - a'' - a'^2)\,\theta^0 \wedge \theta^1\\ \Omega^0_{\ 2} &= -\frac{a'\,\mathrm{e}^{-2b}}{r}\,\theta^0 \wedge \theta^2\\ \Omega^0_{\ 3} &= -\frac{a'\,\mathrm{e}^{-2b}}{r}\,\theta^0 \wedge \theta^3\\ \Omega^1_{\ 2} &= \frac{b'\,\mathrm{e}^{-2b}}{r}\,\theta^1 \wedge \theta^2\\ \Omega^1_{\ 3} &= \frac{b'\,\mathrm{e}^{-2b}}{r}\,\theta^1 \wedge \theta^3\\ \Omega^2_{\ 3} &= \frac{1-\mathrm{e}^{-2b}}{r^2}\,\theta^2 \wedge \theta^3, \quad \Omega_{\mu\nu} := g_{\mu\lambda}\,\Omega^\lambda_{\ \nu} = -\Omega_{\nu\mu}.\end{aligned} \tag{3.1.7}$$

From this, one reads off the components of the Riemann tensor with respect to the basis $\theta^\alpha$. For the Einstein tensor one easily finds

$$\begin{aligned}G^0_{\ 0} &= \frac{1}{r^2} - \mathrm{e}^{-2b}\left(\frac{1}{r^2} - \frac{2b'}{r}\right)\\ G^1_{\ 1} &= \frac{1}{r^2} - \mathrm{e}^{-2b}\left(\frac{1}{r^2} + \frac{2a'}{r}\right)\\ G^2_{\ 2} = G^3_{\ 3} &= -\mathrm{e}^{-2b}\left(a'^2 - a'\,b' + a'' + \frac{a'-b'}{r}\right),\\ &\text{all other} \quad G_{\mu\nu} = 0.\end{aligned} \tag{3.1.8}$$

We now solve the vacuum equations. $G_{00} + G_{11} = 0$ implies that $a' + b' = 0$, and hence $a + b = 0$, since $a, b \to 0$ asymptotically. We then find from $G_{00} = 0$ that

$$e^{-2b}(2b'/r - 1/r^2) + 1/r^2 = 0,$$

or

$$(r\,\mathrm{e}^{-2b})' = 1\,,$$

which implies that $\mathrm{e}^{-2b} = 1 - 2m/r$, where $m$ is an integration constant. We thus obtain the *Schwarzschild solution*

$$g = \left(1 - \frac{2m}{r}\right) dt^2 - \frac{dr^2}{1 - 2m/r} - r^2 (d\vartheta^2 + \sin^2\vartheta\, d\varphi^2)\,. \tag{3.1.9}$$

---

**Exercise:** Allow the functions $a(t, r)$ and $b(t, r)$ in (3.1.1) to be time dependent, and show that then the components of the Einstein tensor with respect to the basis (3.1.2) are given by ($\dot{a} = da/dt$, etc.)

$$\begin{aligned} G^0_0 &= \frac{1}{r^2} - \mathrm{e}^{-2b}\left(\frac{1}{r^2} - \frac{2b'}{r}\right) \\ G^1_1 &= \frac{1}{r^2} - \mathrm{e}^{-2b}\left(\frac{1}{r^2} + \frac{2a'}{r}\right) \\ G^2_2 = G^3_3 &= -\,\mathrm{e}^{-2b}\left(a'^2 - a'\,b' + a'' + \frac{a' - b'}{r}\right) \\ &\quad - \mathrm{e}^{-2a}(-\,\dot{b}^2 + \dot{a}\,\dot{b} - \ddot{b}) \\ G^1_0 &= -\frac{2\dot{b}}{r}\,\mathrm{e}^{-a-b}\,; \end{aligned} \tag{3.1.10}$$

all other $G^\mu_\nu = 0$.

---

*The Birkhoff Theorem*

We now solve the vacuum equations for these expressions. The equation $G^1_0 = 0$ implies that $b$ is independent of time and hence it follows from $G_{00} = 0$ that $b$ has the same form as in the static case. From $G_{00} + G_{11} = 0$ we again obtain the condition $a' + b' = 0$; in this case, however, we can only conclude that

$$a = -\,b + f(t)\,.$$

The other vacuum equations are then all satisfied and the metric reads

$$g = \mathrm{e}^{2f(t)}\left(1 - \frac{2m}{r}\right) dt^2 - \left[\frac{dr^2}{1 - 2m/r} + r^2 (d\vartheta^2 + \sin^2\vartheta\, d\varphi^2)\right].$$

If we introduce a new time coordinate

$$t' = \int \mathrm{e}^{f(t)}\, dt\,,$$

we again obtain the Schwarzschild metric (3.1.9). *For $r > 2m$, a spherically symmetric vacuum field is thus necessarily static.*

We determine the integration constant $m$ in (3.1.9) by comparison with the Newtonian limit at large distances. In this region, we must have $g_{00} \simeq 1 + 2\phi$, $\phi = -GM/r$. Hence

$$m = GM/c^2. \tag{3.1.11}$$

We shall now show that the integration constant $M$ is also equal to the total energy $P^0$.

For this purpose, we write (3.1.9) in nearly Lorentzian coordinates. Let

$$\varrho = \tfrac{1}{2}[r - m + (r^2 - 2mr)^{1/2}].$$

Then

$$r = \varrho(1 + m/2\varrho)^2. \tag{3.1.12}$$

Substitution into (3.1.9) results in

$$g = \left(\frac{1 - m/2\varrho}{1 + m/2\varrho}\right)^2 dt^2 - \left(1 + \frac{m}{2\varrho}\right)^4 (d\varrho^2 + \varrho^2\, d\vartheta^2 + \varrho^2 \sin^2\vartheta\, d\varphi^2). \tag{3.1.13}$$

If we now set

$$x^1 = \varrho \sin\vartheta \cos\varphi, \quad x^2 = \varrho \sin\vartheta \sin\varphi, \quad x^3 = \varrho\cos\vartheta,$$

then the Schwarzschild metric has the form

$$g = h^2(|\boldsymbol{x}|)\, dt^2 - f^2(|\boldsymbol{x}|)\, d\boldsymbol{x}^2 \tag{3.1.14}$$

with

$$h(r) = \frac{1 - m/2r}{1 + m/2r}, \quad f(r) = \left(1 + \frac{m}{2r}\right)^2. \tag{3.1.15}$$

One finds the following connection forms

$$\omega^{0j} = \frac{h'}{f}\frac{x^j}{r}\, dt, \quad \omega^{jk} = \frac{f'}{fr}(x^j\, dx^k - x^k\, dx^j) \tag{3.1.16}$$

with respect to the orthonormal tetrad

$$(\theta^\alpha) = (h\, dt, f\, dx^i).$$

We now compute $P^0$ from (2.6.9):

$$\begin{aligned} P^0 &= -\frac{1}{16\pi G} \oint \omega^{jk} \wedge \eta^0{}_{jk} = -\frac{1}{16\pi G}\, \varepsilon_{0jkl} \oint \omega^{jk} \wedge \theta^l \\ &= \frac{1}{8\pi G}\, \varepsilon_{0jkl} \oint \frac{f'}{r}\, x^k\, dx^j \wedge dx^l. \end{aligned}$$

Let us integrate over the surface of a large sphere. Then

$$\varepsilon_{0jkl}\, x^k\, dx^j \wedge dx^l = -x^k\, \varepsilon_{kjl}\, dx^j \wedge dx^l = -r\, 2r^2\, d\Omega,$$

where $d\Omega$ denotes the solid angle element. This gives, as expected,

$$P^0 = -\frac{1}{4\pi G} \lim_{R\to\infty} \int_{r=R} f' r^2 \, d\Omega = M.$$

It is easy to verify that $P^i = 0$. Furthermore, one obtains a vanishing angular momentum from (2.6.24).

The Schwarzschild solution (3.1.9) has an apparent singularity at

$$r = R_S := 2GM/c^2. \tag{3.1.17}$$

$R_S$ is the so-called *Schwarzschild radius*. Schwarzschild himself was quite disturbed by this "singularity". For this reason, he subsequently investigated the solution of Einstein's field equations for a spherically symmetric static mass distribution having constant energy density. He showed that the radius of such a configuration must be $> 9R_S/8$. He was extremely satisfied by this result, since it showed that the singularity is not relevant (for the special case being considered). However, somewhat later, in 1923, Birkhoff proved that a spherically symmetric vacuum solution of Einstein's equations is necessarily *static* for $r > R_S$. Hence the exterior field for a nonstatic, spherically symmetric mass distribution is necessarily the Schwarzschild solution for $r > R_S$. The lower bound $\frac{9}{8}R_S$ is no longer valid for a nonstatic situation; hence it is necessary to investigate in more detail what is going on in the vicinity of the Schwarzschild sphere $r = R_S$. We shall do this in Sect. 3.7. It will turn out that there is no singularity at $r = R_S$; the coordinate system being used is simply not applicable there. The Schwarzschild sphere has nevertheless physical significance (as a horizon). When $r < R_S$, the solution is no longer static.

### *Geometric Meaning of the Spatial Part of the Schwarzschild Metric*

We shall now give a geometrical illustration of the spatial part of the metric (3.1.9). Consider the two dimensional submanifold ($\vartheta = \frac{\pi}{2}$, $t = \text{const}$) and represent this as a surface of rotation in three-dimensional Euclidean space $E^3$. This submanifold has the metric

$$G = \frac{dr^2}{1 - 2m/r} + r^2 \, d\varphi^2. \tag{3.1.18}$$

On the other hand, a surface of rotation in $E^3$ has the metric

$$G = z'^2 \, dr^2 + (dr^2 + r^2 \, d\varphi^2) = (1 + z'^2) \, dr^2 + r^2 \, d\varphi^2,$$

where $z(r)$ describes the surface. (We use cylindrical coordinates.) If we require this to agree with (3.1.18), then

$$dz/dr = [2m/(r - 2m)]^{1/2}.$$

Integration gives $z=[8m(r-2m)]^{1/2}+\text{const}$. If we set the integration constant equal to zero, we obtain a paraboloid of revolution:

$$z^2=8m(r-2m). \tag{3.1.19}$$

---

**Exercises:**

1) Consider a spherical cavity inside a spherically symmetric non-rotating matter distribution. Show that there the metric is flat. (This remark justifies certain Newtonian considerations in cosmology.)
2) Generalize the Schwarzschild solution for Einstein's equations with a cosmological constant.
3) Determine the solution representing space-time outside a spherically symmetric charged body carrying an electric charge (but no spin or magnetic dipole moment). The result is the *Reissner-Nordstrøm* solution:

$$g=\left(1-\frac{2m}{r}+\frac{G\,e^2}{r^2}\right)dt^2-\left(1-\frac{2m}{r}+\frac{G\,e^2}{r^2}\right)^{-1}dr^2-r^2\,(d\vartheta^2+\sin^2\vartheta\,d\varphi^2),$$

where $m/G$ represents the gravitational mass and $e$ the electric charge of the body ($c=1$).
4) Consider a static Lorentz manifold $(M, g)$. We know that locally $M$ is isometric to a direct product $\mathbb{R}\times N$, where $(N, h)$ is a 3-dimensional Riemannian manifold. If $\{\theta^i\}$ denotes an orthonormal tetrad of $N$, $g$ can be written as

$$g=\varphi^2\,dt^2-\delta_{ik}\,\theta^i\,\theta^k\,,$$

where $\varphi$ is a smooth function on $N$. (More accurately, we should write $\pi^*\,\theta^i$ instead of $\theta^i$, where $\pi$ denotes the projection of $\mathbb{R}\times N$ onto the second factor.)

Thus $\{\theta^0=\varphi\,dt,\,\theta^i\}$ is locally an orthonormal tetrad of $(M, g)$.
a) Compute the components of the Ricci-tensor of $(M, g)$ relative to this basis. The result is

$$R_{00}=\frac{1}{\varphi}\overset{(3)}{\Delta}\varphi,\quad R_{0i}=0$$

$$R_{ij}=-\frac{1}{\varphi}\,\varphi_{\|ij}+\overset{(3)}{R}_{ij}.$$

The index (3) refers to quantities of the 3-dimensional Riemannian manifold $N$ and the double stroke denotes the covariant derivative in $(N, h)$. Thus the vacuum field equations are

$$\overset{(3)}{\Delta}\varphi=0$$

$$\overset{(3)}{R}_{ik}=\frac{1}{\varphi}\,\varphi_{\|ik},$$

which imply $\overset{(3)}{R}=0$.

b) Rewrite these equations in terms of the conformal metric $\gamma = \varphi^2 h$ of $N$ and show that they are equivalent to

$$\operatorname{Ric}[\gamma] = 2\, dU \otimes dU, \quad \text{where} \quad \varphi =: e^U. \tag{*}$$

A static metric thus has locally the form

$$g = e^{2U}\, dt^2 - e^{-2U}\, \gamma_{ik}\, dx^i\, dx^k$$

and the field equations are equivalent to (*). Prove that (*) implies

$$\Delta[\gamma]\, U = 0.$$

[Use the contracted Bianchi identity for $(N, h)$.]

**Remark:** If $N$ is diffeomorphic to $\mathbb{R}^3$, then the Laplace equation for $U$ implies $U = 0$ if $U$ vanishes at infinity. Hence $\varphi = 1$ and $(N, \gamma)$ is Ricci flat: $\operatorname{Ric}[\gamma] = 0$. From this one can conclude (see [5], Volume I, p. 292 and Theorem 7.10) that $(N, \gamma)$ is isometric to the 3-dimensional Euclidean space. A static vacuum manifold with spatial sections diffeomorphic to $\mathbb{R}^3$ is thus isometric to Minkowski space.

c) Verify (*) for the Schwarzschild solution.

---

## 3.2 Equation of Motion in a Schwarzschild Field

We consider a test body in a Schwarzschild field. Its geodesic equation of motion is the Euler equation for the Lagrangian $\mathscr{L} = \frac{1}{2}\, g_{\mu\nu}\, \dot{x}^\mu\, \dot{x}^\nu$, which is given by

$$2\mathscr{L} = (1 - 2\,m/r)\, \dot{t}^2 - \frac{\dot{r}^2}{1 - 2m/r} - r^2(\dot{\vartheta}^2 + \sin^2\vartheta\, \dot{\varphi}^2) \tag{3.2.1}$$

for the Schwarzschild metric (3.1.9) (the dot denotes differentiation with respect to proper time).

Obviously, along the orbit

$$2\mathscr{L} = 1. \tag{3.2.2}$$

We consider first the $\vartheta$-equation

$$(r^2\, \dot{\vartheta})^{\cdot} = r^2 \sin\vartheta \cos\vartheta\, \dot{\varphi}^2.$$

This implies that if the motion of the test body is initially in the equatorial plane $\vartheta = \frac{\pi}{2}$ (and hence $\dot{\vartheta} = 0$), then $\vartheta \equiv \frac{\pi}{2}$. We may therefore take $\vartheta = \frac{\pi}{2}$ without loss of generality. We then have

$$2\mathscr{L} = (1 - 2m/r)\, \dot{t}^2 - \frac{\dot{r}^2}{1 - 2m/r} - r^2\, \dot{\varphi}^2. \tag{3.2.3}$$

The variables $\varphi$ and $t$ are cyclic. Hence

$$-\frac{\partial \mathscr{L}}{\partial \dot{\varphi}} = r^2 \dot{\varphi} = \text{const} =: L \tag{3.2.4}$$

$$\frac{\partial \mathscr{L}}{\partial \dot{t}} = \dot{t}(1 - 2m/r) = \text{const} =: E. \tag{3.2.5}$$

Inserting (3.2.4) and (3.2.5) into (3.2.2) gives

$$(1 - 2m/r)^{-1} E^2 - (1 - 2m/r)^{-1} \dot{r}^2 - L^2/r^2 = 1. \tag{3.2.6}$$

From this we obtain

$$\dot{r}^2 + V(r) = E^2 \tag{3.2.7}$$

with the *effective potential*

$$V(r) = (1 - 2m/r)(1 + L^2/r^2). \tag{3.2.8}$$

**Remark:** The conservation laws (3.2.4) and (3.2.5) are based on the following general fact. Let $\gamma(\tau)$ be a geodesic with tangent vector $u$ and let $\xi$ be a Killing field. Then

$$(u, \xi) = \text{const} \quad \text{along } \gamma. \tag{3.2.9}$$

In fact

$$\begin{aligned} u(u, \xi) &= (\nabla_u u, \xi) + (u, \nabla_u \xi) \\ &= u^\mu u^\nu \xi_{\mu;\nu} = \tfrac{1}{2} u^\mu u^\nu (\xi_{\mu;\nu} + \xi_{\nu;\mu}) = 0. \end{aligned}$$

For the Schwarzschild metric, $\partial/\partial t$ and $\partial/\partial\varphi$ are Killing fields. The corresponding conservation laws (3.2.9) agree with (3.2.4) and (3.2.5) along $\gamma$:

$$(u, \partial/\partial t) = u^t \left(\frac{\partial}{\partial t}, \frac{\partial}{\partial t}\right) = g_{00} u^t = \left(1 - \frac{2m}{r}\right) \dot{t} = \text{const}$$

$$(u, \partial/\partial\varphi) = u^\varphi \left(\frac{\partial}{\partial\varphi}, \frac{\partial}{\partial\varphi}\right) = g_{\varphi\varphi} u^\varphi = -r^2 \dot{\varphi} = \text{const}.$$

In the following, we are primarily interested in the orbit $r(\varphi)$. Now

$$r' = \dot{r}/\dot{\varphi}$$

(the prime denotes differentiation with respect to $\varphi$), and this implies

$$\dot{r} = r' \dot{\varphi} = r' L/r^2.$$

Using this in (3.2.7) gives

$$r'^2 L^2/r^4 = E^2 - V(r).$$

Now let $u = 1/r$ ($r' = -u'/u^2$). In terms of this variable, we have

$$L^2 u'^2 = E^2 - (1 - 2m u)(1 + L^2 u^2)$$

or

$$u'^2 + u^2 = \frac{E^2 - 1}{L^2} + \frac{2m}{L^2} u + 2m u^3 . \tag{3.2.10}$$

We now differentiate (3.2.10) with respect to $\varphi$, and obtain

$$2u' u'' + 2u u' = \frac{2m}{L^2} u' + 6m u' u^2 .$$

Hence either $u' = 0$ (circular motion) or

$$u'' + u = \frac{m}{L^2} + 3m u^2 . \tag{3.2.11}$$

At this point we can make a comparison with the Newtonian theory. In this, the Lagrangian for a gravitational potential $\phi(r)$ is given by

$$\mathcal{L} = \frac{1}{2}\left[\left(\frac{dr}{dt}\right)^2 + r^2\left(\frac{d\varphi}{dt}\right)^2\right] + \phi(r) .$$

Since $\varphi$ is cyclic, we have $r^2\, d\varphi/dt =: L = \text{const}$, and the radial equation is

$$\frac{d^2 r}{dt^2} = r\left(\frac{d\varphi}{dt}\right)^2 - \phi'(r) .$$

Now

$$dr/dt = \frac{dr}{d\varphi}\frac{d\varphi}{dt} = r' L/r^2 = -L u'$$

$$d^2 r/dt^2 = -L u'' \frac{d\varphi}{dt} = -L^2 u'' u^2 .$$

After some rearrangement we obtain

$$u'' + u = -\frac{1}{L^2}\phi'/u^2 . \tag{3.2.12}$$

For the special case $\phi = -GM/r$, we have

$$u'' + u = GM/L^2 . \tag{3.2.13}$$

Equation (3.2.11) contains the additional term $3m u^2$. This "perturbation" is small, since

$$\frac{3m u^2}{m/L^2} = 3u^2 L^2 = 3\frac{1}{r^2}(r^2 \dot\varphi)^2 \simeq 3\left(r\frac{d\varphi}{dt}\right)^2 \Big/ c^2$$

$$\simeq 3v_\perp^2/c^2 \simeq 7.7 \times 10^{-8} \text{ for Mercury}$$

(here $v_\perp$ is the velocity component perpendicular to the radius vector).

According to (3.2.12), we may regard (3.2.11) as a Newtonian equation of motion for the potential

$$\phi(r) = -GM/r - m L^2/r^3 . \tag{3.2.14}$$

**Exercise:** Sketch the effective potential (3.2.8) for various values of $L/m$ and show the following:
a) For $L/m < 2\sqrt{3}$ any incoming particle falls toward $r = 2m$.
b) The most tightly bound, stable circular orbit is at $r = 6m$ with $L/m = 2\sqrt{3}$ and has a fractional binding energy of $1 - \sqrt{8/9}$.
c) Any particle with $E \geqq 1$ will be pulled into $r = 2m$ if $2\sqrt{3} < L/m < 4$.

## 3.3 Advance of the Perihelion of a Planet

We shall now examine the orbit equation (3.2.11), treating the term $3mu^2$ as a small perturbation. In the Newtonian approximation, the orbit is a Kepler ellipse

$$u = p^{-1}(1 + e\cos\varphi)\,, \tag{3.3.1}$$

where $e$ is the eccentricity and

$$p = a(1 - e^2) = L^2/m\,. \tag{3.3.2}$$

We now insert this into the perturbation term and obtain from (3.2.11) to a first approximation

$$u'' + u = m L^{-2} + 3m^3 L^{-4}(1 + e\cos\varphi)^2\,. \tag{3.3.3}$$

For planets, the last term on the right hand side is very small.

Particular solutions of the following three equations

$$u'' + u = \begin{cases} A \\ A\cos\varphi \\ A\cos^2\varphi \end{cases} \tag{3.3.4}$$

are given by

$$u_1 = \begin{cases} A \\ \frac{1}{2}A\,\varphi\sin\varphi \\ \frac{1}{2}A - (A/6)\cos 2\varphi\,. \end{cases} \tag{3.3.5}$$

The constant terms in (3.3.5) are uninteresting, since they simply change the parameters of the unperturbed orbit (3.3.1). The periodic perturbation is also quite unobservable. However, the second term in (3.3.5) leads to a *secular* change. In second order we have

$$\begin{aligned} u &= m L^{-2}(1 + e\cos\varphi + 3m^2 L^{-2} e\,\varphi\sin\varphi) \\ &\simeq m L^{-2}[1 + e\cos(1 - 3m^2 L^{-2})\,\varphi]\,. \end{aligned} \tag{3.3.6}$$

According to (3.3.6), $r$ is a periodic function of $\varphi$ with period

$$2\pi(1 - 3m^2/L^2)^{-1} > 2\pi\,.$$

Using (3.3.2) we find the perihelion anomaly $\Delta\varphi$ to be

$$\Delta\varphi = 2\pi(1-3m^2/L^2)^{-1} - 2\pi \simeq 6\pi m^2/L^2 = 6\pi m a^{-1}(1-e^2)^{-1}. \quad (3.3.7)$$

This effect of GR is most pronounced when the orbit's semimajor axis $a$ is small and/or the eccentricity is large. In addition, it is easier to determine the precise position of the perihelion observationally for large eccentricities than for small ones. In the solar system, *Mercury* provides the most favorable case. Here one obtains

$$\Delta\varphi_{\text{Einstein}} = 42.98'' \text{ per century.} \quad (3.3.8)$$

This agrees with the results obtained by radar ranging measurements to better than $\frac{1}{2}$%. The situation is considerably less favorable for the other planets. For Mercury, the "observed" 43″ per century are the remainder left over after subtracting the Newtonian perturbations to the orbit which are due to the presence of the other planets and which amount to about 500″ per century. A further Newtonian perturbation could be caused by a solar quadrupole moment. We now calculate this possible contribution.

The Newtonian potential exterior to a mass distribution having the density $\varrho(x)$ is

$$\phi(\boldsymbol{x}) = -G\int \frac{\varrho(\boldsymbol{x}')}{|\boldsymbol{x}-\boldsymbol{x}'|}\, d^3x' .$$

For $r > r'$ we have

$$\frac{1}{|\boldsymbol{x}-\boldsymbol{x}'|} = 4\pi \sum_{l=0}^{\infty}\sum_{m=-l}^{+l} \frac{1}{2l+1}\left(\frac{r'}{r}\right)^l \frac{1}{r}\, Y^*_{lm}(\hat{\boldsymbol{x}}')\, Y_{lm}(\hat{\boldsymbol{x}}) .$$

Hence $\phi$ can be expanded in multipoles

$$\phi(\boldsymbol{x}) = -4\pi G \sum_{l,m} \frac{Q^*_{lm}}{2l+1}\, \frac{1}{r^{l+1}}\, Y_{lm}(\hat{\boldsymbol{x}}) , \quad (3.3.9)$$

where

$$Q_{lm} := \int \varrho(\boldsymbol{x}')\, r'^l\, Y_{lm}(\hat{\boldsymbol{x}}')\, d^3x' . \quad (3.3.9')$$

Now suppose that $\varrho(\boldsymbol{x})$ is azimuthally symmetric, and symmetric under reflections at the $(x, y)$-plane. Then $Q_{lm} = 0$ for $m \neq 0$. The monopole contribution (with $Y_{00} = 1/\sqrt{4\pi}$) is equal to $-GM_\odot/r$. Due to mirror symmetry, the dipole contribution vanishes:

$$\int \varrho(\boldsymbol{x}') \underbrace{r'\, Y_{10}(\hat{\boldsymbol{x}}')}_{\propto\, z'}\, d^3x' = 0 .$$

The remaining terms give

$$\phi = -\frac{GM_\odot}{r}\left\{1 - \sum_{l=2}^{\infty} J_l \left(\frac{R_\odot}{r}\right)^l P_l(\cos\vartheta)\right\} \quad (3.3.10)$$

with

$$J_l = \frac{-1}{M_\odot R_\odot^l} \int \varrho(\boldsymbol{x}')\, r'^l P_l(\cos\vartheta')\, d^3x' \,. \tag{3.3.11}$$

It is sufficient to keep only the quadrupole term. For $\vartheta = \frac{\pi}{2}$, we have

$$\phi = -\frac{GM_\odot}{r} - \frac{1}{2}\frac{GM_\odot J_2 R_\odot^2}{r^3} \,. \tag{3.3.12}$$

Note that $J_2$ is positive for an oblate Sun.

This has the same form as (3.2.14), provided we make the substitution $m \to \frac{1}{2} GM_\odot L^{-2} J_2 R_\odot^2$ in the second term.

We then obtain, instead of (3.3.7), using (3.3.2),

$$\varDelta\varphi\,|_{\text{Quad}} = \frac{6\pi}{a(1-e^2)}\frac{1}{2}J_2\frac{GM_\odot R_\odot^2}{L^2} = \frac{6\pi m}{a(1-e^2)}\frac{1}{2}J_2\frac{R_\odot^2}{\left(\frac{GM_\odot}{c}\right)^2 a(1-e^2)}$$

which means that

$$\varDelta\varphi_{\text{Quad}} = \frac{1}{2}J_2\frac{R_\odot^2/(GM_\odot/c^2)}{a(1-e^2)}\varDelta\varphi_{\text{Einstein}} \,. \tag{3.3.13}$$

From this one sees that $\varDelta\varphi_{\text{Quad}}$ and $\varDelta\varphi_{\text{Einstein}}$ depend differently on $a$ and on $e$.

For the case of Mercury, we have numerically (in arc seconds per century)

$$\varDelta\varphi_{\text{Einstein}} + \varDelta\varphi_{\text{Quad}} = 42.98 + 0.013\,(J_2/10^{-7}) \,. \tag{3.3.14}$$

The work of *Hill* et al. [121] shows that $J_2 < 0.5 \times 10^{-5}$. Thus the quadrupole moment of the Sun seems to be small enough not to spoil the beautiful agreement between theory and experiment. For a uniformly rotating Sun with the observed surface angular velocity, one would estimate $J_2 \sim 10^{-7}$. Satellite experiments are planned by NASA with highly eccentric orbits close to the Sun, which will make a more precise measurement of the solar quadrupole moment possible.

For recent developments which led to the result

$$J_2 = (-1.4 \pm 1.5) \times 10^{-6}$$

see the note added in Sect. 3.5.

## 3.4 Bending of Light Rays

For light rays, Eq. (3.2.2) is replaced by $\mathscr{L} = 0$. Instead of (3.2.6), one then obtains

$$(1-2m/r)^{-1}E^2 - (1-2m/r)^{-1}\dot{r}^2 - L^2/r^2 = 0 \,, \tag{3.4.1}$$

and the equation above (3.2.10) becomes

$$L^2 u'^2 = E^2 - (1 - 2m\,u)\,L^2 u^2 .$$

The orbit equation for light rays is thus

$$u'^2 + u^2 = E^2/L^2 + 2m\,u^3 . \tag{3.4.1'}$$

Differentiation results in

$$u'' + u = 3m\,u^2. \tag{3.4.2}$$

[Compare this with (3.2.11).] The right hand side is very small:

$$3m\,u^2/u = 3R_s/2r \lesssim R_s/R_\odot \sim 10^{-6} .$$

If we neglect this, we have (see Fig. 3.1)

$$u = b^{-1}\sin\varphi . \tag{3.4.3}$$

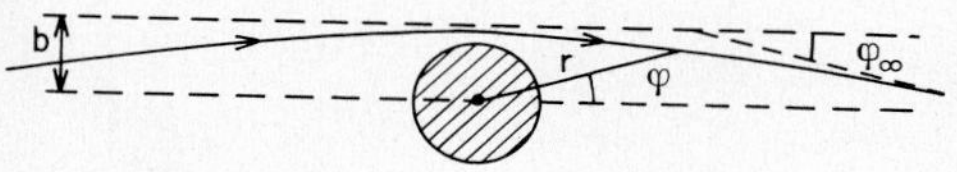


**Fig. 3.1**

Inserting this into the right hand side of (3.4.2) gives

$$u'' + u = 3m\,b^{-2}(1 - \cos^2\varphi) , \tag{3.4.4}$$

which has the particular solution

$$u_1 = 3m\,b^{-2}(1 + \tfrac{1}{3}\cos 2\varphi)/2 .$$

In second order we thus have

$$u = b^{-1}\sin\varphi + \frac{3m}{2b^2}(1 + \tfrac{1}{3}\cos 2\varphi) . \tag{3.4.5}$$

For large $r$ (small $u$), $\varphi$ is small, and we may take $\sin\varphi \simeq \varphi$ and $\cos\varphi \simeq 1$. In the limit $u \to 0$, $\varphi$ approaches $\varphi_\infty$, with

$$\varphi_\infty = -2m/b .$$

The total deflection $\delta$ is equal to $2\,|\varphi_\infty|$; thus

$$\delta = 4m/b = 4GM/c^2 b = 2R_s/b . \tag{3.4.6}$$

For the sun we have

$$\delta = 1.75''\,R_\odot/b . \tag{3.4.7}$$

One also obtains this result from the linearized theory (see Chap. 4). Thus, the bending of light rays does not depend on the nonlinearity of the theory; by contrast, the precession of the perihelion does depend on this, as is obvious from (3.3.7).

Historically, this prediction of Einstein was first tested during the solar eclipse of March 29, 1919. Eddington and Dyson organized two expeditions, one to the Brazilian city Sobral, and one to the island Principe in Portugese Africa.

The effect of the deflection of light is observed as an apparent outward shift in the position of the stars during the eclipse (see Fig. 3.2). This shift can be determined by photographing the stars in the vicinity of the sun during the eclipse and later the same stars at night. Then one compares the two photographs.

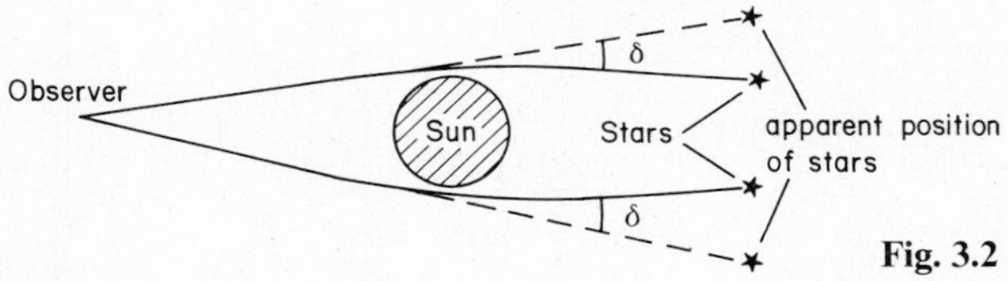


**Fig. 3.2**

The results of the first two expeditions were

$$\delta = \begin{cases} 1.98 \pm 0.16'' & \text{(Sobral)} \\ 1.61 \pm 0.40'' & \text{(Principe)} . \end{cases}$$

This result made headlines in most newspapers. On October 11, 1919, the participants in the physics colloquium in Zürich (Debye, Weyl, ...) sent Einstein a postcard with the verse:

„Alle Zweifel sind entschwunden,
Endlich ist es nun gefunden:
Das Licht, das läuft natürlich krumm
Zu Einsteins allergrösstem Ruhm!"

Since then, numerous measurements of the bending of light have been performed during solar eclipses. The results show considerable scatter.

Since 1969, substantial improvements have been made using radio astronomy. Every year the quasar 3C 279 is eclipsed by the sun on October 8, and thus the deflection of radio waves emitted from this quasar, relative to those from the quasar 3C 273, which is 10° away, can be measured. Similar measurements can be performed on the group 0111+02, 0119+11 and 0116+08.

The increasing precision which has been attained is shown in Fig. 3.3. Here $\frac{1}{2}(1+\gamma)$ is the ratio of the observed deflection (after all corrections) to Einstein's prediction. The most accurate results to date are due to *Fomalont* and *Sramek* [26]. Before a comparison with GR can be made, it is necessary to correct for the additional deflection caused by the solar corona. For radio waves, this deflection depends on frequency as $\omega^{-2}$ (see the following exercise). By observing at two frequencies (2695 and 8085 MHz), the contributions due to the solar corona and the Earth's ionosphere can be determined very precisely.

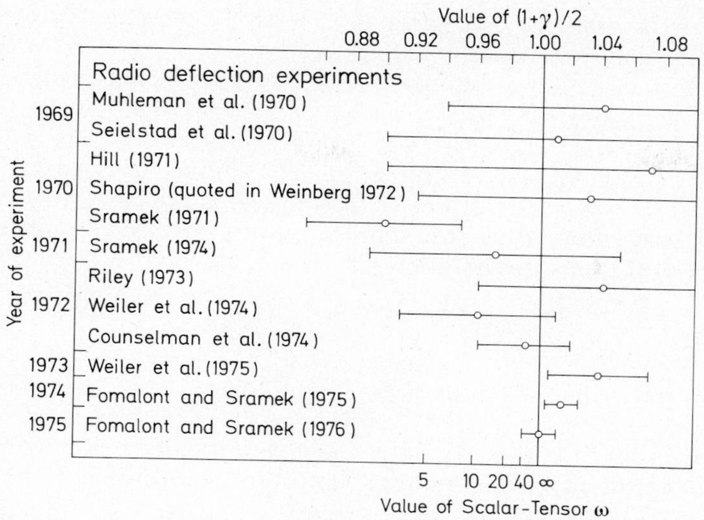


**Fig. 3.3**

Long baseline and very long baseline (VLBI) interferometric techniques have the capability in principle of measuring angular separations and changes in angles as small as $3 \times 10^{-4}$ seconds of arc. This will permit considerably higher accuracy in future measurements.

---

**Exercise** (Deflection by the solar corona): Compute the deflection of radio waves by the solar corona for the following electron density (for $r/R_\odot > 2.5$):

$$n_e(r) = \frac{A}{(r/R_\odot)^6} + \frac{B}{(r/R_\odot)^2} \qquad \begin{aligned} A &= 10^8\ \text{electrons/cm}^3 \\ B &= 10^6\ \text{electrons/cm}^3\,, \end{aligned} \tag{3.4.8}$$

which should be fairly realistic.

*Hint:* The dispersion relation for transverse waves in a plasma (see, e.g. *J. D. Jackson,* Classical Electrodynamics, second edition, Sect. 10.8) is given by

$$\omega^2 = c^2 k^2 + \omega_p^2 . \tag{3.4.9}$$

Here, $\omega_p$ is the plasma frequency, which is related to $n_e$ by

$$\omega_p^2 = \frac{4\pi n_e e^2}{m} . \tag{3.4.10}$$

The corresponding index of refraction is

$$n = c\,k/\omega = \sqrt{1 - \omega_p^2/\omega^2} . \tag{3.4.11}$$

Let $s$ denote the arc length of the light ray $\boldsymbol{x}(s)$. From any optics text one finds that $\boldsymbol{x}(s)$ satisfies the differential equation

$$\frac{d}{ds}\left(n\,\frac{d\boldsymbol{x}}{ds}\right) = \boldsymbol{\nabla} n . \tag{3.4.12}$$

Since the deflection is small, we may calculate it to sufficient accuracy by integrating (3.4.12) along an unperturbed trajectory $y=b$, $-\infty<x<\infty$. Since $n=1$ for $r\to\infty$, the difference in the direction of the asymptotes is given by

$$\left.\frac{d\boldsymbol{x}}{ds}\right|_{\infty}-\left.\frac{d\boldsymbol{x}}{ds}\right|_{-\infty}\simeq\int_{-\infty}^{\infty}\nabla n\,(x\,\boldsymbol{e}_x+b\,\boldsymbol{e}_y)\,dx\,. \tag{3.4.13}$$

The coronal scattering angle $\delta_c$ is (for small $\delta_c$) equal to the left hand side of (3.4.13) multiplied by $\boldsymbol{e}_y$. Since $\nabla n=n'(r)\,\boldsymbol{x}/r$, we have

$$\delta_c\simeq\int_{-\infty}^{+\infty}n'(\sqrt{x^2+b^2})\,\frac{b}{\sqrt{x^2+b^2}}\,dx\,. \tag{3.4.14}$$

Compute this integral using the saddle point method.

---

It has become usual to parametrize the static, spherically symmetric vacuum field independent of theory. In "isotropic" coordinates, one then has, in place of (3.1.13) and (3.1.14)

$$g=\left[1-2\frac{m}{r}+2\beta\left(\frac{m}{r}\right)^2+\ldots\right]dt^2-\left(1+2\gamma\frac{m}{r}+\ldots\right)d\boldsymbol{x}^2\,. \tag{3.4.15}$$

$\beta$ and $\gamma$ are the so-called *Eddington-Robertson parameters.* According to (3.1.14), we have in GR

$$\beta=\gamma=1\,. \tag{3.4.16}$$

For a systematic presentation of the so-called Parametrized Post-Newtonian (PPN) formalism and an analysis of solar system tests within this framework, we refer to [67].

---

**Exercise:** Derive the following generalizations for the deflection of light rays and the advance of the perihelion from the metric (3.4.15):

$$\delta=\tfrac{1}{2}(1+\gamma)\,\delta_{GR} \tag{3.4.17}$$

$$\Delta\varphi=(2-\beta+2\gamma)/3\cdot\Delta\varphi_{GR}\,. \tag{3.4.18}$$

*Hint:* Use Fermat's principle to derive (3.4.17) (see Sect. 1.7).

---

From (3.4.18) it is clear that the precession of the perihelion is sensitive to nonlinearities in the theory (the parameter $\beta$).

## 3.5 Time Delay of Radar Echoes

A relatively new test of GR consists of the determination of the time delay of radar signals which are transmitted from the Earth through a region near the Sun to another planet or satellite and then reflected back to the Earth. We now compute the time delay for such a circuit.

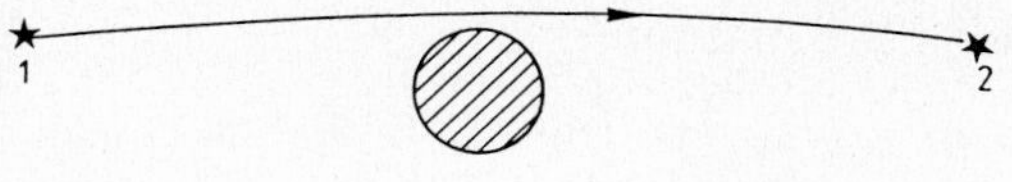


**Fig. 3.4**

Suppose that a radar signal is transmitted from a point $(r_1,\ \vartheta = \frac{\pi}{2},\ \varphi_1)$ to another point $(r_2,\ \vartheta = \frac{\pi}{2},\ \varphi_2)$, as indicated in Fig. 3.4.

We first compute the *coordinate time* $t_{12}$ required. From (3.4.1) we have

$$\dot{r}^2 = E^2 - (1 - 2m/r)\, L^2/r^2 . \tag{3.5.1}$$

Using (3.2.5) we can write

$$\dot{r} = (dr/dt)\, \dot{t} = \frac{dr}{dt} \frac{E}{1 - 2m/r} .$$

Inserting this into (3.5.1) gives

$$(1 - 2m/r)^{-3}\, (dr/dt)^2 = (1 - 2m/r)^{-1} - \left(\frac{L}{E}\right)^2 \frac{1}{r^2} . \tag{3.5.2}$$

At the distance of closest approach to the Sun at $r = r_0$, $dr/dt$ vanishes, so that

$$(L/E)^2 = r_0^2/(1 - 2m/r_0) .$$

If we insert this in (3.5.2), we obtain

$$(1 - 2m/r)^{-3} \left(\frac{dr}{dt}\right)^2 + \left(\frac{r_0}{r}\right)^2 \frac{1}{1 - 2m/r_0} - \frac{1}{1 - 2m/r} = 0 . \tag{3.5.3}$$

According to (3.5.3), the coordinate time which the light requires to go from $r_0$ to $r$ (or the reverse) is

$$t\,(r, r_0) = \int_{r_0}^{r} \frac{dr}{1 - 2m/r} \frac{1}{\sqrt{1 - \dfrac{1 - 2m/r}{1 - 2m/r_0} \left(\dfrac{r_0}{r}\right)^2}} . \tag{3.5.4}$$

Hence (for $|\varphi_1 - \varphi_2| > \pi/2$),

$$t_{12} = t\,(r_1, r_0) + t\,(r_2, r_0) . \tag{3.5.5}$$

We shall treat the quantity $2m/r$ appearing in the integrand of (3.5.4) as small, thus obtaining

$$t\,(r, r_0) \simeq \int_{r_0}^{r} \left\{ (1 + 2m/r) \left[ 1 - \left( 1 + \frac{2m}{r_0} - \frac{2m}{r} \right) \left( \frac{r_0}{r} \right)^2 \right]^{-1/2} \right\} dr .$$

The expression in the square bracket is equal to

$$1-\left(\frac{r_0}{r}\right)^2 - 2m\left(\frac{1}{r_0}-\frac{1}{r}\right)\left(\frac{r_0}{r}\right)^2 = \left[1-\left(\frac{r_0}{r}\right)^2\right]\left(1-\frac{2m\left(\frac{1}{r_0}-\frac{1}{r}\right)r_0^2}{r^2\left[1-\left(\frac{r_0}{r}\right)^2\right]}\right)$$

$$= \left[1-\left(\frac{r_0}{r}\right)^2\right]\left(1-\frac{2m\, r_0}{r\,(r+r_0)}\right).$$

Hence

$$t\,(r, r_0) \simeq \int_{r_0}^{r}\left(1-\frac{r_0^2}{r^2}\right)^{-1/2}\left(1+\frac{2m}{r}+\frac{m\, r_0}{r\,(r+r_0)}\right)dr\,.$$

This integral is elementary, and is equal to

$$t\,(r, r_0) \simeq \sqrt{r^2-r_0^2} + 2m\ln\left(\frac{r+\sqrt{r^2-r_0^2}}{r_0}\right) + m\left(\frac{r-r_0}{r+r_0}\right)^{1/2}. \tag{3.5.6}$$

For the circuit from point 1 to point 2 and back, it is clear that one should introduce the following quantity as the *delay in the coordinate time:*

$$\begin{aligned}\Delta t &:= 2\left[t\,(r_1, r_0) + t\,(r_2, r_0) - \sqrt{r_1^2-r_0^2} - \sqrt{r_2^2-r_0^2}\right]\\ &= 4m\ln\left(\frac{(r_1+\sqrt{r_1^2-r_0^2})\,(r_2+\sqrt{r_2^2-r_0^2})}{r_0^2}\right)\\ &\quad + 2m\left(\sqrt{\frac{r_1-r_2}{r_1+r_0}}+\sqrt{\frac{r_2-r_0}{r_2+r_0}}\right).\end{aligned} \tag{3.5.7}$$

This time is not observable, but it gives an idea of the magnitude and the behavior of the effects of general relativity.

For a round trip from the Earth to Mars and back, we have ($r_0 \ll r_1, r_2$):

$$\Delta t \cong 4m\left(\ln\frac{4r_1 r_2}{r_0^2}+1\right) \tag{3.5.8}$$

$$(\Delta t)_{\max} \simeq 72\ \text{km} \simeq 240\ \mu\text{s}\,. \tag{3.5.9}$$

In order to give an idea of the experimental possibilities, we mention that the error in the time measurement of a circuit during the Viking mission was only about 10 ns.

As we have mentioned, the coordinate time delay (3.5.7) cannot be directly measured. For this, it would be necessary to know $(r_1^2-r_0^2)^{1/2} + (r_2^2-r_0^2)^{1/2}$ extremely precisely (to better than 1 km for a 1% measurement). The various radial coordinates in general use [see, e.g. (3.1.11)] already differ by an amount of the order of $GM_\odot/c^2 \simeq 1.5$ km. For this reason, the theoretical analysis and practical evaluation of the data from a time delay experiment is extremely complicated. One must carry out the measurements of the circuit times over a period of several months and fit the data to a complicated model which describes the motion of the transmitter and receiver very precisely. The post-Newtonian corrections to the equations of motion must be taken into account. In every reasonably realistic model, there is a set of parameters which are not known well enough. It is necessary to include all of these "uninteresting" parameters, along with the relativistic parameter $\gamma$, in a simultaneous fit.

The first ("passive") experiments were performed with radar echoes from Mercury and Venus. The rather complicated topography of these planets limited the precision attainable in these experiments. In addition, the reflected signals were very weak.

The ("active") radio observations from satellites, in which the reflected signals are shifted in frequency by a transponder, suffer from the difficulty that the satellites are subject to strong, nongravitational perturbations, such as radiation pressure, loss of gas, etc.

In all the experiments, time delays due to the solar corona provide an additional source of error.

In spite of all these difficulties, the Viking mission (see [27]) provided a very precise determination of $\gamma$:

$$\gamma = 1 \pm 0.002\,. \tag{3.5.10}$$

An important factor in the attainment of such amazing precision is the fact that during the Viking mission two transponders landed on Mars, while two others continued to orbit the planet. The orbiting transponders transmitted in both the S-band and the X-band (unfortunately, the two "landers" could only transmit at S-band frequencies). It was thus possible to determine the rather large time delays (up to 100 μs) due to the solar corona fairly precisely (since these are proportional to $\nu^{-2}$). Since two of the transponders are fixed on Mars, the determination of the orbits is more precise than was possible with Mariner 6, 7, and 9. One also discovered that the corona varies considerably with time, and hence that the time delays cannot be determined sufficiently precisely with a parametrized theoretical model.

---

**Exercise:** Using the electron density (3.4.8), compute the time delay caused by the solar corona.

*Hint:* The group velocity corresponding to the dispersion relation $\omega = (c^2 k^2 + \omega_p^2)^{1/2}$ is

$$v_g = \partial\omega/\partial k = \frac{k\,c^2}{\sqrt{k^2 c^2 + \omega_p^2}} \simeq (1 - \tfrac{1}{2}\,\omega_p^2/\omega^2)\,c\,.$$

For the computation of the time delay, one may use the unperturbed straight line path for the light ray. This can be seen with Fermat's principle, according to which the actual path of the ray minimizes the transit time (hence the error is second order in the perturbation).

---

*Note Added*

Recently, R. Hellings reported [122] new analysis of radar-ranging data that have led to the following remarkable improvements:

$$\begin{aligned} \beta-1 &= (-2.9 \pm 3.1) \times 10^{-3} \\ \gamma-1 &= (-0.7 \pm 1.7) \times 10^{-3} \\ \alpha_1 &= (0.21 \pm 0.19) \times 10^{-3} . \end{aligned} \tag{3.5.11}$$

(For the parameter $\alpha_1$, see [14], Chapt. 39.) In the same fit the quadrupole moment of the Sun was determined to be

$$J_2 = (-1.4 \pm 1.5) \times 10^{-6} , \tag{3.5.12}$$

which is inconsistent with [123], and the rate of change of the gravitational constant turns out to be bounded by

$$\dot{G}/G = (0.2 \pm 0.4) \times 10^{-11}\,\mathrm{yr}^{-1} . \tag{3.5.13}$$

Unfortunately, no data of comparable quality can be expected for many years as the Mars Viking lander is now dead.

## 3.6 Geodetic Precession

Suppose that a gyroscope with spin $S$ moves along a geodesic. Thus, $S$ is parallel transported (see Sect. 1.10.1). As a special case, we consider circular motion in the plane $\vartheta = \frac{\pi}{2}$ in the Schwarzschild field and compute the corresponding spin precession. We shall consider the precession of the spin for general motion in the post-Newtonian approximation again in Chap. 5.

In the following, we calculate relative to the basis (3.1.2) and its dual basis, which we denote as usual by $e_\alpha$. The four-velocity $u$ and $S$ satisfy the equations

$$(S, u) = 0 , \qquad \nabla_u S = 0 \qquad \nabla_u u = 0 . \tag{3.6.1}$$

The components of $\nabla_u S$ are given by [see (5.54) in Part I]

$$(\nabla_u S)^\mu = \dot{S}^\mu + \omega^\mu_\beta(u)\, S^\beta = 0 . \tag{3.6.2}$$

For circular motion with $\vartheta = \frac{\pi}{2}$, we obviously have $u^1 = u^2 = 0$. If we now use the connection forms (3.1.6), we obtain

$$\dot{S}^0 = -\,\omega^0_\beta(u)\, S^\beta = -\,\omega^0_1(u)\, S^1 = -\,a'\, \mathrm{e}^{-b}\, \theta^0(u)\, S^1 = b'\, \mathrm{e}^{-b}\, u^0\, S^1 .$$

One finds the other equations in a similar manner. The result is

$$\begin{aligned}
\dot S^0 &= b' \,\mathrm{e}^{-b} u^0 S^1\\
\dot S^1 &= b' \,\mathrm{e}^{-b} u^0 S^0 + \frac{1}{r}\,\mathrm{e}^{-b} u^3 S^3\\
\dot S^2 &= 0\\
\dot S^3 &= -\frac{1}{r}\,\mathrm{e}^{-b} u^3 S^1 .
\end{aligned} \tag{3.6.3}$$

For $\nabla_u u = 0$, we obtain in place of the second equation in (3.6.3)

$$0 = b' \,\mathrm{e}^{-b} (u^0)^2 + \frac{1}{r}\,\mathrm{e}^{-b} (u^3)^2 ,$$

which implies that

$$(u^0/u^3)^2 = -\frac{1}{b' r} . \tag{3.6.4}$$

The following normalized vectors are perpendicular to $u$:

$$\bar e_1 = e_1 , \qquad \bar e_2 = e_2 , \qquad \bar e_3 = u^3 e_0 + u^0 e_3 . \tag{3.6.5}$$

When expanded in terms of these, $S$ has the form

$$S = \bar S^i \bar e_i .$$

Obviously

$$S^0 = u^3 \bar S^3 , \qquad S^1 = \bar S^1 , \qquad S^2 = \bar S^2 , \qquad S^3 = u^0 \bar S^3 . \tag{3.6.6}$$

We now rewrite (3.6.3) in terms of the $\bar S^i$, with the result (we now drop the bars)

$$\dot S^1 = -b' \,\mathrm{e}^{-b} \frac{u^0}{u^3} S^3 , \qquad \dot S^2 = 0 , \qquad \dot S^3 = b' \,\mathrm{e}^{-b} \frac{u^0}{u^3} S^1 . \tag{3.6.7}$$

In order to obtain (3.6.7), we used (3.6.4) and $(u^0)^2 - (u^3)^2 = 1$.

We now replace the $\dot S^i$ in (3.6.7) by derivatives with respect to coordinate time. Note that (3.1.2) implies $u^0 = \mathrm{e}^{-b} \dot t$; hence $dS^i/dt = \dot S^i/\dot t = \dot S^i \,\mathrm{e}^{-b}/u^0$, and

$$\frac{dS^1}{dt} = -\frac{b'}{u^3}\,\mathrm{e}^{-2b} S^3 , \qquad \frac{dS^2}{dt} = 0 , \qquad \frac{dS^3}{dt} = \frac{b'}{u^3}\,\mathrm{e}^{-2b} S^1 . \tag{3.6.8}$$

From (3.1.2) we have $u^3 = r\dot\varphi$, and hence $\omega := d\varphi/dt = \dot\varphi/\dot t = u^3 \,\mathrm{e}^{-b}/u^0 r$ or, together with (3.6.4),

$$\omega^2 = \left(\frac{u^3}{u^0}\right)^2 \frac{1}{r^2}\,\mathrm{e}^{-2b} = -\frac{1}{r} b' \,\mathrm{e}^{-2b} = \frac{1}{2r} (\mathrm{e}^{-2b})' .$$

However

$$\mathrm{e}^{-2b} = 1 - 2m/r \tag{3.6.9}$$

and hence

$$\omega^2 = m/r^3 . \tag{3.6.10}$$

This is Kepler's third law.

The precession frequency $\Omega$ in (3.6.8) satisfies

$$\begin{aligned}\Omega^2 &= \frac{(b')^2\,\mathrm{e}^{-4b}}{(u^3)^2} = (b')^2\,\mathrm{e}^{-4b}\left[-1-\frac{1}{b'\,r}\right] = -\,b'\left(b'+\frac{1}{r}\right)\mathrm{e}^{-4b}\\ &= \omega^2\,\mathrm{e}^{-2b}\,(r\,b'+1) = \omega^2\,\mathrm{e}^{-2b}\left[-\frac{m}{r}\,\frac{1}{1-2m/r}+1\right]\\ &= \omega^2\,\mathrm{e}^{-2b}\,\frac{1-3m/r}{1-2m/r},\end{aligned}$$

which means that $\Omega^2 = e^2\,\omega^2$, with

$$e^2 = 1 - 3m/r . \tag{3.6.11}$$

We write (3.6.8) in the form

$$\frac{d}{dt}\boldsymbol{S} = \boldsymbol{\Omega}\times\boldsymbol{S}, \qquad \boldsymbol{\Omega} = (0,\, e\,\omega,\, 0) . \tag{3.6.12}$$

In the Newtonian limit $\boldsymbol{\Omega} = (0, \omega, 0)$. We thus see that in the three-dimensional space oriented toward the center and perpendicular to the direction of motion, $\boldsymbol{S}$ precesses retrograde about an axis perpendicular to the plane of the orbit, with frequency $e\,\omega < \omega$. After one complete orbit, the projection of $\boldsymbol{S}$ onto the plane of the orbit has advanced by an angle

$$2\pi(1-e) =: \omega_s\,\frac{2\pi}{\omega}, \tag{3.6.13}$$

(the reader should convince himself of the sign).

The precession frequency $\omega_s = \omega\,(1-e)$ is given by

$$\omega_s = (m/r^3)^{1/2}\,[1-(3m/r)^{1/2}] \simeq (GM/r^3)^{1/2}\,\frac{3GM}{2r}$$

or

$$\omega_s \simeq \frac{3}{2}\,\frac{(GM)^{3/2}}{r^{5/2}} . \tag{3.6.14}$$

**Remark:** The square root in the expression for $\omega_s$ is always well defined, since there are no circular orbits for $r < 3m$ (prove this).

Taking the Earth as the central mass, one obtains

$$\omega_s \simeq 8.4\left(\frac{R_\oplus}{r}\right)^{5/2} \text{arcs/yr} . \tag{3.6.15}$$

This will probably be measurable in the near future. It is planned to launch a satellite containing a system of gyroscopes. In addition to the geodetic precession (3.6.15), there is a further small contribution from GR which is due to the Earth's rotation. This will be discussed in Chap. 5 (see also Sect. 1.10.4).

**Remarks:**

1) The Earth is a natural gyroscope. After one "sidereal year" ($\Delta\varphi = 2\pi$) the projection of the Earth's axis on the ecliptic has advanced by the angle

$$2\pi(1-e) = 0.019'' .$$

The sideral year is characterized by the return of the Sun to the same location with respect to the fixed stars. The pole axis has then moved with respect to the fixed stars. However, this effect is masked by other perturbations of the Earth's axis.

2) Another natural gyroscope is the binary pulsar 1913+16. As we shall see in Chap. 5, its general relativistic precession amounts to about 1° per year!

## 3.7 Gravitational Collapse and Black Holes (Part 1)

In Chap. 6 we shall show that, in spite of the uncertainties in the equation of state, the mass of nonrotating, cold (neutron) stars is bounded by a few solar masses. Here we mention only a few important results (see also [28]).

Let $\varrho$ denote the mass-energy density, and assume that the equation of state is known for $\varrho \leqq \varrho^*$, where $\varrho^*$ is not significantly larger than the nuclear matter density $\varrho_0 = 2.8 \times 10^{14}\,\mathrm{g/cm^3}$. The mass of a nonrotating neutron star is bounded by

$$M/M_\odot \lesssim 6.75\,(\varrho_0/\varrho^*)^{1/2} , \tag{3.7.1}$$

assuming only that the equation of state above $\varrho^*$ obeys $\varrho > 0$ and $\partial p/\partial\varrho > 0$.

If in addition, one assumes that the speed of sound is less than the speed of light, so that $\partial p/\partial\varrho < c^2$, the bound (3.7.1) can be tightened:

$$M/M_\odot \leqq 4.0\,(\varrho_0/\varrho^*)^{1/2} . \tag{3.7.2}$$

"Realistic" equations of state give typical values of $M_{\max}$ around $2\,M_\odot$.

Of course, it is possible for a supercritical mass to exist in a temporary equilibrium, if pressure is built up by means of thermonuclear processes. Sooner or later, however, the nuclear energy sources will be exhausted, and thus a permanent equilibrium is not possible, unless sufficient mass can be "blown off" (for example in a supernova explosion). Otherwise, the star will collapse to a black hole.

In this section, we shall consider spherically symmetric collapse and spherically symmetric black holes. Many of the qualitatively important properties reveal themselves even in this simple situation. Spherically symmetric collapse is particularly simple, since we already know the gravitational field outside of the star. In this region, it depends only on the total mass of the collapsing object, and not on any dynamical details of the implosion, as we have already shown for $r > 2m$ in Sect. 3.1 (Birkhoff theorem). In the following section, we shall extend this result to the case $r \leqq 2m$.

### 3.7.1 The Kruskal Continuation of the Schwarzschild Solution

The Schwarzschild solution

$$g = (1 - 2m/r)\, dt^2 - \left[\frac{dr^2}{1 - 2m/r} + r^2\,(d\vartheta^2 + \sin^2\vartheta\, d\varphi^2)\right] \tag{3.7.3}$$

has an apparant singularity at $r = 2m$. We shall see that (3.7.3) becomes singular at $r = 2m$ only because the chosen coordinate system loses its applicability[2] at $r = 2m$. A first hint of this behavior comes from the observation that, with respect to the orthonormal basis (3.1.2), the Riemann tensor (3.1.7) is finite at $r = 2m$. A typical component is

$$R^1_{212} = R^1_{313} = e^{-2b}\frac{b'}{r} \sim \frac{1}{r^3}\,.$$

Hence at $r = 2m$, the tidal forces remain finite.

Before we continue the Schwarzschild solution beyond $r = 2m$, let us consider timelike radial geodesics in the vicinity of the horizon. For $L = 0$, we obtain from (3.2.7) and (3.2.8) the equation of motion

$$\dot{r}^2 = 2m/r + E^2 - 1\,. \tag{3.7.4}$$

Suppose that a radially falling particle is at rest at $r = R$, so that $2m/R = 1 - E^2$ $(E < 1)$. From (3.7.4) we have

$$d\tau = \left(\frac{2m}{r} - \frac{2m}{R}\right)^{-1/2} dr\,. \tag{3.7.5}$$

---

[2] It is easy to construct coordinate singularities. For example, consider in $\mathbb{R}^2$ the metric $ds^2 = dx^2 + dy^2$ and introduce in place of $x$ the coordinate $\xi = x^3/3$. The transformed metric is given by

$$ds^2 = (3\,\xi)^{-4/3}\, d\xi^2 + dy^2\,,$$

which is "singular" at $\xi = 0$.

One can easily find the parametric representation (cycloid)

$$r = \frac{R}{2}(1+\cos\eta) \tag{3.7.6}$$

$$\tau = \left(\frac{R^3}{8m}\right)^{1/2}(\eta + \sin\eta)\,. \tag{3.7.7}$$

Nothing in particular happens at $r = 2m$. For $\eta = 0$, we have $r = R$ and $\tau = 0$. The proper time for free fall to $r = 0$ $(\eta = \pi)$ is $\tau = (\pi/2)\,(R^3/2m)^{1/2}$.

On the other hand, consider $r$ as a function of the coordinate time $t$. Together with (3.2.5) we obtain

$$\dot r = \frac{dr}{dt}\dot t = \frac{dr}{dt}\frac{E}{1-2m/r}\,. \tag{3.7.8}$$

Let us use the radial coordinate

$$r^* = r + 2m\ln\left(\frac{r}{2m} - 1\right). \tag{3.7.9}$$

Since

$$\frac{dr^*}{dt} = \frac{1}{1-2m/r}\frac{dr}{dt},$$

it follows from (3.7.8) that

$$\dot r = E\frac{dr^*}{dt}\,. \tag{3.7.10}$$

If we now insert this in (3.7.4), we obtain

$$\left(E\frac{dr^*}{dt}\right)^2 = E^2 - 1 + \frac{2m}{r}\,. \tag{3.7.11}$$

We now determine an approximate solution in the vicinity of $r = 2m$. For $r \downarrow 2m$, we have $r^* \to -\infty$ and the right hand side of (3.7.11) approaches $E^2$. Hence, for $r \simeq 2m$,

$$dr^*/dt \simeq -1$$

$$r^* = 2m + 2m\ln(r/2m - 1) \simeq -t + \text{const}\,.$$

Thus, we obtain

$$r \simeq 2m + \text{const}\; e^{-t/2m}\,. \tag{3.7.12}$$

From this it is clear that one arrives at the Schwarzschild radius only after infinitely long coordinate time.

It is also possible to express $t$ in terms of the parameter $\eta$. From (3.2.5), (3.7.6) and (3.7.7), we have

$$dt = \frac{E\,d\tau}{1-2m/r} = E\frac{d\tau}{d\eta}\frac{1}{1-2m/r}\,d\eta$$
$$= (1-2m/R)^{1/2}\left(\frac{R^3}{8m}\right)^{1/2}\frac{1+\cos\eta}{1-4m\,[R\,(1+\cos\eta)]^{-1}}\,d\eta\,.$$

With the help of integral tables, one finds

$$t/2m = \ln\left|\frac{(R/2m-1)^{1/2}+\tan\eta/2}{(R/2m-1)^{1/2}-\tan\eta/2}\right| + \left(\frac{R}{2m}-1\right)^{1/2}\left[\eta+\frac{R}{4m}(\eta+\sin\eta)\right]. \tag{3.7.13}$$

Here, the constant of integration was chosen such that $t=0$ for $\eta=0$ ($r=R$). For $\tan\eta/2 \to (R/2m-1)^{1/2}$, we have $r\to 2m$, $t\to\infty$ (see Fig. 3.5). Next we consider radial lightlike directions for the metric (3.7.3) (for $r>2m$ and $r<2m$). For these we have $ds=0$ and

$$dr/dt = \pm\,(1-2m/r)\,.$$

As $r\downarrow 2m$, the opening angle of the light cone becomes increasingly narrow, as in Fig. 3.6.

The previous discussion indicates that the use of the coordinates $r$ and $t$ is limited. By using the proper time, it is possible to describe events which only occur after $t=\infty$.

When $r>2m$, $t$ is a distinguished timelike coordinate. The Schwarzschild solution is *static* and the corresponding Killing field $K$ is just $K=\partial/\partial t$. [From Sect. 1.9, the corresponding 1-form $K^\flat$ is given by $K^\flat=(K,K)\,dt$.] Since the coordinate $t$ is adapted to the Killing field $K$, it is uniquely defined, up to an additive constant.

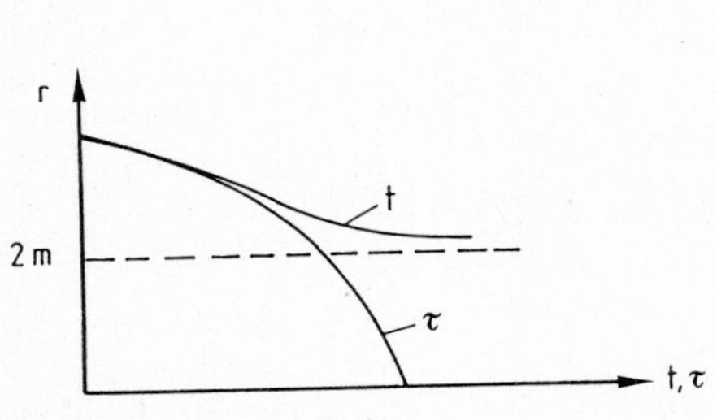


**Fig. 3.5**

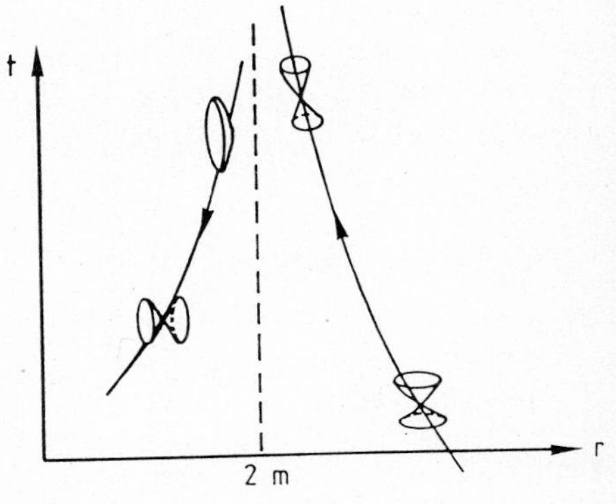


**Fig. 3.6**

**Exercise:** Show that for $r > 2m$, only one timelike hypersurface orthogonal Killing field exists.

We now wish to continue the Schwarzschild manifold $(r > 2m)$ in such a way that the Einstein vacuum field equations are always satisfied. This was done most simply by *Kruskal* in 1960.

We wish to avoid having the light cones contract in a singular manner for $r \downarrow 2m$. Hence we transform to new coordinates $(u, v)$, for which the metric has the following form

$$g = f^2(u, v)(dv^2 - du^2) - r^2(d\vartheta^2 + \sin^2\vartheta\, d\varphi^2)\,. \tag{3.7.14}$$

For radially emitted light rays we then have $(du/dv)^2 = 1$ when $f^2 \neq 0$. The metrics of the two-dimensional submanifolds ($\vartheta$ = const, $\varphi$ = const) are then conformally equivalent to the two-dimensional Minkowski metric $dv^2 - du^2$.

We can easily write down the differential equations for the desired transformation ($\vartheta$ and $\varphi$ are unchanged).

From

$$g_{\alpha\beta} = \frac{\partial x'^{\mu}}{\partial x^{\alpha}} \frac{\partial x'^{\nu}}{\partial x^{\beta}} g'_{\mu\nu}$$

it follows that

$$1 - 2m/r = f^2\left[\left(\frac{\partial v}{\partial t}\right)^2 - \left(\frac{\partial u}{\partial t}\right)^2\right]$$

$$-(1 - 2m/r)^{-1} = f^2\left[\left(\frac{\partial v}{\partial r}\right)^2 - \left(\frac{\partial u}{\partial r}\right)^2\right]$$

$$0 = \frac{\partial u}{\partial t}\frac{\partial u}{\partial r} - \frac{\partial v}{\partial t}\frac{\partial v}{\partial r}\,.$$

The signs of $u$ and $v$ are not determined by these equations.

For simplicity we again introduce $r^*$ as in (3.7.9). Furthermore, let

$$F(r^*) := \frac{1 - 2m/r}{f^2(r)}\,.$$

Here we have assumed that it is possible to find a function $f$ which depends only on $r$. The transformations can now be rewritten as follows:

$$\left(\frac{\partial v}{\partial t}\right)^2 - \left(\frac{\partial u}{\partial t}\right)^2 = F(r^*) \tag{3.7.15}$$

$$\left(\frac{\partial v}{\partial r^*}\right)^2 - \left(\frac{\partial u}{\partial r^*}\right)^2 = -F(r^*) \tag{3.7.16}$$

$$\frac{\partial u}{\partial t} \cdot \frac{\partial u}{\partial r^*} = \frac{\partial v}{\partial t}\frac{\partial v}{\partial r^*}\,. \tag{3.7.17}$$

We now construct the linear combinations (3.7.15) + (3.7.16) $\pm 2 \times$ (3.7.17) with the result

$$\left(\frac{\partial v}{\partial t}+\frac{\partial v}{\partial r^*}\right)^2=\left(\frac{\partial u}{\partial t}+\frac{\partial u}{\partial r^*}\right)^2 \tag{3.7.18}$$

$$\left(\frac{\partial v}{\partial t}-\frac{\partial v}{\partial r^*}\right)^2=\left(\frac{\partial u}{\partial t}-\frac{\partial u}{\partial r^*}\right)^2 . \tag{3.7.19}$$

When extracting square roots, we choose the positive roots in the first equation and the negative roots in the second (this avoids vanishing of the Jacobi determinants). We then obtain

$$\frac{\partial v}{\partial t}=\frac{\partial u}{\partial r^*}, \qquad \frac{\partial v}{\partial r^*}=\frac{\partial u}{\partial t} . \tag{3.7.20}$$

This gives

$$\frac{\partial^2 u}{\partial t^2}-\frac{\partial^2 u}{\partial r^{*2}}=0, \qquad \frac{\partial^2 v}{\partial t^2}-\frac{\partial^2 v}{\partial r^{*2}}=0 . \tag{3.7.21}$$

The general solution of these equations is

$$v=h\,(r^*+t)+g\,(r^*-t) \tag{3.7.22}$$

$$u=h\,(r^*+t)-g\,(r^*-t) . \tag{3.7.23}$$

We now insert this into (3.7.15–17). Equation (3.7.17) is satisfied identically, while (3.7.15) and (3.7.16) lead to a further equation:

$$-4h'\,(r^*+t)\,g'\,(r^*-t)=F\,(r^*) . \tag{3.7.24}$$

For the moment assume that $r>2m$ and hence that $F(r^*)>0$. Differentiation of (3.7.24) with respect to $r^*$ (resp. $t$) leads to

$$\frac{F'(r^*)}{F(r^*)}=\frac{h''(r^*+t)}{h'(r^*+t)}+\frac{g''(r^*-t)}{g'(r^*-t)} \tag{3.7.25}$$

$$0=\frac{h''(r^*+t)}{h'(r^*+t)}-\frac{g''(r^*-t)}{g'(r^*-t)} . \tag{3.7.26}$$

Adding these gives

$$[\ln F\,(r^*)]'=2\,(\ln h')'\,(r^*+t) . \tag{3.7.27}$$

We may regard $r^*$ and $y=r^*+t$ as independent variables. In Eq. (3.7.27), both sides must then be equal to the same constant $2\,\eta$. By introducing suitable constants of integration, we then have

$$h\,(y)=\tfrac{1}{2}\,\mathrm{e}^{\eta y}, \qquad F\,(r^*)=\eta^2\,\mathrm{e}^{2\eta r^*} . \tag{3.7.28}$$

From (3.7.26) we also find

$$g\,(y)=-\tfrac{1}{2}\,\mathrm{e}^{\eta y} . \tag{3.7.29}$$

The relative sign of $h$ and $g$ is determined by $F > 0$ and (3.7.24).

Summarizing, we have

$$u = h\,(r^* + t) - g\,(r^* - t) = \tfrac{1}{2}\,e^{\eta(r^*+t)} + \tfrac{1}{2}\,e^{\eta(r^*-t)}$$

$$= e^{\eta r^*} \cosh(\eta t) = \left(\frac{r}{2m} - 1\right)^{2m\eta} e^{\eta r} \cosh(\eta t)\,,$$

which means that

$$u = \left(\frac{r}{2m} - 1\right)^{2m\eta} e^{\eta r} \cosh(\eta t)\,.$$

Similarly

$$v = \left(\frac{r}{2m} - 1\right)^{2m\eta} e^{\eta r} \sinh(\eta t)\,.$$

Furthermore,

$$f^2 = \frac{2m}{\eta^2 r}\left(\frac{r}{2m} - 1\right)^{1-4m\eta} e^{-2\eta r}\,.$$

We now choose $\eta$ such that $f^2 \neq 0$ for $r = 2m$. This requires $\eta = 1/4m$. In this manner, we are led to the *Kruskal transformation:*

$$u = \sqrt{\frac{r}{2m} - 1}\; e^{r/4m} \cosh\left(\frac{t}{4m}\right) \qquad (3.7.30)$$

$$v = \sqrt{\frac{r}{2m} - 1}\; e^{r/4m} \sinh\left(\frac{t}{4m}\right). \qquad (3.7.31)$$

By construction, the metric expressed in terms of these new coordinates has the form (3.7.14) with

$$f^2 = \frac{32\,m^3}{r}\, e^{-r/2m}\,. \qquad (3.7.32)$$

**Discussion:**

1) Up to now we have done nothing but make a change of coordinates. In the $(u, v)$ plane, the Schwarzschild region $r > 2m$ corresponds to the shaded quadrant $u > |v|$ of Fig. 3.7.

The lines $r = \text{const}$ in the $(r, t)$ plane correspond to hyperbolas. This follows from

$$u^2 - v^2 = \left(\frac{r}{2m} - 1\right) e^{r/2m}\,, \qquad \frac{v}{u} = \tanh\left(\frac{t}{4m}\right). \qquad (3.7.33)$$

In the limit $r \to 2m$ these approach the lines at 45° in Fig. 3.7. According to (3.7.33), the lines $t = \text{const}$ correspond to radial lines through the origin, as shown in Fig. 3.8.

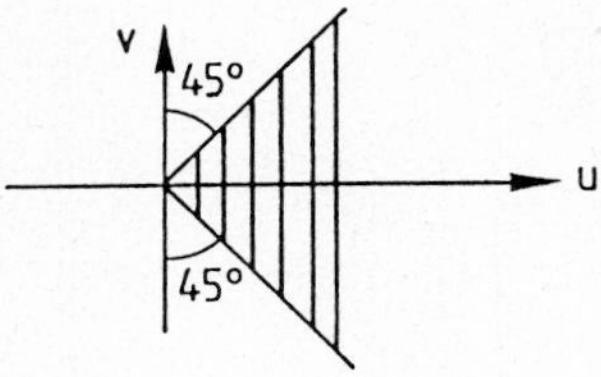


Fig. 3.7

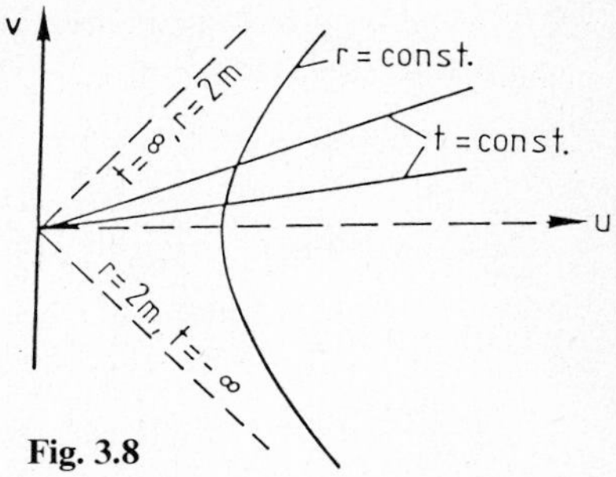


Fig. 3.8

2) However, the transformed metric

$$g = f^2(u, v)(dv^2 - du^2) - r^2(u, v)\, d\Omega^2, \tag{3.7.34}$$

in which $r$ is implicitly defined as a function of $u$ and $v$ by (3.7.33), is regular in a larger region than the quadrant $u > |v|$. The region in which (3.7.34) is nonsingular is bounded by the hyperbolas $v^2 - u^2 = 1$, where, from (3.7.33) we have $r = 0$ and hence, by (3.7.32), $f^2$ becomes singular. When $v^2 - u^2 < 1$, $r$ is a uniquely defined function of $u$ and $v$, since the right hand side of the first equation in (3.7.33) is a monotonic function when $r > 0$; its derivative with respect to $r$ is proportional to $r\,e^{r/2m} > 0$ for $r > 0$. If we write (3.7.33) as $u^2 - v^2 = g(r/2m)$, we have the graph of $g$ as shown in Fig. 3.9. We have thus isometrically embedded the original Schwarzschild manifold in a larger Lorentz manifold, which is known as the *Schwarzschild-Kruskal* manifold. The Ricci tensor vanishes on the extended manifold. This is a result of the next two remarks.

3) In the derivation of the Kruskal transformation, we arbitrarily chose the sign of $h$ to be positive (and hence $g$ to be negative). We could just as well have chosen the reverse. This would be equivalent to the transformation $(u, v) \to (-u, -v)$. Thus the regions I and III in Fig. 3.10 are isometric.

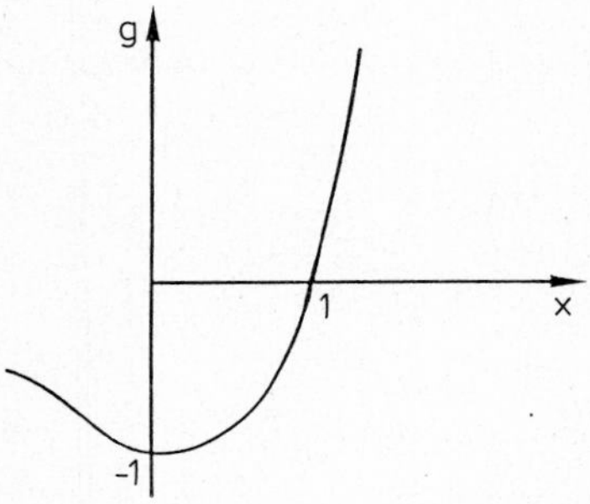


Fig. 3.9

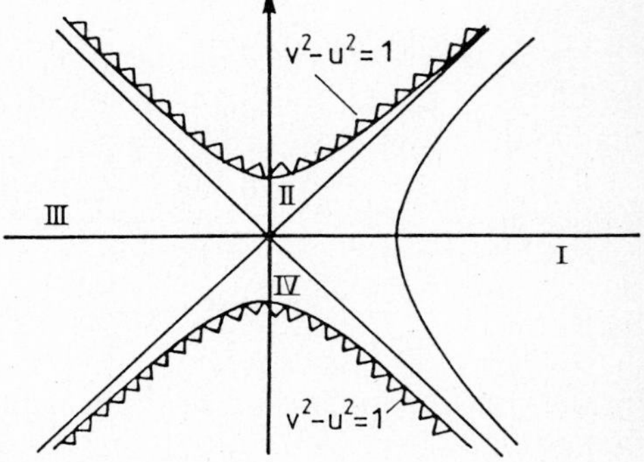


Fig. 3.10

4) We now consider the original Schwarzschild solution (3.7.3) for $r<2m$. This also satisfies the vacuum equations, as is easily shown by direct computation (see also the appendix to this chapter). In this region, $r$ behaves like a time coordinate and $t$ like a space coordinate. This manifold can be mapped isometrically onto region II of Fig. 3.10. In order to see this, consider the derivation of the Kruskal transformation again. For $0<r<2m$, we now have $F(r^*)<0$, and hence the relative sign of $g$ and $h$ must be positive. If we choose both to be positive, we obtain

$$\begin{aligned} u &= \sqrt{1-r/2m}\, \mathrm{e}^{r/4m} \sinh\left(\frac{t}{4m}\right) \\ v &= \sqrt{1-r/2m}\, \mathrm{e}^{r/4m} \cosh\left(\frac{t}{4m}\right). \end{aligned} \tag{3.7.35}$$

The transformed metric again has the form (3.7.34), and $f$ is also given in terms of $r$ by (3.7.32).

The image of $(0<r<2m)$ under the transformation (3.7.35) is precisely the region II. The inverse of (3.7.35) is given by

$$v^2-u^2=(1-r/2m)\,\mathrm{e}^{r/2m}, \qquad u/v=\tanh(t/4m). \tag{3.7.36}$$

Taking $g$ and $h$ both negative is equivalent to the transformation $(u, v) \rightarrow (-u, -v)$, so that the regions II and IV are mutually isometric.

5) Along the hyperbolas $v^2-u^2=1$, the metric is truly *singular.* For example, here the invariant $R_{\alpha\beta\gamma\delta}R^{\alpha\beta\gamma\delta}$ diverges. The Kruskal extension is *maximal*, which means that every geodesic can either be extended to arbitrary large values of the affine parameter, or it runs into the singularity $v^2-u^2=1$ for a finite value of this parameter.

6) Causal relations in the Schwarzschild-Kruskal manifold are easy to survey. Radial light rays are lines at 45°, as in a Minkowski diagram. Observers in I and III (where the metric is static) can receive signals from IV and send them to II. However, any particle which enters region II must run into the singularity at $r=0$ within a finite proper time. A particle in region IV must have come from the singularity at a finite previous proper time. A causal connection between regions I and III is not possible.

The singularity in the future is shielded from distant observers in regions I and III by an *event horizon.* The boundary of the region which is causally connected with distant observers in regions I and III is given by the surfaces $r=2m$. Signals which are sent from IV to I (or III) will already have reached I at $t=-\infty$. It is probably not possible to observe such light.

7) In Sect. 3.7.2 we shall see that the full Kruskal extension is not relevant in astrophysics.

8) For a geometric visualization of the Schwarzschild-Kruskal manifold, we represent the two-dimensional surface $(v=0,\ \vartheta=\frac{\pi}{2})$ as a

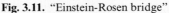

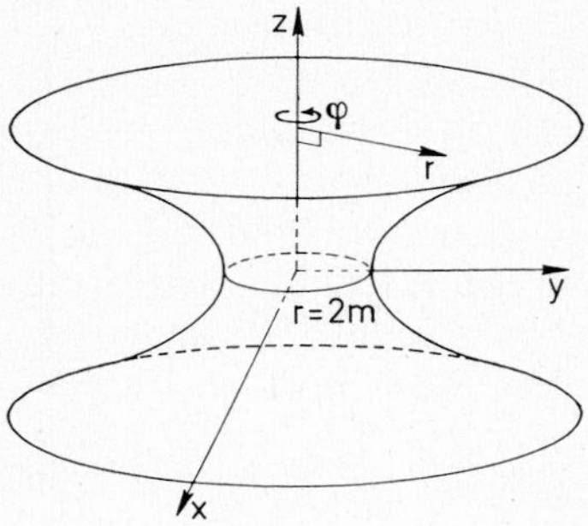


**Fig. 3.11.** "Einstein-Rosen bridge"

surface of rotation in the three-dimensional space (Fig. 3.11). The upper part of the embedding diagram corresponds to $u > 0$ (I) and the lower part to $u < 0$ (III). The reader should consider how the funnel-like structure ("Schwarzschild throat", "Einstein-Rosen bridge") changes for slices with increasing $v$.

9) The Schwarzschild-Kruskal manifold is static in regions I and III, but is dynamical in regions II and IV. In these regions, the Killing field $K = \partial/\partial t$ becomes spacelike. (At the horizon, $K$ is lightlike: $(K, K) = 1 - 2m/r = 0$.) There are no observers at rest in regions II and IV.

10) In the Appendix to this chapter, we shall prove the following generalization of the Birkhoff theorem:

*Every spherically symmetric solution of Einstein's field equations in vacuum is isometric to a part of the Schwarzschild-Kruskal manifold.*

### *Eddington-Finkelstein Coordinates*

For spherically symmetric gravitational collapse, only the regions I and II of the Kruskal diagram are relevant. For this part we shall introduce other, frequently used coordinates, which are originally due to Eddington and which were rediscovered by Finkelstein. They are related to the original Schwarzschild coordinates $(t, r)$ in (3.7.3) by

$$\begin{aligned} &r = r', \quad \vartheta = \vartheta', \quad \varphi = \varphi', \\ &t = t' - 2m \ln\left(\frac{r}{2m} - 1\right) \quad \text{for } r > 2m, \\ &t = t' - 2m \ln\left(1 - \frac{r}{2m}\right) \quad \text{for } r < 2m. \end{aligned} \tag{3.7.37}$$

They are related to the Kruskal coordinates by

$$\begin{aligned} u &= \tfrac{1}{2}\, e^{r/4m}\left(e^{t'/4m} + \frac{r-2m}{2m}\, e^{-t'/4m}\right) \\ v &= \tfrac{1}{2}\, e^{r/4m}\left(e^{t'/4m} - \frac{r-2m}{2m}\, e^{-t'/4m}\right). \end{aligned} \tag{3.7.38}$$

Thus one immediately finds

$$\frac{r-2m}{2m}\,e^{r/2m}=u^2-v^2\,,\qquad e^{t'/2m}=\frac{r-2m}{2m}\,\frac{u+v}{u-v}\,. \tag{3.7.39}$$

Using these equations, one can readily convince oneself that the transformation (3.7.38) is regular for $u>-v$ (i.e. in the region ($\mathrm{I}\cup\mathrm{II}$). Inserting (3.7.37) into (3.7.3) gives the metric

$$g=\left(1-\frac{2m}{r}\right)dt'^2-\left(1+\frac{2m}{r}\right)dr^2-\frac{4m}{r}\,dt'\,dr-r^2\,d\Omega^2\,. \tag{3.7.40}$$

This form is valid in the region $\mathrm{I}\cup\mathrm{II}$. The metric coefficients expressed in terms of the Eddington-Finkelstein coordinates are independent of $t'$; the price one pays is the existence of nondiagonal terms. We now determine the light cones for (3.7.40). For radially propagating light rays the condition $ds^2=0$ results in

$$(dt'+dr)\left[\left(1-\frac{2m}{r}\right)dt'-\left(1+\frac{2m}{r}\right)dr\right]=0\,. \tag{3.7.41}$$

Hence the radial lightlike directions are given by

$$dr/dt'=-1\,,\qquad dr/dt'=\frac{r-2m}{r+2m}\,. \tag{3.7.42}$$

The radial null geodesics corresponding to the first of these equations are given by

$$t'+r=\text{const}\,.$$

These are the straight lines in Fig. 3.12. The second equation of (3.7.42) shows that the tangents to the null geodesics of the other family have the following properties:

$$dr/dt'\to-1\quad\text{for } r\to 0\,,\qquad dr/dt'\to 0\quad\text{for } r\to 2m. \tag{3.7.43}$$

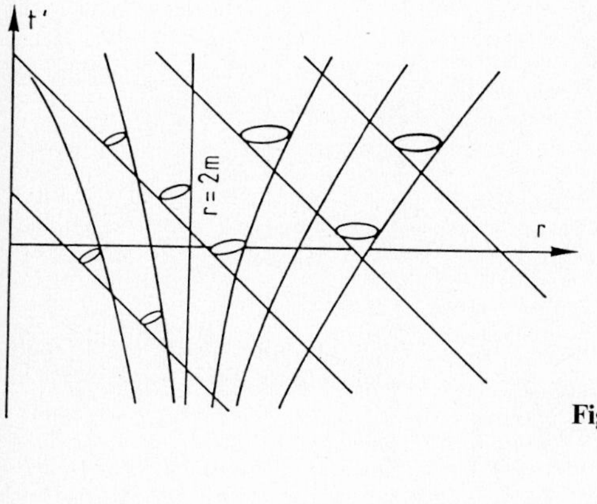


**Fig. 3.12**

The second of these properties implies that the geodesics do not cross the surface $r = 2m$. Hence we obtain Fig. 3.12. One can easily show that nonradial null geodesics and also timelike directions have $dr/dt'$ between the two values in (3.7.42). Hence the light cones have the form shown in Fig. 3.12.

### 3.7.2 Spherically Symmetric Collapse to a Black Hole

We now consider the spherically symmetric, catastrophic collapse of a supercritical mass. Since, according to the generalized Birkhoff theorem, the exterior field is a region of the Kruskal manifold, the collapse is most easily visualized in Kruskal coordinates, as in Fig. 3.13. In Fig. 3.14 we show the same process in Eddington-Finkelstein coordinates. From these two figures we may draw the following conclusions:

1) Equilibrium is no longer possible when the stellar radius becomes smaller than the radius of the horizon, since the world lines of the stellar surface must lie inside the light cones. Collapse to a singularity cannot be avoided. At some point in the vicinity of the singularity, GR will probably no longer be valid, since quantum effects will become important.

2) If a signal is emitted from inside the horizon, it will not reach a distant observer. The stellar matter is literally cut off from the outside world. All light rays also fall into the singularity. Thus, the horizon is the boundary of the region which is causally connected to a distant observer. In general, it is defined by this property. Thus, the event horizon acts like a one-way membrane through which energy and information can pass to the interior, but not to the exterior. The existence of horizons, or causal boundaries, in our universe is a remarkable consequence of GR. The singularity is on the other side of the horizon,

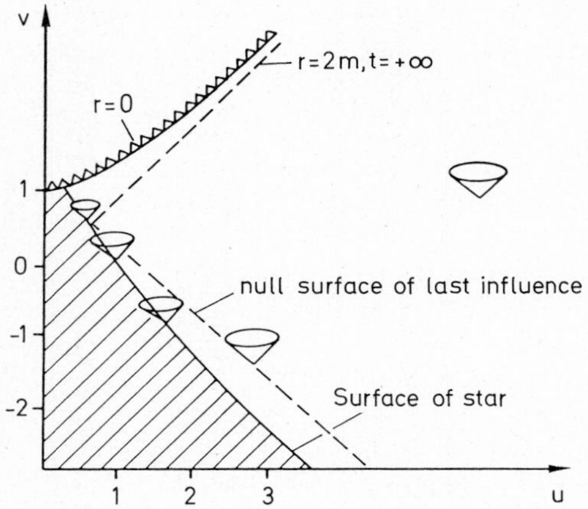


**Fig. 3.13.** Spherically symmetric collapse in Kruskal coordinates

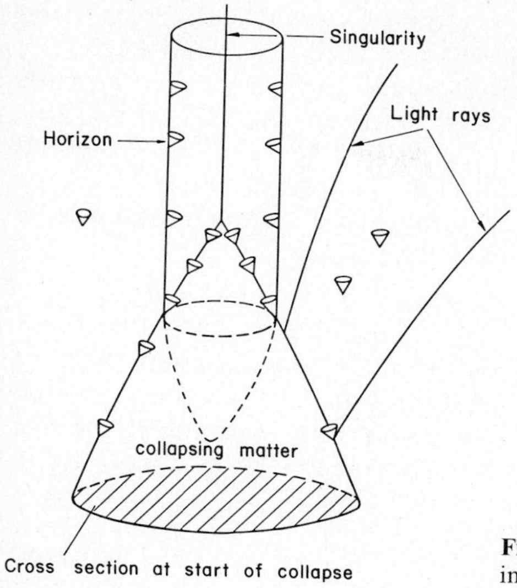


**Fig. 3.14.** Space-time diagram of a collapsing star in Eddington-Finkelstein coordinates

and hence has no causal connection to an external observer; he cannot "see" it. It is suspected that this is true for all "realistic" singularities ("cosmic censorship").

3) An observer on the surface of a collapsing star will not notice anything peculiar when the horizon is crossed. *Locally* the space-time geometry is the same as it is elsewhere. Furthermore, for very large masses the tidal forces at the horizon are harmless (see below). Thus the horizon is a *global* phenomenon of the space-time manifold. Note the "null surface of last influence" shown in Fig. 3.13.

4) An external observer far away from the star will see it reach the horizon only after an infinitely long time. As a result of the gravitational time dilation, the star "freezes" at the Schwarzschild horizon. However, in practice, the star will suddenly become invisible, since the red shift will start to increase exponentially (see below), and the luminosity decreases correspondingly. The characteristic time for this to happen is $\tau \sim R_s/c \simeq 10^{-5}(M/M_\odot)$ s; for $M \sim M_\odot$ this is extremely short. Afterwards, we are dealing with a "black hole". It makes sense to call the horizon the surface, and the external geometry the gravitational field of the black hole. The interior is not relevant for astrophysics. When observed from a distance, the exterior field looks exactly like that of any massive object.

Black holes act like cosmic vacuum cleaners. For example, if a black hole is part of a tightly bound binary system, it can suck up matter from its partner (if this happens, for instance, to be a giant star), and heat it to such a degree that a strong x-ray source results. There is good reason to believe that the x-ray source Cygnus $X-1$ arises from this mechanism. We shall examine the evidence more closely in Chap. 6.

Astrophysicists are becoming increasingly convinced that gigantic black holes, with perhaps $10^9 M_\odot$, may exist in the centers of very active galaxies. The accretion of matter by these black holes would provide a relatively natural explanation for the huge amounts of energy which are set free in a rather small volume. A test of this hypothesis by means of detailed observation remains in the future.

*Red Shift for a Distant Observer*

Suppose that a transmitter approaches the Schwarzschild horizon radially with four velocity $V$. The emitted signals (frequency $\omega_e$) are received by a distant observer at rest (with four velocity $U$) at the frequency $\omega_0$ (see Fig. 3.15). If $k$ is the wave vector, then, according to (1.9.17), we have

$$1+z=\frac{\omega_e}{\omega_0}=\frac{(k,V)}{(k,U)}. \qquad (3.7.44)$$

We use the "retarded" time

$$u=t-r^*, \qquad r^*=r+2m\ln\left(\frac{r}{2m}-1\right). \qquad (3.7.45)$$

In terms of the coordinates $u$, $r$, $\vartheta$, $\varphi$ the metric then reads

$$g=(1-2m/r)\,du^2+2\,du\,dr-r^2\,d\Omega^2. \qquad (3.7.46)$$

For a radially emitted light ray we have $u$, $\vartheta$, $\varphi=\text{const}$. Hence $k$ is proportional to $\partial/\partial r$. Now the Lagrangian for radial geodesics is given by

$$\mathscr{L}=\tfrac{1}{2}(1-2m/r)\,\dot{u}^2+\dot{u}\,\dot{r}.$$

Since $u$ is cyclic, we have

$$p_u:=\partial\mathscr{L}/\partial\dot{u}=(1-2m/r)\,\dot{u}+\dot{r}=\text{const}. \qquad (3.7.47)$$

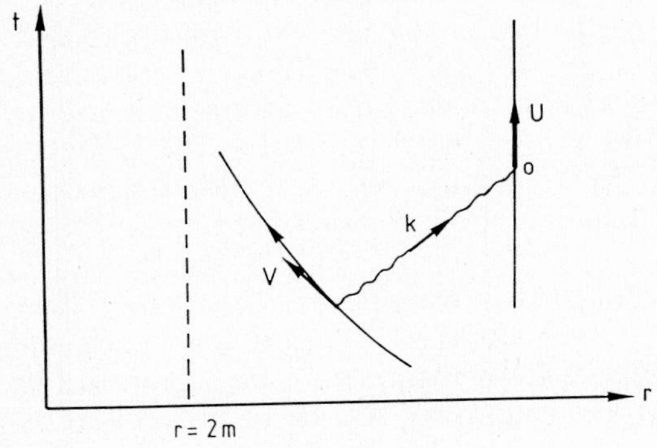


**Fig. 3.15**

This shows that for radial light rays ($\dot{u}=0$), $\dot{r}=p_u=$ const. Hence we have, in terms of the coordinates $(u, r, \vartheta, \varphi)$,

$$k = \text{const}\,\partial/\partial r\,. \tag{3.7.48}$$

Together with (3.7.44) this gives

$$1+z=\frac{(V, \partial/\partial r)}{(U, \partial/\partial r)}\,. \tag{3.7.49}$$

For $V$ we have

$$V=\dot{u}\frac{\partial}{\partial u}+\dot{r}\frac{\partial}{\partial r}$$

and hence

$$(V, \partial/\partial r)=\dot{u}\,g_{ur}=\dot{u}\,.$$

Similarly, we have for a distant observer, $(U, \partial/\partial r)=\dot{t}\simeq 1$, and thus

$$1+z=\dot{u}=\dot{t}-(1-2m/r)^{-1}\dot{r}\,. \tag{3.7.50}$$

We set

$$E:=\dot{t}\,(1-2m/r)\,. \tag{3.7.51}$$

For a radial geodesic, $E$ would be constant. However, we shall not assume free fall. Since $2\mathscr{L}=1$, i.e.

$$(1-2m/r)\,\dot{t}^2-\frac{\dot{r}^2}{1-2m/r}=1\,, \tag{3.7.52}$$

we have

$$\dot{r}^2+(1-2m/r)=E^2\,. \tag{3.7.53}$$

It then follows from (3.7.50) and (3.7.51) that

$$1+z=\frac{E-\dot{r}}{1-2m/r}=(1-2m/r)^{-1}\,[E+(E^2-1+2m/r)^{1/2}]\,. \tag{3.7.54}$$

The observer has $U\simeq\partial/\partial u$, which means that $u$ is the proper time, which we also denote by $\tau_0$. Keeping (3.7.54) in mind, we are interested in $r_e$ as a function of $\tau_0$. From (3.7.50) we have

$$dr_e/d\tau_0=dr_e/du=(\dot{r}/\dot{u})_e=\dot{r}/(1+z)\,.$$

If we set (3.7.54) into this expression and use (3.7.53) we obtain, for $r=r_e$

$$dr/d\tau_0=-(1-2m/r)\,(E^2-1+2m/r)^{1/2}/[E+(E^2-1+2m/r)^{1/2}]\,. \tag{3.7.55}$$

Now $r_e(\tau_e)$ has qualitatively the form shown in Fig. 3.16, which means that after a finite proper time the transmitter reaches the Schwarzschild horizon (e.g., in free fall). From (3.7.53) we see that for $r\simeq 2m$, $E$ is

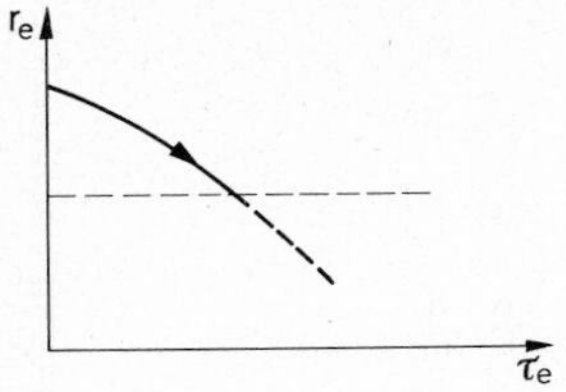


Fig. 3.16

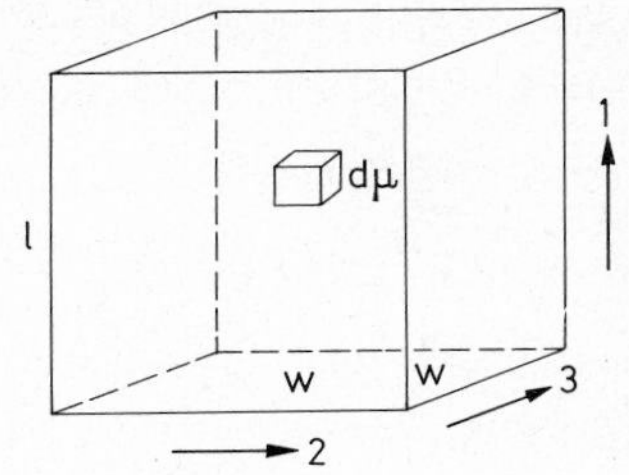


Fig. 3.17

finite, nonzero, and slowly varying. It then follows from (3.7.55) that in the vicinity of $r = 2m$

$$dr/d\tau_0 \simeq -\tfrac{1}{2}(1 - 2m/r) \simeq -\frac{r-2m}{4m}$$

or

$$r - 2m \simeq \text{const}\ e^{-\tau_0/4m}. \tag{3.7.56}$$

From (3.7.54) we have the asymptotic expression

$$1 + z \simeq \frac{4mE}{r-2m}. \tag{3.7.57}$$

Setting (3.7.56) into (3.7.57) gives the previously mentioned result

$$1 + z \propto e^{\tau_0/4m}. \tag{3.7.58}$$

*Fate of an Observer on the Surface of the Star*

As the stellar radius decreases, an observer on the surface of the star experiences constantly increasing tidal forces. Since his feet are attracted more strongly than his head, he experiences a longitudinal stress. At the same time, he is compressed from the sides (lateral stress). We now calculate these forces.

As a first step, we compute the components of the Riemann tensor relative to the basis (3.1.2). If we insert

$$e^{2a} = e^{-2b} = 1 - 2m/r \tag{3.7.59}$$

into the curvature forms (3.1.7), we obtain immediately

$$\begin{aligned} R_{0101} &= -R_{2323} = 2mr^{-3} \\ R_{1212} &= R_{1313} = -R_{0202} = -R_{0303} = m/r^3. \end{aligned} \tag{3.7.60}$$

All other components which are not determined by (3.7.60) and symmetry properties vanish.

The falling observer will use his rest system as a frame of reference. For $r > 2m$, this is reached by a special Lorentz transformation in the radial direction, which transforms $e_0$ into the four velocity $U$. Remarkably, the components of the Riemann tensor with respect to this new system are unchanged. The "boost" has the form

$$\Lambda = \left(\begin{array}{cc|c} \cosh\alpha & \sinh\alpha & 0 \\ \sinh\alpha & \cosh\alpha & \\ \hline & 0 & 1 \end{array}\right).$$

From (3.7.60), we then obtain, for example,

$$\begin{aligned} R'_{0101} &= \Lambda_0{}^{\mu}\Lambda_1{}^{\nu}\Lambda_0{}^{\sigma}\Lambda_1{}^{\varrho}R_{\mu\nu\sigma\varrho} \\ &= R_{0101}(\underset{\substack{\uparrow \\ (0101)}}{\cosh^4\alpha} - \underset{\substack{\uparrow \\ (1001),(0110)}}{2\cosh^2\alpha\,\sinh^2\alpha} + \underset{\substack{\uparrow \\ (1010)}}{\sinh^4\alpha}) = R_{0101}\,. \end{aligned}$$

One proves the invariance of the other components in a similar manner. This invariance depends on the particular structure of the Schwarzschild geometry.

We now follow the observer in quadrant II. For this, we need some results from the Appendix to this chapter. According to (A.11), the metric there has the form (A.1) with

$$e^{2b} = e^{-2a} = 2m/t - 1\,, \qquad R = t\,. \tag{3.7.61}$$

If one inserts these functions into the curvature forms (A.5) with respect to the basis (A.2), one obtains the same expressions as we had for $r > 2m$, except that $r$ and $t$ are interchanged. If we denote the timelike variable $t$ in (3.7.61) by $r$, then the expressions (3.7.60) are always valid for the local rest system.

The equation for geodesic deviation implies that the distance between two freely falling test bodies increases according to (2.1.10):

$$\ddot{n}^i = R^i_{00j}\,n^j\,, \tag{3.7.62}$$

where the dot denotes differentiation with respect to the proper time. Using (3.7.60) we thus obtain for our problem

$$\begin{aligned} \ddot{n}^1 &= \frac{2m}{r^3}\,n^1 \\ \ddot{n}^2 &= -\frac{m}{r^3}\,n^2 \\ \ddot{n}^3 &= -\frac{m}{r^3}\,n^3\,. \end{aligned} \tag{3.7.63}$$

We now consider a square prism having mass $\mu$, height $l$ and length = width = $w$, as in Fig. 3.17. We now ask what forces must be present in the body in order to avoid having its various parts move along diverging (or converging) geodesics.

From (3.7.63) we see that the 1,2,3-directions define the principal axes of the stress tensor (the 2- and 3-directions are equivalent).

The longitudinal stress is calculated as follows: According to (3.7.63) an element $d\mu$ of the body at a height $h$ above the center of mass (the distance is measured along $e_1$) would be accelerated relative to the center of mass with $b = 2m\,r^{-3}h$, if it were free to move. In order to prevent this, a force

$$dK = b\,d\mu = \frac{2m}{r^3}\,h\,d\mu$$

must act on the element. This force contributes to the stress through the horizontal plane through the center of mass. The total force through this plane is

$$K = \int_0^{l/2} \frac{2m\,h}{r^3}\,\frac{\mu}{l\,w^2}\,w^2\,dh$$

or

$$K = \frac{1}{4}\,\frac{\mu\,m\,l}{r^3}.$$

The longitudinal stress $T^l$ is $T^l = -K/w^2$, so that

$$T^l = -\frac{1}{4}\,\frac{\mu\,m\,l}{r^3\,w^2}. \tag{3.7.64}$$

Similarly, using (3.7.63), one finds that the lateral stress $T_\perp$ is

$$T_\perp = +\frac{1}{8}\,\frac{\mu\,m}{l\,r^3}. \tag{3.7.65}$$

As a numerical example, take $\mu = 75$ kg, $l = 1.8$ m and $w = 0.2$ m. We find

$$\begin{aligned} T^l &\simeq -1.1\times 10^{14}\,M/M_\odot\,(r/1\,\text{km})^{-3}\,\text{N/m}^2 \\ T_\perp &\simeq +0.7\times 10^{12}\,M/M_\odot\,(r/1\,\text{km})^{-3}\,\text{N/m}^2. \end{aligned} \tag{3.7.66}$$

For comparison, recall the 1 atmosphere $= 1.013\times 10^5\,\text{N/m}^2$.

When the mass is very large, the tidal forces are even at the Schwarzschild horizon perfectly tolerable.

After the body is finally torn apart, its constituents (electrons and nucleons) move along geodesics. For $r < 2m$, these have the qualitative form indicated in Fig. 3.18. The spatial coordinate $t$ is nearly constant as $r \to 0$ (the reader should provide an analytic discussion). Thus, when the upper (lower) end has the radial coordinate $t_2$ ($t_1$), the length of the body approaches

$$L = g_{tt}(r)^{1/2}(t_2 - t_1) \simeq \left(\frac{2m}{r}\right)^{1/2}(t_2 - t_1) \propto r^{-1/2} \propto (\tau_{\text{collapse}} - \tau)^{-1/3} \to \infty$$

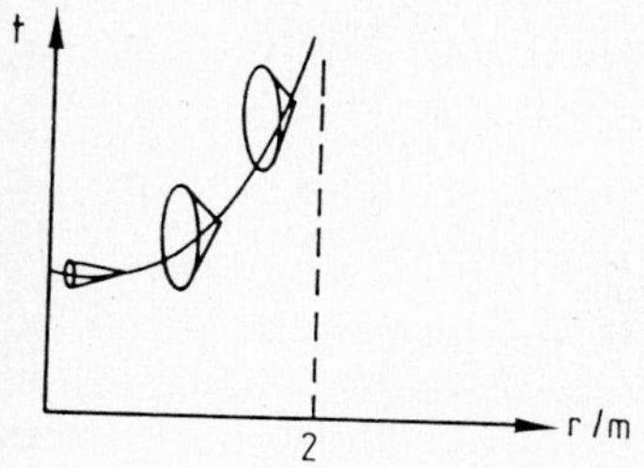


**Fig. 3.18.** Qualitative form of geodesics for $r < 2m$

as $r \to 0$. Here, $\tau$ is the proper time

$$\tau = -\int^{r} \left(\frac{2m}{r} - 1\right)^{-1/2} dr$$

and $\tau_{\text{collapse}} = \tau|_{r=0}$. For $r \simeq 0$, we have $\tau_{\text{collapse}} - \tau \propto r^{3/2}$. For the surface area we find

$$A = [g_{\vartheta\vartheta}(r)\, g_{\varphi\varphi}(r)]^{1/2}\, \Delta\vartheta\, \Delta\varphi \propto r^2 \propto (\tau_{\text{collapse}} - \tau)^{4/3} \to 0 .$$

The volume behaves according to

$$\text{Vol} = AL \propto r^{3/2} \alpha (\tau_{\text{collapse}} - \tau) \to 0 .$$

It is thus not advisable to fall into a black hole.

*Stability of the Schwarzschild Black Hole*

The Schwarzschild solution is only physically relevant if it is stable agains small perturbations.

This a very difficult problem. Some insight can be expected from a *linear* stability analysis. It turns out ([68], [69]) that the frequencies $\omega_n$ of the normal modes in the time dependent factors $\exp(-\mathrm{i}\,\omega_n t)$ are all real, because the $\omega_n^2$ can be shown to be the eigenvalues of a self-adjoint positive operator of the Schrödinger type. (The radial parts of the normal modes are the corresponding eigenfunctions.)

From this we cannot conclude anything about the stability of the nonlinear problem (as is well known from simple examples of ordinary nonlinear differential equations).

---

**Exercise:** Introduce Eddington-type coordinates for the Reissner-Nordström solution (see Exercises in Sect. 3.1) for $m > |e|\sqrt{G}$ such that the metric is regular for all positive values of $r$. Discuss the radial null geodesics and show that there is a horizon. Investigate radial timelike geodesics for neutral test particles and show that they cannot fall into the singularity at $r = 0$. Is there a simple physical explanation of this fact?

---

# Appendix: Spherically Symmetric Gravitational Fields

In this appendix, we consider spherically symmetric fields and prove, among other things, the generalized Birkhoff theorem.

**Definition:** A Lorentz manifold $(M, g)$ is *spherically symmetric* provided it admits the group SO(3) as an isometry group, in such a way that the group orbits are two-dimensional spacelike surfaces. The group orbits are necessarily surfaces of constant positive curvature.

## *1. General Form of the Metric*

Let $q \in M$ and let $\Omega(q)$ be the orbit through $q$. The (one-dimensional) subgroup of SO(3) which leaves $q$ invariant is the so-called stabilizer $G_q$. We denote the set of all geodesics passing through $q$ which are perpendicular to $\Omega(q)$ by $N(q)$. Locally these form a two dimensional surface [also denoted by $N(q)$] which is invariant under $G_q$. A vector $v \in T_q$ which is perpendicular to $\Omega(q)$ is transformed into another such vector by $G_q$, for if $u \in T_q(M)$ is tangent to $\Omega(q)$, and if $g \in G_q$, then $(u, g \cdot v) = (g^{-1} \cdot u, v) = 0$, since $g^{-1} \cdot u$ is also tangent to $\Omega(q)$.

Now consider a point $p \in N(q)$ and the directions [vectors in $T_p(M)$] perpendicular to $N(q)$. The group $G_q$ permuts these directions, since $N(q)$ remains invariant under $G_q$. On the other hand, $G_q$ leaves the orbit $\Omega(p)$ invariant, and hence this must be perpendicular to $N(q)$, as indicated in Fig. A1. We can now define locally a bijective mapping $\phi$: $\Omega(q) \to \Omega(p)$ between two group orbits as follows: $\phi$ maps the point $q$ into the point $\Omega(p) \cap N(q)$. This mapping is $G_q$-invariant, since $q$, as well as $\Omega(p)$ and $N(q)$ are invariant under $G_q$.

Therefore vectors of $T_q\Omega(q)$ having the same length are transformed by $\phi$ into vectors of $T_{\phi(q)}\Omega(p)$ having also the same length. Furthermore, since all points in $\Omega(q)$ are equivalent, one obtains the same magnification factors for all points in $\Omega(q)$; this means that the mapping $\phi$ is *conformal.*

Thus we can introduce local coordinates $(t, r, \vartheta, \varphi)$ as follows. Let the group orbits be surfaces of constant $t$ and $r$; the surfaces $N$ orthogonal to these are the surfaces $(\vartheta, \varphi = \text{const})$. Since $\phi$ is conformal, and

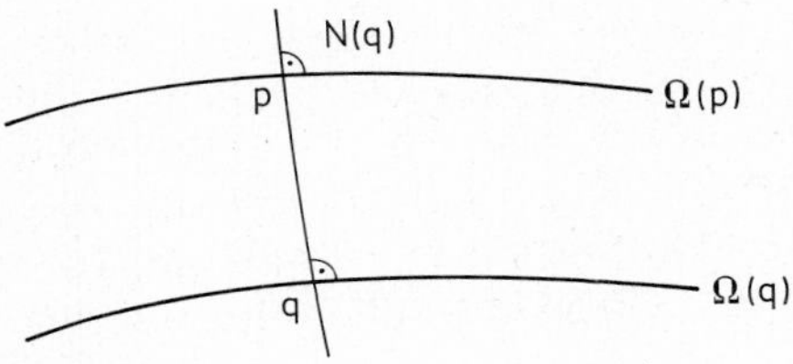

**Fig. A1**

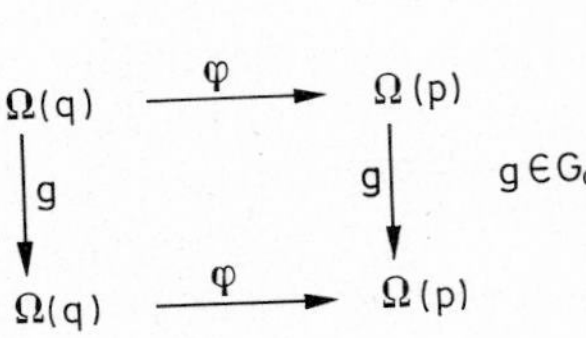

**Fig. A2**

since the orbits are surfaces of constant positive curvature, the metric has the form

$$ds^2 = d\tau^2(t, r) - R^2(t, r)(d\vartheta^2 + \sin^2\vartheta\, d\varphi^2),$$

where $d\tau^2$ is an indefinite metric in $t$ and $r$. In other words, $d\tau^2$ has the form

$$d\tau^2 = a^2\, dt^2 + 2\, a\, b\, dt\, dr - c^2\, dr^2.$$

Now the 1-form in two variables $a\,dt + b\,dr$ always has an integrating denominator: $a\,dt + b\,dr = A\,dt'$. In terms of the coordinates $(t', r)$, $d\tau^2$ has the form $d\tau^2 = A^2\, dt'^2 - B^2\, dr^2$, and the curves $\{t' = \text{const}\}$ and $\{r = \text{const}\}$ are mutually orthogonal. Thus, it is always possible to introduce coordinates such that the metric has the form

$$g = e^{2a(t,r)}\, dt^2 - [e^{2b(t,r)}\, dr^2 + R^2(t, r)(d\vartheta^2 + \sin^2\vartheta\, d\varphi^2)]. \tag{A.1}$$

At this point, one still has the freedom to choose either $t$ or $r$ arbitrarily in the surfaces $N$. In the following, we shall make use of this fact in several different ways.

We will need the connection forms and curvature forms corresponding to the metric (A.1), as well as the Ricci and Einstein tensors not only in this appendix, but also in the following chapters. We compute these relative to the orthonormal basis

$$\theta^0 = e^a\, dt, \quad \theta^1 = e^b\, dr, \quad \theta^2 = R\, d\vartheta, \quad \theta^3 = R \sin\vartheta\, d\varphi. \tag{A.2}$$

Exterior differentiation of the $\theta^\mu$ results immediately in (here the prime denotes $\partial/\partial r$ and the dot denotes $\partial/\partial t$):

$$\begin{aligned}
d\theta^0 &= a'\, e^{-b}\, \theta^1 \wedge \theta^0 \\
d\theta^1 &= \dot{b}\, e^{-a}\, \theta^0 \wedge \theta^1 \\
d\theta^2 &= \frac{\dot{R}}{R}\, e^{-a}\, \theta^0 \wedge \theta^2 + \frac{R'}{R}\, e^{-b}\, \theta^1 \wedge \theta^2 \\
d\theta^3 &= \frac{\dot{R}}{R}\, e^{-a}\, \theta^0 \wedge \theta^3 + \frac{R'}{R}\, e^{-b}\, \theta^1 \wedge \theta^3 + \frac{1}{R} \cot\vartheta\, \theta^2 \wedge \theta^3 .
\end{aligned} \tag{A.3}$$

Comparison with the first structure equation leads to

$$\begin{aligned}
\omega^0_1 &= \omega^1_0 = a'\, e^{-b}\, \theta^0 + \dot{b}\, e^{-a}\, \theta^1 \\
\omega^0_2 &= \omega^2_0 = \frac{\dot{R}}{R}\, e^{-a}\, \theta^2 \\
\omega^0_3 &= \omega^3_0 = \frac{\dot{R}}{R}\, e^{-a}\, \theta^3 \\
\omega^2_1 &= -\, \omega^1_2 = \frac{R'}{R}\, e^{-b}\, \theta^2
\end{aligned} \tag{A.4}$$

$$\omega^3_1 = -\,\omega^1_3 = \frac{R'}{R}\,e^{-b}\,\theta^3$$

$$\omega^3_2 = -\,\omega^2_3 = \frac{1}{R}\cot\vartheta\,\theta^3.$$

One may verify immediately that the first structure equations are satisfied.

The curvature forms are obtained from the second structure equations by a straightforward calculation. The result is

$$\begin{aligned}
\Omega^0_1 &= E\,\theta^0 \wedge \theta^1\\
\Omega^0_2 &= \tilde{E}\,\theta^0 \wedge \theta^2 + H\,\theta^1 \wedge \theta^2\\
\Omega^0_3 &= \tilde{E}\,\theta^0 \wedge \theta^3 + H\,\theta^1 \wedge \theta^3\\
\Omega^1_2 &= \tilde{F}\,\theta^1 \wedge \theta^2 - H\,\theta^0 \wedge \theta^2\\
\Omega^1_3 &= \tilde{F}\,\theta^1 \wedge \theta^3 - H\,\theta^0 \wedge \theta^3\\
\Omega^2_3 &= F\,\theta^2 \wedge \theta^3,
\end{aligned}\tag{A.5}$$

where

$$\begin{aligned}
E &:= e^{-2a}(\dot{b}^2 - \dot{a}\,\dot{b} + \ddot{b}) - e^{-2b}(a'^2 - a'\,b' + a'')\\
\tilde{E} &:= \frac{1}{R}\,e^{-2a}(\ddot{R} - \dot{a}\,\dot{R}) - \frac{1}{R}\,e^{-2b}\,a'\,R'\\
H &:= \frac{1}{R}\,e^{-a-b}(\dot{R}' - a'\,\dot{R} - \dot{b}\,R')\\
F &:= \frac{1}{R^2}(1 - R'^2\,e^{-2b} + \dot{R}^2\,e^{-2a})\\
\tilde{F} &:= \frac{1}{R}\,e^{-2a}\,\dot{b}\,\dot{R} + \frac{1}{R}\,e^{-2b}(b'\,R' - R'').
\end{aligned}\tag{A.6}$$

One obtains the Ricci tensor immediately from (A.5):

$$\begin{aligned}
R_{00} &= -\,E - 2\tilde{E}\\
R_{01} &= -\,2H\\
R_{02} &= R_{03} = 0\\
R_{11} &= E + 2\tilde{F}\\
R_{12} &= R_{13} = 0\\
R_{22} &= R_{33} = \tilde{E} + \tilde{F} + F, \quad R_{23} = 0.
\end{aligned}\tag{A.7}$$

The Riemann curvature scalar is then given by

$$R = -\,2(E + F) - 4(\tilde{E} + \tilde{F}),\tag{A.8}$$

and hence the Einstein tensor is given by

$$
\begin{aligned}
&G^0_0 = F + 2\tilde{F}\\
&G^0_1 = -2H\\
&G^0_2 = G^0_3 = 0\\
&G^1_1 = 2\tilde{E} + F\\
&G^1_2 = G^1_3 = 0\\
&G^2_2 = G^3_3 = \tilde{E} + \tilde{F} + E\\
&G^2_3 = 0.
\end{aligned}
\tag{A.9}
$$

*2. The Generalized Birkhoff Theorem*

We now consider spherically symmetric solutions of the vacuum field equations, and prove the following

**Theorem:** Every ($C^2-$) solution of the Einstein field equations in vacuum which is spherically symmetric in an open subset $U$ is locally isometric to a portion of the Schwarzschild-Kruskal solution.

*Proof:* The local solution depends on the nature of the surfaces $\{R(t, r) = \text{const}\}$. We first consider the case
a) The surfaces $\{R = \text{const}\}$ are timelike in $U$ and $dR \neq 0$. In this case, we may choose $R(t, r) = r$. If we now use $\dot{R} = 0$ and $R' = 1$ in $G^0_1 = 0$, we find from (A.6) and (A.9) that $\dot{b} = 0$. This information, together with $G_{00} + G_{11} = 0$, i.e. $\tilde{E} - \tilde{F} = 0$, implies $a' + b' = 0$. Hence $a(t, r) = -b(r) + f(t)$. Thus, by choosing a new time coordinate, one can arrange things such that $a$ is also independent of time. The equation $G_{00} = 0$ $(F + 2\tilde{F} = 0)$ then gives $(r\,\mathrm{e}^{-2b})' = 1$, or $\mathrm{e}^{-2b} = 1 - 2m/r$. Thus, for case a) we obtain the Schwarzschild solution for $r > 2m$:

$$
g = (1 - 2m/r)\,dt^2 - \frac{dr^2}{1 - 2m/r} - r^2(d\vartheta^2 + \sin^2\vartheta\, d\varphi^2). \tag{A.10}
$$

b) The surface $\{R = \text{const}\}$ are spacelike in $U$ and $dR \neq 0$. In this case, we may choose $R(t, r) = t$. Now the equation $G^0_1 = 0$ gives the condition $a' = 0$, so that $a$ depends only on $t$. We now obtain $\dot{a} + \dot{b} = 0$ from $G_{00} + G_{11} = 0$, and we can introduce a new $r$-coordinate such that $b$ also depends only on $t$. Finally we obtain from $G_{00} = 0$ the equation $(t\,\mathrm{e}^{-2a})^{\cdot} = -1$, or

$$
\mathrm{e}^{-2a} = \mathrm{e}^{2b} = \frac{2m}{t} - 1, \quad R = t, \quad (t < 2m). \tag{A.11}
$$

This is the Schwarzschild solution for $r < 2m$, for if we interchange the variables $r$ and $t$, we obtain (A.10) with $r < 2m$.

c) Suppose that $(dR, dR) = 0$ in $U$. Consider first the special case, that $R$ is constant in $U$. Then $G_{00} = 0$ implies $R = \infty$. We may therefore assume that $dR$ is lightlike and not equal to zero. The coordinates $(t, r)$ can be chosen such that $R(t, r) = t - r$. If $dR$ is lightlike, then we must have $a = b$. $G_1^0 = 0$ is then automatically satisfied and from $G_{00} + G_{11} = 0$ we obtain the condition $a' = -\dot{a}$. Hence $a$ is a function of $t - r$. $G_{00} = 0$ then implies the ridiculous condition $t^{-1} - r = 0$. Case c) is therefore incompatible with the field equations.
d) Suppose that $\{R = \text{const}\}$ is spacelike in some part of $U$ and timelike in another part.

In this case, one obtains the solutions a), resp. b). As in Sect. 3.7.2, one can join these smoothly along the surfaces $(dR, dR) = 0$, and thus obtains a portion of the Schwarzschild-Kruskal solution.

*3. Spherically Symmetric Solutions for a Fluid*

We now consider the spherically symmetric field of a (not necessarily static) fluid, whose four velocity $U$ is assumed to be invariant with respect to the isotropy group SO(3). Hence $U$ is perpendicular to the orbits and the integral curves lie in the surfaces $N$ (see Sect. 3.1). We may choose $t$ such that $U$ is proportional to $\partial/\partial t$. Since $(U, U) = 1$, we have

$$U = e^{-a}\,\partial/\partial t. \tag{A.12}$$

We could have chosen the time coordinate such that $U = \partial/\partial t$, but then the $(t, r)$ portion of the metric would not necessarily be orthogonal, but would have the form

$$d\tau^2 = dt^2 + 2g_{tr}\,dt\,dr + g_{rr}\,dr^2. \tag{A.13}$$

For an ideal fluid with $p = 0$ (incoherent dust), the field equations imply $\nabla_U U = 0$. The $r$-component of this equation implies $\Gamma^r_{tt} = 0$, or, in other words, $\partial g_{tr}/\partial t = 0$. Let us therefore choose new coordinates

$$t' = t + \int^r g_{tr}(r)\,dr$$
$$r' = r.$$

Then the metric has the form (the primes have been suppressed)

$$g = dt^2 - [\mathrm{e}^{2b}\,dr^2 + R^2(t, r)\,d\Omega^2] \tag{A.14}$$

and

$$U = \frac{\partial}{\partial t}. \tag{A.15}$$

However, this is only possible when $p = 0$!

Of course, for a *static* situation, $U$ [in (A.12)] should be proportional to a timelike Killing field, which we can choose to be $\partial/\partial t$ (by a suitable choice of coordinates). In this case, $R$, $a$, and $b$ are independent of $t$. In the basis (A.2), the energy-momentum tensor has the form

$$T^{\mu\nu} = (p+\varrho)\, U^{\mu} U^{\nu} - p\, g^{\mu\nu} = \begin{pmatrix} \varrho & & & 0 \\ & -p & & \\ & & -p & \\ 0 & & & -p \end{pmatrix}. \tag{A.16}$$

If we had $R' = 0$, then, because $G_{00} - G_{11} = 8\pi G(\varrho + p)$, we would have to require $\varrho + p = 0$. This is excluded for a physical fluid. We may therefore choose $R(r) = r$, and the metric has the form

$$g = e^{2a(r)}\, dt^2 - [e^{2b(r)}\, dr^2 + r^2 (d\vartheta^2 + \sin^2\vartheta\, d\varphi^2)]. \tag{A.17}$$

---

**Exercise:** Let $(M, g)$ be a spherically symmetric, stationary Lorentz manifold. Show that a timelike Killing field $K$ exists locally, which is perpendicular to the orbits $\Omega(q)$ of SO(3). In this case, the $t$-coordinate in the surfaces $N$ can be choosen such that $K = \partial/\partial t$. Then all the functions in (A.1) are independent of $t$. If we exclude the degenerate case $R' = 0$, the metric can again be written in the form (A.17).

---

# Chapter 4. Weak Gravitational Fields

## 4.1 The Linearized Theory of Gravity

In this chapter, we consider systems for which the gravitational field (at least in a certain region of space-time) is nearly flat. Thus a coordinate system exists for which

$$g_{\mu\nu} = \eta_{\mu\nu} + h_{\mu\nu}, \quad |h_{\mu\nu}| \ll 1. \tag{4.1.1}$$

For example, in the solar system, we have $|h_{\mu\nu}| \sim |\phi|/c^2 \lesssim GM_\odot/c^2 R_\odot \sim 10^{-6}$. However, the field can vary rapidly with time, as is the case for weak gravitational waves.

For such fields, we expand the field equations in powers of $h_{\mu\nu}$ and keep only the linear terms. The Ricci tensor is given, up to quadratic terms, by

$$R_{\mu\nu} = \partial_\lambda \Gamma^\lambda_{\nu\mu} - \partial_\nu \Gamma^\lambda_{\lambda\mu}. \tag{4.1.2}$$

Furthermore

$$\Gamma^\alpha_{\mu\nu} = \tfrac{1}{2}\eta^{\alpha\beta}(h_{\mu\beta,\nu} + h_{\beta\nu,\mu} - h_{\mu\nu,\beta}) = \tfrac{1}{2}(h^\alpha_{\mu,\nu} + h^\alpha_{\nu,\mu} - h_{\mu\nu}^{,\alpha}). \tag{4.1.3}$$

We use the convention that indices are raised or lowered with $\eta^{\mu\nu}$ or $\eta_{\mu\nu}$, respectively.

Using (4.1.3) gives for the Ricci curvature

$$R_{\mu\nu} = \tfrac{1}{2}[h^\lambda_{\mu,\lambda\nu} - \Box h_{\mu\nu} - h^\lambda_{\lambda,\mu\nu} + h^\lambda_{\nu,\lambda\mu}] \tag{4.1.4}$$

and for the scalar Riemann curvature

$$R = h^{\lambda\sigma}_{,\lambda\sigma} - \Box h, \tag{4.1.5}$$

where

$$h := h^\lambda_\lambda = \eta^{\alpha\beta} h_{\alpha\beta}. \tag{4.1.6}$$

Thus, in the linear approximation, the Einstein tensor is given by

$$2G_{\mu\nu} = -\Box h_{\mu\nu} - h_{,\mu\nu} + h^\lambda_{\mu,\lambda\nu} + h^\lambda_{\nu,\lambda\mu} + \eta_{\mu\nu}\Box h - \eta_{\mu\nu} h^{\lambda\sigma}_{,\lambda\sigma}. \tag{4.1.7}$$

The contracted Bianchi identity reduces to

$$G^{\mu\nu}_{,\nu} = 0, \tag{4.1.8}$$

as one also sees by direct computation. As a consequence of (4.1.8) and the field equations $G_{\mu\nu} = 8\pi G T_{\mu\nu}$, or

$$\Box h_{\mu\nu} + h_{,\mu\nu} - h^{\lambda}_{\mu,\lambda\nu} - h^{\lambda}_{\nu,\lambda\mu} - \eta_{\mu\nu} \Box h + \eta_{\mu\nu} h^{\lambda\sigma}_{,\lambda\sigma} = -16\pi G T_{\mu\nu} \tag{4.1.9}$$

we have

$$T^{\mu\nu}_{,\nu} = 0 . \tag{4.1.10}$$

Thus in this approximation, although $T_{\mu\nu}$ produces a gravitational field, this does not react back on the source. [For example, if we consider incoherent dust, $T^{\mu\nu} = \varrho u^\mu u^\nu$, $(\varrho u^\mu)_{,\mu} = 0$ (4.1.10) implies $u^\nu u^\mu_{,\nu} = 0$ or $du^\mu/ds = 0$, so that the integral curves of $u^\mu$ are straight lines.]

As the starting point of an iterative procedure, one can, for weak fields, determine $h_{\mu\nu}$ from the field equations (4.1.9) with the "flat" $T_{\mu\nu}$ and compute the reaction on physical systems by setting $g_{\mu\nu} = \eta_{\mu\nu} + h_{\mu\nu}$ in the equations of Sect. 1.4. This procedure makes sense only if the reaction is small. We shall discuss some specific examples further on.

In the following, it will be useful to use the quantity

$$\gamma_{\mu\nu} = h_{\mu\nu} - \tfrac{1}{2} \eta_{\mu\nu} h ; \tag{4.1.11}$$

$h_{\mu\nu}$ can then be obtained from

$$h_{\mu\nu} = \gamma_{\mu\nu} - \tfrac{1}{2} \eta_{\mu\nu} \gamma , \quad \gamma := \gamma^{\lambda}_{\lambda} . \tag{4.1.11'}$$

The field equations written in terms of $\gamma_{\mu\nu}$ read

$$-\Box \gamma_{\mu\nu} - \eta_{\mu\nu} \gamma^{,\alpha\beta}_{\alpha\beta} + \gamma^{\;\;,\alpha}_{\mu\alpha,\nu} + \gamma^{\;\;,\alpha}_{\nu\alpha,\mu} = 16\pi G T_{\mu\nu} . \tag{4.1.12}$$

We may regard the linearized theory as a Lorentz covariant field theory (for the field $h_{\mu\nu}$) in flat space. If we consider a global Lorentz transformation, we obtain immediately from the transformation properties of $g_{\mu\nu}$ with respect to coordinate transformations, that $h_{\mu\nu}$ transforms as a tensor with respect to the Lorentz group. Thus, if

$$x^\mu = \Lambda^\mu_{\ \nu} x'^\nu , \quad \Lambda^T \eta \Lambda = \eta , \quad \eta := \operatorname{diag}(1, -1, -1, -1)$$

then

$$\eta_{\alpha\beta} + h'_{\alpha\beta} = \frac{\partial x^\mu}{\partial x'^\alpha} \frac{\partial x^\nu}{\partial x'^\beta} g_{\mu\nu} = \Lambda^\mu_{\ \alpha} \Lambda^\nu_{\ \beta} (\eta_{\mu\nu} + h_{\mu\nu})$$
$$= \eta_{\alpha\beta} + \Lambda^\mu_{\ \alpha} \Lambda^\nu_{\ \beta} h_{\mu\nu} ,$$

so that

$$h'_{\alpha\beta} = \Lambda^\mu_{\ \alpha} \Lambda^\nu_{\ \beta} h_{\mu\nu} . \tag{4.1.13}$$

As is the case in electrodynamics, there is a *gauge group*. The linearized Einstein tensor (4.1.7) is *invariant* with respect to gauge transformations of the form (here $\xi^\mu$ is a vector field)

$$h_{\mu\nu} \to h_{\mu\nu} + \xi_{\mu,\nu} + \xi_{\nu,\mu} \tag{4.1.14}$$

as one can easily verify by direct computation. This gauge transformation can be regarded as an *infinitesimal coordinate transformation.* In fact, let

$$x'^{\mu} = x^{\mu} + \xi^{\mu}(x), \tag{4.1.15}$$

where $\xi^{\mu}(x)$ is an infinitesimal vector field [it must be infinitesimal to avoid inconsistency with (4.1.1)].

This transformation induces infinitesimal changes in all the fields. With the exception of the gravitational field, which is described by the infinitesimal $h_{\mu\nu}$, it is possible to neglect these changes in the linearized theory.

From the transformation law

$$g_{\mu\nu}(x) = \underbrace{\frac{\partial x'^{\sigma}}{\partial x^{\mu}}}_{[\delta^{\sigma}_{\mu} + \xi^{\sigma}{}_{,\mu}]} \cdot \underbrace{\frac{\partial x'^{\varrho}}{\partial x^{\nu}}}_{[\delta^{\varrho}_{\nu} + \xi^{\varrho}{}_{,\nu}]} g'_{\sigma\varrho}(x')$$

we obtain

$$g_{\mu\nu}(x) = g'_{\mu\nu}(x') + g'_{\sigma\nu}(x')\, \xi^{\sigma}_{,\mu} + g'_{\mu\sigma}(x')\, \xi^{\sigma}_{,\nu}.$$

To first order in $\xi^{\mu}$ and $h^{\mu\nu}$ we then have

$$h_{\mu\nu}(x) = h'_{\mu\nu}(x) + \xi_{\mu,\nu} + \xi_{\nu,\mu}.$$

Gauge invariance can be expressed in other terms as follows: Previously, we emphasized that a $g$-field is equivalent to every field which can be obtained from it by a diffeomorphism. In particular, $g$ is equivalent to $\phi_s^* g$, where $\phi_s$ denotes the flow of some vector field $X$. For small $s$ we have

$$\phi_s^* g = g + s L_X g + O(s^2).$$

If we express this in terms of $h := g - \eta$, and set $\xi = s X$, this means

$$h \to h + L_\xi g + O(s^2) = h + L_\xi \eta + L_\xi h + O(s^2).$$

Thus, it is clear that for weak fields $h$ and $h + L_\xi \eta$ are equivalent for an infinitesimal vector field $\xi$. Now

$$(L_\xi \eta)_{\mu\nu} = \xi_{\mu,\nu} + \xi_{\nu,\mu}.$$

One demonstrates the gauge invariance of the linearized Riemann tensor $R^{(1)}$ as follows: For the Riemann tensor $R[g]$ we have generally $\phi_s^*(R[g]) = R[\phi_s^* g]$. Hence

$$L_X(R[g]) = \frac{d}{ds}\bigg|_{s=0} (R[g + s L_X g]). \tag{*}$$

If we now set $g = \eta$ on both sides, we find $0 = R^{(1)}(L_X \eta)$, and thus

$$R^{(1)}(h + L_X \eta) = R^{(1)}(h)$$

for every vector field $X$.

**Remarks:** For small perturbations $h$ around some curved background metric $g$, the gauge transformations are

$$h \to h + L_\xi g \quad \text{or} \quad h_{\mu\nu} \to h_{\mu\nu} + \xi_{\mu;\nu} + \xi_{\nu;\mu},$$

where the semicolon denotes covariant derivatives with respect to the background metric.

If $g$ satisfies the vacuum equations $\mathrm{Ric}[g] = 0$, then the linearized Ricci tensor is gauge invariant. This follows as before from

$$L_X(\mathrm{Ric}[g]) = \left.\frac{d}{ds}\right|_{s=0} \mathrm{Ric}[g + s L_X g].$$

We can easily find an explicit expression for the linearized Ricci tensor $\delta R_{\mu\nu}[h]$. The arguments that led to (2.3.5) show that

$$\delta R_{\mu\nu} = (\delta \Gamma^{\alpha}_{\mu\nu})_{;\alpha} - (\delta \Gamma^{\alpha}_{\mu\alpha})_{;\nu}.$$

This is known as the *Palatini identity*. In addition, we have (2.3.20),

$$\delta \Gamma^{\mu}_{\alpha\beta} = \tfrac{1}{2} g^{\mu\nu} [h_{\nu\alpha;\beta} + h_{\nu\beta;\alpha} - h_{\alpha\beta;\nu}].$$

Combining the two equations gives

$$\delta R_{\mu\nu}[h] = \tfrac{1}{2} g^{\alpha\beta} [h_{\beta\mu;\nu\alpha} - h_{\alpha\beta;\mu\nu} + h_{\beta\nu;\mu\alpha} - h_{\mu\nu;\beta\alpha}]$$

[compare this with (4.1.4)].

The linearized field equations

$$\delta R_{\mu\nu}[h] = 8\pi G [\delta T_{\mu\nu} - \tfrac{1}{2} g_{\mu\nu} g^{\alpha\beta} \delta T_{\alpha\beta} + \tfrac{1}{2} g_{\mu\nu} h_{\alpha\beta} T^{\alpha\beta} - \tfrac{1}{2} h_{\mu\nu} T^{\lambda}_{\lambda}]$$

are invariant under infinitesimal diffeomorphisms

$$h_{\mu\nu} \to h_{\mu\nu} + (L_\xi h)_{\mu\nu}, \quad T_{\mu\nu} \to T_{\mu\nu} + (L_\xi T)_{\mu\nu}$$

or

$$h_{\mu\nu} \to h_{\mu\nu} + \xi_{\mu;\nu} + \xi_{\nu;\mu}$$
$$T_{\mu\nu} \to T_{\mu\nu} + T^{\lambda}_{\mu} \xi_{\lambda;\nu} + T^{\lambda}_{\nu} \xi_{\lambda;\mu} + T_{\mu\nu;\lambda} \xi^{\lambda}.$$

They describe the propagation of small disturbances on a given space-time background, satisfying the field equations with energy-momentum tensor $T_{\mu\nu}$.

---

**Exercises:**

1. Compute the Riemann tensor in the linearized theory and show that it is invariant under the gauge transformations (4.1.14).
2. Show that the vacuum field equations (4.1.9) can be derived from the Lagrangian

$$\mathscr{L} = \tfrac{1}{4} h_{\mu\nu,\sigma} h^{\mu\nu,\sigma} - \tfrac{1}{2} h_{\mu\nu,\sigma} h^{\sigma\nu,\mu} - \tfrac{1}{4} h^{,\sigma} + \tfrac{1}{2} h_{,\sigma} h^{\nu\sigma}_{,\nu}.$$

For this, show that

$$\delta \int_D \mathscr{L}\, d^4x = \int_D G^{\mu\nu}\, \delta h_{\mu\nu}\, d^4x \tag{4.1.17}$$

for variations $\delta h_{\mu\nu}$ which vanish on the boundary of $D$.

3. Show that the Lagrangian (4.1.16) changes only by a divergence under gauge transformations (4.1.14). Thus if $\xi^\mu$ vanishes on the boundary of $D$, the action integral over $D$ is invariant under gauge transformations. Using this, derive the Bianchi identity (4.1.8) from (4.1.17).

---

**Remark:** Since $\mathscr{L}$ changes if the gauge is changed, the canonical energy-momentum tensor (from SR) is *not gauge invariant.* This is a reflection of the nonlocalizability of the gravitational energy. Even in the linearized theory, it is not possible to find gauge invariant expressions for the energy and momentum densities.

We now make use of gauge invariance to simplify the field equations. One can always find a gauge such that

$$\gamma^{\alpha\beta}_{,\beta} = 0 \quad \text{(Hilbert gauge).} \tag{4.1.18}$$

*Proof:* When

$$h_{\mu\nu} \to h_{\mu\nu} + \xi_{\mu,\nu} + \xi_{\nu,\mu}$$

then

$$\gamma_{\mu\nu} \to \gamma_{\mu\nu} + \xi_{\mu,\nu} + \xi_{\nu,\mu} - \eta_{\mu\nu}\, \xi^{\lambda}_{,\lambda}. \tag{4.1.19}$$

Thus

$$\gamma^{\mu\nu}_{,\nu} \to \gamma^{\mu\nu}_{,\nu} + \Box\, \xi^\mu + \xi^{\nu,\mu}_{,\nu} - \xi^{\lambda,\mu}_{,\lambda} = \Box\, \xi^\mu + \gamma^{\mu\nu}_{,\nu}\,.$$

If $\gamma^{\mu\nu}_{,\nu} \neq 0$, we merely need to choose $\xi^\mu$ to be a solution of

$$\Box\, \xi^\mu = -\gamma^{\mu\nu}_{,\nu}.$$

Such solutions always exist (retarded potentials).

For the Hilbert gauge, (4.1.12) has the simple form

$$\Box\, \gamma_{\mu\nu} = -16\,\pi\, G T_{\mu\nu}. \tag{4.1.20}$$

The most general solution of (4.20), with subsidiary condition (4.1.18) is

$$\gamma_{\mu\nu} = -16\pi\, G\, D_{\mathrm{R}} * T_{\mu\nu} + \text{solution of the homogeneous equation,} \tag{4.1.21}$$

where $D_R$ is the retarded Green's function

$$D_{\mathrm{R}}(x) = \frac{1}{4\pi\,|\boldsymbol{x}|}\, \delta(x^0 - |\boldsymbol{x}|)\, \theta(x^0)\,.$$

Note that

$$\partial_\nu (D_{\mathrm{R}} * T^{\mu\nu}) = D_{\mathrm{R}} * \partial_\nu T^{\mu\nu} = 0\,,$$

so that the first term of (4.1.21) satisfies the Hilbert condition (4.1.18). We shall discuss the solutions to the homogeneous equation (waves) in Sect. 4.3.

The retarded solution is given by

$$\gamma_{\mu\nu}(x) = -4G\int \frac{T_{\mu\nu}(x^0 - |\boldsymbol{x}-\boldsymbol{x}'|, \boldsymbol{x}')}{|\boldsymbol{x}-\boldsymbol{x}'|}\, d^3x' . \tag{4.1.22}$$

We interpret this field as the field resulting from the source, while the second term in (4.1.21) represents gravitational waves coming from infinity. As in electrodynamics, we conclude that gravitational effects propagate at the speed of light.

## 4.2 Nearly Newtonian Gravitational Fields

We now consider nearly Newtonian sources, with $T_{00} \gg |T_{0j}|, |T_{ij}|$ and such small velocities that retardation effects are negligible. In this case (4.1.22) becomes

$$\gamma_{00} = 4\phi, \quad \gamma_{0j} = \gamma_{ij} = 0, \tag{4.2.1}$$

where $\phi$ is the Newtonian potential

$$\phi = -G\int \frac{T_{00}(t, \boldsymbol{x}')}{|\boldsymbol{x}-\boldsymbol{x}'|}\, d^3x' . \tag{4.2.2}$$

For the metric

$$g_{\mu\nu} = \eta_{\mu\nu} + h_{\mu\nu} = \eta_{\mu\nu} + (\gamma_{\mu\nu} - \tfrac{1}{2}\eta_{\mu\nu}\gamma) \tag{4.2.3}$$

we have from (4.1.23) and $\gamma = 4\phi$

$$g_{00} = 1 + 2\phi, \quad g_{0i} = 0 \quad \text{and} \quad g_{ij} = -(1-2\phi)\,\delta_{ij}, \tag{4.2.4}$$

or

$$g = (1+2\phi)\, dt^2 - (1-2\phi)(dx^2 + dy^2 + dz^2). \tag{4.2.5}$$

At large distances from the source, the monopole contribution to (4.2.2) dominates and we obtain

$$g = (1-2m/r)\, dt^2 - (1+2m/r)(dx^2 + dy^2 + dz^2), \tag{4.2.6}$$

where, as usual, $m = GM/c^2$.

The errors in the metric (4.2.5) are:

(i) Terms of order $\phi^2$ are missing. These arise as a result of the nonlinearity of the theory.

(ii) $\gamma_{0j}$ vanishes up to terms of order $v\,\phi$, where $v \sim |T_{0j}|/T_{00}$ is a typical velocity of the source.

(iii) $\gamma_{ij}$ vanishes, up to terms of order $\phi\, |T_{ij}|/T_{00}$.

In the solar system, all of these errors are of order $10^{-12}$, while $\phi \sim 10^{-6}$.

**Remark:** Fermat's principle (1.6.9) and (4.2.5) imply that light rays behave in almost Newtonian fields exactly as in an optical medium with index of refraction

$$n = 1 - 2\,\phi. \tag{4.2.7}$$

**Exercises:**

1. Show that the metric (4.2.6) implies the correct deflection of light rays.
2. Show that the metric (4.2.6) gives $\frac{4}{3}$ times the Einstein value for the precession of the perihelion. This demonstrates that the precession of the perihelion is sensitive to nonlinearities in the theory.

## 4.3 Gravitational Waves in the Linearized Theory

We now consider the linearized theory in vacuum. In the Hilbert gauge, the field equations then read [see (4.1.20)]

$$\Box\,\gamma_{\mu\nu} = 0\,. \tag{4.3.1}$$

For the free field, it is possible to find, even within the Hilbert gauge class, a gauge such that we also have

$$\gamma = 0\,. \tag{4.3.2}$$

*Proof:* From (4.1.19), a change of gauge has the effect

$$\gamma_{\mu\nu} \to \gamma_{\mu\nu} + \xi_{\mu,\nu} + \xi_{\nu,\mu} - \eta_{\mu\nu}\,\xi^{\lambda}_{,\lambda}$$

and hence

$$\gamma \to \gamma - 2\,\xi^{\lambda}_{,\lambda}$$

as well as

$$\gamma^{\mu\nu}_{,\nu} \to \gamma^{\mu\nu}_{,\nu} + \Box\,\xi^{\mu}\,.$$

In order to satisfy the Hilbert condition $\gamma^{\mu\nu}_{,\nu} = 0$, we must have

$$\Box\,\xi^{\mu} = 0\,. \tag{4.3.3}$$

If $\gamma \neq 0$, then we must seek a $\xi^{\mu}$ which satisfies (4.3.3) and which also satisfies $2\,\xi^{\mu}_{,\mu} = \gamma$. Equation (4.3.3) then requires the consistency condition $\Box\,\gamma = 0$, which is generally satisfied only in vacuum. However, this condition is also *sufficient*. We shall show that if $\phi$ is a scalar field such that $\Box\,\phi = 0$, then there exists a vector field $\xi^{\mu}$ with the properties $\Box\,\xi^{\mu} = 0$, $\xi^{\mu}_{,\mu} = \phi$.

*Construction:* Let $\eta^{\mu}$ be a solution of $\eta^{\mu}_{,\mu} = \phi$. Such a solution exists; one may take $\eta_{\mu} = \Lambda_{,\mu}$, with $\Box\Lambda = \phi$. Now let $\zeta^{\mu} = \Box\,\eta^{\mu}$.

Since $\zeta^{\mu}_{,\mu}=0$, there exists[1] an antisymmetric tensor field $f^{\mu\nu}=-f^{\nu\mu}$, such that $\zeta^{\mu}=f^{\mu\nu}_{,\nu}$. Now let $\sigma^{\mu\nu}=-\sigma^{\nu\mu}$ be a solution of $\Box\sigma^{\mu\nu}=f^{\mu\nu}$. Then

$$\xi^{\mu}:=\eta^{\mu}-\sigma^{\mu\nu}_{,\nu}$$

satisfies the equations

$$\xi^{\mu}_{,\mu}=\eta^{\mu}_{,\mu}=\phi$$
$$\Box\xi^{\mu}=\Box\eta^{\mu}-\Box\sigma^{\mu\nu}_{,\nu}=\zeta^{\mu}-f^{\mu\nu}_{,\nu}=0\,.$$

For the gauge class determined by (4.1.18) and (4.3.1), only gauge transformation which satisfy the additional conditions

$$\Box\xi^{\mu}=0 \quad \text{and} \quad \xi^{\mu}_{,\mu}=0 \tag{4.3.4}$$

are allowed. Within this gauge class we have $\gamma_{\mu\nu}=h_{\mu\nu}$.

*Plane Waves*

The most general solution of (4.3.1) can be represented as a superposition of plane waves

$$h_{\mu\nu}=\mathrm{Re}\,\{\varepsilon_{\mu\nu}\,\mathrm{e}^{-\mathrm{i}(k,x)}\}\,. \tag{4.3.5}$$

The field equations (4.3.1) are satisfied provided

$$k^2:=(k,k)=0\,. \tag{4.3.6}$$

The Hilbert condition $\gamma^{\mu\nu}_{,\nu}=h^{\mu\nu}_{,\nu}=0$ implies

$$k_{\mu}\,\varepsilon^{\mu}_{\nu}=0\,, \tag{4.3.7}$$

and (4.3.2) gives

$$\varepsilon^{\mu}_{\mu}=0\,. \tag{4.3.8}$$

The matrix $\varepsilon_{\mu\nu}$ is called the *polarisation tensor*. The conditions (4.3.7) and (4.3.8) imply that at most five components are independent. We shall now prove that, due to the remaining gauge invariance, only two of them are independent.

Under a gauge transformation belonging to

$$\xi^{\mu}(x)=\mathrm{Re}\,\{\mathrm{i}\,\varepsilon^{\mu}\,\mathrm{e}^{-\mathrm{i}(k,x)}\} \tag{4.3.9}$$

$\varepsilon_{\mu\nu}$ changes according to

$$\varepsilon_{\mu\nu}\to\varepsilon_{\mu\nu}+k_{\mu}\,\varepsilon_{\nu}+k_{\nu}\,\varepsilon_{\mu}\,. \tag{4.3.10}$$

Now consider a wave which propagates in the positive $z$-direction:

$$k^{\mu}=(k,0,0,k)\,. \tag{4.3.11}$$

[1] This follows from Poincaré's Lemma, written in terms of the codifferential. See Sect. 4.6.3 of Part I.

From (4.3.7) we have

$$\varepsilon_{0\nu} = \varepsilon_{3\nu} \Rightarrow \varepsilon_{00} = \varepsilon_{30} = \varepsilon_{03} = \varepsilon_{33}$$

$$\varepsilon_{01} = \varepsilon_{31}\,, \qquad \varepsilon_{02} = \varepsilon_{32}$$

and (4.3.8) implies

$$\varepsilon_{00} - \varepsilon_{11} - \varepsilon_{22} - \varepsilon_{33} = 0\,,$$

so that

$$\varepsilon_{11} + \varepsilon_{22} = 0\,.$$

These relations permit us to express all components in terms of

$$\varepsilon_{00},\ \varepsilon_{11},\ \varepsilon_{01},\ \varepsilon_{02},\ \varepsilon_{12}\,. \tag{4.3.12}$$

In particular, we have

$$\varepsilon_{03} = \varepsilon_{00}\,, \quad \varepsilon_{13} = \varepsilon_{01}\,, \quad \varepsilon_{22} = -\,\varepsilon_{11}\,, \quad \varepsilon_{23} = \varepsilon_{02}\,, \quad \varepsilon_{33} = \varepsilon_{00}\,. \tag{4.3.13}$$

For a gauge transformation (4.3.10), we have

$$\begin{aligned} &\varepsilon_{00} \to \varepsilon_{00} + 2k\,\varepsilon_0, \quad \varepsilon_{11} \to \varepsilon_{11}, \quad \varepsilon_{01} \to \varepsilon_{01} + k\,\varepsilon_1 \\ &\varepsilon_{02} \to \varepsilon_{02} + k\,\varepsilon_2, \qquad \varepsilon_{12} \to \varepsilon_{12}\,. \end{aligned} \tag{4.3.14}$$

If we require that (4.3.9) satisfies the conditions (4.3.4), then $k^\mu\,\varepsilon_\mu = 0$, or, in this case, $\varepsilon_0 = \varepsilon_3$. It then follows from (4.3.14) that one can choose $\varepsilon^\mu$ such that only $\varepsilon_{12}$ and $\varepsilon_{11} = -\,\varepsilon_{22}$ do not vanish. As is the case for light, one has *only two linearly independent polarization states*. Under a rotation about the $z$-axis, these transform according to

$$\varepsilon'_{\mu\nu} = R^\alpha_\mu\,R^\beta_\nu\,\varepsilon_{\alpha\beta}\,,$$

where

$$R(\varphi) = \left(\begin{array}{c|ccc} 1 & & 0 & \\ \hline & \cos\varphi & \sin\varphi & 0 \\ 0 & -\sin\varphi & \cos\varphi & 0 \\ & 0 & 0 & 1 \end{array}\right).$$

One easily finds

$$\begin{aligned} \varepsilon'_{11} &= \quad \varepsilon_{11}\cos 2\varphi + \varepsilon_{12}\sin 2\varphi \\ \varepsilon'_{12} &= -\,\varepsilon_{11}\sin 2\varphi + \varepsilon_{12}\cos 2\varphi \end{aligned}$$

or, for

$$\begin{aligned} \varepsilon_\pm &:= \varepsilon_{11} \mp \mathrm{i}\,\varepsilon_{12} \\ \varepsilon'_\pm &= \mathrm{e}^{\pm 2\mathrm{i}\varphi}\,\varepsilon_\pm\,. \end{aligned}$$

The polarization states $\varepsilon_\pm$ have *helicity* $\pm 2$ (left, resp. right handed circular polarization).

**Exercise:** Work out the analogous considerations for electrodynamics.

*Traceless, Transverse Gauge*

In a gauge for which only $\varepsilon_{12}$ and $\varepsilon_{11} = -\varepsilon_{22}$ do not nonvanish, we obviously have

$$h_{\mu 0} = 0\,, \qquad h_{kk} = 0\,, \qquad h_{kj,j} = 0\,; \tag{4.3.15}$$

here and henceforth $h_{kk} := \sum_{k=1}^{3} h_{kk}$!

We now consider a *general* gravitational wave $h_{\mu\nu}$ in the linearized theory. For each plane wave of the Fourier decomposition, we choose the special gauge (4.3.15). One can achieve this by first choosing the special gauge $h^{\mu\nu}_{\;\;,\nu} = 0$, $h^{\mu}_{\mu} = 0$, and then carrying out a superposition of gauge transformations of the form (4.3.9), i.e. with

$$\xi^{\mu}(x) = \mathrm{Re}\,\{\textstyle\int \mathrm{i}\,\varepsilon^{\mu}(k)\,\mathrm{e}^{-\mathrm{i}(k,x)}\,d^4k\}\,.$$

Since the gauge conditions (4.3.15) are all linear, they will then be satisfied for the general wave under consideration. We have thus shown that the gauge conditions (4.3.15) can be satisfied for an arbitrary gravitational wave. If a symmetric tensor satisfies (4.3.15), we call it a traceless, transverse (TT) tensor, and the gauge (4.3.15) is known as the traceless transverse gauge.

Note that in the gauge (4.3.15), only the $h_{ij}$ are nonvanishing, and hence we have *only six* wave equations

$$\Box\, h_{ij} = 0\,. \tag{4.3.16}$$

We now compute the linearized Riemann tensor. In general, we have

$$R^{\mu}_{\;\sigma\nu\varrho} = \partial_{\nu}\,\Gamma^{\mu}_{\varrho\sigma} - \partial_{\varrho}\,\Gamma^{\mu}_{\nu\sigma} + \text{quadratic terms in } \Gamma.$$

Using (4.1.3) for the Christoffel symbols, we obtain ($R_{\mu\sigma\nu\varrho} = g_{\mu\lambda}\,R^{\lambda}_{\;\sigma\nu\varrho}$)

$$\begin{aligned} R_{\mu\sigma\nu\varrho} &= \tfrac{1}{2}\,(h_{\sigma\mu,\nu\varrho} + h_{\varrho\mu,\sigma\nu} - h_{\varrho\sigma,\mu\nu}) - \tfrac{1}{2}\,(h_{\sigma\mu,\nu\varrho} + h_{\nu\mu,\sigma\varrho} - h_{\nu\sigma,\mu\varrho}) \\ &= \tfrac{1}{2}\,(h_{\nu\sigma,\mu\varrho} + h_{\varrho\mu,\sigma\nu} - h_{\nu\mu,\sigma\varrho} - h_{\varrho\sigma,\mu\nu})\,. \end{aligned} \tag{4.3.17}$$

We consider in particular the components

$$R_{i0j0} = R_{0i0j} = -R_{i00j} = -R_{0ij0}\,.$$

In the TT-gauge we have

$$R_{i0j0} = \tfrac{1}{2}\,(h_{0j,0i} + h_{i0,j0} - h_{ij,00} - h_{00,ij}) = -\tfrac{1}{2}\,h_{ij,00}\,. \tag{4.3.18}$$

Since the Riemann tensor is gauge invariant, one sees from (4.3.18) that $h_{\mu\nu}$ cannot be reduced to fewer components than is possible in the TT-gauge.

*Geodetic Deviation for a Gravitational Wave*

As in Sect. 2.1 we consider a collection (congruence) of freely falling test bodies. The separation vector $n$ between neighboring geodesics satisfies (2.1.11) relative to an orthonormal frame $\{e_i\}$ perpendicular to $\dot{\gamma}$, which is parallel displaced along an arbitrary geodesic $\gamma(\tau)$ of the congruence, namely

$$\frac{d^2}{d\tau^2}\boldsymbol{n} = K\boldsymbol{n}, \tag{4.3.19}$$

where

$$K_{ij} = R^i_{00j}. \tag{4.3.20}$$

There exists a TT-*coordinate system*, in which $h_{\mu\nu}$ satisfy the TT conditions (4.3.15), for which, to leading order in the $h_{ij}$ we have, along $\gamma$

$$\partial/\partial x^i = e_i \quad (i = 1, 2, 3), \qquad \partial/\partial t = e_0 := \dot{\gamma}. \tag{4.3.21}$$

This system can be constructed in two steps. One first constructs a local inertial system along $\gamma$ (in which $t = \tau$), as in Sect. 1.10.5. This satisfies (4.3.21). Then one performs an infinitesimal coordinate transformation, such that the TT conditions (4.3.15) are satisfied.

Thus, according to (4.3.18) and (4.3.19), we have (since $t$ is equal to the proper time $\tau$ along $\gamma$, to first order in the $h_{ij}$)

$$\frac{d^2 n^i}{dt^2} = -\frac{1}{2}\frac{\partial^2 h_{ij}}{\partial t^2} n^j. \tag{4.3.22}$$

This equation describes the oscillations of test bodies near $\gamma$ induced by weak gravitational waves. For example, if the particles were at rest relative to one another before the wave arrives ($\boldsymbol{n} = \boldsymbol{n}_{(0)}$ for $h_{ij} = 0$), then integration of (4.3.22) results in

$$n^i(\tau) \simeq n^i_{(0)} - \tfrac{1}{2} h_{ij}(\gamma(\tau))\, n^j_{(0)}. \tag{4.3.23}$$

We now consider the special case of a plane wave propagating in the $z$-direction. In the TT-gauge the nonvanishing components are

$$h_{xx} = -h_{yy} = A(t - z)$$
$$h_{xy} = h_{yx} = B(t - z). \tag{4.3.24}$$

The particle will not oscillate if it is moving in the direction of propagation ($\boldsymbol{n}_{(0)} = (0, 0, a)$), since then $h_{ij}n^j_{(0)} = 0$. Hence only *transverse oscillations are possible.* We shall now discuss these for various polarizations of the incident plane wave. The transverse components of the displacement vector $\boldsymbol{n}$ satisfy

$$\ddot{n}_\perp = K_\perp n_\perp, \tag{4.3.25}$$

where

$$K_\perp = -\tfrac{1}{2}\begin{pmatrix} \ddot{h}_{xx} & \ddot{h}_{xy} \\ \ddot{h}_{yx} & \ddot{h}_{yy} \end{pmatrix}, \qquad K_\perp^T = K_\perp, \qquad \mathrm{Tr}\, K_\perp = 0\,. \tag{4.3.26}$$

According to (4.3.23), an approximate solution is

$$n_\perp \simeq n_\perp^{(0)} - \tfrac{1}{2}\begin{pmatrix} h_{xx} & h_{xy} \\ h_{yx} & h_{yy} \end{pmatrix} n_\perp^{(0)}\,. \tag{4.3.27}$$

If we now transform the matrix on the right hand side to principal axes,

$$-\tfrac{1}{2}\begin{pmatrix} h_{xx} & h_{xy} \\ h_{yx} & h_{yy} \end{pmatrix} = R\begin{pmatrix} \Omega & 0 \\ 0 & -\Omega \end{pmatrix} R^T, \tag{4.3.28}$$

we have, with

$$n_\perp =: R\begin{pmatrix} \xi \\ \eta \end{pmatrix}, \qquad n_\perp^{(0)} = R\begin{pmatrix} \xi_0 \\ \eta_0 \end{pmatrix} \tag{4.3.29}$$

the motion

$$\begin{aligned} \xi &\simeq \xi_0 + \Omega(t)\, \xi_0 \\ \eta &\simeq \eta_0 - \Omega(t)\, \eta_0\,. \end{aligned} \tag{4.3.30}$$

Thus the point $(\xi, \eta)$ performs a "quadrupole oscillation" about the point $(\xi_0, \eta_0)$.

We now consider the special case of a *periodic* plane wave propagating in the $z$-direction

$$\begin{aligned} h_{xx} &= -h_{yy} = \mathrm{Re}\,\{A_1\, \mathrm{e}^{-\mathrm{i}\omega(t-z)}\} \\ h_{xy} &= h_{yx} = \mathrm{Re}\,\{A_2\, \mathrm{e}^{-\mathrm{i}\omega(t-z)}\}\,. \end{aligned} \tag{4.3.31}$$

If $A_2 = 0$, the principal axes conincide with the $x$- and $y$-axes, as can be seen from (4.3.28). If $A_1 = 0$, the principal axes are rotated by 45°. For obvious reasons, we say that the wave is *linearly polarized* in these two

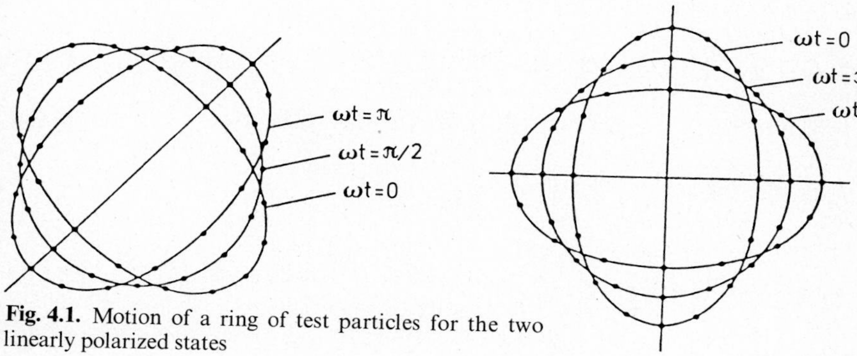


**Fig. 4.1.** Motion of a ring of test particles for the two linearly polarized states

cases. If $A_2 = \pm \mathrm{i} A_1$, the wave is said to be (right, resp. left) circularly polarized. For the upper sign, the principal axes rotate in the positive direction (counter clockwise for a wave approaching the reader), and in the negative direction for the lower sign. The principal axes rotate at the frequency $\omega/2$.

Figure 4.1 shows the motion of a ring of test particles moving about a central particle in the transverse plane, for the two linearly polarized states.

**Example.** Consider a pulse corresponding to the radiation of the energy equivalent to 10% of a solar mass by a supernova event at the center of our galaxy. If the duration of the pulse is 1 ms, the flux will be $\sim 10^{10}\,\mathrm{erg\,cm^{-2}\,s^{-1}}$. From this we conclude (see Exercise on p. 235) that the dimensionless strain amplitude is about $10^{-17}$. This would induce in the Earth-Moon separation a change of $\sim 10^{-7}$ cm!

## 4.4 The Gravitational Field at Large Distances from the Source

We now consider a stationary gravitational field far away from an arbitrary distribution of isolated sources and determine the leading terms of its expansion in powers of $1/r$.

For a stationary field we choose an adapted coordinate system, with respect to which the $g_{\mu\nu}$ are independent of time. It is possible to choose the asymptotic behavior to be nearly Lorentzian:

$$g_{\mu\nu} = \eta_{\mu\nu} + h_{\mu\nu}, \qquad |h_{\mu\nu}| \ll 1 .$$

As a first step, we ignore the nonlinearity of the field equations in the asymptotic region. There we have, in the Hilbert gauge,

$$\Delta \gamma_{\mu\nu} = 0, \qquad \gamma_{\mu\nu} := h_{\mu\nu} - \tfrac{1}{2}\eta_{\mu\nu} h . \tag{4.4.1}$$

Every solution of the Poisson equation $\Delta\phi = 0$ can be expanded in terms of spherical harmonics as

$$\phi(\boldsymbol{x}) = \sum_{l=0}^{\infty} \sum_{m=-l}^{+l} (a_{lm} r^l + b_{lm} r^{-(l+1)}) Y_{lm}(\hat{\boldsymbol{x}}) .$$

For our problem, the $a_{lm}$ vanish and we may take the $\gamma_{\mu\nu}$ to have the form

$$\gamma_{00} = \frac{A^0}{r} + \frac{B^j n^j}{r^2} + O\left(\frac{1}{r^3}\right), \qquad n^j = x^j/r$$

$$\gamma_{0j} = \frac{A^j}{r} + \frac{B^{jk} n^k}{r^2} + O\left(\frac{1}{r^3}\right), \qquad \gamma_{jk} = \frac{A^{jk}}{r} + \frac{B^{jkl} n^l}{r^2} + O\left(\frac{1}{r^3}\right), \tag{4.4.2}$$

where $A^{jk} = A^{kj}$ and $B^{jkl} = B^{kjl}$.

From the Hilbert condition $\gamma^{\mu\nu}_{,\nu} = 0$, we obtain

$$A^j = 0\,, \qquad A^{jk} = 0$$
$$B^{jk}(\delta^{jk} - 3n^j n^k) = 0\,, \quad B^{jkl}(\delta^{kl} - 3n^k n^l) = 0\,. \tag{4.4.3}$$

We now decompose $B^{jk}$ and $B^{jkl}$ in the useful form

$$B^{jk} = \delta^{jk} B + S^{jk} + A^{jk}\,, \tag{4.4.4}$$

where $B$ is the trace of $B^{jk}$, $S^{jk}$ is symmetric and traceless, $S^{jj} = 0$, and $A^{jk} = \varepsilon^{jkl} F^l$ is antisymmetric.

The gauge conditions (4.4.3) imply

$$(\delta^{jk} B + S^{jk} + A^{jk})(\delta^{jk} - 3n^j n^k) = -3S^{jk} n^j n^k = 0\,,$$

which means that

$$S^{jk} = 0\,.$$

---

**Exercise:** Show that the $B^{jkl}$ can be written in an analogous manner as follows (parentheses indicate symmetrization in the enclosed indices):

$$B^{jkl} = \delta^{jk} A^l + C^{(j}\delta^{k)l} + \varepsilon^{ml(j} E^{k)m} + S^{jkl}\,,$$

where $E^{km}$ is symmetric and traceless and $S^{jkl}$ is completely antisymmetric and traceless in all pairs of indices. From (4.4.3) we then have

$$C^j = -2A^j\,, \qquad E^{km} = S^{jkl} = 0\,.$$

---

If one now uses these results in (4.4.2), one obtains

$$\gamma_{00} = \frac{A^0}{r} + \frac{B^j n^j}{r^2} + O\left(\frac{1}{r^3}\right)$$
$$\gamma_{0j} = \frac{\varepsilon^{jkl} n^k F^l}{r^2} + \frac{B n^j}{r^2} + O\left(\frac{1}{r^3}\right) \tag{4.4.5}$$
$$\gamma_{jk} = \frac{\delta^{jk} A^l n^l - A^j n^k - A^k n^j}{r^2} + O\left(\frac{1}{r^3}\right).$$

Within the Hilbert gauge, one can choose $B$ in $\gamma_{0j}$ and $A^j$ in $\gamma_{jk}$ to vanish. This is achieved with the gauge transformation defined by the vector field $\xi_j = -A^j/r$ and $\xi_0 = B/r$. Then (4.4.5) becomes

$$\gamma_{00} = \frac{A^0}{r} + \frac{(B^j + A^j)\, n^j}{r^2} + O\left(\frac{1}{r^3}\right)$$
$$\gamma_{0j} = \frac{\varepsilon^{jkl} n^k F^l}{r^2} + O\left(\frac{1}{r^3}\right) \tag{4.4.6}$$
$$\gamma_{ik} = O(1/r^3)\,.$$

Finally, we shift the origin of the coordinate system such that

$$x^j_{\text{new}} = x^j_{\text{old}} - (B^j + A^j)/A^0 .$$

The second term in $\gamma_{00}$ then vanishes. Rewriting everything in terms of $h_{\mu\nu}$ then gives

$$\begin{aligned} h_{00} &= -\frac{2m}{r} + O\left(\frac{1}{r^3}\right) \\ h_{0i} &= 2G\,\varepsilon_{ijk}\frac{S^j x^k}{r^3} + O\left(\frac{1}{r^3}\right) \\ h_{ij} &= -\frac{2m}{r}\,\delta_{ij} + O\left(\frac{1}{r^3}\right) . \end{aligned} \tag{4.4.7}$$

The constants $m$ and $S^j$ will be interpreted further on.

Up to now we have considered only the linear approximation in the asymptotic region. The dominant nonlinear terms must be proportional to the square, $(m/r)^2$, of the dominant linear terms. In order to calculate them, it is simplest to consider the Schwarzschild solution in appropriate coordinates and expand it in powers of $m/r$. (In Chap. 5, we will obtain the same result in the post-Newtonian approximation.)

First of all, we note that the Hilbert gauge condition is the linearized form of the harmonic gauge condition $\delta dx = 0$. According to Sect. 4.6.3 of Part I, this can be written as

$$(\sqrt{-g}\; g^{\mu\nu})_{,\nu} = 0 . \tag{4.4.8}$$

Now

$$\sqrt{-g} \simeq 1 + \tfrac{1}{2}h \quad \text{and} \quad g^{\mu\nu} \simeq \eta^{\mu\nu} - h^{\mu\nu} ,$$

and hence

$$\sqrt{-g}\; g^{\mu\nu} \simeq \eta^{\mu\nu} - \gamma^{\mu\nu} . \tag{4.4.9}$$

Thus, in the linear approximation, (4.4.8) does in fact reduce to $\gamma^{\mu\nu}_{,\nu} = 0$.

For this reason, we transform the Schwarzschild solution

$$ds^2 = (1 - 2m/r)\, dt^2 - \frac{dr^2}{1 - 2m/r} - r^2\,(d\vartheta^2 + \sin^2\vartheta\, d\varphi^2)$$

to harmonic coordinates. These are constructed using the results of the following exercise.

---

**Exercise:** Insert the ansatz

$$x^1 = R(r)\sin\vartheta\cos\varphi , \quad x^2 = R(r)\sin\vartheta\sin\varphi , \quad x^3 = R(r)\cos\vartheta ,$$
$$x^0 = t$$

into the gauge condition $\delta dx = 0$ and obtain the following differential equation for $R(r)$:

$$\frac{d}{dr}\left[r^2(1-2m/r)\frac{dR}{dr}\right]-2R=0\,.$$

A useful solution is $R=r-m$.

---

The transformed Schwarzschild metric then reads, if now $r=(x^i x^i)^{1/2}$,

$$\begin{aligned}g_{00}&=\frac{1-m/r}{1+m/r}\\ g_{i0}&=0\\ g_{ij}&=-(1+m/r)^2\,\delta_{ij}-(m/r)^2\,\frac{1+m/r}{1-m/r}\frac{x^i x^j}{r^2}\,.\end{aligned}\tag{4.4.10}$$

The quadratic terms in $g_{00}$ are $g_{00}^{(2)}=2\,(m/r)^2$. We thus obtain the expansion

$$\begin{aligned}g_{00}&=1-\frac{2m}{r}+2\frac{m^2}{r^2}+O\left(\frac{1}{r^3}\right)\\ g_{i0}&=2G\,\varepsilon_{ijk}\frac{S^j x^k}{r^3}+O\left(\frac{1}{r^3}\right)\\ g_{ij}&=-\left(1+\frac{2m}{r}\right)\delta_{ij}+O\left(\frac{1}{r^2}\right).\end{aligned}\tag{4.4.11}$$

We shall also obtain this result in the post-Newtonian approximation in Chap. 5.

We can now compute the energy and angular momentum of the system, using the asymptotic expressions (4.4.11). We use the expressions (2.6.19) and (2.6.24) for these quantities in terms of flux integrals. In our coordinates, the connection forms are given by (4.1.1)

$$\omega^\alpha_\beta=\Gamma^\alpha_{\gamma\beta}\,dx^\gamma=\tfrac{1}{2}\,(h^\alpha_{\gamma,\beta}+h^\alpha_{\beta,\gamma}-h_{\gamma\beta}^{;\alpha})\,dx^\gamma\,.$$

According to (2.6.19) the total momentum is given by

$$\begin{aligned}P^\varrho&=-\frac{1}{16\pi G}\oint\sqrt{-g}\;\omega_{\alpha\beta}\wedge\eta^{\alpha\beta\varrho}\\ &=-\frac{1}{16\pi G}\,\varepsilon^{\varrho\alpha\beta}{}_\sigma\oint\tfrac{1}{2}\,(h_{\alpha\gamma,\beta}-h_{\gamma\beta,\alpha})\,dx^\gamma\wedge dx^\sigma\\ &=\frac{1}{16\pi G}\,\varepsilon^{\varrho\alpha\beta}{}_\sigma\oint h_{\beta\gamma,\alpha}\,dx^\gamma\wedge dx^\sigma\,.\end{aligned}$$

Only $P^0$ does not vanish:

$$P^0=\frac{1}{16\pi G}\,\varepsilon^{0ij}{}_l\oint h_{jk,i}\,dx^k\wedge dx^l\,.$$

According to (4.4.11), we have

$$h_{jk,i} = \frac{2m}{r^3} x^i \delta_{jk} .$$

If we now integrate over a large spherical surface, we then obtain (with $M = m/G$):

$$P^0 = \frac{M}{8\pi} \int \frac{1}{r^3} \underbrace{x^i \varepsilon_{ijk}\, dx^j \wedge dx^k}_{2r^3\, d\Omega} = M .$$

Thus, as expected, we have

$$P^0 = m/G . \tag{4.4.12}$$

According to (2.6.24), the angular momentum is given by the following flux integral:

$$J^{\varrho\alpha} = \frac{1}{16\pi G} \oint \sqrt{-g}\; [(x^\varrho \eta^{\alpha\beta\gamma} - x^\alpha \eta^{\varrho\beta\gamma}) \wedge \omega_{\beta\gamma} + \eta^{\varrho\alpha}] . \tag{4.4.13}$$

For a "surface at infinity", the first term is

$$\begin{aligned} J^{\varrho\alpha}_{(1)} &= \frac{1}{16\pi G} \oint \sqrt{-g}\; (x^\varrho \eta^{\alpha\beta\gamma} - x^\alpha \eta^{\varrho\beta\gamma}) \wedge \omega_{\beta\gamma} \\ &= \frac{1}{2} \frac{1}{16\pi G} \oint (x^\varrho \eta^{\alpha\beta\gamma} - x^\alpha \eta^{\varrho\beta\gamma}) \wedge (h_{\beta\sigma,\gamma} - h_{\sigma\gamma,\beta})\, dx^\sigma \\ &= \frac{1}{16\pi G} \oint (x^\varrho \varepsilon^{\alpha\beta\gamma}{}_\mu - \varrho \leftrightarrow \alpha]\, h_{\beta\sigma,\gamma}\, dx^\mu \wedge dx^\sigma . \end{aligned}$$

The spatial components are

$$\begin{aligned} J^{ri}_{(1)} &= \frac{1}{16\pi G} \varepsilon^{i0k}{}_l \oint x^r h_{0j,k}\, dx^l \wedge dx^j - (r \leftrightarrow i) \\ &= -\frac{1}{16\pi G} \varepsilon_{ikl} \oint (x^r h_{0j,k} - i \leftrightarrow r)\, dx^l \wedge dx^j . \end{aligned}$$

The components of the angular momentum vector are given by

$$J_s = \tfrac{1}{2} \varepsilon_{ris} J^{ri} .$$

For these, we obtain

$$\begin{aligned} J^{(1)}_s &= -\frac{1}{16\pi G} \underbrace{\varepsilon_{ris}\varepsilon_{ikl}}_{-(\delta_{rk}\delta_{sl} - \delta_{rl}\delta_{ks})} \oint x^r h_{0j,k}\, dx^l \wedge dx^j \\ &= -\frac{1}{16\pi G} \oint [-x^k h_{0j,k}\, dx^s \wedge dx^j + h_{0j,s} \overbrace{x^l dx^l}^{\frac{1}{2} d(r^2)} \wedge dx^j] . \end{aligned}$$

The second term does not contribute to the surface integral over a large sphere, and hence

$$J_s^{(1)} = \frac{1}{16\pi G} \oint x^k \underbrace{h_{0j,k}}_{2G\,\varepsilon_{jmn} S^m \underbrace{\left(\frac{x^n}{r^3}\right)_{,k} x^k}_{x^n/r^3 - 3\frac{x^n x^k}{r^5} x^k = -2\frac{x^n}{r^3}}} dx^s \wedge dx^j$$

$$= -\frac{1}{4\pi} \oint \varepsilon_{jmn} S^m \frac{x^n}{r^3} dx^s \wedge dx^j .$$

After an integration by parts, we have

$$J_s^{(1)} = \frac{1}{4\pi} \int \frac{x^s}{r^3} \varepsilon_{jmn} S^m dx^n \wedge dx^j = \frac{1}{4\pi} 2 \oint S^m \frac{x^s}{r^3} r^2 \hat{x}^m d\Omega$$

$$= \frac{1}{4\pi} 2 S^m \oint \hat{x}^s \hat{x}^m d\Omega = \frac{2}{3} S^s$$

or

$$J_k^{(1)} = \frac{2}{3} S^k . \tag{4.4.14}$$

The second term of (4.4.13) contributes

$$J_{(2)}^{ik} = \frac{1}{16\pi G} \oint \frac{1}{2!} \varepsilon^{ik}{}_{\alpha\beta} dx^\alpha \wedge dx^\beta$$

$$= \frac{1}{16\pi G} \oint 2 \frac{1}{2!} \varepsilon^{ikol} g_{or} g_{ls} dx^r \wedge dx^s$$

$$= \frac{1}{16\pi G} \oint \varepsilon_{ikl} \delta_{ls} g_{or} dx^r \wedge dx^s$$

or

$$J_l^{(2)} = \frac{1}{16\pi G} \oint g_{or} dx^r \wedge dx^l = \frac{1}{8\pi} S^m \oint \varepsilon_{rmn} \frac{x^n}{r^3} dx^r \wedge dx^l$$

$$= \frac{1}{8\pi} S^m \oint \frac{x^l}{r^3} \varepsilon_{rmn} dx^n \wedge dx^r = \frac{1}{8\pi} S^m \oint \frac{x^l}{r} 2 \hat{x}^m d\Omega = \frac{1}{3} S^l .$$

Hence

$$J_k^{(2)} = \tfrac{1}{3} S^k . \tag{4.4.15}$$

The total angular momentum is the sum of (4.4.14) and (4.4.15):

$$J_k = S^k . \tag{4.4.16}$$

We have thus shown that the parameters $S^j$ in (4.4.11) are the components of the system's angular momentum.

On the other hand, we have representations of $P^0$ and $J_k$ in terms of volume integrals [see (2.6.17) and (2.6.18)]. For the particular case that the field is weak *everywhere*, these reduce to

$$P^0 \simeq \int_{x^0=\text{const}} *T_0 \simeq \int_{x^0=\text{const}} T^{00}\, d^3x \tag{4.4.17}$$

and

$$J^{ij} \simeq \int_{x^0=\text{const}} \sqrt{-g}\,(x^i *T^j - x^j *T^i) \simeq \int_{x^0=\text{const}} (x^i T^{j0} - x^j T^{i0})\, d^3x\,. \tag{4.4.18}$$

The angular momentum $S^k$ of a system can be measured with a gyroscope. Suppose this is far away from the source and at rest in the coordinate system in which (4.4.11) is valid. The gyroscope then rotates relative to the basis $\partial/\partial x^i$ with the angular velocity [see (1.10.33)]

$$\boldsymbol{\Omega} \simeq -\tfrac{1}{2}\nabla \times \boldsymbol{g}, \quad \text{where } \boldsymbol{g} = (g_{01}, g_{02}, g_{03})\,.$$

If we now use (4.4.11) for the $g_{i0}$, we obtain

$$\boldsymbol{\Omega} = \frac{G}{r^3}\left[\boldsymbol{S} - 3\frac{(\boldsymbol{S}\cdot\boldsymbol{x})\,\boldsymbol{x}}{r^2}\right] \tag{4.4.19}$$

for the precession frequency of a gyroscope relative to an asymptotically Lorentzian system (the "fixed stars").

---

**Exercise:** Consider a stationary asymptotically flat metric of an isolated system in asymptotically Minkowskian coordinates such that the stationary Killing field has the form $K = \partial/\partial x^0$ and the metric coefficients are given by (4.4.11). Show that the total mass $M$ can be expressed in a coordinate independent manner by the *Komar formula:*

$$8\pi GM = \lim_{\infty} \oint K^{[\mu;\nu]}\, dS_{\mu\nu}, \tag{4.4.20}$$

where the integral is taken over a spacelike 2-surface $S$ in the limit as this surface is taken out to arbitrarily large asymptotic distances. The measure-valued 2-form $dS_{\mu\nu}$ corresponds to the 2-forms $\eta_{\mu\nu}$ of $S$ (i.e. $X^\mu Y^\nu dS_{\mu\nu}$ is the measure on $S$ corresponding to $i_X i_Y \eta$ for vector fields $X$ and $Y$ which are linearly independent in every point of $S$).
*Solution:* From $K_\mu = g_{\mu o}$ we have $K_{[\mu,\nu]} = \frac{1}{2}(g_{\mu o,\nu} - g_{\nu o,\mu})$ and

$$\begin{aligned}\lim_{\infty}\oint_S K_{[\mu,\nu]}\,\eta^{\mu\nu} &= \lim_{\infty}\oint_S h_{\mu o,\nu}\tfrac{1}{2}\varepsilon^{\mu\nu}{}_{\lambda\sigma}\,dx^\lambda\wedge dx^\sigma\\ &= \lim_{\infty}\tfrac{1}{2}\varepsilon^{0j}{}_{kl}\oint_S h_{00,j}\,dx^k\wedge dx^l\\ &= \lim_{\infty}\tfrac{1}{2}\oint_S \frac{2m}{r^3}\,x^j\,\varepsilon_{jkl}\,dx^k\wedge dx^l = 8\pi m\,.\end{aligned}$$

---

**Remark:** With Stokes' theorem one can convert the integral (4.4.20) into a volume integral (plus surface integrals over horizons). For this one needs

$$d(K^{[\mu;\nu]}\eta_{\mu\nu}) = D(K^{\mu;\nu}\eta_{\mu\nu}) = DK^{\mu;\nu}\wedge\eta_{\mu\nu}$$
$$= K^{\mu;\nu}{}_{;\lambda}\,\theta^\lambda\wedge\eta_{\mu\nu} = (K^{\mu;\nu}{}_{;\nu} - K^{\nu;\mu}{}_{;\nu})\,\eta_\mu\,.$$

The Killing equation implies

$$K^{\nu;\mu}{}_{;\nu} = K^{\nu}{}_{;\nu}{}^{;\mu} + R^\mu_\nu K^\nu = R^\mu_\nu K^\nu$$

and

$$K^{\mu;\nu}{}_{;\nu} = -K^{\nu;\mu}{}_{;\nu} = -R^\mu_\nu K^\nu\,.$$

Thus we obtain

$$d(K^{[\mu;\nu]}\eta_{\mu\nu}) = -2R^\mu_\nu K^\nu\eta_\mu. \tag{4.4.21}$$

If there are no horizons (4.4.20) and (4.4.21) imply

$$4\pi GM = -\int_\Sigma R^\mu_\nu K^\nu d\sigma_\mu\,, \tag{4.4.22}$$

where $\Sigma$ is a spacelike hypersurface extending to infinity and $d\sigma_\mu$ is the measure-valued 1-form corresponding to $\eta_\mu$ on $\Sigma$ (i.e. $X^\mu d\sigma_\mu$ is the measure corresponding to the volume form $i_X\eta$ on $\Sigma$ for every non-vanishing vector field $X$).

On the right hand side of (4.4.22) one can use the field equations to replace $R^\mu_\nu$ by $8\pi G(T^\mu_\nu - \frac{1}{2}\delta^\mu_\nu T)$.

## 4.5. Emission of Gravitational Radiation

In this section we investigate, in the framework of the linearized theory, the energy radiated in the form of gravitational waves by a time-dependent source. For this, we need the solution (4.1.22)

$$\gamma_{\mu\nu}(t,\boldsymbol{x}) = -4G\int\frac{T_{\mu\nu}(t-|\boldsymbol{x}-\boldsymbol{x}'|,\boldsymbol{x}')}{|\boldsymbol{x}-\boldsymbol{x}'|} \tag{4.5.1}$$

in the wave zone. We shall expand in terms of both the retardation within the source *and* in terms of $|\boldsymbol{x}'|/|\boldsymbol{x}| \ll 1$.

To lowest order, we have

$$\gamma^{\mu\nu}(t,\boldsymbol{x}) = -\frac{4G}{r}\int T^{\mu\nu}(t-r,\boldsymbol{x}')\,d^3x'\,, \tag{4.5.2}$$

where $r = |\boldsymbol{x}|$. First of all, we transform the integral on the right hand side for spacelike indices. For this, we use $T^{\mu\nu}{}_{,\nu} = 0$, which implies the following identity:

$$\int T^{kl}\, d^3x = \frac{1}{2}\frac{\partial^2}{\partial t^2}\int T^{00}\, x^k\, x^l\, d^3x\,. \tag{4.5.3}$$

*Proof:* We have

$$0 = \int x^k\, \partial_\nu\, T^\nu_\mu\, d^3x = \frac{\partial}{\partial t}\int x^k\, T^0_\mu\, d^3x + \int x^k\, \partial_l\, T^l_\mu\, d^3x\,.$$

Integration by parts in the last term gives

$$\int T^k_\mu\, d^3x = \frac{\partial}{\partial t}\int T^0_\mu x^k\, d^3x\,. \tag{4.5.4}$$

As a result of Gauss' theorem, we also have

$$\begin{aligned}\frac{\partial}{\partial t}\int T^{00}\, x^k\, x^l\, d^3x &= \int \partial_\nu (T^{\nu 0}\, x^k\, x^l)\, d^3x\\ &= \int T^{\nu 0}\, \partial_\nu (x^k\, x^l)\, d^3x\\ &= \int T^{k0}\, x^l + T^{l0})\, d^3x\,.\end{aligned}$$

Together with (4.5.4) this implies

$$\frac{1}{2}\frac{\partial^2}{\partial t^2}\int T^{00}\, x^k\, x^l\, d^3x = \frac{1}{2}\frac{\partial}{\partial t}\int (T^{k0}\, x^l + T^{l0}\, x^k)\, d^3x = \int T^{kl}\, d^3x\,.$$

Using $T^{00} \simeq \varrho$, together with (4.5.2) and (4.5.3), we obtain

$$\gamma^{kl}(t, \boldsymbol{x}) = -\frac{2G}{r}\left[\frac{\partial^2}{\partial t^2}\int \varrho(\boldsymbol{x}')\, x'^k\, x'^l\, d^3x'\right]_{t-r}. \tag{4.5.5}$$

We can write this in terms of the quadrupole tensor

$$Q_{kl} = \int (3x'^k\, x'^l - r'^2\, \delta_{kl})\, \varrho(\boldsymbol{x}')\, d^3x' \tag{4.5.6}$$

as

$$\gamma_{kl}(t, x) = -\frac{2G}{r}\frac{1}{3}\left[\frac{\partial^2}{\partial t^2}Q_{kl} + \delta_{kl}\frac{\partial^2}{\partial t^2}\int r'^2 \varrho(\boldsymbol{x}')\, d^3x'\right]_{t-r}. \tag{4.5.7}$$

This allows us to compute the "energy flux". At large distances from the source, the wave can be considered to be (locally) a plane wave. If the wave propagates in the $x^1$-direction, the energy current in the $x^1$-direction is

$$t^{01} = \frac{1}{16\pi G}[(\dot{h}_{23})^2 + \tfrac{1}{4}(\dot{h}_{22} - \dot{h}_{33})^2], \tag{4.5.8}$$

as will be shown in the next exercise. Now $h_{23} = \gamma_{23}$ and $h_{22} - h_{33} = \gamma_{22} - \gamma_{33}$. Hence the second term in (4.5.7) does not contribute to the radiation and the first term gives in the $x^1$-direction

$$t^{01} = \frac{G}{36\pi}\frac{1}{r^2}[(\dddot{Q}_{23})^2 + \tfrac{1}{4}(\dddot{Q}_{22} - \dddot{Q}_{33})^2]. \tag{4.5.9}$$

Using $Q_{kk}=0$, and letting $\boldsymbol{n}=(1,0,0)$ be the unit vector in the $x^1$-direction, we may also write

$$t^{0s}\,n^s=\frac{G}{36\,\pi}\frac{1}{r^2}\left[\tfrac{1}{2}\dddot{Q}_{kl}\dddot{Q}_{kl}-\dddot{Q}_{kl}\dddot{Q}_{km}\,n^l\,n^m+\tfrac{1}{4}\,(\dddot{Q}_{kl}\,n^k\,n^l)^2\right]. \tag{4.5.10}$$

This expression is, of course, valid for an arbitrary direction $\boldsymbol{n}$. The energy radiated per unit time per solid angle in the direction $\boldsymbol{n}$ is equal to $r^2\,t^{0s}\,n^s$, so that

$$\frac{dI}{d\Omega}=\frac{G}{36\,\pi}\left[\tfrac{1}{2}\dddot{Q}_{kl}\dddot{Q}_{kl}-\dddot{Q}_{kl}\dddot{Q}_{km}\,n^l\,n^m+\tfrac{1}{4}\,(\dddot{Q}_{kl}\,n^k\,n^l)^2\right]. \tag{4.5.11}$$

We are interested in the total energy radiated. Using the average values

$$\begin{aligned}&\frac{1}{4\pi}\int n^l\,n^m\,d\Omega=\tfrac{1}{3}\,\delta_{lm}\\&\frac{1}{4\pi}\int n^k\,n^l\,n^m\,n^r\,d\Omega=\frac{1}{15}\,(\delta_{kl}\,\delta_{mr}+\delta_{km}\,\delta_{lr}+\delta_{kr}\,\delta_{lm})\end{aligned} \tag{4.5.12}$$

we obtain for the energy loss (first derived by Einstein in 1917)

$$-\frac{dE}{dt}=\frac{G}{45\,c^5}\,\dddot{Q}_{kl}\,\dddot{Q}_{kl}. \tag{4.5.13}$$

In contrast to electrodynamics, the lowest multipole is quadrupole radiation. This, of course, is due to the spin 2 nature of the gravitational field. The derivation of the quadrupole formula (4.5.13) in the framework of the linearized theory is possibly misleading. Up to now, there is no really satisfactory derivation of this result based on a systematic approximation scheme which permits an estimate of the errors. It is entirely possible that the validity of (4.5.13) is better than its derivation. This is indicated by observations of the binary pulsar 1913 + 16, as discussed below and in Sect. 5.6.

The validity of Einstein's quadrupole formula is at present a source of heated debate. For the current status of the discussion, we refer the reader to [169, 170].

---

**Exercise:** *Energy and Momentum of a Plane Wave.* Consider (in the linearized theory) a not necessarily harmonic wave $h_{\mu\nu}(x^1-t)$ propagating in the $x^1$-direction. Show that it is possible to find a gauge within the Hilbert gauge class such that at most $h_{23}$ and $h_{22}=-h_{33}$ do not vanish. Compute the energy-momentum complex $t^{\mu\nu}_{L-L}$ in this particular gauge.

*Solution:* From the Hilbert condition $\gamma^\nu_{\mu,\nu}=0$, we have $\gamma^{0\prime}_\mu=\gamma^{1\prime}_\mu$. Expect for an irrelevant integration constant (we are only interested in the variable fields), this equation is also valid for the fields. Hence

$$\gamma^0_0=\gamma^1_0\quad\gamma^0_1=\gamma^1_1,\quad\gamma^0_2=\gamma^1_2,\quad\gamma^0_3=\gamma^1_3$$

and thus $\gamma^0_0=-\gamma^1_1$.

The independent components are thus $\gamma_{00}, \gamma_{02}, \gamma_{03}, \gamma_{22}, \gamma_{23}, \gamma_{33}$.

By introducing a gauge transformation with $\xi^{\mu}(x^1 - t)$ (so that $\Box \xi^{\mu} = 0$), we have $-\xi^{\lambda}_{,\lambda} = \xi'_0 + \xi'_1$ and thus obtain the transformation rules

$$\gamma_{00} \to \gamma_{00} - \xi'_0 + \xi'_1,\ \gamma_{02} \to \gamma_{02} - \xi'_2,\ \gamma_{03} \to \gamma_{03} - \xi'_3,$$

$$\gamma_{22} \to \gamma_{22} - \xi'_0 - \xi'_1,\ \gamma_{23} \to \gamma_{23},\ \gamma_{33} \to \gamma_{33} - \xi'_0 - \xi'_1. \tag{4.5.14}$$

Among these, $\gamma_{23}$ and $\gamma_{22} - \gamma_{33}$ remain invariant. Obviously it is possible to make $\gamma_{00}, \gamma_{02}, \gamma_{03}$, and $\gamma_{22} + \gamma_{33}$ vanish by a suitable choice of $\xi_{\mu}$.

We now compute the energy-momentum 3-forms (2.6.9) in this gauge. We shall perform the calculation with respect to an orthonormal basis. In the linearized theory, such a basis has the form

$$\theta^{\alpha} = dx^{\alpha} + \varphi^{\alpha}{}_{\beta}\, dx^{\beta}, \tag{4.5.15}$$

where

$$\varphi_{\alpha\beta} + \varphi_{\beta\alpha} = h_{\alpha\beta}. \tag{4.5.16}$$

Equation (4.5.16) determines only the symmetric part of $\varphi_{\alpha\beta}$. Under infinitesimal local Lorentz transformations, $\varphi_{\alpha\beta}$ is changed by an antisymmetric contribution. We may thus choose $\varphi_{\alpha\beta}$ to be symmetric, i.e. $\varphi_{\alpha\beta} = \frac{1}{2} h_{\alpha\beta}$. Since

$$d\theta^{\alpha} = \varphi^{\alpha}{}_{\beta,\gamma}\, dx^{\gamma} \wedge dx^{\beta} \simeq \varphi^{\alpha}{}_{\beta,\gamma}\, \theta^{\gamma} \wedge \theta^{\beta}$$

one finds, to first order in $\varphi_{\alpha\beta}$, for the connection forms

$$\omega^{\alpha}{}_{\beta} = (\varphi^{\alpha}{}_{\gamma,\beta} - \varphi_{\beta\gamma}{}^{,\alpha})\, dx^{\gamma}. \tag{4.5.17}$$

For our special plane wave, the $\varphi_{\alpha\beta}$ are functions only of $u = x^1 - t$, and hence

$$\varphi_{\alpha\beta,\gamma} = \varphi'_{\alpha\beta}\, n_{\gamma},\ (n_{\gamma}) := (-1, +1, 0, 0)\,.$$

Therefore

$$\omega_{\alpha\beta} = \phi_{\alpha}\, n_{\beta} - \phi_{\beta}\, n_{\alpha}, \quad \phi_{\alpha} := \varphi'_{\alpha\gamma}\, dx^{\gamma}. \tag{4.5.18}$$

Since only $\varphi_{23}$ and $\varphi_{22} = -\varphi_{33}$ do not vanish, we note that

$$n^{\alpha}\, \phi_{\alpha} = 0, \quad n^{\alpha}\, n_{\alpha} = 0. \tag{4.5.19}$$

From this, it follows that $\omega_{\alpha\beta} \wedge \omega^{\alpha}{}_{\gamma} = 0$ and the first term of the Landau-Lifshitz forms (2.6.9),

$$*t^{\alpha}_{L-L} = -\frac{1}{16\pi G}\, \eta^{\alpha\beta\gamma\delta} (\omega_{\sigma\beta} \wedge \omega^{\sigma}{}_{\gamma} \wedge \theta_{\delta} - \omega_{\beta\gamma} \wedge \omega_{\sigma\delta} \wedge \theta^{\sigma})\,,$$

vanishes. Using this and (4.5.18), we obtain

$$*t^{\alpha}_{L-L} = \frac{1}{16\pi G}\, 2\eta^{\alpha\beta\gamma\delta}\, \phi_\beta\, \underbrace{n_\gamma \wedge (\phi_\sigma n_\delta - \phi_\delta n_\sigma)}_{\text{symm. in } \gamma,\,\delta} \wedge \theta^\sigma$$

$$= -\frac{1}{8\pi G}\,\eta^{\alpha\beta\gamma\delta}\,\phi_\beta \wedge \phi_\delta \wedge n_\gamma \underbrace{n_\sigma \theta^\sigma}_{du},$$

so that

$$*t^{\alpha}_{L-L} = -\frac{1}{8\pi G}\, n_\gamma\, \eta^{\alpha\beta\gamma\delta}\, \phi_\beta \wedge \phi_\delta \wedge du\,. \tag{4.5.20}$$

In particular, we have for $\alpha = 0$:

$$*t^{0}_{L-L} = \frac{1\cdot 2}{8\pi G}\, n_1\, \eta^{0213}\, \phi_2 \wedge \phi_3 \wedge du$$

$$= -\frac{1}{4\pi G}\, \phi_2 \wedge \phi_3 \wedge du$$

$$= -\frac{1}{4\pi G}\, [\varphi'_{22}\, dx^2 + \varphi'_{23}\, dx^3] \wedge [\varphi'_{32}\, dx^2 + \varphi'_{33}\, dx^3] \wedge du$$

$$= -\frac{1}{4\pi G}\, [\varphi'_{22}\,\varphi'_{33} - (\varphi'_{23})^2]\, dx^2 \wedge dx^3 \wedge du\,.$$

This gives, with $\theta^\mu \wedge *t^\alpha = t^{\alpha\mu}\,\eta$,

$$t^{00}_{L-L} = t^{01}_{L-L} = \frac{1}{16\pi G}\, [(h'_{23})^2 + \tfrac{1}{4}\,(h'_{22} - h'_{33})^2]. \tag{4.5.21}$$

Here we used $h'_{22} = -\,h'_{33}$. These expressions are invariant under gauge transformations of the type (4.5.14).

---

*Radiation from a Binary Star System*

As an example, we compute the radiation loss from a double star system. We shall perform the calculation using the units $G = c = 1$. Let the masses be $m_1$ and $m_2$. The semimajor axis $a$ and the eccentricity $e$ are related to the total energy $E$ ($E < 0$) and angular momentum $L$ by

$$a = -\,m_1 m_2 / 2E \tag{4.5.22}$$

$$e^2 = 1 + \frac{2EL^2(m_1 + m_2)}{m_1^3\, m_2^3}\,. \tag{4.5.23}$$

Therefore

$$\frac{da}{dt} = \frac{m_1 m_2}{2E^2}\frac{dE}{dt} \tag{4.5.24}$$

$$e\,\frac{de}{dt} = \frac{m_1 + m_2}{m_1^3\, m_2^3}\left(L^2 \frac{dE}{dt} + 2EL\,\frac{dL}{dt}\right). \tag{4.5.25}$$

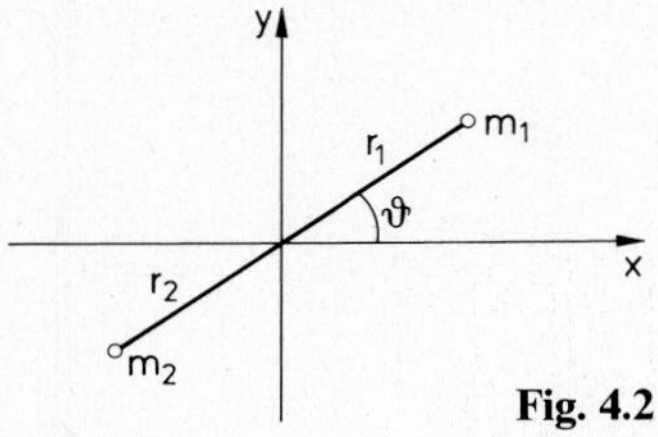


Fig. 4.2

Let $r$ be the distance between the two masses. Then (see Fig. 4.2)

$$r = \frac{a(1-e^2)}{1+e\cos\vartheta} \tag{4.5.26}$$

and

$$r_1 = \frac{m_2}{m_1+m_2}\, r$$

$$r_2 = \frac{m_1}{m_1+m_2}\, r\,.$$

The components of the moment of inertia tensor

$$I_{kl} = \int \varrho(\mathbf{x})\, x^k\, x^l\, d^3x$$

satisfy

$$I_{xx} = m_1 x_1^2 + m_2 x_2^2 = \frac{m_1 m_2}{m_1+m_2}\, r^2 \cos^2\vartheta$$

$$I_{yy} = \frac{m_1 m_2}{m_1+m_2}\, r^2 \sin^2\vartheta$$

$$I_{xy} = \frac{m_1 m_2}{m_1+m_2}\, r^2 \sin\vartheta\cos\vartheta$$

$$I := I_{xx} + I_{yy} = \frac{m_1 m_2}{m_1+m_2}\, r^2\,. \tag{4.5.27}$$

The angular momentum is $L = m_1 m_2/(m_1+m_2)\, r^2\, \dot\vartheta$ and hence we have, from (4.5.22) and (4.5.23)

$$\dot\vartheta = \frac{[(m_1+m_2)\, a(1-e^2)]^{1/2}}{r^2}\,. \tag{4.5.28}$$

From this and (4.5.26) it follows that

$$\dot r = e\sin\vartheta \left(\frac{m_1+m_2}{a(1-e^2)}\right)^{1/2}. \tag{4.5.29}$$

We now compute the time derivatives of the $I_{kl}$, simplifying the resulting expressions using (4.5.26), (4.5.28) and (4.5.29). The result is

$$\dot{I}_{xx} = \frac{-2m_1 m_2}{[(m_1+m_2)\, a\,(1-e^2)]^{1/2}}\, r \cos\vartheta \sin\vartheta \tag{4.5.30}$$

$$\ddot{I}_{xx} = \frac{-2m_1 m_2}{a(1-e^2)} (\cos 2\vartheta + e\cos^3\vartheta) \tag{4.5.31}$$

$$\dddot{I}_{xx} = \frac{2m_1 m_2}{a(1-e^2)} (2\sin 2\vartheta + 3e\cos^2\vartheta \sin\vartheta)\,\dot{\vartheta} \tag{4.5.32}$$

$$\dot{I}_{yy} = \frac{2m_1 m_2}{[(m_1+m_2)\, a\,(1-e^2)]^{1/2}}\, r(\sin\vartheta\cos\vartheta + e\sin\vartheta) \tag{4.5.33}$$

$$\ddot{I}_{yy} = \frac{2m_1 m_2}{a(1-e^2)} (\cos 2\vartheta + e\cos\vartheta + e\cos^3\vartheta + e^2) \tag{4.5.34}$$

$$\dddot{I}_{yy} = \frac{-2m_1 m_2}{a(1-e^2)} (2\sin 2\vartheta + e\sin\vartheta + 3e\cos^2\vartheta\sin\vartheta)\,\dot{\vartheta} \tag{4.5.35}$$

$$\dot{I}_{xy} = \frac{m_1 m_2}{[(m_1+m_2)\, a\,(1-e^2)]^{1/2}}\, r(\cos^2\vartheta - \sin^2\vartheta + e\cos\vartheta) \tag{4.5.36}$$

$$\ddot{I}_{xy} = \frac{-2m_1 m_2}{a(1-e^2)} (\sin 2\vartheta + e\sin\vartheta + e\sin\vartheta\cos^2\vartheta) \tag{4.5.37}$$

$$\dddot{I}_{xy} = \frac{-2m_1 m_2}{a(1-e^2)} (2\cos 2\vartheta - e\cos\vartheta + 3e\cos^3\vartheta)\,\dot{\vartheta} \tag{4.5.38}$$

$$\dddot{I} = \dddot{I}_{xx} + \dddot{I}_{yy} = \frac{-2m_1 m_2}{a(1-e^2)}\, e\sin\vartheta\,\dot{\vartheta}\,. \tag{4.5.39}$$

Equation (4.5.13) can be written in the form

$$-\frac{dE}{dt} = \frac{1}{5}(\dddot{I}_{kl}\dddot{I}_{kl} - \tfrac{1}{3}\dddot{I}^2) = \tfrac{1}{5}(\dddot{I}^2_{xx} + 2\,\dddot{I}^2_{xy} + \dddot{I}^2_{yy} - \tfrac{1}{3}\dddot{I}^2)\,.$$

Inserting the explicit expressions gives

$$-\frac{dE}{dt} = \frac{8m_1^2 m_2^2}{15\,a^2(1-e^2)^2}[12(1+e\cos\vartheta)^2 + e^2\sin^2\vartheta]\,\dot{\vartheta}^2.$$

We now average the energy loss over a period. According to Kepler's third law, the orbital period is

$$T = \frac{2\pi a^{3/2}}{(m_1+m_2)^{1/2}}\,. \tag{4.5.40}$$

Hence

$$-\left\langle\frac{dE}{dt}\right\rangle=\frac{1}{T}\int_0^T-\frac{dE}{dt}\,dt=\frac{1}{T}\int_0^{2\pi}-\frac{dE}{dt}\frac{1}{\dot{\vartheta}}\,d\vartheta$$

$$=\frac{32}{5}\frac{m_1^2m_2^2(m_1+m_2)}{a^5(1-e^2)^{7/2}}\left(1+\frac{73}{24}e^2+\frac{37}{96}e^4\right). \tag{4.5.41}$$

From this and (4.5.24) we obtain

$$\left\langle\frac{da}{dt}\right\rangle=\frac{2a^2}{m_1m_2}\left\langle\frac{dE}{dt}\right\rangle$$

$$=-\frac{64}{5}\frac{m_1m_2(m_1+m_2)}{a^3(1-e^2)^{7/2}}\left(1+\frac{73}{24}e^2+\frac{37}{96}e^4\right). \tag{4.5.42}$$

From (4.5.40) we then find for the change in the orbital period

$$\dot{T}/T=\frac{3}{2}\dot{a}/a=-\frac{96}{5}\frac{1}{a^4}m_1m_2(m_1+m_2)f(e), \tag{4.5.43}$$

where

$$f(e)=\left(1+\frac{73}{24}e^2+\frac{37}{96}e^4\right)(1-e^2)^{-7/2}. \tag{4.5.44}$$

If we replace $a$ by $T$ in (4.5.43), we also have

$$\dot{T}/T=-\frac{96}{5}\frac{m_1m_2}{(T/2\pi)^{8/3}(m_1+m_2)^{1/3}}f(e). \tag{4.5.45}$$

In recent years, it has become possible to verify this expression by observations of a remarkable binary system, which consists of a pulsar (PSR 1913 + 16) and a compact partner, which is either a white dwarf, a neutron star, or a black hole. (A detailed discussion of this system will be given in Sect. 5.6.) After five years of observation, one has succeeded in determining not only the Keplerian parameters, but also the post-Keplerian parameters, such as the perihelion precession. From these, one finds (see Sect. 5.6) that $m_1\simeq m_2\simeq 1.4\,M_\odot$. If one inserts this in (4.5.45), as well as

$$T=27906.980\,(2)\ s,\quad e=0.617,$$

one obtains

$$\dot{T}_{\text{theor}}=(-2.403\pm0.005)\times10^{-12}. \tag{4.5.46}$$

The observed value is [29]

$$\dot{T}_{\text{exp}}=(-2.30\pm0.22)\times10^{-12}. \tag{4.5.47}$$

**Exercises:**

1. Compute the radiated energy (4.5.41) of the system in erg/s.
2. Compare the quadrupole formula (4.5.11) with the quadrupole contribution to electromagnetic radiation. Use this to estimate, in a naive linearized quantum theory of gravity, the graviton emissing rate for the $3d \to 1s$ transition in hydrogen. Compare the corresponding lifetime with the age of the universe.

*Result:* The transition rate is ($\hbar = c = 1$):

$$\Gamma_{\text{grav}}(3d \to 1s) = \frac{1}{2^3 \cdot 3^2 \cdot 5} \alpha^6 G m_e^3$$

corresponding to a lifetime $\tau \simeq 0.5 \times 10^{32}$ yr.

# Chapter 5. The Post-Newtonian Approximation

For most astrophysical systems (except black holes) the Newtonian theory provides a good first approximation. This is certainly true for the planetary system but also for more exotic objects like the binary pulsar PSR 1913 + 16. In such situations, it suffices to study the effects predicted by general relativity in a consistent post-Newtonian approximation, i.e. at a level of approximation in which gravitational radiation plays no rôle.

## 5.1 The Field Equations in the Post-Newtonian Approximation

The small parameter, $\varepsilon$, in this approximation is

$$\varepsilon \sim \bar{v}/c \sim (G\bar{M}/c^2\,\bar{r})^{1/2}, \tag{5.1.1}$$

where $\bar{M}$, $\bar{r}$, and $\bar{v}$ are typical values of masses, separations, and velocities of the bodies involved. (In the second relation we have used the virial theorem.)

From the results of Sect. 4.2 it follows that the metric has the following expansion in $\varepsilon$:

$$\begin{aligned} g_{00} &= 1 + \overset{(2)}{g}_{00} + \overset{(4)}{g}_{00} + \overset{(6)}{g}_{00} + \dots \\ g_{ij} &= -\delta_{ij} + \overset{(2)}{g}_{ij} + \overset{(4)}{g}_{ij} + \dots \\ g_{0i} &= \overset{(3)}{g}_{0i} + \overset{(5)}{g}_{0i} + \dots \end{aligned} \tag{5.1.2}$$

with ($c = 1$)

$$\overset{(2)}{g}_{00} = 2\,\Phi, \quad \overset{(2)}{g}_{ij} = 2\,\Phi\,\delta_{ij} \tag{5.1.3}$$

and where $\overset{(n)}{g}_{\mu\nu}$ denotes the term in $g_{\mu\nu}$ of order $\varepsilon^n$. The last line of (5.1.2) contains only odd powers of $\varepsilon$ because $g_{0i}$ must change sign under the time-reversal transformation $t \to -t$. At the moment, (5.1.2) together with (5.1.3) should be considered as a reasonable ansatz, which has to be justified by showing that it leads to consistent solutions of the field equations. We shall see that a consistent post-Newtonian limit requires determination of $g_{00}$ correct through $O(\varepsilon^4)$, $g_{0i}$ through $O(\varepsilon^3)$ and $g_{ij}$ through $O(\varepsilon^2)$. From $g^{\mu\lambda} g_{\lambda\nu} = \delta^\mu_\nu$ we find easily

$$\begin{aligned} g^{00} &= 1 + \overset{(2)}{g}{}^{00} + \overset{(4)}{g}{}^{00} + \overset{(6)}{g}{}^{00} + \dots \\ g^{ij} &= -\delta^{ij} + \overset{(2)}{g}{}^{ij} + \overset{(4)}{g}{}^{ij} + \dots \\ g^{0i} &= \overset{(3)}{g}{}^{0i} + \overset{(5)}{g}{}^{0i} + \dots \end{aligned} \tag{5.1.4}$$

with

$$\overset{(2)}{g}{}^{00} = -\overset{(2)}{g}_{00}, \quad \overset{(2)}{g}{}^{ij} = -\overset{(2)}{g}_{ij}, \quad \overset{(3)}{g}{}^{0i} = \overset{(3)}{g}_{0i}, \text{ etc.} \tag{5.1.5}$$

In computing the Christoffel symbols

$$\Gamma^\mu_{\nu\lambda} = \tfrac{1}{2} g^{\mu\varrho} (g_{\varrho\nu,\lambda} + g_{\varrho\lambda,\nu} - g_{\nu\lambda,\varrho})$$

one should regard

$$\partial/\partial x^i \sim 1/\bar{r}, \qquad \partial/\partial x^0 \sim \bar{v}/\bar{r} \sim \varepsilon/\bar{r} .$$

Inserting the expansion for $g_{\mu\nu}$ and $g^{\mu\nu}$ leads to

$$\Gamma^\mu_{\nu\lambda} = \overset{(2)}{\Gamma}{}^\mu_{\nu\lambda} + \overset{(4)}{\Gamma}{}^\mu_{\nu\lambda} + \dots \quad \text{for} \quad \Gamma^i_{00}, \Gamma^i_{jk}\ \Gamma^0_{0i} \tag{5.1.6}$$

$$\Gamma^\mu_{\nu\lambda} = \overset{(3)}{\Gamma}{}^\mu_{\nu\lambda} + \overset{(5)}{\Gamma}{}^\mu_{\nu\lambda} + \dots \quad \text{for} \quad \Gamma^i_{0j}\ \Gamma^0_{00}\ \Gamma^0_{ij}, \tag{5.1.7}$$

where $\overset{(n)}{\Gamma}{}^\mu_{\nu\lambda}$ denotes the term in $\Gamma^\mu_{\nu\lambda}$ of order $\varepsilon^n/\bar{r}$. The explicit expressions which will be needed are

$$\begin{aligned} \overset{(2)}{\Gamma}{}^i_{00} &= \tfrac{1}{2} \overset{(2)}{g}_{00,i} \\ \overset{(4)}{\Gamma}{}^i_{00} &= \tfrac{1}{2} \overset{(4)}{g}_{00,i} - \overset{(3)}{g}_{0i,0} + \tfrac{1}{2} \overset{(2)}{g}_{ij} \overset{(2)}{g}_{00,j} \\ \overset{(3)}{\Gamma}{}^i_{0j} &= -\tfrac{1}{2} [\overset{(3)}{g}_{i0,j} + \overset{(2)}{g}_{ij,0} - \overset{(3)}{g}_{0j,i}] \\ \overset{(2)}{\Gamma}{}^i_{jk} &= -\tfrac{1}{2} [\overset{(2)}{g}_{ij,k} + \overset{(2)}{g}_{ik,j} - \overset{(2)}{g}_{jk,i}] \\ \overset{(3)}{\Gamma}{}^0_{00} &= \tfrac{1}{2} \overset{(2)}{g}_{00,0} \\ \overset{(2)}{\Gamma}{}^0_{0i} &= \tfrac{1}{2} \overset{(2)}{g}_{00,i} . \end{aligned} \tag{5.1.8}$$

Next we calculate the Ricci-tensor:

$$R_{\mu\nu} = \Gamma^{\alpha}_{\mu\nu,\alpha} - \Gamma^{\alpha}_{\mu\alpha,\nu} + \Gamma^{\varrho}_{\mu\nu}\Gamma^{\alpha}_{\varrho\alpha} - \Gamma^{\varrho}_{\mu\alpha}\Gamma^{\alpha}_{\varrho\nu}\,.$$

From (5.1.6) and (5.1.7) we get the expansions

$$\begin{aligned} R_{00} &= \overset{(2)}{R}_{00} + \overset{(4)}{R}_{00} + \ldots \\ R_{0i} &= \overset{(3)}{R}_{0i} + \overset{(5)}{R}_{0i} + \ldots \\ R_{ij} &= \overset{(2)}{R}_{ij} + \overset{(4)}{R}_{ij} + \ldots \end{aligned} \tag{5.1.9}$$

with the following expressions of $\overset{(n)}{R}_{\mu\nu}$ (term of order $\varepsilon^n/\bar{r}^2$ of $R_{\mu\nu}$) in terms of the Christoffel symbols

$$\begin{aligned} \overset{(2)}{R}_{00} &= \overset{(2)}{\Gamma}{}^{i}_{00,i} \\ \overset{(4)}{R}_{00} &= \overset{(4)}{\Gamma}{}^{i}_{00,i} - \overset{(3)}{\Gamma}{}^{i}_{0i,0} + \overset{(2)}{\Gamma}{}^{i}_{00}\overset{(2)}{\Gamma}{}^{j}_{ij} - \overset{(2)}{\Gamma}{}^{0}_{0i}\overset{(2)}{\Gamma}{}^{i}_{00} \\ \overset{(3)}{R}_{0i} &= \overset{(2)}{\Gamma}{}^{0}_{0i,0} + \overset{(3)}{\Gamma}{}^{j}_{0i,j} - \overset{(3)}{\Gamma}{}^{0}_{00,i} - \overset{(3)}{\Gamma}{}^{j}_{0j,i} \\ \overset{(2)}{R}_{ij} &= \overset{(2)}{\Gamma}{}^{k}_{ij,k} - \overset{(2)}{\Gamma}{}^{0}_{i0,j} - \overset{(2)}{\Gamma}{}^{k}_{ik,j}\,. \end{aligned} \tag{5.1.10}$$

Inserting expressions (5.1.8) for the Christoffel symbols gives

$$\overset{(2)}{R}_{00} = \tfrac{1}{2}\Delta\overset{(2)}{g}_{00} \tag{5.1.11}$$

$$\begin{aligned} \overset{(4)}{R}_{00} = {} & \tfrac{1}{2}\Delta\overset{(4)}{g}_{00} - \overset{(3)}{g}_{0i,0i} + \tfrac{1}{2}\overset{(2)}{g}_{ii,00} + \tfrac{1}{2}\overset{(2)}{g}_{ij}\overset{(2)}{g}_{00,ij} + \tfrac{1}{2}\overset{(2)}{g}_{ij,j}\overset{(2)}{g}_{00,i} \\ & - \tfrac{1}{4}\overset{(2)}{g}_{00,i}\overset{(2)}{g}_{jj,i} - \tfrac{1}{4}\overset{(2)}{g}_{00,i}\overset{(2)}{g}_{00,i} \end{aligned} \tag{5.1.12}$$

$$\overset{(3)}{R}_{0i} = \tfrac{1}{2}\overset{(2)}{g}_{jj,0i} - \tfrac{1}{2}\overset{(3)}{g}_{j0,ij} - \tfrac{1}{2}\overset{(2)}{g}_{ij,0j} + \tfrac{1}{2}\Delta\overset{(3)}{g}_{0i} \tag{5.1.13}$$

$$\overset{(2)}{R}_{ij} = -\tfrac{1}{2}\overset{(2)}{g}_{00,ij} + \tfrac{1}{2}\overset{(2)}{g}_{kk,ij} - \tfrac{1}{2}\overset{(2)}{g}_{ik,kj} - \tfrac{1}{2}\overset{(2)}{g}_{kj,ki} + \tfrac{1}{2}\Delta\overset{(2)}{g}_{ij}\,. \tag{5.1.14}$$

We have the freedom to impose four "gauge conditions". A useful choice is the following: We set $g_{\alpha\beta} = \eta_{\alpha\beta} + h_{\alpha\beta}$ and require

$$h_{0k,k} - \tfrac{1}{2}h_{kk,0} = O\,(\varepsilon^5/\bar{r}) \tag{5.1.15}$$

$$h_{jk,k} + \tfrac{1}{2}h_{,j} = O\,(\varepsilon^4/\bar{r})\,, \tag{5.1.16}$$

where $h := h_{\alpha\beta}\eta^{\alpha\beta} = h_{00} - h_{ii}$.

Inserting (5.1.2), these gauge conditions read explicitely

$$\overset{(3)}{g}_{0k,k} - \tfrac{1}{2}\overset{(2)}{g}_{kk,0} = 0 \tag{5.1.17}$$

$$\tfrac{1}{2}\overset{(2)}{g}_{00,i} + \overset{(2)}{g}_{ij,j} - \tfrac{1}{2}\overset{(2)}{g}_{jj,i} = 0\,. \tag{5.1.18}$$

The second equation is automatically satisfied by (5.1.3). [In arriving at (5.1.3) we had already imposed the gauge condition (5.1.18).] We differentiate (5.1.18) with respect to $x^k$, obtaining

$$\tfrac{1}{2}\overset{(2)}{g}_{00,ik} + \overset{(2)}{g}_{ij,jk} - \tfrac{1}{2}\overset{(2)}{g}_{jj,ik} = 0\,.$$

Interchanging $i$ and $k$ and adding the two equations gives

$$\overset{(2)}{g}_{00,ik} + \overset{(2)}{g}_{ij,jk} + \overset{(2)}{g}_{kj,ji} - \overset{(2)}{g}_{jj,ik} = 0\,. \tag{5.1.19}$$

Hence $\overset{(2)}{R}_{ij}$ in (5.1.14) simplifies to

$$\overset{(2)}{R}_{ij} = \tfrac{1}{2}\Delta\overset{(2)}{g}_{ij}\,. \tag{5.1.20}$$

Next we use (5.1.17) to simplify $\overset{(4)}{R}_{00}$. Equation (5.1.17) implies

$$\overset{(3)}{g}_{0i,i0} - \tfrac{1}{2}\overset{(2)}{g}_{ii,00} = 0$$

and thus

$$\begin{aligned}\overset{(4)}{R}_{00} = {} & \tfrac{1}{2}\Delta\overset{(4)}{g}_{00} + \tfrac{1}{2}\overset{(2)}{g}_{ij}\overset{(2)}{g}_{00,ij} + \tfrac{1}{2}\overset{(2)}{g}_{ij,j}\overset{(2)}{g}_{00,i} \\ & - \tfrac{1}{4}\overset{(2)}{g}_{00,i}\overset{(2)}{g}_{00,i} - \tfrac{1}{4}\overset{(2)}{g}_{00,i}\overset{(2)}{g}_{jj,i}\,.\end{aligned} \tag{5.1.21}$$

Here we insert (5.1.3) and obtain finally

$$\overset{(4)}{R}_{00} = \tfrac{1}{2}\Delta\overset{(4)}{g}_{00} + 2\Phi\,\Delta\Phi - 2\,(\nabla\Phi)^2\,. \tag{5.1.22}$$

If we also use the gauge condition (5.1.17) in (5.1.13) we obtain

$$\overset{(3)}{R}_{0i} = \tfrac{1}{4}\overset{(2)}{g}_{jj,0i} - \tfrac{1}{2}\overset{(2)}{g}_{ij,0j} + \tfrac{1}{2}\Delta\overset{(3)}{g}_{0i} \tag{5.1.23}$$

or, using (5.1.3)

$$\overset{(3)}{R}_{0i} = \tfrac{1}{2}\Delta\overset{(3)}{g}_{0i} + \tfrac{1}{2}\Phi_{,0i}\,. \tag{5.1.24}$$

The energy-momentum tensor has the expansion

$$\begin{aligned} T^{00} &= \overset{(0)}{T}{}^{00} + \overset{(2)}{T}{}^{00} + \ldots \\ T^{i0} &= \overset{(1)}{T}{}^{i0} + \overset{(3)}{T}{}^{i0} + \ldots \\ T^{ij} &= \overset{(2)}{T}{}^{ij} + \overset{(4)}{T}{}^{ij} + \ldots\,, \end{aligned} \tag{5.1.25}$$

where $\overset{(n)}{T}{}^{\mu\nu}$ denotes the term in $T^{\mu\nu}$ of order $\varepsilon^n \bar{M}/\bar{r}^3$. For the field equations we need

$$S_{\mu\nu} := T_{\mu\nu} - \tfrac{1}{2}\,g_{\mu\nu}T^{\lambda}_{\lambda}\,. \tag{5.1.26}$$

Using (5.1.2) one finds from (5.1.25) the following expansions

$$\begin{aligned} S_{00} &= \overset{(0)}{S}_{00} + \overset{(2)}{S}_{00} + \dots \\ S_{i0} &= \overset{(1)}{S}_{i0} + \overset{(3)}{S}_{i0} + \dots \\ S_{ij} &= \overset{(0)}{S}_{ij} + \overset{(2)}{S}_{ij} + \dots \end{aligned} \tag{5.1.27}$$

with

$$\begin{aligned} \overset{(0)}{S}_{00} &= \tfrac{1}{2} \overset{(0)}{T}{}^{00} \\ \overset{(2)}{S}_{00} &= \tfrac{1}{2} [\overset{(2)}{T}{}^{00} + 2 \overset{(2)}{g}_{00} \overset{(0)}{T}{}^{00} + \overset{(2)}{T}{}^{ii}] \\ \overset{(1)}{S}_{0i} &= - \overset{(1)}{T}{}^{0i} \\ \overset{(0)}{S}_{ij} &= \tfrac{1}{2} \delta_{ij} \overset{(0)}{T}{}^{00} . \end{aligned} \tag{5.1.28}$$

With our expressions for $\overset{(n)}{R}_{\mu\nu}$ and $\overset{(n)}{S}_{\mu\nu}$ the field equations

$$R_{\mu\nu} = 8 \pi G S_{\mu\nu}$$

read (note $G\bar{M}/\bar{r} \sim \varepsilon^2$):

$$\Delta \overset{(2)}{g}_{00} = 8 \pi G \overset{(0)}{T}{}^{00} \tag{5.1.29}$$

$$\begin{aligned} \Delta \overset{(4)}{g}_{00} = &- \overset{(2)}{g}_{ij} \overset{(2)}{g}_{00,ij} - \overset{(2)}{g}_{ij,j} \overset{(2)}{g}_{00,i} + \tfrac{1}{4} \overset{(2)}{g}_{00,i} \overset{(2)}{g}_{00,i} + \tfrac{1}{4} \overset{(2)}{g}_{00,i} \overset{(2)}{g}_{jj,i} \\ &+ 8 \pi G [\overset{(2)}{T}{}^{00} + 2 \overset{(2)}{g}_{00} \overset{(0)}{T}{}^{00} + \overset{(2)}{T}{}^{ii}] \end{aligned} \tag{5.1.30}$$

$$\Delta \overset{(3)}{g}_{0i} = - \tfrac{1}{2} \overset{(2)}{g}_{jj,0i} + \overset{(2)}{g}_{ij,0j} - 16 \pi G \overset{(1)}{T}{}^{i0} \tag{5.1.31}$$

$$\Delta \overset{(2)}{g}_{ij} = 8 \pi G \delta_{ij} \overset{(0)}{T}{}^{00} . \tag{5.1.32}$$

The second order equations (5.1.29) and (5.1.32) are, of course, satisfied by (5.1.3) with

$$\Phi = - G \int \frac{\overset{(0)}{T}{}^{00} (t, \boldsymbol{x}')}{|\boldsymbol{x} - \boldsymbol{x}'|} d^3x' . \tag{5.1.33}$$

According to (5.1.22) and (5.1.24) the field equations in third and fourth order are

$$\Delta \overset{(3)}{g}_{0i} = - 16 \pi G \overset{(1)}{T}{}^{i0} - \Phi_{,0i} \tag{5.1.34}$$

$$\Delta \overset{(4)}{g}_{00} = 8 \pi G [\overset{(2)}{T}{}^{00} + 4 \Phi \overset{(0)}{T}{}^{00} + \overset{(2)}{T}{}^{ii}] - 4 \Phi \Delta \Phi + 4 (\boldsymbol{\nabla} \Omega)^2 . \tag{5.1.35}$$

We use $\Delta\Phi = 4\pi G \overset{(0)}{T}{}^{00}$ and

$$(\nabla\Omega)^2 = \tfrac{1}{2}\,\Delta\,(\Phi^2) - \Phi\,\Delta\Phi$$

to rewrite (5.1.35) in the following form

$$\Delta\,(\overset{(4)}{g}_{00} - 2\Phi^2) = 8\pi G\,(\overset{(2)}{T}{}^{00} + \overset{(2)}{T}{}^{ii})\,. \tag{5.1.36}$$

We define the potential $\psi$ by

$$\overset{(4)}{g}_{00} = 2\Phi^2 + 2\psi\,. \tag{5.1.37}$$

$\psi$ satisfies the equation

$$\Delta\psi = 4\pi G\,(\overset{(2)}{T}{}^{00} + \overset{(2)}{T}{}^{ii})\,. \tag{5.1.38}$$

Since $\overset{(4)}{g}_{00}$ must vanish at infinity, the solution is

$$\psi = -G\int \frac{d^3x'}{|\boldsymbol{x}-\boldsymbol{x}'|}\,[\overset{(2)}{T}{}^{00}(\boldsymbol{x}', t) + \overset{(2)}{T}{}^{ii}(\boldsymbol{x}', t)]\,. \tag{5.1.39}$$

We define the potentials $\zeta_i$ and $\chi$ through

$$\zeta_i(\boldsymbol{x}, t) := -4G\int \frac{d^3x'}{|\boldsymbol{x}-\boldsymbol{x}'|}\,\overset{(1)}{T}{}^{i0}(\boldsymbol{x}', t) \tag{5.1.40}$$

$$\chi(\boldsymbol{x}, t) := -\frac{G}{2}\int d^3x'\,|\boldsymbol{x}-\boldsymbol{x}'|\,\overset{(0)}{T}{}^{00}(\boldsymbol{x}', t)\,. \tag{5.1.41}$$

They satisfy

$$\Delta\zeta_i = 16\pi G\,\overset{(1)}{T}{}^{i0} \tag{5.1.42}$$

$$\Delta\chi = \Phi\,. \tag{5.1.43}$$

From (5.1.34) we obtain

$$\overset{(3)}{g}_{i0} = -\zeta_i - \frac{\partial^2\chi}{\partial t\,\partial x^i}\,. \tag{5.1.44}$$

The gauge condition (5.1.17) reads [with (5.1.3, 42, 43)]

$$\nabla\cdot\zeta + 4\frac{\partial\Phi}{\partial t} = 0\,. \tag{5.1.45}$$

We shall see below that this condition is satisfied by virtue of the conservation conditions satisfied by $T^{\mu\nu}$.

We also express the Christoffel symbols (5.1.8) in terms of $\Phi$, $\zeta_i$, $\chi$, and $\psi$:

$$\begin{aligned}
\overset{(2)}{\Gamma}{}^i_{00} &= \frac{\partial \Phi}{\partial x^i} \\
\overset{(4)}{\Gamma}{}^i_{00} &= \frac{\partial}{\partial x^i}(2\Phi^2+\psi)+\frac{\partial \zeta_i}{\partial t}+\frac{\partial^3 \chi}{\partial t^2\,\partial x^i} \\
\overset{(3)}{\Gamma}{}^i_{0j} &= -\delta_{ij}\frac{\partial \Phi}{\partial t}+\frac{1}{2}\left(\frac{\partial \zeta_i}{\partial x^j}-\frac{\partial \zeta_j}{\partial x^i}\right) \\
\overset{(3)}{\Gamma}{}^0_{00} &= \frac{\partial \Phi}{\partial t} \\
\overset{(2)}{\Gamma}{}^0_{0i} &= \frac{\partial \Phi}{\partial x^i} \\
\overset{(2)}{\Gamma}{}^i_{jk} &= -\delta_{ij}\frac{\partial \Phi}{\partial x^k}-\delta_{ik}\frac{\partial \Phi}{\partial x^j}+\delta_{jk}\frac{\partial \Phi}{\partial x^i}\,.
\end{aligned} \tag{5.1.46}$$

Now we show that $T^{\mu\nu}_{;\nu}=0$ implies the gauge condition (5.1.45). This "conservation law" reads more explicitly

$$T^{\mu\nu}_{,\nu} = -\Gamma^\mu_{\nu\lambda}\,T^{\lambda\nu}-\Gamma^\nu_{\nu\lambda}\,T^{\mu\lambda}\,. \tag{5.1.47}$$

Since all $\Gamma$'s are at least of order $\varepsilon^2/\bar{r}$, we obtain in lowest order

$$\frac{\partial \overset{(0)}{T}{}^{00}}{\partial t}+\frac{\partial \overset{(1)}{T}{}^{i0}}{\partial x^i}=0\,. \tag{5.1.48}$$

This implies, together with (5.1.42)

$$\Delta\left(\nabla\cdot\boldsymbol{\zeta}+4\frac{\partial \Phi}{\partial t}\right)=0\,.$$

Since $\Phi$ and $\zeta_i$ vanish at infinity we conclude that (5.1.45) holds.

Note that in the next approximation we obtain from (5.1.47) and (5.1.46)

$$\frac{\partial \overset{(1)}{T}{}^{i0}}{\partial t}+\frac{\partial \overset{(2)}{T}{}^{ij}}{\partial x^j}=-\frac{\partial \Phi}{\partial x^i}\overset{(0)}{T}{}^{00}\,. \tag{5.1.49}$$

There are no other conservation laws which involve *only* terms in $T^{\mu\nu}$ needed to calculate the fields in the post-Newtonian approximation. Furthermore, the conservation laws (5.1.48) and (5.1.49) involve $g_{\mu\nu}$ only through the Newtonian potential.

For a particle in an external post-Newtonian field ($\Phi$, $\zeta_i$, $\chi$, $\psi$) the equation of motion follows from

$$\delta \int \left(\frac{d\tau}{dt}\right) dt = 0 .$$

With $v^i = dx^i/dt$ we have

$$\begin{aligned}(d\tau/dt)^2 &= g_{\mu\nu}\,(dx^\mu/dt)\,(dx^\nu/dt)\\ &= 1 - v^2 + \overset{(2)}{g}_{00} + \overset{(4)}{g}_{00} + 2\,\overset{(3)}{g}_{0i}\,v^i + \overset{(2)}{g}_{ij}\,v^i v^j .\end{aligned}$$

This gives for $L := 1 - d\tau/dt$

$$\begin{aligned}L &= \tfrac{1}{2}\,v^2 + \tfrac{1}{8}\,(v^2)^2 - \tfrac{1}{2}\,\overset{(2)}{g}_{00} - \tfrac{1}{2}\,\overset{(4)}{g}_{00}\\ &\quad - \overset{(3)}{g}_{0i}\,v^i - \tfrac{1}{2}\,\overset{(2)}{g}_{ij}\,v^i v^j + \tfrac{1}{8}\,(\overset{(2)}{g}_{00})^2 - \tfrac{1}{4}\,\overset{(2)}{g}_{00}\,v^2\\ &= \tfrac{1}{2}\,v^2 - \Phi + \tfrac{1}{8}\,(v^2)^2 - \tfrac{1}{2}\,\Phi^2 - \psi - \tfrac{3}{2}\,\Phi\,v^2 + v^i\left(\zeta_i + \frac{\partial^2\chi}{\partial t\,\partial x^i}\right). \qquad (5.1.50)\end{aligned}$$

The Euler equations for the Lagrangian function $L$ give the equations of motion for a test particle. Explicitly they read in 3-dimensional notation

$$\begin{aligned}\frac{d\boldsymbol{v}}{dt} &= -\nabla(\Phi + 2\,\Phi^2 + \psi) - \frac{\partial\boldsymbol{\zeta}}{\partial t} - \frac{\partial^2}{\partial t^2}\,\nabla\chi + \boldsymbol{v}\times(\nabla\times\boldsymbol{\zeta})\\ &\quad + 3\,\boldsymbol{v}\,\frac{\partial\Phi}{\partial t} + 4\,\boldsymbol{v}\,(\boldsymbol{v}\cdot\nabla)\,\Phi - v^2\,\nabla\Phi . \qquad (5.1.51)\end{aligned}$$

## 5.2 Asymptotic Fields

Far away from an isolated distribution of energy and momentum we can replace

$$\frac{1}{|\boldsymbol{x} - \boldsymbol{x}'|} = \frac{1}{r} + \frac{\boldsymbol{x}\cdot\boldsymbol{x}'}{r^3}\,\ldots ,$$

where $r = |\boldsymbol{x}|$. We then obtain

$$\Phi = -\frac{G\overset{(0)}{M}}{r} - \frac{G\,\boldsymbol{x}\cdot\overset{(0)}{\boldsymbol{D}}}{r^3} + O\left(\frac{1}{r^3}\right) \qquad (5.2.1)$$

$$\zeta_i = -4\,\frac{G\overset{(1)}{P^i}}{r} - 2G\,\frac{x^j}{r^3}\,\overset{(1)}{J}_{ji} + O\left(\frac{1}{r^3}\right) \qquad (5.2.2)$$

$$\psi = -\frac{G\overset{(2)}{M}}{r} - \frac{G\,\boldsymbol{x}\cdot\overset{(2)}{\boldsymbol{D}}}{r^3} + O\left(\frac{1}{r^3}\right), \qquad (5.2.3)$$

where

$$\overset{(0)}{M} := \int \overset{(0)}{T}{}^{00}\, d^3x \tag{5.2.4}$$

$$\overset{(0)}{\boldsymbol{D}} := \int \boldsymbol{x}\, \overset{(0)}{T}{}^{00}\, d^3x \tag{5.2.5}$$

$$\overset{(1)}{P}{}^i := \int \overset{(1)}{T}{}^{i0}\, d^3x \tag{5.2.6}$$

$$\overset{(1)}{J}_{ij} := 2 \int x^i\, \overset{(1)}{T}{}^{j0}\, d^3x \tag{5.2.7}$$

$$\overset{(2)}{M} := \int (\overset{(2)}{T}{}^{00} + \overset{(2)}{T}{}^{ii})\, d^3x \tag{5.2.8}$$

$$\overset{(2)}{\boldsymbol{D}} := \int \boldsymbol{x} (\overset{(2)}{T}{}^{00} + \overset{(2)}{T}{}^{ii})\, d^3x . \tag{5.2.9}$$

From the mass-conservation equation (5.1.48) it follows that

$$\frac{d\overset{(0)}{M}}{dt} = 0 \tag{5.2.10}$$

$$\frac{d\overset{(0)}{\boldsymbol{D}}}{dt} = \overset{(1)}{\boldsymbol{P}} . \tag{5.2.11}$$

Let us now specialize to a time independent energy-momentum tensor. Then (5.1.48) reduces to

$$(\partial/\partial x^i)\, \overset{(1)}{T}{}^{i0} = 0 .$$

With partial integrations we obtain from this equation

$$0 = \int x^i \frac{\partial}{\partial x^j} \overset{(1)}{T}{}^{j0}\, d^3x = -\overset{(1)}{P}{}^i$$

$$0 = 2 \int x^i x^j \frac{\partial}{\partial x^k} \overset{(1)}{T}{}^{k0}\, d^3x = -\overset{(1)}{J}_{ij} - \overset{(1)}{J}_{ji} .$$

This shows that $\overset{(1)}{\boldsymbol{P}}$ vanishes and that $\overset{(1)}{J}_{ij}$ is antisymmetric. Hence

$$\overset{(1)}{J}_{ij} = \varepsilon_{ijk} \overset{(1)}{J}_k , \tag{5.2.12}$$

where

$$\overset{(1)}{J}_k = \tfrac{1}{2} \varepsilon_{ijk} \overset{(1)}{J}_{ij} = \int d^3x\, \varepsilon_{ijk}\, x^i\, \overset{(1)}{T}{}^{j0} \tag{5.2.13}$$

is the angular momentum vector. We conclude from these results and (5.2.2) that

$$\zeta = \frac{2G}{r^3} (\boldsymbol{x} \times J) + O\left(\frac{1}{r^3}\right). \tag{5.2.14}$$

In order to bring the metric into a convenient form, we perform a gauge transformation

$$t_{\text{old}} = t_{\text{new}} + \frac{\partial \chi}{\partial t}, \quad \boldsymbol{x}_{\text{old}} = \boldsymbol{x}_{\text{new}}. \tag{5.2.15}$$

This leads to

$$g_{00} \to g_{00} + 2\frac{\partial^2 \chi}{\partial t^2}, \quad g_{0i} \to g_{0i} + \frac{\partial^2 \chi}{\partial t\, \partial x^i}. \tag{5.2.16}$$

In the *new gauge* we obtain from (5.1.44) and (5.1.37)

$$\overset{(3)}{g}_{0i} = -\zeta_i \tag{5.2.17}$$

$$\overset{(4)}{g}_{00} = 2\,\Phi^2 + 2\,\psi + 2\frac{\partial^2 \chi}{\partial t^2} \tag{5.2.18}$$

and hence, up to fourth order

$$\begin{aligned} g_{00} &= 1 + 2\,\Phi + 2\,\psi + 2\frac{\partial^2 \chi}{\partial t^2} + 2\,\phi^2 + \ldots \\ &= 1 + 2\left(\Phi + \psi + \frac{\partial^2 \chi}{\partial t^2}\right) + 2\left(\Phi + \psi + \frac{\partial^2 \chi}{\partial t^2}\right)^2 + O(\varepsilon^6). \end{aligned} \tag{5.2.19}$$

For the physically significant field $\Phi + \psi + \partial^2\chi/\partial t^2$ we have from (5.1.38) and (5.1.43)

$$\Delta\left(\Phi + \psi + \frac{\partial^2 \chi}{\partial t^2}\right) = 4\pi G\left(\overset{(0)}{T}{}^{00} + \overset{(2)}{T}{}^{00} + \overset{(2)}{T}{}^{ii} + \frac{1}{4\pi G}\,\frac{\partial^2 \Phi}{\partial t^2}\right)$$

and hence asymptotically

$$\Phi + \psi + \frac{\partial^2 \chi}{\partial t^2} = -\frac{GM}{r} - \frac{G\,\boldsymbol{x}\cdot\boldsymbol{D}}{r^3} + O\left(\frac{1}{r^3}\right) \tag{5.2.20}$$

with[1]

$$M := \int \left(\overset{(0)}{T}{}^{00} + \overset{(2)}{T}{}^{00} + \overset{(2)}{T}{}^{ii}\right) d^3x = \overset{(0)}{M} + \overset{(2)}{M} \tag{5.2.21}$$

$$\boldsymbol{D} := \int \boldsymbol{x}\left(\overset{(0)}{T}{}^{00} + \overset{(2)}{T}{}^{00} + \overset{(2)}{T}{}^{ii} + \frac{1}{4\pi G}\,\frac{\partial^2 \Phi}{\partial t^2}\right) d^3x. \tag{5.2.22}$$

We can write (5.2.20) in the form

$$\Phi + \psi + \frac{\partial^2 \chi}{\partial t^2} = -\frac{GM}{|\boldsymbol{x} - \boldsymbol{D}/M|} + O\left(\frac{1}{r^3}\right). \tag{5.2.23}$$

---

[1] The term $\partial^2\Phi/\partial t^2$ does not contribute to $M$ because it equals $-\frac{1}{4}\nabla\cdot(\partial\zeta/\partial t)$ [see (5.1.45)], and hence vanishes upon integration.

This shows that we can always choose the coordinate system such that $\boldsymbol{D}=0$. Then the metric has the form [use (5.2.19, 23, 17 and 14)]

$$
\begin{aligned}
g_{00} &= 1-\frac{2GM}{r}+2\frac{G^2M^2}{r^2}+O\left(\frac{1}{r^3}\right)\\
g_{0i} &= 2G\,\varepsilon_{ijk}\frac{J^j x^k}{r^3}+O\left(\frac{1}{r^3}\right)\\
g_{ij} &= -\left(1+\frac{2GM}{r}\right)\delta_{ij}+O\left(\frac{1}{r^3}\right).
\end{aligned}
\tag{5.2.24}
$$

This agrees with (4.4.11). (See also the discussion of these results in Sect. 4.4.)

## 5.3 The Post-Newtonian Potentials for a System of Point Particles

From our experience with electrodynamics, it is not surprising that point particles are incompatible with Einstein's field equations. On the level of the first post-Newtonian approximation, there are, however, no serious difficulties. All divergences can be absorbed by mass renormalizations. Furthermore, the resulting Einstein-Infeld-Hoffmann equations can also be derived by starting with extended bodies. But even then, higher order post-Newtonian approximations suffer from divergences.

In other words, even for extended bodies, there exists no formal approximation scheme in which each order is well-defined. (For the present status we refer to [169].)

In special relativity the energy-momentum tensor of a system of point particles is given by

$$T^{\mu\nu}(\boldsymbol{x},t)=\sum_a m_a\int\frac{dx_a^\mu}{d\tau_a}\frac{dx_a^\nu}{d\tau_a}\,\delta^4(x-x_a(\tau_a))\,d\tau_a.$$

Since $\sqrt{-g}\,d^4x$ is an invariant measure, it follows that $\delta^4\,(x-y)/\sqrt{-g}$ is an invariant distribution. From this we conclude that the general relativistic energy-momentum tensor for a system of point particles is

$$
\begin{aligned}
T^{\mu\nu}(\boldsymbol{x},t) &= \frac{1}{\sqrt{-g}}\sum_a m_a\int\frac{dx_a^\mu}{d\tau_a}\frac{dx_a^\nu}{d\tau_a}\,\delta^4(x-x_a(\tau_a))\,d\tau_a\\
&= \frac{1}{\sqrt{-g}}\sum_a m_a\frac{dx_a^\mu}{dt}\frac{dx_a^\nu}{dt}\left(\frac{d\tau_a}{dt}\right)^{-1}\delta^3(\boldsymbol{x}-\boldsymbol{x}_a(t)).
\end{aligned}
\tag{5.3.1}
$$

One easily finds the expansion

$$-g=1+\overset{(2)}{g}+\overset{(4)}{g}+\dots \tag{5.3.2}$$

with

$$\overset{(2)}{g} = \overset{(2)}{g}_{00} - \overset{(2)}{g}_{ii} = -4\,\Phi\,. \tag{5.3.3}$$

Using this in (5.3.1) and $d\tau/dt = 1 - L$ (with $L$ given by (5.1.50)) gives

$$\begin{aligned}
\overset{(0)}{T}{}^{00} &= \sum_a m_a\,\delta^3(\boldsymbol{x} - \boldsymbol{x}_a)\\
\overset{(2)}{T}{}^{00} &= \sum_a m_a\left(\tfrac{1}{2}\,\boldsymbol{v}_a^2 + \Phi\right)\delta^3(\boldsymbol{x} - \boldsymbol{x}_a)\\
\overset{(1)}{T}{}^{i0} &= \sum_a m_a\,v_a^i\,\delta^3(\boldsymbol{x} - \boldsymbol{x}_a)\\
\overset{(2)}{T}{}^{ij} &= \sum_a m_a\,v_a^i\,v_a^j\,\delta^3(\boldsymbol{x} - \boldsymbol{x}_a)\,.
\end{aligned} \tag{5.3.4}$$

One verifies easily that the conservation laws (5.1.48) and (5.1.49) are satisfied, provided each particle obeys the Newtonian equations of motion

$$\frac{d\boldsymbol{v}_a}{dt} = -\nabla\Phi(\boldsymbol{x}_a)\,. \tag{5.3.5}$$

Clearly we have

$$\Phi(\boldsymbol{x}, t) = -G\sum_a \frac{m_a}{|\boldsymbol{x} - \boldsymbol{x}_a|}\,. \tag{5.3.6}$$

From (5.1.38) and (5.3.4) we obtain

$$\Delta\psi = 4\pi G\sum_a m_a\left(\Phi_a' + \tfrac{3}{2}\,\boldsymbol{v}_a^2\right)\delta^3(\boldsymbol{x} - \boldsymbol{x}_a)\,.$$

On the right hand side we replaced the undefined Newtonian potential at the position $\boldsymbol{x}_a$ by

$$\Phi_a' := -G\sum_{b \neq a} \frac{m_b}{|\boldsymbol{x}_b - \boldsymbol{x}_a|}\,. \tag{5.3.7}$$

This can be interpreted as a mass renormalization. We obtain

$$\psi = -G\sum_a \frac{m_a\Phi_a'}{|\boldsymbol{x} - \boldsymbol{x}_a|} - \frac{3G}{2}\sum_a \frac{m_a\,\boldsymbol{v}_a^2}{|\boldsymbol{x} - \boldsymbol{x}_a|}\,. \tag{5.3.8}$$

From (5.1.40) and (5.3.4) we obtain

$$\zeta_i = -4G\sum_a \frac{m_a\,v_a^i}{|\boldsymbol{x} - \boldsymbol{x}_a|} \tag{5.3.9}$$

and (5.1.41) gives

$$\chi = -\frac{G}{2}\sum_a m_a\,|\boldsymbol{x} - \boldsymbol{x}_a|\,. \tag{5.3.10}$$

Using (5.3.9) and (5.3.10) we find from (5.1.44)

$$\overset{(3)}{g}_{0i} = \frac{G}{2} \sum_a \frac{m_a}{|\boldsymbol{x} - \boldsymbol{x}_a|} [7 v_a^i + (\boldsymbol{v}_a \cdot \boldsymbol{n}_a)\, n_a^i], \tag{5.3.11}$$

where $\boldsymbol{n}_a = (\boldsymbol{x} - \boldsymbol{x}_a)/|\boldsymbol{x} - \boldsymbol{x}_a|$.

*Summary*

For later reference, we summarize the important formulae. The small parameter in the PN-approximation is

$$\varepsilon \sim \bar{v}/c \sim (G\bar{M}/\bar{r}\, c^2)^{1/2}.$$

Metric:

$$\begin{aligned} g_{00} &= 1 + \overset{(2)}{g}_{00} + \overset{(4)}{g}_{00} + \overset{(6)}{g}_{00} + \ldots \\ g_{ij} &= -\delta_{ij} + \overset{(2)}{g}_{ij} + \overset{(4)}{g}_{ij} + \ldots \\ g_{0i} &= \overset{(3)}{g}_{0i} + \overset{(5)}{g}_{0i} + \ldots . \end{aligned} \tag{5.3.12}$$

Only the terms up to the dashed boundary are taken into account in the first PN-approximation. Their values are

$$\begin{aligned} \overset{(2)}{g}_{00} &= 2\Phi, \qquad \overset{(4)}{g}_{00} = 2\Phi^2 + 2\psi \\ \overset{(2)}{g}_{ij} &= 2\delta_{ij}\Phi, \qquad \overset{(3)}{g}_{0i} = -\zeta_i - \frac{\partial^2 \chi}{\partial t\, \partial x^i}, \end{aligned} \tag{5.3.13}$$

where

$$\Phi = -G \sum_a \frac{m_a}{|\boldsymbol{x} - \boldsymbol{x}_a|} \tag{5.3.14}$$

$$\zeta_i = -4G \sum_a \frac{m_a v_a^i}{|\boldsymbol{x} - \boldsymbol{x}_a|} \tag{5.3.15}$$

$$\chi = -\frac{G}{2} \sum_a m_a |\boldsymbol{x} - \boldsymbol{x}_a| \tag{5.3.16}$$

$$\psi = -G \sum_a \frac{m_a \Phi'_a}{|\boldsymbol{x} - \boldsymbol{x}_a|} - 3G \sum_a \frac{m_a v_a^2}{|\boldsymbol{x} - \boldsymbol{x}_a|} \tag{5.3.17}$$

with

$$\Phi'_a = -G \sum_{b \neq a} m_b / |\boldsymbol{x}_a - \boldsymbol{x}_b|. \tag{5.3.18}$$

From this (5.3.11) follows, i.e.

$$\overset{(3)}{g}_{0i} = \frac{G}{2} \sum_a \frac{m_a}{|\boldsymbol{x} - \boldsymbol{x}_a|} [7 v_a^i + (\boldsymbol{v}_a \cdot \boldsymbol{n}_a)\, n_a^i]. \tag{5.3.19}$$

## 5.4 The Einstein-Infeld-Hoffmann Equations

The Lagrangian $L_a$ of particle $a$ in the field of the other particles is, according to (5.1.50), (5.3.14, 17, and 19)

$$\begin{aligned} L_a = & \frac{1}{2} v_a^2 + \frac{1}{8} v_a^4 + G \sum_{b \neq a} \frac{m_b}{r_{ab}} \\ & - \frac{1}{2} G^2 \sum_{b, c \neq a} \frac{m_b m_c}{r_{ab} r_{ac}} - G^2 \sum_{b \neq a} \sum_{c \neq a, b} \frac{m_b m_c}{r_{ab} r_{bc}} \\ & + \frac{3}{2} v_a^2 G \sum_{b \neq a} \frac{m_b}{r_{ab}} + \frac{3}{2} G \sum_{b \neq a} \frac{m_b v_b^2}{r_{ab}} \\ & - \frac{G}{2} \sum_{b \neq a} \frac{m_b}{r_{ab}} [7 \boldsymbol{v}_a \cdot \boldsymbol{v}_b + (\boldsymbol{v}_a \cdot \boldsymbol{n}_{ab})(\boldsymbol{v}_b \cdot \boldsymbol{n}_{ab})] , \end{aligned} \tag{5.4.1}$$

where

$$r_{ab} := | \boldsymbol{x}_a - \boldsymbol{x}_b | , \qquad \boldsymbol{n}_{ab} := (\boldsymbol{x}_a - \boldsymbol{x}_b)/r_{ab} .$$

The total Lagrangian $L$ of the $N$-body system must be a symmetric expression of $(m_a, \boldsymbol{x}_a, \boldsymbol{v}_a; a = 1, 2, \ldots, N)$ with the property that $\lim_{m_a \to 0} L/m_a = L_a$. This latter condition just says that in the limit $m_a \to 0$ particle a moves on a geodesic in the field of the other particles.

One verifies easily that $L$ is uniquely given by the following expression

$$\begin{aligned} L = & \sum_a \frac{1}{2} m_a v_a^2 + \sum_a \frac{1}{8} m_a v_a^4 + \frac{1}{2} \sum_{\substack{a, b \\ a \neq b}} \frac{m_a m_b}{r_{ab}} \\ & + \frac{3G}{2} \sum_a m_a v_a^2 \sum_{b \neq a} \frac{m_b}{r_{ab}} \\ & - \sum_{\substack{a, b \\ a \neq b}} \frac{G m_a m_b}{4 r_{ab}} [7 \boldsymbol{v}_a \cdot \boldsymbol{v}_b + (\boldsymbol{v}_a \cdot \boldsymbol{n}_{ab})(\boldsymbol{v}_b \cdot \boldsymbol{n}_{ab})] \\ & - \frac{G^2}{2} \sum_a \sum_{b \neq a} \sum_{c \neq a} \frac{m_a m_b m_c}{r_{ab} r_{ac}} . \end{aligned} \tag{5.4.2}$$

The corresponding Euler equations are the *Einstein-Infeld-Hoffmann* (EIH) equations. They read

$$\begin{aligned} \dot{\boldsymbol{v}}_a = - \sum_{b \neq a} m_b (\boldsymbol{x}_{ab}/r_{ab}^3) \Bigg[ & 1 - 4 \sum_{c \neq a} m_c / r_{ac} \\ & + \sum_{c \neq a, b} m_c \left( - \frac{1}{r_{bc}} + \boldsymbol{x}_{ab} \cdot \boldsymbol{x}_{bc} / 2 r_{bc}^3 \right) - 5 m_a / r_{ab} \\ & + v_a^2 - 4 \boldsymbol{v}_a \cdot \boldsymbol{v}_b + 2 v_b^2 - \frac{3}{2} (\boldsymbol{v}_b \cdot \boldsymbol{x}_{ab} / r_{ab})^2 \Bigg] \end{aligned}$$

$$-\frac{7}{2}\sum_{b\neq a}(m_b/r_{ab})\sum_{c\neq a,b} m_c\,\boldsymbol{x}_{bc}/r_{bc}^3$$
$$+\sum_{b\neq a} m_b\,(\boldsymbol{x}_{ab}/r_{ab}^3)\cdot(4\,\boldsymbol{v}_a-3\,\boldsymbol{v}_b)\,(\boldsymbol{v}_a-\boldsymbol{v}_b)\,. \tag{5.4.3}$$

*The Two-body Problem in the Post-Newtonian Approximation*

For two particles (5.4.2) reduces to ($r := r_{12}$, $\boldsymbol{n} := \boldsymbol{n}_{12}$)

$$L=\frac{m_1}{2}\,\boldsymbol{v}_1^2+\frac{m_2}{2}\,\boldsymbol{v}_2^2+\frac{G\,m_1\,m_2}{r}+\frac{1}{8}\,(m_1\,\boldsymbol{v}_1^4+m_2\,\boldsymbol{v}_2^4)$$
$$+\frac{G\,m_1\,m_2}{2\,r}\,[3\,(\boldsymbol{v}_1^2+\boldsymbol{v}_2^2)-7\,\boldsymbol{v}_1\cdot\boldsymbol{v}_2-(\boldsymbol{v}_1\cdot\boldsymbol{n})\,(\boldsymbol{v}_2\cdot\boldsymbol{n})]$$
$$-\frac{G^2}{2}\,\frac{m_1\,m_2\,(m_1+m_2)}{r^2}\,. \tag{5.4.4}$$

The corresponding EIH equations imply that the center of mass

$$\boldsymbol{X}=(m_1^*\,\boldsymbol{x}_1+m_2^*\,\boldsymbol{x}_2)/(m_1^*+m_2^*) \tag{5.4.5}$$

with

$$m_a^* := m_a+\frac{1}{2}\,m_a\,\boldsymbol{v}_a^2-\frac{1}{2}\,\frac{m_a\,m_b}{r_{ab}}\,,\qquad a\neq b \tag{5.4.6}$$

is not accelerated

$$\frac{d^2\boldsymbol{X}}{dt^2}=0\,. \tag{5.4.7}$$

If we choose $\boldsymbol{X}=0$, then

$$\boldsymbol{x}_1=\left[\ \frac{m_2}{m}+\frac{\mu\,\delta m}{2m^2}\left(\boldsymbol{v}^2-\frac{m}{r}\right)\right]\boldsymbol{x}$$
$$\boldsymbol{x}_2=\left[-\frac{m_1}{m}+\frac{\mu\,\delta m}{2m^2}\left(\boldsymbol{v}^2-\frac{m}{r}\right)\right]\boldsymbol{x}\,, \tag{5.4.8}$$

where

$$\boldsymbol{x} := \boldsymbol{x}_1-\boldsymbol{x}_2\,,\qquad \boldsymbol{v} := \boldsymbol{v}_1-\boldsymbol{v}_2\,,\qquad m := m_1+m_2\,,$$
$$\delta m := m_1-m_2\,,\qquad \mu := m_1\,m_2/m\,. \tag{5.4.9}$$

For the relative motion we obtain from (5.4.4) with (5.4.8), after dividing by $\mu$:

$$L\ =L_0+L_1$$
$$L_0=\frac{1}{2}\,\boldsymbol{v}^2+\frac{G\,m}{r} \tag{5.4.10}$$

$$L_1 = \frac{1}{8}\left(1 - \frac{3\mu}{m}\right) \boldsymbol{v}^4 + \frac{G\,m}{2r}\left[3\,\boldsymbol{v}^2 + \frac{\mu}{m}\,\boldsymbol{v}^2 + \frac{\mu}{m}\,(\boldsymbol{v}\cdot\boldsymbol{x}/r)^2\right] - \frac{G^2 m^2}{2r^2}\,.$$

The corresponding Euler equation is

$$\dot{\boldsymbol{v}} = -\frac{G\,m}{r^3}\,\boldsymbol{x}\left[1 - \frac{G\,m}{r}\,(4 + 2\,\mu/m) + (1 + 3\,\mu/m)\,\boldsymbol{v}^2 - (3\,\mu/2\,m)\left(\frac{\boldsymbol{v}\cdot\boldsymbol{x}}{r}\right)^2\right]$$
$$+ \frac{G\,m}{r^3}\,\boldsymbol{v}\,(\boldsymbol{v}\cdot\boldsymbol{x})\,(4 - 2\,\mu/m)\,. \tag{5.4.11}$$

In the Newtonian limit we choose the solution corresponding to a Keplerian orbit in the plane $z = 0$ with periastron on the $x$-axis. In standard notations we have for $G = 1$:

$$\boldsymbol{x} = r\,(\cos\phi, \sin\phi, 0) \tag{5.4.12}$$

$$r = \frac{p}{1 + e\cos\phi} \tag{5.4.13}$$

$$r^2\,\frac{d\phi}{dt} = \sqrt{m\,p}\,. \tag{5.4.14}$$

The post-Newtonian solution is obtained in the following way: We write

$$r^2\,\frac{d\phi}{dt} = |\,\boldsymbol{x}\times\boldsymbol{v}\,| = \sqrt{m\,p}\,\,(1 + \delta h) \tag{5.4.15}$$

$$\boldsymbol{v} = \frac{d\boldsymbol{x}}{dt} = \sqrt{\frac{m}{p}}\,\,(-\sin\phi, e + \cos\phi, 0) + \delta\boldsymbol{v}. \tag{5.4.16}$$

Substituting (5.4.15) into the identity

$$\frac{d}{dt}\left(r^2\,\frac{d\phi}{dt}\right) \equiv |\,\boldsymbol{x}\times\dot{\boldsymbol{v}}\,|$$

and using (5.4.11) for $\dot{\boldsymbol{v}}$ gives

$$\frac{d}{dt}\,\delta h = \frac{m}{r^3}\left(4 - \frac{2\mu}{m}\right)|\,\boldsymbol{v}\cdot\boldsymbol{x}\,| = \left(4 - \frac{2\mu}{m}\right)\sqrt{m/p}\,\frac{m}{r^2}\,e\sin\phi$$
$$= -\frac{m\,e}{p}\,(4 - 2\,\mu/m)\,(\cos\phi)^{\bullet}\,,$$

where we have used $e\sin\phi = -\,(\cos\phi)^{\bullet}\,r^2/\sqrt{m\,p}$. Hence

$$r^2\,\frac{d\phi}{dt} = \sqrt{m\,p}\left[1 - \frac{m\,e}{p}\,(4 - 2\,\mu/m)\cos\phi\right]. \tag{5.4.17}$$

Inserting (5.4.16) into (5.4.11), one finds by simple integration

$$\begin{aligned} \boldsymbol{v} = \sqrt{m/p} \Big( & - \sin\phi\, \boldsymbol{e}_x + (e + \cos\phi)\, \boldsymbol{e}_y \\ & + \frac{m}{p} \Big\{ \boldsymbol{e}_x \Big[ - 3\, e\, \phi + (3 - \mu/m) \sin\phi - (1 + 21\, \mu/8\, m)\, e^2 \sin\phi \\ & + \frac{1}{2} (1 - 2\mu/m)\, e \sin 2\phi - (\mu/8\, m)\, e^2 \sin 3\phi \Big] \\ & + \boldsymbol{e}_y \Big[ - (3 - \mu/m) \cos\phi - (3 - 31\, \mu/8\, m)\, e^2 \cos\phi \\ & - \frac{1}{2} (1 - 2\mu/m)\, e \cos 2\phi + (\mu/8\, m)\, e^2 \cos 3\phi \Big] \Big\} \Big) . \end{aligned} \tag{5.4.18}$$

If one substitutes (5.4.17) and (5.4.18) into the identity

$$\frac{d}{d\phi} \frac{1}{r} \equiv - \frac{1}{r^2\, d\phi/dt} (\boldsymbol{x} \cdot \boldsymbol{v}/r)$$

and integrates with respect to $\phi$, then one finds for the orbit again formula (5.4.13) with

$$\begin{aligned} p/r = 1 + e \cos\phi + \frac{m}{p} \Big[ & - (3 - \mu/m) + (1 + 9\, \mu/4\, m)\, e^2 \\ & + \frac{1}{2} (7 - 2\mu/m)\, e \cos\phi + 3\, e\, \phi \sin\phi - (\mu/4\, m)\, e^2 \cos 2\phi \Big] . \end{aligned} \tag{5.4.19}$$

The next to last coefficient of $m/p$ gives the periastron motion

$$\delta\phi = \frac{6\pi\, m}{p} . \tag{5.4.20}$$

This is the same expression as for the Schwarzschild solution, but now $m$ denotes the *sum* of the two masses.

The results of this section are important in the analysis of the binary pulsar PSR 1913+16 (see Sect. 5.6).

The result (5.4.20) can be obtained faster with the Hamilton-Jacobi theory. The Hamiltonian corresponding to (5.4.4) is

$$H = H_0 + H_1 , \tag{5.4.21}$$

where $H_0$ is the Hamilton function corresponding to the unperturbed problem

$$H_0 = \frac{1}{2 m_1} \boldsymbol{p}_1^2 + \frac{1}{2 m_2} \boldsymbol{p}_2^2 - \frac{G\, m_1 m_2}{r} \tag{5.4.22}$$

and

$$H_1 = - L_1 . \tag{5.4.23}$$

In (5.4.23) $L_1$ has to be expressed in terms of the positions and momenta. (It is sufficient to use $\boldsymbol{p}_a = m_a \boldsymbol{v}_a$.) Equation (5.4.23) follows from a general result of perturbation theory[2]. Inserting (5.4.4) gives

$$\begin{aligned} H_1 = & -\frac{1}{8}\left(\frac{\boldsymbol{p}_1^4}{m_1^3}+\frac{\boldsymbol{p}_2^4}{m_2^3}\right) \\ & -\frac{G}{2r}\left[3\left(\frac{m_2}{m_1}\boldsymbol{p}_1^2+\frac{m_1}{m_2}\boldsymbol{p}_2^2\right)-7\boldsymbol{p}_1\cdot\boldsymbol{p}_2-(\boldsymbol{p}_1\cdot\boldsymbol{n})(\boldsymbol{p}_2\cdot\boldsymbol{n})\right] \\ & +\frac{G^2}{2}\,\frac{m_1 m_2 (m_1+m_2)}{r^2}\,. \end{aligned} \tag{5.4.24}$$

For the relative motion we have $\boldsymbol{p}_1 = -\boldsymbol{p}_2 =: \boldsymbol{p}$ and

$$\begin{aligned} H_{\text{rel}} = & \frac{1}{2}\left(\frac{1}{m_1}+\frac{1}{m_2}\right)\boldsymbol{p}^2-\frac{G m_1 m_2}{r}-\frac{\boldsymbol{p}^4}{8}\left(\frac{1}{m_1^3}+\frac{1}{m_2^3}\right) \\ & -\frac{G}{2r}\left[3\boldsymbol{p}^2\left(\frac{m_2}{m_1}+\frac{m_1}{m_2}\right)+7\boldsymbol{p}^2+(\boldsymbol{p}\cdot\boldsymbol{n})^2\right]+\frac{G^2 m_1 m_2 (m_1+m_2)}{r^2}\,. \end{aligned} \tag{5.4.25}$$

We use polar angles and consider the motion in the plane $\theta = \frac{\pi}{2}$. Replacing $\boldsymbol{p}^2$ by $p_r^2 + p_\phi^2/r^2$, with $p_\phi := L = \text{const}$, the equation $H(p_r, r) = E$ becomes[3]

$$\begin{aligned} H_{\text{rel}}(p_r, r) = & \frac{1}{2}\left(\frac{1}{m_1}+\frac{1}{m_2}\right)\left(p_r^2+\frac{L^2}{r^2}\right)-\frac{G m_1 m_2}{r} \\ & -\frac{1}{8}\left(\frac{1}{m_1^3}+\frac{1}{m_2^3}\right)\left(\frac{2 m_1 m_2}{m_1+m_2}\right)^2\left(E+\frac{G m_1 m_2}{r}\right)^2 \\ & -\frac{G}{2r}\left\{3\left(\frac{m_2}{m_1}+\frac{m_1}{m_2}\right)+7\right\}\frac{2 m_1 m_2}{m_1+m_2}\left(E+\frac{G m_1 m_2}{r}\right) \\ & -\frac{G}{2r}p_r^2+G^2\,\frac{m_1 m_2 (m_1+m_2)}{2r^2} = E\,. \end{aligned} \tag{5.4.26}$$

The solution of this equation for $p_r$ has the form

$$p_r^2 = -\frac{L^2}{r^2}+\frac{L^2}{r^2}\,\frac{G}{2r}\,\frac{2 m_1 m_2}{m_1+m_2}+A+\frac{B}{r}+\frac{C}{r^2}\,. \tag{5.4.27}$$

---

[2] Let $(q, \dot{q}, \lambda)$ be a Lagrangian which depends on a parameter $\lambda$. The canonical momenta are $p = \partial L/\partial \dot{q}$ and it is assumed that these equations can uniquely be solved for $\dot{q} = \phi(q, p, \lambda)$ for every $\lambda$. Now $H(p, q, \lambda) = p\,\phi(q, p, \lambda) - L(q, \phi(q, p, \lambda), \lambda)$ and hence

$$\frac{\partial H}{\partial \lambda} = p\frac{\partial \phi}{\partial \lambda} - \frac{\partial L}{\partial \dot{q}}\frac{\partial \phi}{\partial \lambda} - \frac{\partial L}{\partial \lambda} = -\frac{\partial L}{\partial \lambda}\,.$$

[3] In the perturbation terms we can replace $\boldsymbol{p}^2$ by $\dfrac{2 m_1 m_2}{m_1+m_2}\left(E+\dfrac{G m_1 m_2}{r}\right)$.

We choose a new radial variable $r'$ such that

$$\frac{L^2}{r^2}\left(1-\frac{G}{r}\frac{m_1 m_2}{m_1+m_2}\right)=\frac{L^2}{r'^2}\,.$$

Up to higher orders

$$r'=r+\frac{G}{2}\frac{m_1 m_2}{m_1+m_2}\,. \tag{5.4.28}$$

The term in the expression for $p_r^2$ proportional to $1/r'^2$ is equal to $-L^2+BGm_1 m_2/2\,(m_1+m_2)+C$, whereby only the lowest order terms in $B$ have to be taken into account. A simple calculation shows that (with new constants $A$, $B$)

$$p_r^2=A+\frac{B}{r'}-(L^2-6\,G^2 m_1^2 m_2^2)\frac{1}{r'^2}\,.$$

The Hamilton-Jacobi function $S$ is

$$S=S_r+S_\phi-E\,t=S_r+L\,\phi-E\,t \tag{5.4.29}$$

with

$$S_r:=\int\sqrt{A+B/r-(L^2-6\,G^2 m_1^2 m_2^2)/r^2}\;dr\,. \tag{5.4.30}$$

The orbit equation follows from $\partial S/\partial L=\text{const}$, or

$$\phi+\partial S_r/\partial L=\text{const}\,.$$

From the first to the second periastron we have the change

$$\Delta\phi=-\frac{\partial}{\partial L}\Delta S_r\,. \tag{5.4.31}$$

Without the correction term to $L^2$ in (5.4.30) the orbit would be a Kepler ellipse and

$$-\frac{\partial}{\partial L}\Delta S_r^{(0)}=\Delta\phi^{(0)}=2\pi\,. \tag{5.4.32}$$

Since

$$\begin{aligned}\Delta S_r&=\Delta S_r^{(0)}-6\,G^2 m_1^2 m_2^2\,\frac{\partial}{\partial(L^2)}\Delta S_r^{(0)} \\ &=\Delta S_r^{(0)}-\frac{3\,G^2 m_1^2 m_2^2}{L}\frac{\partial}{\partial L}\Delta S_r^{(0)}=\Delta S_r^{(0)}+\frac{6\pi\,G^2 m_1^2 m_2^2}{L}\end{aligned} \tag{5.4.33}$$

it follows from (5.4.31, 32, and 33)

$$\begin{aligned}\Delta\phi&=2\pi-\frac{\partial}{\partial L}\left(\frac{6\pi\,G^2 m_1^2 m_2^2}{L}\right) \\ &=2\pi+\frac{6\pi\,G^2 m_1^2 m_2^2}{L^2}\,.\end{aligned}$$

Hence the periastron advance is

$$\delta\phi = \frac{6\pi G^2 m_1^2 m_2^2}{L^2} \equiv \frac{6\pi G (m_1 + m_2)}{c^2 a (1 - e^2)}, \tag{5.4.34}$$

which agrees with (5.4.20).

For the binary pulsar PSR 1913 + 16 the measured periastron shift is (see Sect. 5.6)

$$\dot{\omega} = 4.226 \pm 0.002 \text{ deg/yr} . \tag{5.4.35}$$

The general-relativistic prediction for $\dot{\omega}$ is, using the known values of the orbital elements

$$\dot{\omega}_{\mathrm{GR}} = 2.11 \left(\frac{m_1 + m_2}{M_\odot}\right)^{2/3} \text{deg/yr} . \tag{5.4.36}$$

If $\dot{\omega}_{\text{observed}} = \dot{\omega}_{\mathrm{GR}}$, then

$$m_1 + m_2 = 2.85\, M_\odot . \tag{5.4.37}$$

## 5.5 Precession of a Gyroscope in the PN-Approximation

The equation of transport for the gyroscope spin is [see (1.10.8)]

$$\nabla_u S = -(S, a)\, u, \qquad (S, u) = 0, \quad a := \nabla_u u . \tag{5.5.1}$$

We are interested in the change of the components of $S$ relative to a comoving frame. We first establish the relation between these components ($\mathscr{S}$) and the coordinate components $S^\mu$ [in the gauge (5.1.17), (5.1.18)]. We calculate up to the third order. The metric is, to sufficient accuracy [see (5.3.13)]:

$$g = (1 + 2\phi)\, dt^2 - (1 - 2\phi)\, \delta_{ij}\, dx^i dx^j + 2 h_i\, dt\, dx^i , \tag{5.5.2}$$

where

$$h_i = -\zeta_i - \frac{\partial^2 \chi}{\partial t\, \partial x^i} . \tag{6.5.3}$$

It is useful to introduce the orthonormal basis of 1-forms

$$\begin{aligned} \tilde{\theta}^0 &= (1 + \phi)\, dt + h_i\, dx^i \\ \tilde{\theta}^j &= (1 - \phi)\, dx^j . \end{aligned} \tag{5.5.4}$$

The comoving frame $\hat{\theta}^\mu$ is obtained by a boost. We need the 3-velocity $\tilde{v}_j$ of the gyroscope with respect to $\tilde{\theta}^\mu$. If $\tilde{u}^\mu$ denotes the corresponding 4-velocity, we have, using (5.5.4)

$$\tilde{v}_j = \frac{\tilde{u}^j}{\tilde{u}^0} = \frac{\langle \tilde{\theta}^j, u\rangle}{\langle \tilde{\theta}^0, u\rangle} = \frac{(1 - \phi)\, u^j}{(1 + \phi)\, u^0 + h_i u^i} \simeq (1 - 2\phi) \frac{u^j}{u^0} .$$

Hence

$$\tilde{v}_j = (1 - 2\phi)\, v_j\,, \tag{5.5.5}$$

where $v_j$ is the coordinate velocity

The relation between $\tilde{S}^\mu := \langle \tilde{\theta}^\mu, S \rangle$ and $\mathscr{S}$ is, to a sufficient approximation

$$\tilde{S}^\mu = (\tilde{v}_j \mathscr{S}^j, \mathscr{S}^i + \tfrac{1}{2}\tilde{v}_i (\tilde{v}_j \mathscr{S}^j))\,,$$

but $\tilde{S}^i = \langle \tilde{\theta}^i, S \rangle = (1 - \phi)\, S^i$.

Hence, to the required accuracy

$$S^i = (1 + \phi)\mathscr{S}^i + \tfrac{1}{2} v_i (v_k \mathscr{S}^k)\,.$$

The covariant components $S_i$ are obtained by multiplication with $-(1 - 2\phi)$:

$$S_i = (1 - \phi)\mathscr{S}_i + \tfrac{1}{2} v_i (v_k \mathscr{S}_k)\,, \tag{5.5.7}$$

note that $\mathscr{S}_i = -\mathscr{S}^i$. Next we derive an equation of motion for $S_i$ and translate this with (5.5.7) into an equation for $\mathscr{S}$. In components, (5.5.1) reads

$$\frac{dS_\mu}{d\tau} = \Gamma^\lambda_{\mu\nu} S_\lambda u^\nu - a^\lambda S_\lambda g_{\mu\nu} u^\nu\,. \tag{5.5.8}$$

We set $\mu = i$, multiply with $d\tau/dt$ and use

$$S_0 = -v_i S_i\,, \tag{5.5.9}$$

which follows from $(S, u) = 0$. Thus

$$\begin{aligned}\frac{dS_i}{dt} &= \Gamma^j_{i0} S_j - \Gamma^0_{i0} v_j S_j + \Gamma^j_{ik} v_k S_j - \Gamma^0_{ik} v_k v_j S_j \\ &\quad - (g_{0i} + g_{ik} v_k)(-a^0 v_j S_j + a^j S_j)\,.\end{aligned} \tag{5.5.10}$$

Up to third order we obtain (with $a_0 = -v_i a_i$):

$$\frac{dS_i}{dt} = [\overset{(3)}{\Gamma^j_{i0}} - \overset{(2)}{\Gamma^0_{i0}} v_j + \overset{(2)}{\Gamma^j_{ik}} v_k]\, S_j + v_i a^j S_j\,. \tag{5.5.11}$$

The Christoffel symbols are given in (5.1.46). We find for $\boldsymbol{S} = (S_1, S_2, S_3)$

$$\begin{aligned}\frac{d}{dt}\boldsymbol{S} &= \tfrac{1}{2}\boldsymbol{S} \times (\nabla \times \boldsymbol{\zeta}) - \boldsymbol{S}\frac{\partial \phi}{\partial t} - 2(\boldsymbol{v}\cdot\boldsymbol{S})\nabla\phi - \boldsymbol{S}(\boldsymbol{v}\cdot\nabla\phi) \\ &\quad + \boldsymbol{v}(\boldsymbol{S}\cdot\nabla\phi) + \boldsymbol{v}(\boldsymbol{a}\cdot\boldsymbol{S})\,,\end{aligned} \tag{5.5.12}$$

where $\boldsymbol{a} = (a^1, a^2, a^3)$.

To the required order, we can invert (5.5.7)

$$\mathscr{S} = (1 + \phi)\,\boldsymbol{S} - \tfrac{1}{2}\boldsymbol{v}(\boldsymbol{v}\cdot\boldsymbol{S})\,. \tag{5.5.13}$$

We know that $\mathscr{S}^2 = \text{const}$ [see (1.10.22)]. The rate of change of $\mathscr{S}$ is given to third order by

$$\dot{\mathscr{S}} = \dot{\boldsymbol{S}} + \boldsymbol{S}\left(\frac{\partial \phi}{\partial t} + \boldsymbol{v} \cdot \nabla \phi\right) - \tfrac{1}{2}\, \dot{\boldsymbol{v}}\, (\boldsymbol{v} \cdot \boldsymbol{S}) - \tfrac{1}{2}\, \boldsymbol{v}\, (\dot{\boldsymbol{v}} \cdot \boldsymbol{S}).$$

Here we can use $\dot{\boldsymbol{v}} \cong -\nabla\phi + \boldsymbol{a}$:

$$\dot{\mathscr{S}} = \dot{\boldsymbol{S}} + \boldsymbol{S}\left(\frac{\partial \phi}{\partial t} + \boldsymbol{v} \cdot \nabla \phi\right) + \tfrac{1}{2}\, \nabla\phi\, (\boldsymbol{v} \cdot \boldsymbol{S}) + \tfrac{1}{2}\, \boldsymbol{v}\, (\boldsymbol{S} \cdot \nabla\phi)$$
$$- \tfrac{1}{2}\, \boldsymbol{a}\, (\boldsymbol{v} \cdot \boldsymbol{S}) - \tfrac{1}{2}\, \boldsymbol{v}\, (\boldsymbol{S} \cdot \boldsymbol{a})\,. \tag{5.5.14}$$

Now we substitute (5.5.12), and find to the required order

$$\dot{\mathscr{S}} = \boldsymbol{\Omega} \times \mathscr{S} \tag{5.1.15}$$

with the precession angular velocity

$$\boldsymbol{\Omega} = -\tfrac{1}{2}\, (\boldsymbol{v} \times \boldsymbol{a}) - \tfrac{1}{2}\, \nabla \times \boldsymbol{\zeta} - \tfrac{3}{2}\, \boldsymbol{v} \times \nabla\phi\,. \tag{5.5.16}$$

The first term is just the Thomas precession. This term is not present for a geodesic motion. The third term is the geodetic precession, while the second term gives the Lense-Thirring precession [see (1.10.33) and the end of Sect. 4.4].

*Gyroscope in Orbit Around the Earth*

As a first application of (5.5.16) we consider a gyroscope placed in a circular orbit about the Earth. We first need the potential $\boldsymbol{\zeta}$. A system which is at rest and spherically symmetric, but which rotates with angular frequency $\boldsymbol{\omega}\,(r)$, has the momentum density

$$\overset{(1)}{T}{}^{i0}\,(\boldsymbol{x}, t) = \overset{(0)}{T}{}^{00}\,(r)\,[\boldsymbol{\omega}\,(r) \times \boldsymbol{x}]_i\,. \tag{5.5.17}$$

Using this in (5.1.40) leads to

$$\boldsymbol{\zeta}\,(\boldsymbol{x}) = -4G \int \frac{d^3x'}{|\boldsymbol{x} - \boldsymbol{x}'|}\, \boldsymbol{\omega}\,(r') \times \boldsymbol{x}'\, \overset{(0)}{T}{}^{00}\,(r')\,. \tag{5.5.18}$$

The solid-angle integral is

$$\int d\Omega' \frac{\boldsymbol{x}'}{|\boldsymbol{x} - \boldsymbol{x}'|} = \begin{cases} \dfrac{4\pi r'^2}{3 r^3}\, \boldsymbol{x} & \text{for } r' < r \\[2ex] \dfrac{4\pi}{3 r'}\, \boldsymbol{x} & \text{for } r' > r\,. \end{cases} \tag{5.5.19}$$

Thus the field outside the sphere is

$$\boldsymbol{\zeta}\,(\boldsymbol{x}) = \frac{16\pi G}{3 r^3}\, \boldsymbol{x} \times \int \boldsymbol{\omega}\,(r')\, \overset{(0)}{T}{}^{00}\,(r')\, r'^4\, dr'\,. \tag{5.5.20}$$

On the other hand, the angular momentum is

$$\begin{aligned}\boldsymbol{J} &= \int \{\boldsymbol{x}' \times (\boldsymbol{\omega}(r') \times \boldsymbol{x}')\} \overset{(0)}{T}{}^{00}(r')\, d^3x' \\ &= \int \{r'^2 \boldsymbol{\omega}(r') - \boldsymbol{x}'(\boldsymbol{x}' \cdot \boldsymbol{\omega}(r'))\} \overset{(0)}{T}{}^{00}(r')\, d^3x' \\ &= \frac{8\pi}{3} \int \boldsymbol{\omega}(r') \overset{(0)}{T}{}^{00}(r')\, r'^4\, dr' .\end{aligned} \tag{5.5.21}$$

Comparing this with (5.5.20) shows that

$$\boldsymbol{\zeta}(\boldsymbol{x}) = \frac{2G}{r^3}(\boldsymbol{x} \times \boldsymbol{J}) . \tag{5.5.22}$$

This agrees with (5.2.14) but now the formula holds everywhere outside the sphere.

Inserting (5.5.22) and $\phi = -GM/r$ into (5.5.16) gives

$$\boldsymbol{\Omega} = \frac{G}{r^3}\left(-\boldsymbol{J} + \frac{3(\boldsymbol{J} \cdot \boldsymbol{x})\boldsymbol{x}}{r^2}\right) + \frac{3GM\, \boldsymbol{x} \times \boldsymbol{v}}{2r^3} . \tag{5.5.23}$$

The last term, which is independent of $\boldsymbol{J}$, represents the geodetic precession.

If $\boldsymbol{n}$ denotes the normal to the plane of the orbit, we have

$$\boldsymbol{v} = -\left(\frac{GM}{r^3}\right)^{1/2} \boldsymbol{x} \times \boldsymbol{n} \tag{5.5.24}$$

and the precession rate, averaged over a revolution, is

$$\langle \boldsymbol{\Omega} \rangle = \frac{G}{2r^3}\{\boldsymbol{J} - \boldsymbol{n}(\boldsymbol{n} \cdot \boldsymbol{J})\} + \frac{3(GM)^{3/2}\boldsymbol{n}}{2r^{5/2}} . \tag{5.5.25}$$

For the Earth and $r \simeq R_\oplus$ we get

$$\frac{\text{Lense-Thirring}}{\text{geodetic}} \simeq \frac{J_\oplus G}{3(M_\oplus G)^{3/2} R_\oplus^{1/2}} = 6.5 \times 10^{-3} . \tag{5.5.26}$$

This shows that the main effect is a precession around the orbital angular momentum with an averaged angular velocity

$$|\langle \boldsymbol{\Omega} \rangle| \simeq \frac{3GM_\oplus^{3/2}}{2r^{5/2}} \simeq 8.4\left(\frac{R_\oplus}{r}\right)^{5/2} \text{sec/yr} . \tag{5.5.27}$$

This may be measurable in the coming years.

*Precession of the Binary Pulsar*

As an interesting application of (5.5.16) we determine $\boldsymbol{\Omega}$ for the binary pulsar. The field of the companion (indexed by 2) is [see (5.3.14) and (5.3.15)]

$$\phi(\boldsymbol{x}) = -\frac{G\,m_2}{|\boldsymbol{x}-\boldsymbol{x}_2|}\,, \qquad \boldsymbol{\zeta}(\boldsymbol{x},t) = -4\,G\,m_2\frac{\boldsymbol{v}_2}{|\boldsymbol{x}-\boldsymbol{x}_2|}\,. \tag{5.5.28}$$

Using (5.5.16) we find

$$\boldsymbol{\Omega} = -2\,G\,m_2\frac{\boldsymbol{x}\times\boldsymbol{v}_2}{r^3} + \frac{3}{2}\,G\,m_2\,\frac{\boldsymbol{x}\times\boldsymbol{v}_1}{r^3}\,,$$

where now

$$\boldsymbol{x} := \boldsymbol{x}_1 - \boldsymbol{x}_2\,, \qquad r := |\boldsymbol{x}|\,.$$

Since

$$\boldsymbol{v}_1 = \frac{m_2}{m_1+m_2}\,\dot{\boldsymbol{x}}\,, \qquad \boldsymbol{v}_2 = -\frac{m_1}{m_1+m_2}\,\dot{\boldsymbol{x}}\,,$$

we obtain

$$\boldsymbol{\Omega} = (\boldsymbol{L}/\mu)\left\{2\,G\,m_2\frac{m_1}{m_1+m_2} + \frac{3}{2}\,G\,m_2\frac{m_2}{m_1+m_2}\right\}\frac{1}{r^3}\,, \tag{5.5.29}$$

where $\boldsymbol{L}$ is the angular momentum:

$$\boldsymbol{L} = \mu\,\boldsymbol{x}\times\dot{\boldsymbol{x}}\,, \qquad \mu = \frac{m_1\,m_2}{m_1+m_2}\,.$$

We average $\boldsymbol{\Omega}$ over a period:

$$\langle\boldsymbol{\Omega}\rangle = \hat{\boldsymbol{L}}\,\{\ldots\}\,\frac{L}{\mu}\,\frac{1}{T}\int_0^{2\pi}\frac{1}{r^3}\,\frac{1}{\dot\phi}\,d\phi\,.$$

Using

$$L = \mu\,r^2\,\dot\phi\,, \qquad r = \frac{a\,(1-e^2)}{1+e\cos\phi}\,,$$

we find

$$\begin{aligned}\boldsymbol{\Omega} &= \hat{\boldsymbol{L}}\,\{\ldots\}\,\frac{1}{T}\int_0^{2\pi}\frac{1}{r}\,d\phi\\ &= \hat{\boldsymbol{L}}\,\{\ldots\}\,\frac{1}{a\,(1-e^2)}\,\frac{1}{T}\int_0^{2\pi}(1+e\cos\phi)\,d\phi\,.\end{aligned}$$

Thus

$$\langle\boldsymbol{\Omega}\rangle = \hat{\boldsymbol{L}}\,\frac{3\pi\,G\,m_2}{T}\,\frac{1}{a\,(1-e^2)}\Bigg\{\underbrace{\frac{m_2}{m_1+m_2}}_{\substack{\text{geodetic}\\ \text{precession}}} + \underbrace{\frac{4}{3}\,\frac{m_1}{m_1+m_2}}_{\substack{\text{Lense-}\\ \text{Thirring}}}\Bigg\}\,. \tag{5.5.30}$$

We compare this with the periastron motion

$$\dot\omega = \frac{6\pi\,G\,(m_1+m_2)}{T\,a\,(1-e^2)\,c^2} \tag{5.5.31}$$

and obtain

$$|\langle \boldsymbol{\Omega} \rangle| / \dot{\omega} = \frac{1}{2} \frac{m_2\left(1 + \frac{1}{3}\frac{m_1}{m_1 + m_2}\right)}{m_1 + m_2} . \tag{5.5.32}$$

For $m_1 \simeq m_2 \simeq (m_1 + m_2)/2$ we have

$$|\langle \boldsymbol{\Omega} \rangle| / \dot{\omega} \simeq \frac{7}{24} . \tag{5.5.33}$$

There is some chance that this is measurable.

## 5.6 The Binary Pulsar

At several occasions we have compared general relativity effects with observational data of the binary system containing the pulsar PSR 1913+16. In the present section, we describe this marvelous system in some detail.

The binary pulsar was discovered by *Hulse* and *Taylor* in summer, 1974 [70] in a systematic search for new pulsars. The nominal pulse period of the pulsar is 59 ms. This short period was observed to be periodically shifted, which proves that the pulsar is a member of a binary system with an orbital period of 7.75 hours. With Kepler's third law and reasonable masses, one concludes from this that the system is rather narrow, having a diameter of roughly $1 R_\odot$. Correspondingly, the velocity of the pulsar is $\sim 10^{-3} c$ and it moves through a relatively strong gravitational field ($GM/c^2 r \sim 10^{-6}$). These numbers show that several special and general relativistic effects should be observable. This has indeed been achieved in the meantime with increasing accuracy.

The measurements yield enough information to determine all parameters of the system, and allow in addition a test of GR, provided the companion star has a negligible quadrupole moment and tidal interactions are small enough. We shall see that these conditions for a "clean" relativistic system are probably fulfilled.

### 5.6.1 Pulse Arrival-Time Data and their Analysis

The observations of the pulsar have been made with the giant 305 m Arecibo radiotelescope in Puerto Rico. The data were obtained at frequencies near 430 and 1410 MHz. The observing procedure and the analysis of the data involve the following main steps.

1) Measurement of the absolute arrival times of the pulses with very good time resolution (the time resolution at 430 MHz has been improved to 43 μs). PSR 1913 + 16 is one of the weakest pulsars (≈ 5 m Jy at 430 MHz). For this reason, individual pulses cannot be observed, so

synchronous averaging over several minutes must be used to yield an average pulse profile. (For the steady improvement in accuracy, see Fig. 1 in [71].) The resulting mean pulse profile is then fitted by the method of least squares to a long term average "standard profile", to obtain the precise arrival time. It is very fortunate that PSR 1913 + 16 has made no discontinuous jumps (glitches) or noise since it has been discovered. Hopefully, it will continue to behave so well.

2) The pulse arrival times are now corrected from the location of the observatory to the barycenter of the solar system. In this step, one has to include relativistic clock corrections to account for the annual motion of the Earth around the Sun.

3) The dispersion delay due to the interstellar plasma is removed. This can be computed from the measurements at two different frequencies. Clearly, one must use the frequencies at which the signal propagates through the interstellar medium, rather than the observed frequency, which is Doppler shifted by the Earth's motion. The dispersion measure is rather large $DM = 168.77\ \mathrm{cm}^{-3}\,\mathrm{pc}$, leading to differential time delays of 70 ms across a typical bandwidth of 4 MHz at 430 MHz. This is more than the pulsar period. For this reason, the dispersion of the signals had to be compensated from the very beginning. The large dispersion measure indicates that the pulsar is quite far away ($\approx 5\,\mathrm{kpc}$).

4) In a next step, one must express the previously corrected arrival times in terms of the proper times $t_p$ measured at the pulsar. This is done in the post-Newtonian approximation, using in particular the results of Sect. 5.4 for the two-body problem. The details of this step, in which GR enters in an important way are described in [72].

5) The pulsar phase $\phi$ as a function of the proper time $t_p$ is expanded as follows

$$\phi = \phi_0 + \nu t_p + \tfrac{1}{2}\dot{\nu} t_p^2 + \tfrac{1}{6}\ddot{\nu} t_p^3. \tag{5.6.1}$$

Here $\phi_0$ is a constant and $\nu$, $\dot{\nu}$, $\ddot{\nu}$ denote the rotation frequency of the pulsar and its first and second time derivatives. From experience with other pulsars, one can confidently neglect higher derivatives of $\nu$. Together with the previous steps, one obtains the integer $\phi$ in terms of the measured arrival times of the pulses and a large number (up to 20) of quantities, some of which are listed in Table 1.

6) The parameters in this relation are determined by an iterative procedure. Initial estimates of the parameters are used to calculate the deviation of $\phi$ from the next integer. These so-called phase residuals are then minimized in a least-squares fit to determine improved values of the model parameters. A very good fit to the data is obtained, as can be seen from the postfit residuals (see [71]). There are no systematic phase-dependent trends exceeding about 20 μs.

Table 1 shows the values of the most interesting parameters, determined up to 1982 [71].

**Table 1.** Parameter estimates from pulse arrival times (taken from 71)]

| | |
|---|---|
| Right ascension (1950.0) | $19^{\text{h}}\,13^{\text{m}}\,12.47^{\text{s}}$ |
| Declination (1950.0) | $16°\,01'\,08.2''$ |
| Period $P$ | 0.059029952709 s |
| $\dot{P}$ | $8.628 \times 10^{-18}$ s/s |
| $\ddot{P}$ | $(-58 \pm 1200) \times 10^{-30}\ \text{s}^{-1}$ |
| Projected semimajor axis of the pulsar orbit $a_p \sin i$ | 2.34186 lt-s |
| Excentricity $e$ | 0.61714 |
| Orbital period $P_b$ | 27906.9816 s |
| Rate of advance of periastron passage $\dot{\omega}$ | 4.226 deg $\text{yr}^{-1}$ |
| Variable part of the gravitational redshift and transverse Doppler shift $\gamma$ | 0.00438 ($\pm$ 0.00024) s |
| Rate of change of the orbital period $\dot{P}_b$ | $(-2.30 \pm 0.22) \times 10^{-12}$ s/s |

## 5.6.2 Relativistic Effects

In the analysis of the data the periastron shift is left as a free param- eter. If the binary system is "clean", it should be given by GR Eq. (5.4.34), or, using Kepler's third law, by

$$\dot{\omega}_{\text{GR}} = \frac{3\,G^{2/3}}{c^2\,(1-e^2)} \left(\frac{2\pi}{P_b}\right)^{5/3} (m_{\text{p}} + m_{\text{c}})^{2/3}. \qquad (5.6.2$$

Here $m_{\text{p}}$ is the mass of the pulsar and $m_{\text{c}}$ that of the companion. Th quantity $\dot{\omega}_{\text{GR}}$ is of order $(v/c)^2$. The transverse Doppler shift of th signals and the variable part of the gravitational red shift are of th same order. Since both terms have the same dependence on th classical parameters, only their sum can be measured. It is given by th parameter (see [72])

$$\gamma = \frac{G^{2/3}}{c^2}\, e \left(\frac{P_b}{2\pi}\right)^{1/3} m_{\text{c}} \frac{m_{\text{p}} + 2m_{\text{c}}}{(m_{\text{p}} + m_{\text{c}})^{4/3}}\,. \qquad (5.6.3$$

Terms of order $(v/c)^3$ are at present unobservable. From Table 1 w find for (5.6.2) and (5.6.3)

$$\dot{\omega}_{\text{GR}} = 2.11353 \left(\frac{m_{\text{p}} + m_{\text{c}}}{M_\odot}\right)^{2/3} \text{deg yr}^{-1} \qquad (5.6.4$$

$$\gamma = 0.00293696 \left(\frac{m_{\text{c}}}{M_\odot}\right) \left(\frac{m_{\text{p}} + 2\,m_{\text{c}}}{M_\odot}\right) \left(\frac{m_{\text{p}} + m_{\text{c}}}{M_\odot}\right)^{-4/3}. \qquad (5.6.$$

Assuming for the moment that the companion can be treated as a poi mass, we can use (5.6.4) and (5.6.5) to determine from the measur values the masses of the pulsar and the unseen companion. (This is remarkable new application of GR.) The results are

$$m_{\rm p} + m_{\rm c} = 2.8275 \pm 0.0007\ M_\odot$$
$$m_{\rm p} = 1.42 \pm 0.06\ M_\odot$$
$$m_{\rm c} = 1.41 \pm 0.06\ M_\odot .$$

It his way, we obtain the first precise value of the mass of a radio pulsar. It is close to the Chandrasekhar mass limit of a white dwarf (see Chap. 6).

We can derive also all the remaining unknown parameters of the system.

The inclination angle is determined by (use Kepler's third law)

$$\sin i = G^{-1/3} \frac{a_{\rm p} \sin i}{m_{\rm c}} \left(\frac{P_b}{2\pi}\right)^{-2/3} (m_{\rm p} + m_{\rm c})^{2/3} . \tag{5.6.6}$$

The semimajor axis of the relative orbit, pulsar orbit, and companion orbit $a$, $a_{\rm p}$, $a_{\rm c}$ may be calculated from known quantities, using Kepler's third law,

$$a = G^{1/3} \left(\frac{P_b}{2\pi}\right)^{2/3} (m_{\rm p} + m_{\rm c})^{1/3} \tag{5.6.7}$$

$$a_{\rm p} = \frac{m_{\rm c}}{m_{\rm p} + m_{\rm c}} a \tag{5.6.8}$$

$$a_{\rm c} = \frac{m_{\rm p}}{m_{\rm p} + m_{\rm c}} a . \tag{5.6.9}$$

The resulting values are listed, together with the mass determinations, in Table 2.

The spin precession of the pulsar has already been discussed at the end of Sect. 5.5. The magnitude of the spin angular frequency is given by (5.5.32) and its direction is orthogonal to the orbital plane. This effect could show up as a slow change in the observed pulse profile (or as a polarisation change) with an expected rate of about 1 deg $\rm yr^{-1}$. No such changes have been observed in the last few years, indicating that the spin of the pulsar is orthogonal to the orbit plane. (Some fortuitous geometry cannot be excluded.)

**Table 2.** Derived quantities

| | |
|---|---|
| Total mass | $M = 2.8275 \pm 0.0007\ M_\odot$ |
| Pulsar mass | $m_{\rm p} = 1.42 \quad \pm 0.06\ M_\odot$ |
| Companion mass | $m_{\rm c} = 1.41 \quad \pm 0.06\ M_\odot$ |
| Inclination | $\sin i = 0.72 \quad \pm 0.03$ |
| Relative semimajor axis | $a = 6.5011 \pm 0.0005$ lt-s |
| Pulsar semimajor axis | $a_{\rm p} = 3.24 \quad \pm 0.13$ lt-s |
| Companion semimajor axis | $a_{\rm c} = 3.26 \quad \pm 0.13$ lt-s |

### 5.6.3 Gravitational Radiation

We have already seen (Sect. 4.5) that the quadrupole formula agrees remarkably well with the measured values of $\dot{P}_b$, giving for the first time direct evidence of the emission of gravitational waves. This is an effect of order $(v/c)^5$. Recently, a more satisfactory analysis of the relativistic two-body problem has been given [169], [170], in which the equations of motion have been derived up to order $(v/c)^5$. It is quite satisfactory that the effects of order $(v/c)^5$ turned out to be the same as in our naive discussion.

The self-consistency among the measured values of $\dot{\omega}$, $\gamma$, and $\dot{P}_b$ is demonstrated in Fig. 5.1.

An instructive representation of the observed change of orbital period is given in Fig. 5.2.

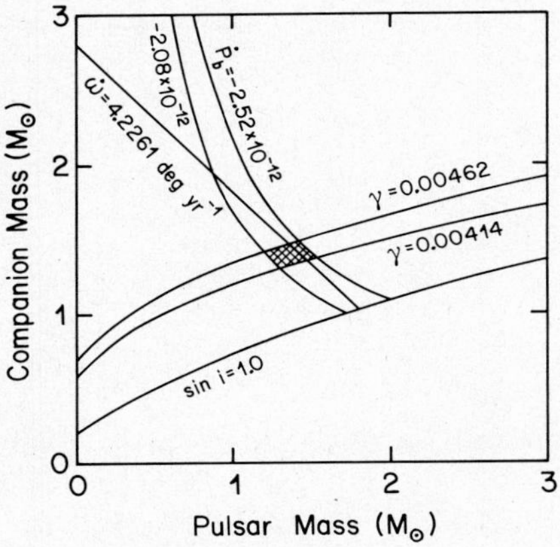


**Fig. 5.1.** Curves in the pulsar mass-companion mass plane corresponding to the measured values of $\dot{\omega}$, $\gamma$, and $\dot{P}_b$, if these are determined by GR. The portion of the plane below $\sin i = 1$ is forbidden by the observed mass function (from [71])

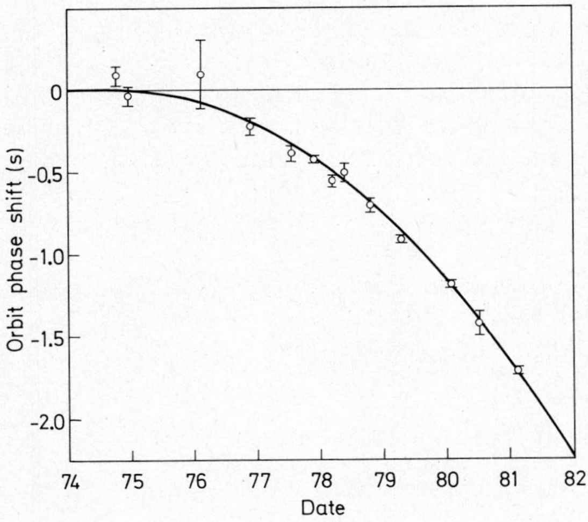


**Fig. 5.2.** Orbital phase residuals from the observed times of periastron passage. For a constant period, the points would lie on a straight line. The curve through the points corresponds to the general relativistic prediction based on the quadrupole formula for the energy loss by gravitational radiation (from [71])

An analysis of the binary pulsar observations in the framework of other gravitational theories can be found in [67], Chap. 12. Many of the proposed theories do not even predict the right sign of the orbital period change due to the emission of gravitational radiation, let alone the proper magnitude. Together with the solar system tests almost no theories of gravitation, except GR, remain viable. (Certain theories with torsion can, however, not be excluded.)

### 5.6.4 The Companion Star

The binary pulsar can only be used as a test system for GR if both stars can be treated as point masses. If not, other Newtonian effects, such as tidal or rotational distortion of the companion and viscous dissipation of orbital energy, are important. This would mask an orbital period decrease due to the emission of gravitational radiation. These classical effects are completely negligible for a neutron star, a black hole, or a slowly rotating white dwarf companion. They may, however, be important if the companion is not a collapsed star.

The companion can certainly not be a main-sequence star, because the system is so narrow and no eclipses of the pulsar are seen. Moreover, tidal deformation of a main-sequence star would result in a periastron advance larger than the observed value [see (5.6.20) below]. The only remaining viable candidates for the companion are a helium star, white dwarf, neutron star, or a black hole. Among these only a helium star of a rapidly rotating white dwarf may not act as a point mass. We proceed to discuss this in some detail.

*Tidal Distortion*

We study first the tidal contribution to the periastron advance.

The distortion of the companion may be computed at any time from the equilibrium conditions, using the instantaneous distance between the two components. This adiabatic approximation is valid, since the characteristic dynamical time scales (pulsation periods) of the companion are much shorter than the orbital period, so that the star is capable of adjusting itself at any time to the instantaneous distorting force.

Since the distortion is treated as a perturbation, we ignore the rotation of the star and add these effects later on.

The total Newtonian potential $\phi$ is the sum $\phi = \phi_p + \phi_c$, where $\phi_p$ is the potential of the pulsar (treated as a point mass) and $\phi_c$ is that of the companion. The basic equations are the hydrostatic equilibrium condition

$$\nabla P = -\varrho \nabla(\phi_p + \phi_c) \tag{5.6.10}$$

and Poisson's equation

$$\Delta \phi_c = 4\pi G \varrho, \tag{5.6.11}$$

where $\varrho$ denotes the density of the companion. In solving these equations, one treats the distortion of the companion as a small perturbation. The details of the calculations are described in Schwarzschild's classic book [73], Sect. 18. From Eqs. (18.13) and (18.16) of this reference, we find for the interaction energy $V$ of the two stars

$$V = -\frac{G m_p m_c}{r}\left[1 + 2k\frac{m_p}{m_c}\left(\frac{R_c}{r}\right)^5\right]. \tag{5.6.12}$$

Here $R_c$ is the unperturbed radius of the companion and $k$ is a constant which is determined by the mass distribution of the unperturbed companion through the Radau equation [see Eqs. (18.17), (18.11), (18.12), and (18.15) in [73]].

We compute now the apsidal motion due to the perturbation $\delta V$ in (5.6.12). (Schwarzschild gives only the result, which is, however, as everywhere (?) else in the literature, in error by a factor of two.) This is most easily done by using the following general formula for the periastron advance $\delta\varphi$ per revolution

$$\delta\varphi = \frac{\partial}{\partial L}\left(\frac{2\mu}{L}\int_0^\pi r^2\,\delta V\, d\varphi\right), \tag{5.6.13}$$

where $\mu$ is the reduced mass and $L$ is the orbital angular momentum of the relative motion.

---

**Exercise:** Prove Eq. (5.6.13). (The solution can be found in Landau-Lifshitz, Mechanics, solution of Exercise 2 in Sect. 15.)

---

In our case we get

$$\delta\varphi = \frac{\partial}{\partial L}\left(-\frac{2\mu}{L}\,2k\,G m_p^2 R_c^5 \int_0^\pi \frac{1}{r^4}\, d\varphi\right).$$

Using standard notation, we have for the unperturbed Keplerian motion (see Sect. 4.5)

$$r = \frac{p}{1 + e\cos\varphi}, \quad p = a(1-e^2) \tag{5.6.14}$$

$$a = -m_c m_p/2E, \quad p = \frac{L^2}{\mu G m_p m_c}, \quad e^2 = 1 + \frac{2EL^2(m_p + m_c)}{G m_p^3 m_c^3}.$$

Using

$$\int_0^\pi (1 + e\cos\varphi)^4\, d\varphi = \pi\left(1 + 3e^2 + \tfrac{3}{8}e^4\right),$$

we obtain

$$\delta\varphi = \frac{\partial}{\partial L}\left[-2\pi\frac{\mu}{L}\,2k\,G m_p^2 R_c^5\left(\frac{\mu G m_p m_c}{L^2}\right)^4\left(1 + 3e^2 + \tfrac{3}{8}e^4\right)\right].$$

With the explicit $L$-dependence of $e^2$ given in (5.6.14), we find

$$\delta\varphi = 30\,\pi\,2k\,\frac{m_p}{m_c}\left(\frac{R_c}{a}\right)^5 \frac{1+\frac{3}{2}e^2+\frac{1}{8}e^4}{(1-e^2)^5}\,, \tag{5.6.15}$$

which is twice as large as the expression we have found in the literature.

If we eliminate $a$ with Kepler's third law and use the observed values for $e$ and $P_b$, we find for the tidal contribution to $\dot{\omega}$

$$\dot{\omega}_{\text{tidal}} = 22.3\,\frac{k}{0.01}\,\frac{m_p}{m_c}\left(\frac{R_c}{0.2\,R_\odot}\right)^5\left(\frac{m_p+m_c}{M_\odot}\right)^{-5/3}\ \text{deg yr}^{-1}. \tag{5.6.16}$$

This formula (up to the mentioned factor of two) has been used in [74].

*Rotational Distortion*

The rotational contribution to $\dot{\omega}$ is also discussed in Ref. [74]. For simplicity, we derive $\dot{\omega}_{\text{rot}}$ only for the case where the companion is rigidly rotating with angular velocity $\omega_c$ around an axis perpendicular to the orbit plane. Then the stationary Euler equation for the companion can be written as

$$\nabla P = -\varrho\,\nabla(\phi_c - \tfrac{1}{2}\boldsymbol{v}^2) \equiv -\varrho\,\nabla(\phi_c + \phi_d)\,.$$

Thus the deformation potential $\phi_d$ is determined by the velocity field $\boldsymbol{v}$,

$$\phi_d(r,\theta) = -\tfrac{1}{2}\boldsymbol{v}^2 = -\tfrac{1}{2}r^2\omega_c^2\sin^2\theta = \tfrac{1}{3}r^2\omega_c^2\,[P_2(\cos\theta)-1]\,,$$

where $\theta$ denotes the polar angle with respect to the rotation axis. If one compares this potential with the tidal distortion potential [Eq. (18.4) in Schwarzschild's book], one can immediately write down the result for $\phi_c$ by an appropriate substitution:

$$\phi_c = -\frac{Gm_c}{r} + \frac{2k}{3}\,\frac{1}{r^3}\,R_c^5\,\omega_c^2\,P_2(\cos\theta)\,.$$

The interaction energy is $V = m_c\,\phi_c(\theta = \frac{\pi}{2})$, i.e.

$$V = -\frac{Gm_p m_c}{r} - \frac{k}{3}\,\frac{1}{r^3}\,R_c^5\,\omega_c^2. \tag{5.6.17}$$

In a similar manner as before, one easily finds

$$\delta\varphi = 6\pi\,k\,\frac{\omega_c^2 R_c^5}{3Gm_c a^2}\,\frac{1}{(1-e^2)^2}\,. \tag{5.6.18}$$

---

**Exercise:** Derive Eq. (5.6.18).

---

The potential (5.6.16) has the form (3.3.10) with

$$J_2 R_c^2 = \frac{2k\,R_c^5\,\omega_c^2}{3Gm_c}\,. \tag{5.6.19}$$

We also note that the definition (3.3.10) of $J_2$ can be expressed as

$$J_2 m_c R_c^2 = C - A, \tag{5.6.20}$$

where $C$ and $A$ are the moments of inertia about the spin axis and about an axis perpendicular to the spin axis, respectively. From (5.6.19) and (5.6.20), we find

$$\frac{C-A}{m_c} = \frac{2k\, R_c^5\, \omega_c^2}{3Gm_c}.$$

Hence we can write (5.6.18) in the form

$$\delta\varphi = 3\pi \frac{C-A}{m_c a^2} \frac{1}{(1-e^2)^2}. \tag{5.6.21}$$

For an arbitrary angle $\theta$ between the spin axis and the normal to the orbit plane, one finds [see Eq. (3.12) of Ref. [74]]:

$$\delta\varphi = 3\pi \frac{C-A}{m_c a^2} (1-e^2)^{-2} (1 - \tfrac{3}{2}\sin^2\theta) \tag{5.6.22}$$

leading to the following rotational contribution of the apsidal motion

$$\dot{\omega}_{\mathrm{rot}} = 0.84\, \alpha_6 \left(\frac{M}{M_\odot}\right)^{-2/3} (1 - \tfrac{3}{2}\sin^2\theta) \quad \mathrm{deg\ yr^{-1}}, \tag{5.6.23}$$

where

$$\alpha_6 = \frac{C-A}{m_c \times 10^6\,\mathrm{km}^2}. \tag{5.6.24}$$

Note that $\dot{\omega}_{\mathrm{rot}}$ can have either sign, depending on the orientation of the spin axis.

The rate of the tidally or rotationally induced periastron advance are both proportional to $R_c^5$ and are thus utterly negligible for the pulsar.

### *Helium Star Companion*

A tidally deformed helium star would induce an $\dot{\omega}_{\mathrm{tidal}}$ comparable to $\dot{\omega}_{\mathrm{GR}}$. Let us for a moment ignore rotational effects. Then the observed value of $\dot{\omega}$ should be equal to $\dot{\omega}_{\mathrm{tidal}} + \dot{\omega}_{\mathrm{GR}}$. Using the values for $k$ in (5.6.20) of model calculations [75], one finds the curve (He star) in Fig. 5.3.

This result, together with the limit $\sin i \leqq 1$ obtained from the mass function[4], shows that the pulsar would be a low mass star, a possibility

[4] The mass function is obtained from the radial velocity curve; see Chap. 8.

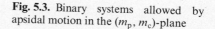

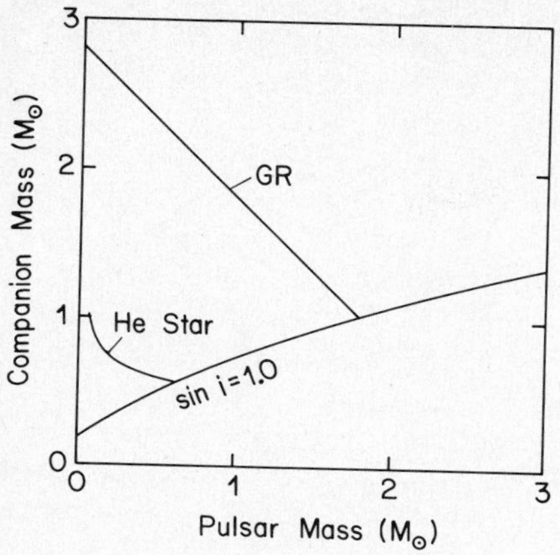


**Fig. 5.3.** Binary systems allowed by apsidal motion in the $(m_p, m_c)$-plane

that is very unlikely on evolutionary grounds (see Chap. 8). Rotation of the helium star, with a spin axis aligned with the orbital angular momentum vector, reduces the mass even further.

For such a low mass system, the gravitational radiation gives only a small contribution to $\dot{P}_b$. It would indeed be fortuitous if viscous dissipation would just give the correct change of orbital energy.

There have been claims of an optical identification [76] of the pulsar companion. A candidate with $m_V = 22.5$, $m_R = 20.9$ has been found which lies $0.13'' \pm 0.21''$ east of the pulsar, and $0.36'' \pm 0.22''$ north of it. The discrepancy in declination is not large enough to reject the hypothesis of a physical association. Further information, such as an optical spectrum, and the presence or absence of periodic Doppler shifts, is needed.

Helium stars have strong, radiation-driven winds. A search for variations in the pulsar's dispersion measure with respect to orbit phase has yielded an upper limit of $0.002\ \mathrm{pc\ cm^{-3}}$, which leads to an upper limit on the mean electron density inside the pulsar's orbit of about $10^5\ \mathrm{cm^{-3}}$. This translates to the stringent restriction for the wind parameters: $\dot{M}/v < 10^{-17}\ M_\odot\ \mathrm{yr^{-1}/km\ s^{-1}}$. From this, we cannot necessarily conclude that the companion is not a helium star because the pulsar radiation might sweep the ionized gas from the system, or the stellar wind might be weaker than expected.

*White Dwarf Companion*

For a white dwarf, we can neglect $\dot{\omega}_{\mathrm{tidal}}$, because the factor $(R_c/0.2\ R_\odot)^5$ in (5.6.20) is typically a few times $10^{-7}$ ($R_c \sim R_\oplus$). Rotational deformations can, however, be important for a rapidly rotating white dwarf. The deformation parameter $\alpha_6$ needed in (5.6.23) has been evaluated

[74] from models of such stars. One finds values for $\alpha_6$ up to $\sim 15$, allowing a rather large region in the $(m_p, m_c)$-plane, satisfying the condition $\dot{\omega} = \dot{\omega}_{GR} + \dot{\omega}_{rot}$ (see Figs. 5 and 6 of Ref. [74]).

One cannot exclude the possibility that viscous dissipation of orbital energy of rotationally deformed white dwarfs could cause an orbit change of the same magnitude as that caused by gravitational radiation. Ionic, electronic, and molecular viscosity are not efficient, but magnetic or turbulent viscosity might be large enough. Unfortunately, we cannot even estimate the order of magnitude of these effects. Again, it is unlikely that they would be just of the right magnitude.

The self-consistency of the interpretation of the binary system with the pulsar PSR 1913 + 16 as a "clean" relativistic system makes it very likely that the companion is a neutron star or a black hole. Some uncertainties remain, however.

Evolutionary scenarios which could lead to the observed configuration will be discussed in Chap. 8.

# Part III

# Relativistic Astrophysics

The term "relativistic astrophysics" was coined in 1963, shortly after the discovery of quasars. The unusual properties of these objects immediately gave rise to quite exotic hypotheses for the time. Even fairly conventional attempts at an explanation clearly indicated that general relativity would probably play a decisive role in understanding quasars.

In the meantime, relativistic astrophysics has grown into an extensive research area which is developing extremely rapidly. The discovery of quasars was soon followed by other important astronomical discoveries. In this last part, we shall be mainly concerned with some theoretical aspects of the physics of compact objects.

# Chapter 6. Neutron Stars

## Introduction

During the course of their evolution, sufficiently massive stars ($10\,M_\odot \lesssim M \lesssim 60\,M_\odot$) acquire an "onion skin" structure. This consists of an iron-nickel core surrounded by concentric shells of $^{28}$Si, $^{16}$O, $^{20}$Ne, $^{12}$C, $^{4}$He, and H. There is also some admixture of $^{24}$Mg, $^{32}$S, and other elements in the inner zones. For a recent evolutionary calculation, see [30].

The "iron" core finally attains a mass of about $1.6\,M_\odot$ with an electron fraction $Y_e$ (the number of electrons/nucleon) of $0.43-0.44$. (The time scales involved are sufficiently long that elements in the vicinity of iron are significantly neutronized via electron capture.) As the temperature rises further, the elements in the iron group will sooner or later be broken up by photodisintegration into $\alpha$-particles. One should keep in mind that the *free* energy is minimized in thermodynamic equilibrium; as the temperature increases, the entropy contribution to $F = U - TS$ becomes important. The dissociation of iron into helium costs a large amount of internal energy and reduces the adiabatic index to below the critical value of $\frac{4}{3}$; the stellar core becomes unstable. (This will be discussed in more detail in Sect. 6.9, p. 337.) During the last evolutionary stages of a massive star the neutrino losses become immense. Before the stellar core starts to collapse, the neutrino luminosity can reach values as high as $(7-8)\times 10^{48}$ erg/s $\sim 2\times 10^{15}\,L_\odot$. (See Sect. 6.9.) This is considerably more than the optical luminosity of an entire galaxy.

At the start of the collapse of the core, the central temperature and density typically have the values (see [30]):

$$T_c = \begin{cases} 8.3\times 10^9\,\text{K} & \text{for} \quad M = 15\,M_\odot \\ 8.3\times 10^9\,\text{K} & \text{for} \quad M = 25\,M_\odot \end{cases}$$

$$\varrho_c = \begin{cases} 6.0\times 10^9\,\text{g/cm}^3 & \text{for} \quad M = 15\,M_\odot \\ 3.5\times 10^9\,\text{g/cm}^3 & \text{for} \quad M = 25\,M_\odot . \end{cases}$$

After the stellar core has become unstable, it collapses to nuclear density and forms a hot ($T \sim 10$ MeV) neutron star (or a black hole). The gravitational energy set free amounts to about $0.1\, M_{\text{core}}\, c^2 \approx 10^{53}$ erg. This is rapidly radiated away, mainly in the form of neutrinos. Gravitational radiation may also play a role at the beginning of a strongly asymmetric collapse. After at most a few minutes, the matter in the interior of the neutron star is practically in its ground state. The temperature is considerably lower than the degeneracy temperature of the electrons, protons, and neutrons, due to the extremely high densities involved. At least in some cases, the stellar envelope explodes as a supernova. A famous example which supports this picture is the Crab nebula which is the relict of a supernova explosion, observed by the Chinese as a "guest star" in 1054 A.D., and which contains a pulsar in its center. The Vela pulsar provides another example. (The current status of supernova theory is described in Sect. 6.9 and pulsars are discussed in Sect. 6.10.)

## 6.1 Order-of-Magnitude Estimates

The general direction of stellar evolution can be understood qualitatively quite simply. Our starting point is the virial theorem [1]

$$E_G + 3 \int P\, dV = 0. \tag{6.1.1}$$

($E_G$ is the gravitational energy). The electron pressure is

$$P_e \simeq P_e(T=0) + n_e k T, \tag{6.1.4}$$

where $n_e$ is the number of electrons per unit volume. This is a simple interpolation between Maxwell-Boltzmann and degenerate behavior.

Now

$$P_e(T=0) = (\gamma - 1)\, u_e(T=0), \tag{6.1.5}$$

where $u_e(T=0)$ is the zero point energy density of the electrons and

$$4/3 \leqq \gamma \leqq 5/3. \tag{6.1.6}$$

[1] The equation for hydrostatic equilibrium reads

$$dP/dr = -\varrho(r)\, GM(r)/r^2, \tag{6.1.2}$$

where $M(r)$ is the mass contained in a concentric sphere of radius $r$. Obviously

$$dM/dr = 4\pi r^2 \varrho. \tag{6.1.3}$$

If we multiply (6.1.2) by $4\pi r^3/3$ and integrate over the star, we obtain $4\pi/3 \int r^3\, dP = -1/3 \int (GM(r)/r)\, dM(r)$, and hence (6.1.1) after an integration by parts.

The lower limit in this inequality corresponds to the extreme relativistic limit and the upper one to the nonrelativistic limit. Furthermore,

$$u_e(T=0) = n_e m_e c^2 [(1+x^2)^{1/2} - 1] \tag{6.1.7}$$

with

$$x = \bar{p}_e / m_e c \simeq \frac{\hbar}{m_e c} n_e^{1/3} = \lambda\!\!\!^{-}_e n_e^{1/3} \simeq \lambda\!\!\!^{-}_e \frac{N_e^{1/3}}{R}; \tag{6.1.8}$$

here $N_e$ is the total number of electrons and $R$ is the radius of the star.

In (6.1.8) we have used the uncertainty relation for the average momentum $\bar{p}_e$ of the electrons and the Pauli principle. At $T=0$ there is exactly one electron in each de Broglie cube having volume $(\hbar/p)^3$. If $N$ denotes the total number of particles (of all kinds), then $N = N_e / Y_e \mu$, where $\mu$ is the average molecular weight, and we obtain from the virial theorem:

$$-\frac{GM^2}{R} + 3Nk\bar{T} + 3(\gamma - 1) N_e m_e c^2 (\sqrt{1+x^2} - 1) \simeq 0. \tag{6.1.9}$$

However $M = N m_N Y_e^{-1}$, where $m_N$ is the nucleon mass. Defining

$$N_0 := \left(\frac{\hbar c}{G m_N^2}\right)^{3/2}, \quad N_{e0}^{1/3} = N_0^{1/3} Y_e, \tag{6.1.10}$$

we obtain from (6.1.8) and (6.1.9)

$$\frac{k\bar{T}}{m_e c^2} \simeq \frac{1}{3} \mu Y_e \left[\left(\frac{N_e}{N_{e0}}\right)^{2/3} x + \left(1 - \sqrt{1+x^2}\right)\right]. \tag{6.1.11}$$

We have replaced a factor $3(\gamma - 1)$ by 1, which is justified by (6.1.6).

These expressions permit us to draw the following interesting conclusions:

(i) With increasing $x$, i.e. with increasing density, $k\bar{T}$ reaches a maximum provided $N_e < N_{e0}$, and then decreases montonically. The value $T=0$ is reached for

$$x \sim 2 (N_e / N_{e0})^{2/3},$$

provided $N_e / N_{e0}$ is sufficiently small, so that $x \ll 1$. From (6.1.8) we then have for the stellar radius

$$R \simeq \lambda\!\!\!^{-}_e N_e^{1/3} \tfrac{1}{2} (N_{e0}/N_e)^{2/3} = \tfrac{1}{2} Y_e^2 \lambda\!\!\!^{-}_e N_0^{1/3} \left(\frac{N_0}{N_e}\right)^{1/3}. \tag{6.1.12}$$

The characteristic length

$$l_e = \lambda\!\!\!^{-}_e N_0^{1/3} = 5 \times 10^8 \text{ cm}, \tag{6.1.13}$$

which appears is comparable to the Earth's radius $R_\oplus = 6.4 \times 10^8$ cm. If the electrons are nonrelativistic, one obtains from (6.1.12) the following relation between mass and radius:

$$R \simeq \tfrac{1}{2} l_e \left(\frac{N_0 m_N}{M}\right)^{1/3} Y_e^{5/3}. \tag{6.1.14}$$

Numerically, we have

$$N_0 m_N = 1.85\, M_\odot . \tag{6.1.15}$$

In the other limiting case, $x \gg 1$, (6.1.11) reduces to

$$x\left[1 - \frac{1}{Y_e^2}(N_e/N_0)^{2/3} + \frac{1}{2x^2} + \ldots\right] \simeq 1 - 3\mu\, Y_e \frac{kT}{m_e c^2} .$$

For $T \simeq 0$, this means that

$$1 - Y_e^{-2}(N_e/N_0)^{2/3} \gtrsim 0$$

or

$$M \lesssim Y_e^2 N_0 m_N . \tag{6.1.16}$$

This is the famous *Chandrasekhar limit.* A more precise calculation [77] results in

$$M_{Ch} = 5.76\, Y_e^2 M_\odot = 1.44\, M_\odot \quad \text{for} \quad Y_e = \tfrac{1}{2} . \tag{6.1.17}$$

The existence of a limiting mass is an extremely important consequence of special relativity and quantum mechanics. (For further discussion, see Sect. 6.8.)
(ii) If $M > M_{Ch}$, the temperature increases without limit. We thus expect that the evolution of the star will encounter an instability.
(iii) If $M < M_{Ch}$ and $x \ll 1$, the maximum temperature is given approximately by

$$(kT/mc^2)_{max} \simeq \frac{1}{6}\mu\, Y_e \left(\frac{N_e}{N_{e0}}\right)^{4/3} = \frac{1}{6}\mu\, Y_e^{-5/3}\left(\frac{M}{N_0 m_N}\right)^{4/3} . \tag{6.1.18}$$

This shows that the maximum temperature is of the order of magnitude $m_e c^2$.

One should note that in the nonrelativistic limit a cold equilibrium is always possible as the density rises.

*Historical Remarks* (see also [31])

By 1925 it was known through the work of *W.* Adams on the binary system of Sirius that Sirius B has the enormous density of about $10^6$ g/cm$^3$. The existence of such compact stars constituted one of the major puzzles of astrophysics until the quantum statistical theory of the electron gas was worked out.

On August 26, 1926, Dirac's paper containing the Fermi-Dirac distribution was communicated by *Fowler* to the Royal Society. On November 3 of that year, Fowler presented his own work to the Royal Society in which he systematically worked out the quantum statistics of identical particles and in the process developed the well-known

Darwin-Fowler method. Shortly thereafter, on December 10, he communicated to the Royal Astronomical Society a new paper with the title "Dense Matter". In this work he showed that the electron gas in Sirius B is almost completely degenerate in the sense of the new Fermi-Dirac statistics. This paper by Fowler concludes with the following words:

"The black-dwarf material is best likened to a single gigantic molecule in its lowest quantum state. On the Fermi Dirac statistics, its high density can be achieved in one and only one way, in virtue of a correspondingly great energy content. But this energy can no more be expended in radiation than the energy of a normal atom or molecule. The only difference between black dwarf matter and a normal molecule is that the molecule can exist in a free state while the black-dwarf matter can only so exist under very high external pressure."

Since Fowler treated the electrons nonrelativistically, he found an equilibrium configuration for every mass (see above). However, since

$$\varrho = \frac{n_e}{Y_e} m_N = \frac{1}{Y_e} (p_F^3/3\pi^2 \hbar^3) m_N \tag{6.1.19}$$

we find

$$\varrho = B x^3, \quad x := p_F/m_e c \tag{6.1.20}$$

$$B := \frac{8\pi m_e^3 c^3 m_N}{3 h^3} \frac{1}{Y_e} = 0.97 \times 10^6 \frac{1}{Y_e} \text{ [g/cm}^3\text{]}. \tag{6.1.21}$$

This shows that the momenta of the electrons in a white dwarf are comparable to $m_e c$ and the electron gas must thus be treated relativistically. This was noticed 1930 by *Chandrasekhar* and he discovered the limiting mass (6.1.17). This number was also independently estimated by *Landau* in 1932. In 1934, Chandrasekhar derived the exact relation between mass and radius for completely degenerate configurations. (This can be found in his classic book [77].) He concluded his paper with the following statement:

"The life-history of a star of small mass must be essentially different from the life-history of a star of large mass. For a star of small mass, the natural white-dwarf stage is an initial step towards complete extinction. A star of large mass cannot pass into the white-dwarf stage and one is left speculating on other possibilities."

This conclusion was not accepted by the leading astrophysicists of the time. A comment by *Eddington* contains the following correct conclusion:

"Chandrasekhar shows that a star of mass greater than a certain limit remains a perfect gas and can never cool down. The star has to go on radiating and radiating and contracting and contracting, until, I suppose, it gets down to a few kilometers radius when gravity becomes strong enough to hold the radiation and the star can at last find peace."

If Eddington had stopped at this point, he would have been the first to predict the existence of black holes. However, he did not take this conclusion seriously, and continues:

"I felt driven to the conclusion that this was almost a *reductio ad absurdum* of the relativistic degeneracy formula. Various accidents may intervene to save the star, but I want more protection than that. I think that there should be a law of nature to prevent the star from behaving in this absurd way."

And here is *Landau* [88] in similar vein:

"For $M > 1.5\,M_\odot$ there exists in the whole quantum theory no cause preventing the system from collapsing to a point. As in reality such masses exist quietly as stars and do not show any such ridiculous tendencies we must conclude that all stars heavier than $1.5\,M_\odot$ certainly possess regions in which the laws of quantum mechanics (and therefore of quantum statistics) are violated."

It is remarkable that *Baade* and *Zwicky* predicted the fate of massive stars as early as 1934, as can be seen in the following citation from their work [Phys. Rev. *45*, 128 (1934)]:

"With all reserve we suggest the view that supernovae represent the transitions from ordinary stars into *neutron stars*, which in their final stages consist of extremely closely packed neutrons."

It is also worth pointing out the recollection of L. Rosenfeld, that on the day in 1932, when the discovery of the neutron became known in Copenhagen, he spent the evening with Bohr and Landau discussing the implications of this important discovery. During the course of the discussion, Landau considered the possibility of the existence of dense stars which consist primarily of neutrons. However, the idea was not published until 1937.

The first model calculations of the properties of neutron stars were performed within the framework of GR by *Oppenheimer* and *Volkoff*. In this work, the matter was taken to be an ideal degenerate neutron gas.

Afterwards, theoretical interest in neutron stars dwindled, since no relevant observations existed. For two decades, Zwicky was one of the very few who took seriously the probable role of neutron stars in the evolution of massive stars.

Interest in the subject was reawakened at the end of the 1950s and beginning of the 1960s. When pulsars, particularly the pulsar at the center of the Crab Nebula, were discovered in 1967, it became clear that neutron stars could be formed in supernova events through the collapse of the stellar core to nuclear densities.

The gross features of a neutron star can be estimated quite easily. One must simply replace the electron mass by the neutron mass in the equations describing a white dwarf, and also take $\mu$ and $Y_e$ equal to one. From (6.1.16) we obtain for the limiting mass

$$M \lesssim N_0 m_N. \tag{6.1.22}$$

If $N/N_0$ is not too close to unity, the radius of a typical neutron star is given, according to (6.1.14), by

$$R \sim \lambda\!\!\!^{-}_N N_0^{1/3} (N_0/N)^{1/3}. \tag{6.1.23}$$

The linear dimensions of a neutron star are thus smaller than those of a white dwarf by a factor $m_e/m_N$. We note also that

$$\lambda_N N_0^{1/3} = 2.7 \times 10^5\,\mathrm{cm}. \tag{6.1.24}$$

In contrast to the white dwarfs, we obtain a *maximum* mass for neutron stars, since the kinetic energy of the neutrons also contributes to the gravitational mass of a neutron star. In the virial theorem one should set

$$E_G \simeq -\frac{GM^2}{R}, \quad M = M_0 + U/c^2, \tag{6.1.25}$$

where $M_0 = N\,m_N$ and $U$ is the zero point energy of the neutrons.

For cold configurations we then obtain

$$-(N/N_0)^{2/3}\,x(1+x^2) + \sqrt{1+x^2} - 1 \simeq 0. \tag{6.1.26}$$

For $x \gg 1$ this gives

$$N/N_0 \simeq 1/x^2, \quad M \simeq N_0\,m_N/x^2. \tag{6.1.27}$$

As a consequence $M$ decreases with increasing $x$ when $x \gg 1$. One finds the maximal mass for $x \simeq 0.8$.

## 6.2 Relativistic Equations for Stellar Structure

If $M \sim N_0\,m_N$, then $R \sim \lambda_N N_0^{1/3}$, and hence $GM/Rc^2 \sim 1$. Thus, GR is important for a quantitative description.

For a static, spherically symmetric star, the metric has the form (see the Appendix to Chap. 3)

$$g = e^{2a(r)}\,dt^2 - [e^{2b(r)}\,dr^2 + r^2(d\vartheta^2 + \sin^2\vartheta\,d\varphi^2)]. \tag{6.2.1}$$

With respect to the orthonormal tetrad

$$\theta^0 = e^a\,dt, \quad \theta^1 = e^b\,dr, \quad \theta^2 = r\,d\vartheta, \quad \theta^3 = \sin\vartheta\,d\varphi \tag{6.2.2}$$

the energy-momentum tensor has the form

$$(T^{\mu\nu}) = \begin{pmatrix} \varrho & & & 0 \\ & P & & \\ & & P & \\ 0 & & & P \end{pmatrix}, \tag{6.2.3}$$

where $\varrho$ is the total mass-energy density and $P$ is the pressure. The Einstein tensor corresponding to (6.2.1) was computed in Sect. 3.1, with the result

$$G_0^0 = \frac{1}{r^2} - e^{-2b}\left(\frac{1}{r^2} - \frac{2b'}{r}\right)$$

$$G_1^1 = \frac{1}{r^2} - e^{-2b}\left(\frac{1}{r^2} + \frac{2a'}{r}\right) \tag{6.2.4}$$

$$G_2^2 = G_3^3 = - e^{-2b}\left(a'^2 - a'\,b' + a'' + \frac{a' - b'}{r}\right)$$

and all other components equal to zero.

The field equations give ($c = 1$)

$$r^{-2} - e^{-2b}\left(\frac{1}{r^2} - \frac{2b'}{r}\right) = 8\pi G\,\varrho \tag{6.2.5}$$

$$r^{-2} - e^{-2b}\left(\frac{1}{r^2} + \frac{2a'}{r}\right) = - 8\pi G P\,. \tag{6.2.6}$$

If we use the notation $u/r := e^{-2b}$, (6.2.5) reads

$$u' = - 8\pi G\,\varrho\,r^2 + 1\,.$$

Integration of this equation gives $u = r - 2GM(r)$, where

$$M(r) = 4\pi \int_0^r \varrho(r')\,r'^2\,dr'\,. \tag{6.2.7}$$

Hence

$$e^{-2b(r)} = 1 - \frac{2GM(r)}{r}\,. \tag{6.2.8}$$

If we subtract (6.2.6) from (6.2.5), we obtain

$$e^{-2b(r)}\,(a' + b') = 4\pi\,G\,(\varrho + P)\,r \tag{6.2.9}$$

and hence

$$a = - b + 4\pi\,G \int_\infty^r dr'\,e^{2b(r')}\,r'\,(\varrho + P)\,. \tag{6.2.10}$$

Thus, if $\varrho$ and $P$ are known, it is possible to determine the gravitational field.

An additional useful relation follows from the "conservation law" $D{*}\,T^\alpha = 0$. Equation (6.2.3) gives

$${*}T^\alpha = q^\alpha\,\eta^\alpha \quad \text{(no sum)}\,, \tag{6.2.11}$$

with $q^0 = \varrho$ and $q^i = -P$. Now

$$\begin{aligned} D\,{*T^\alpha} &= d\,(q^\alpha\,\eta^\alpha) + \sum_\beta \omega^\alpha_\beta \wedge (q^\beta\,\eta^\beta) \\ &= dq^\alpha \wedge \eta^\alpha + \sum_\beta \omega^\alpha_\beta \wedge \eta^\beta\, q^\beta + q^\alpha \underbrace{d\eta^\alpha}_{-\sum_\beta \omega^\alpha_\beta \wedge \eta^\beta} \\ &= dq^\alpha \wedge \eta^\alpha + \sum_\beta \omega^\alpha_\beta \wedge \eta^\beta\,(q^\beta - q^\alpha) \\ &= 0\,. \end{aligned}$$

For $\alpha = 1$, this gives, after making use of the connection forms (3.1.6),

$$dP \wedge \eta^1 = \omega^1_0 \wedge \eta^0\,(\varrho + P)$$

or

$$\frac{dP}{dr}\,\mathrm{e}^{-b}\,\underbrace{\theta^1 \wedge \eta^1}_{-\eta} = a'\,\mathrm{e}^{-b}\,\underbrace{\theta^0 \wedge \eta^0}_{\eta}\,(\varrho + P)$$

so that

$$a' = -\frac{P'}{\varrho + P}\,. \tag{6.2.12}$$

On the other hand, we obtain from (6.2.7, 8, 9)

$$a' = \frac{G}{1 - 2GM(r)/r}\left(\frac{M(r)}{r^2} + 4\pi\,r\,P\right). \tag{6.2.13}$$

If we compare this with (6.2.12), we obtain the *Tolman, Oppenheimer, Volkoff* (TOV) equation

$$-P' = \frac{G\,(\varrho + P)\,[M(r) + 4\pi\,r^3\,P]}{r^2\,[1 - 2\,G\,M(r)/r]}\,. \tag{6.2.14}$$

At the stellar radius $R$, the pressure vanishes. Outside the star, the metric is given by the Schwarzschild solution. The gravitational mass is

$$M = M(R) = \int_0^R \varrho\,4\pi\,r^2\,dr\,. \tag{6.2.15}$$

The TOV equation generalizes the equation

$$-P' = \frac{G\,\varrho\,M(r)}{r^2} \tag{6.2.16}$$

of the Newtonian theory. The pressure gradient towards the center increases for three reasons in GR:

(i) Since pressure also acts as a source of a gravitational field, there is a term proportional to $P$ in addition to $M(r)$.
(ii) Since gravity also acts on $P$, the density $\varrho$ is replaced by $(\varrho + P)$.
(iii) The gravitational force increases faster than $1/r^2$; $1/r^2$ is replaced by $r^{-2}[1 - 2\,G\,M(r)/r]^{-1}$.

These modifications will lead us to the conclusion (see Sect. 6.6) that an arbitrarily massive neutron star cannot exist, even if the equation of state becomes extremely stiff at high densities.

In order to construct a stellar model, one needs an equation of state $P(\varrho)$. If this is given, all properties of the star are determined by the central density $\varrho_c$.

We summarize the relevant equations:

$$-P' = G(\varrho + P)\frac{M(r) + 4\pi r^3 P}{r^2[1 - 2GM(r)/r]} \tag{6.2.17a}$$

$$M' = 4\pi r^2 \varrho \tag{6.2.17b}$$

$$a' = G\frac{M(r) + 4\pi r^3 P}{r^2[1 - 2GM(r)/r]}\,. \tag{6.2.17c}$$

The initial condition

$$M(0) = 0 \tag{6.2.18a}$$

is obtained from (6.2.7). The other two boundary conditions are

$$P(0) = P(\varrho_c) \quad \text{and} \quad e^{a(R)} = 1 - 2GM(R)/R\,. \tag{6.2.18b}$$

*Interpretation of M*

We compare the quantity $M$ appearing in (6.2.15) with the mass $M_0 = N m_N$ [$N$ is the total number of nucleons (baryons) in the star]. If $J$ denotes the baryon current, then

$$N = \int_{t=\text{const}} *J\,. \tag{6.2.19}$$

If $J = J_\mu \theta^\mu$, then $*J = J_\mu \eta^\mu$ and we obtain

$$N = \int_{t=\text{const}} J_0 \eta^0 = \int_{t=\text{const}} J^0 \theta^1 \wedge \theta^2 \wedge \theta^3$$

or, using (6.2.2)

$$N = \int_0^R J^0 4\pi r^2 e^{b(r)} dr\,.$$

The component $J^0$ with respect to the basis (6.2.2) is the baryon number density

$$n := U_\mu J^\mu = J^0 \tag{6.2.20}$$

since the four-velocity $U_\mu = (1, 0, 0, 0)$. Hence

$$N = \int_0^R 4\pi r^2 e^{b(r)} n(r)\, dr = \int_0^R \frac{4\pi r^2}{\sqrt{1 - 2GM(r)/r}} n(r)\, dr\,. \tag{6.2.21}$$

The invariant internal energy density is defined by

$$\varepsilon(r) = \varrho(r) - m_{\mathrm{N}}\, n(r)\,, \tag{6.2.22}$$

and the total internal energy is correspondingly

$$E = M - m_{\mathrm{N}}\, N\,. \tag{6.2.23}$$

We may use (6.2.21) to decompose $E$ as follows:

$$E = T + V\,, \tag{6.2.24}$$

where

$$T = \int_0^R \frac{4\pi r^2}{\sqrt{1 - 2GM(r)/r}}\, \varepsilon(r)\, dr \tag{6.2.25}$$

$$V = \int_0^R 4\pi r^2 \left[1 - \frac{1}{\sqrt{1 - 2GM(r)/r}}\right] \varrho(r)\, dr\,. \tag{6.2.26}$$

In order to find the connection with Newtonian theory, we expand the square roots in (6.2.25) and (6.2.26), assuming that $GM(r)/r \ll 1$. This gives

$$T = \int_0^R 4\pi r^2 \left[1 + \frac{GM(r)}{r} + \dots\right] \varepsilon(r)\, dr \tag{6.2.27}$$

$$V = -\int_0^R 4\pi r^2 \left[\frac{GM(r)}{r} + \frac{3G^2 M^2(r)}{2r^2} + \dots\right] \varrho(r)\, dr\,. \tag{6.2.28}$$

The leading terms in $T$ and $V$ are the Newtonian values for the internal and gravitational energy of the star.

---

**Exercise:** Solve the stellar structure equations for a star having uniform density ($\varrho = \text{const}$) and show that its mass is limited by the inequality

$$M < \tfrac{4}{9}\,(3\pi\varrho)^{-1/2}\,. \tag{6.2.29}$$

---

## 6.3 Stability

An equilibrium solution is physically relevant only if it is also stable.

When investigating the stability of an equilibrium configuration, one usually restricts oneself to a linear analysis. One thus considers time-dependent perturbations and expands all equations which the system satisfies about the equilibrium configuration, keeping only the linear terms. The system is stable provided the frequencies of the corresponding normal modes all have negative imaginary parts. General theorems show that this result is unchanged by nonlinearities. On the other hand, if the imaginary part of the frequency of some normal mode is positive, this mode grows exponentially, and one expects that the system is unstable.

For (cold) spherically symmetric configurations, one considers as a first step only *radial* perturbations, which are also adiabatic. Since hydrodynamic time scales are usually much shorter than the characteristic times for energy transport, this is reasonable. In this case, one obtains an eigenvalue problem [2] (of the Sturm-Liouville type) for $\omega^2$, since the equations for adiabatic perturbations are time reversal invariant (there is no dissipation). The equilibrium is stable provided $\omega_0^2 > 0$ for the lowest mode frequency $\omega_0$.

One can often find *sufficient* criteria for instability by making use of the Rayleigh-Ritz variational principle and simple trial functions.

Sometimes it is possible to decide, simply by examining the equilibrium solution, whether or not it is stable against radial, adiabatic pulsations. Here we prove only the following Proposition 6.3.1 and refer the interested reader to [32].

We consider only cold matter. For a given equation of state, there is then a one-parameter family of equilibrium solutions. We choose the central density $\varrho_c$ as the most suitable parameter.

**Proposition 6.3.1:** At each critical point of the function $M(\varrho_c)$, precisely one radial adiabatic normal mode changes its stability properties; elsewhere, the stability properties do not change.
*"Proof":* Suppose that for a given equilibrium configuration a radial mode changes its stability property (the frequency $\omega$ passes through zero). This implies that there exist infinitesimally nearby equilibrium configurations into which the given one can be transformed, without changing the gravitational mass (energy). Hence if $\omega$ passes through zero we have $M'(\varrho_c) = N'(\varrho_c) = 0$. From Proposition 6.3.2, $N'(\varrho_c) = 0$ already follows from $M'(\varrho_c) = 0$.

**Proposition 6.3.2:** For a radial adiabatic variation of a cold equilibrium solution, we have

$$\delta M = \frac{\varrho + P}{n} e^a \delta N . \tag{6.3.1}$$

In particular, if $\delta M = 0$ then also $\delta N = 0$.
*Proof:* With the notation $m(r) = GM(r)$, it follows from (6.2.15) and (6.2.20) that

$$\delta M = \int_0^\infty 4\pi r^2 \delta\varrho(r)\, dr \tag{6.3.2}$$

---

[2] A detailed derivation of the eigenvalue equation for radial adiabatic pulsations of relativistic stars can be found in Chap. 26 of [14].

$$\delta N = \int_0^\infty 4\pi r^2 \left(1 - \frac{2m(r)}{r}\right)^{-1/2} \delta n(r)\, dr$$
$$+ \int_0^\infty 4\pi r \left(1 - \frac{2m(r)}{r}\right)^{-3/2} n(r)\, \delta m(r)\, dr\,. \tag{6.3.3}$$

For an adiabatic variation we have

$$\delta(\varrho/n) + P\,\delta(1/n) = T\,\delta s = 0 \tag{6.3.4}$$

or

$$\delta n = \frac{n}{P+\varrho}\,\delta\varrho\,. \tag{6.3.5}$$

Obviously (in this proof we take $G = 1$)

$$\delta m(r) = \int_0^r 4\pi r'^2\, \delta\varrho(r')\, dr'\,. \tag{6.3.6}$$

We now insert these last two relations into (6.3.3) and interchange in one term the order of the integrations over $r$ and $r'$, with the result

$$\delta N = \int_0^\infty 4\pi r^2 \left[\left(1 - \frac{2m(r)}{r}\right)^{-1/2} \frac{n(r)}{P(r)+\varrho(r)}\right.$$
$$\left. + \int_r^\infty 4\pi r'\, n(r') \left(1 - \frac{2m(r')}{r'}\right)^{-3/2} dr'\right] \delta\varrho(r)\, dr\,. \tag{6.3.7}$$

The expression in the square bracket is independent of $r$. In order to see this, we differentiate with respect to $r$, obtaining

$$\frac{d}{dr}[\ldots] = \left(\frac{n'}{P+\varrho} - \frac{n\,(P'+\varrho')}{(P+\varrho)^2}\right)\left(1 - \frac{2m(r)}{r}\right)^{-1/2}$$
$$+ \frac{n}{P+\varrho}\left(4\pi r\varrho - \frac{m(r)}{r^2}\right)\left(1 - \frac{2m(r)}{r}\right)^{-3/2}$$
$$- 4\pi r n \left(1 - \frac{2m(r)}{r}\right)^{-3/2}.$$

From (6.3.5) we have, for cold matter,

$$n' = \frac{n}{P+\varrho}\,\varrho'\,. \tag{6.3.8}$$

If we use this, we obtain

$$\frac{d}{dr}[\ldots] = -\frac{n}{P+\varrho}\left(1 - \frac{2m(r)}{r}\right)^{-1/2}$$
$$\cdot \left[\frac{P'}{P+\varrho} + \left(1 - \frac{2m(r)}{r}\right)^{-1} \frac{1}{r^2}\,(4\pi r^3 P + m)\right].$$

According to the TOV equation, the right hand side vanishes.

The (constant) expression in the square bracket in (6.3.7) is most conveniently evaluated at the stellar surface $r = R$ where $P = 0$:

$$[\ldots] = \frac{n(R)}{\varrho(R)}\left(1 - \frac{2M}{R}\right)^{-1/2} = \frac{n}{\varrho + P}\,\mathrm{e}^{-a}\bigg|_{\text{stellar radius}} .$$

It remains to show that $(P + \varrho)\,\mathrm{e}^{a}/n$ is independent of $r$, or that $\ln(P + \varrho) - \ln(n) + a = \text{const}$, which is equivalent to

$$\frac{\varrho' + P'}{\varrho + P} - \frac{n'}{n} + a' = 0 .$$

From (6.2.12) and (6.3.8) we know, however, that this equation is correct.

---

**Exercise:** Study Chap. 26 of [14] and fill in all the details.

---

## 6.4 The Interior of Neutron Stars

### 6.4.1 Qualitative Overview

Figure 6.1 shows a cross section through a typical neutron star. The outer crust is found underneath an atmosphere of just a few meters thickness[3]. It consists of a lattice of completely ionized nuclei and a

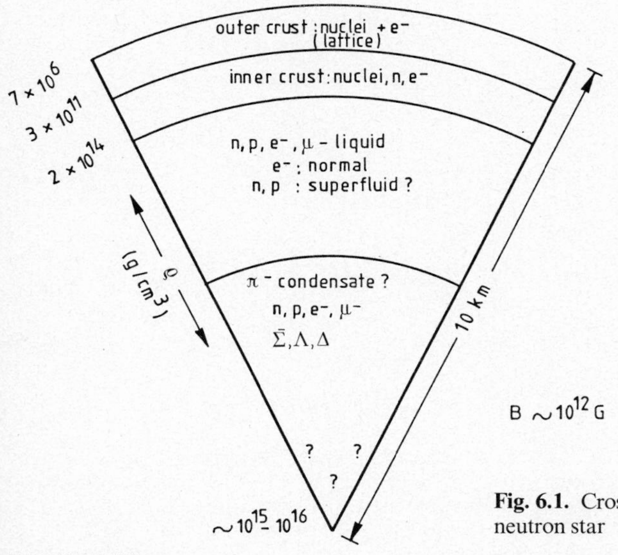


**Fig. 6.1.** Cross sectional view of a typical neutron star

[3] Estimate the pressure scale height for an ideal gas.

highly degenerate relativistic electron gas. In this layer the density $\varrho$ increases from $7\times10^6\,\mathrm{g/cm^3}$ to about $3\times10^{11}\,\mathrm{g/cm^3}$; its thickness is typically about one kilometer. (We shall examine "realistic" density profiles in Sect. 6.5.)

The inner crust starts at $\varrho\sim3\times10^{11}\,\mathrm{g/cm^3}$ and is again about 1 km thick, at which point the density has reached the value $\varrho\sim2\times10^{14}\,\mathrm{g/cm^3}$. In addition to increasingly neutron rich nuclei (determined by $\beta$-equilibrium) and degenerate relativistic electrons, it contains also a degenerate neutron gas, which may be superfluid. We shall examine this possibility in more detail in Sect. 6.7.

The inner fluid consists mainly of neutrons, which are possibly superfluid in a certain density range, together with a few per cent (superconducting?) protons and normal electrons, in order to maintain $\beta$-equilibrium (see Sect. 6.4.2).

The central core is a region within a few km of the center which may contain, in addition to hyperons, a pion condensate (see Sect. 6.7), and, in the case of particularly massive stars, even a quark phase near the center[4].

## 6.4.2 Ideal Mixture of Neutrons, Protons, and Electrons

This brief overview shows that the physics of the interior of a neutron star is extremely complicated. The main problem is to obtain a reliable equation of state.

In order to get a feeling for the orders of magnitude involved, let us first consider an ideal mixture of nucleons and electrons in $\beta$-equilibrium.

The energy density, number density and pressure of an ideal Fermi gas at $T=0$ are given by (here $c=1$):

$$\varrho=\frac{8\pi}{(2\pi\hbar)^3}\int_0^{p_F}\sqrt{p^2+m^2}\,p^2\,dp \tag{6.4.1}$$

$$n=\frac{8\pi}{(2\pi\hbar)^3}\int_0^{p_F}p^2\,dp \tag{6.4.2}$$

$$P=\frac{1}{3}\,\frac{8\pi}{(2\pi\hbar)^3}\int_0^{p_F}\frac{p^2}{\sqrt{p^2+m^2}}\,p^2\,dp\,. \tag{6.4.3}$$

---

[4] One can roughly estimate the density at which the (phase) transition might occur by noting that the density at which nucleons begin to touch is $n_t\simeq(4\pi r_N^3/3)^{-1}$, where $r_N$ is an effective nucleon radius. In terms of nuclear matter density, $n_t/n_0\simeq1.4/r_N^3$, where $r_N$ is measured in fm ($n_0=0.17\,\mathrm{fm^{-3}}$). For $r_N$ as large as 1.0 fm, $n_t$ is as small as $0.24\,\mathrm{fm^{-3}}$; for $r_N=0.5$ fm, $n_t=1.9\,\mathrm{fm^{-3}}$. Thus, one might expect the transition to quark matter to occur somewhere in the range $(2-10)\,n_0$!

The equilibrium reactions

$$e^- + p \rightarrow n + \nu_e$$
$$n \rightarrow p + e^- + \bar{\nu}_e$$

conserve the baryon number density $n_n + n_p$ and maintain charge neutrality $n_e = n_p$.

The condition for chemical equilibrium is

$$\mu_n = \mu_p + \mu_e \,, \tag{6.4.4}$$

where $\mu_i$ ($i$ = n, p, e) are the chemical potentials of the three Fermi gases.

For an *ideal* gas at $T = 0$ we have

$$\mu = \varepsilon_F = \sqrt{p_F^2 + m^2} = \sqrt{\Lambda^2 n^{2/3} + m^2} \,, \tag{6.4.5}$$

where

$$\Lambda := (3\pi^2 \hbar^3)^{1/3} \,. \tag{6.4.6}$$

If we now use this in the equilibrium condition (6.4.4), we can determine the ratio $n_p/n_n$ as a function of $n_n$. A short computation results in

$$\frac{n_p}{n_n} = \frac{1}{8}\left(\frac{1 + \dfrac{2(m_n^2 - m_p^2 - m_e^2)}{\Lambda^2 n_n^{2/3}} + \dfrac{(m_n^2 - m_p^2)^2 - 2m_e^2(m_n^2 + m_p^2) + m_e^4}{\Lambda^4 n_n^{4/3}}}{1 + \dfrac{m_n^2}{\Lambda^2 n_n^{4/3}}}\right)^{3/2} \tag{6.4.7}$$

Let $Q = m_n - m_p$. Since $Q$, $m_e \ll m_n$, we may simplify (6.4.7), obtaining

$$\frac{n_p}{n_n} \simeq \frac{1}{8}\left(\frac{1 + \dfrac{4Q}{m_n}\left(\dfrac{\varrho_0}{m_n n_n}\right)^{2/3} + \dfrac{4(Q^2 - m_e^2)}{m_n^2}\left(\dfrac{\varrho_0}{m_n n_n}\right)^{4/3}}{1 + (\varrho_0/m_n n_n)^{2/3}}\right)^{3/2} , \tag{6.4.8}$$

where

$$\varrho_0 := m_n^4/\Lambda^3 = 6.11 \times 10^{15}\ \mathrm{g/cm^3} \,. \tag{6.4.9}$$

We now calculate the Fermi momentum of the electrons

$$p_{F,e}^2 = \Lambda^2 n_e^{2/3} = \Lambda^2 n_p^{2/3} = m_n^2 \left(\frac{m_n n_n}{\varrho_0}\right)^{2/3} \left(\frac{n_p}{n_n}\right)^{2/3}$$

$$= \frac{\dfrac{m_n^2}{4}\left(\dfrac{m_n n_n}{\varrho_0}\right)^{4/3} + Q\, m_n \left(\dfrac{m_n n_n}{\varrho_0}\right)^{2/3} + Q^2 - m_e^2}{1 + (m_n n_n/\varrho_0)^{2/3}} \,. \tag{6.4.10}$$

We readily see from this that (for $n_n > 0$), $p_{F,e}$ is larger than the maximum momentum of an electron in neutron $\beta$-decay ($p_{max} \simeq \sqrt{Q^2 - m_e^2}$ = 1.19 MeV). Hence the neutrons are stable.

The proton-neutron ratio (6.4.8) is large for small $n_n$ and decreases with increasing $n_n$, reaching a minimum value for

$$\varrho \simeq \varrho_0 \left( \frac{4\,(Q^2 - m_e^2)}{m_n^2} \right)^{3/4} = 1.28 \times 10^{-4}\,\varrho_0 \tag{6.4.11}$$

with

$$(n_p/n_n)_{\min} \simeq \left( \frac{Q + \frac{1}{2}\,(Q^2 - m_e^2)^{1/2}}{m_n} \right)^{3/2} = 0.002\,. \tag{6.4.12}$$

Afterwards $n_p/n_n$ increases monotonically and approaches the asymptotic value 1/8.

As a typical example, we take $\varrho \simeq m_n n_n = 0.107\,\varrho_0$ and find $n_p/n_n = 0.013$ and $p_{F,e} = 105$ MeV/c. Above this density, *muons* are stable.

Note also that $\mu_p \ll \mu_e$ since $p_{F,e} = p_{F,p}$ and thus we have $\mu_n \simeq \mu_e$.

The gravitational mass of a neutron star for an ideal neutron gas is shown as a function of the central density in Fig. 6.2 (*Oppenheimer* and *Volkoff*, 1939).

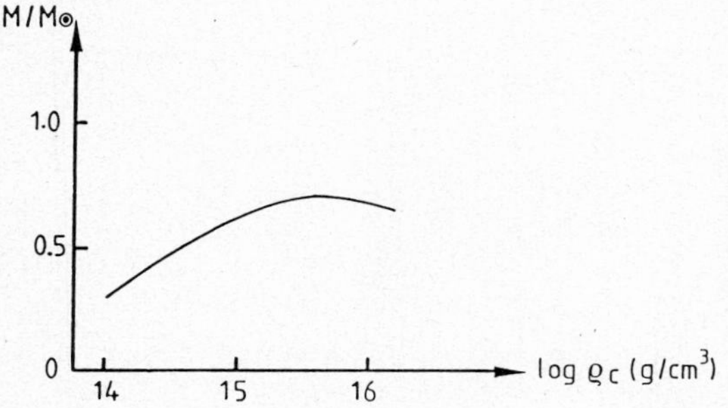


**Fig. 6.2.** $M(\varrho_c)$ for an ideal neutron gas

The maximum mass

$$M_m = 0.71\ M_\odot \tag{6.4.13}$$

is reached for the density

$$\varrho_m = 4 \times 10^{15}\ \mathrm{g/cm^3}\,. \tag{6.4.14}$$

The corresponding radius is

$$R_m = 9.6\ \mathrm{km}\,. \tag{6.4.15}$$

If $\varrho > \varrho_m$, the star is *unstable* according to Proposition 6.3.1. If $\varrho \ll \varrho_m$, a Newtonian stability analysis shows that the star is stable. The adiabatic index is $\frac{5}{3}$, which is well above the critical value $\frac{4}{3}$.

---

**Exercise:** In recent years, one has discovered that galaxies and galactic clusters are considerably more massive than one had believed. It seems that a large part (perhaps even the dominating contribution) of the mass is hidden in extended halos of dark matter. This is probably due to the presence of many low-mass stars with correspondingly low luminosity. However, one cannot exclude the possibility that the galactic halos consist of massive neutrinos. There is some experimental (from tritium decay) and theoretical evidence for a neutrino mass $m_\nu \sim 10$ eV, although this is not yet convincing. In that case, *neutrino stars* might exist.

Show that the mass and radius of a degenerate neutrino star can be obtained from the corresponding expressions for a neutron star if one multiplies them by $(m_n/m_\nu)^2$. This shows that one could easily produce the necessary galactic halos with neutrinos.

## 6.5 Models for Neutron Stars

Realistic models of neutron stars (mass-radius relation, density profile, etc.) require an equation of state which is reliable for all the layers

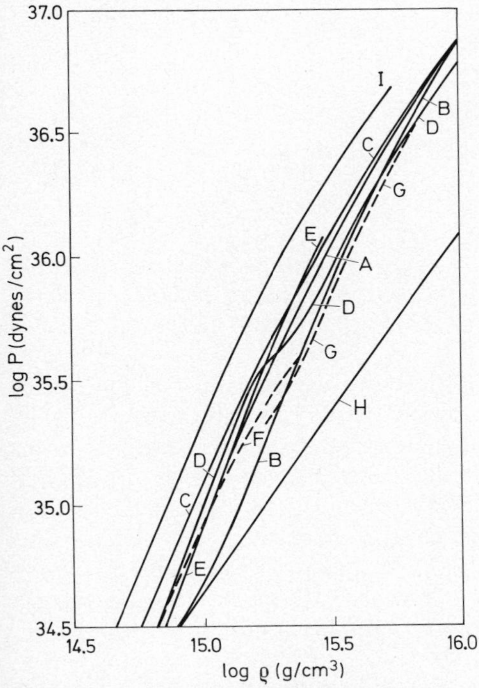


**Fig. 6.3.** Various "realistic" equations of state above nuclear densities (from [35])

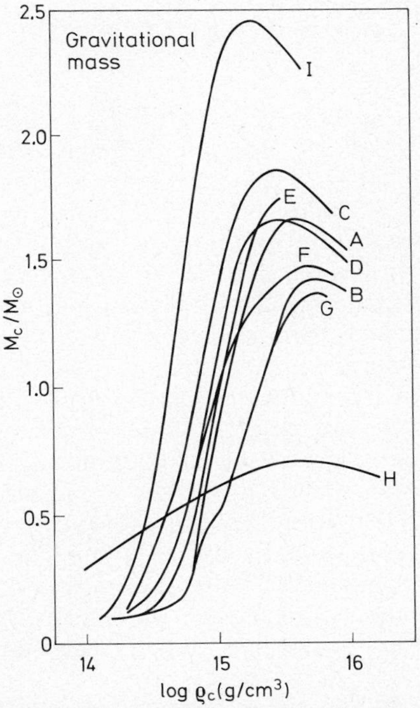


**Fig. 6.4.** Gravitational mass as a function of the central density for the equations of state shown in Fig. 6.3 (from [35])

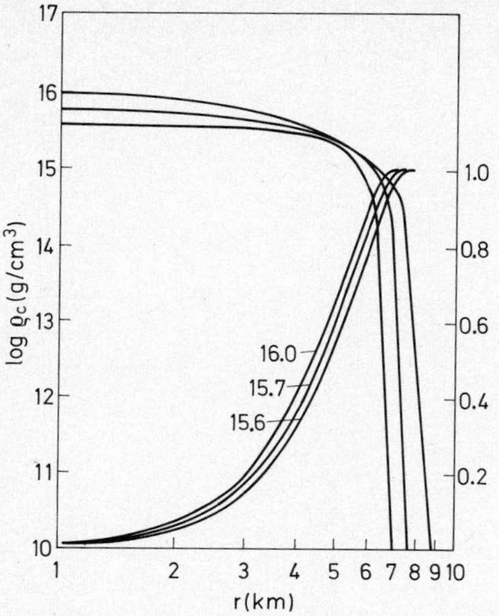


**Fig. 6.5.** Density profile *(left hand scale)* and mass fraction *(right-hand scale)* as function of the distance from the center for equation of state B of Fig. 6.3 (from [35])

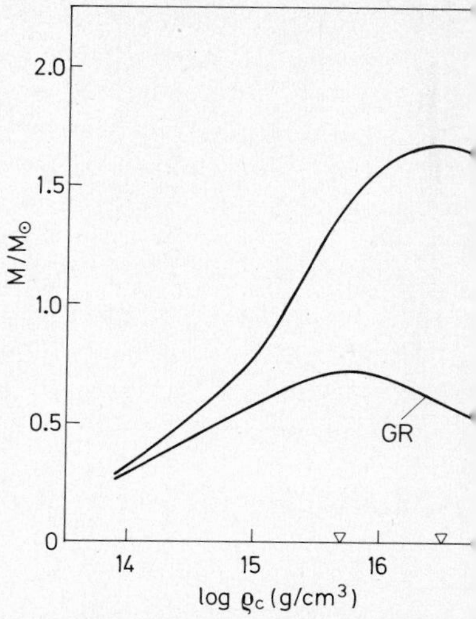


**Fig. 6.6.** Newtonian gravitation compar with general relativity for the equatic of state H of Fig. 6.3 (from [35])

described in Sect. 6.4.1. As the density increases, this becomes an i creasingly difficult problem. For densities in the vicinity of nucle matter density or higher, we have to deal not only with a difficu many-body problem, but also with the fundamental difficulty that o description of the strong interaction is not complete.

It would be out of place here to try to describe all the effort whic has been made to obtain a realistic equation of state. It will suffice refer to the review articles [33] and [34], and to the large number papers cited therein.

As an illustration we show in Fig. 6.3 several "realistic" equations state for densities above nuclear matter density. For comparison, th equation of state for an ideal degenerate neutron gas is also show (curve H). The various equations of state differ considerably from eac other, which indicates the degree of uncertainty. Figure 6.4 shows th gravitational mass as a function of central density $\varrho_c$ for the variou equations of state shown in Fig. 6.3. As before, only the increasin branches represent stable configurations. In all cases, the maximu mass is less than 2.5 $M_\odot$ (see also the more recent results in [36]).

Typical density profiles (for equation of state B) are shown in Fi 6.5. The different predictions of the Newtonian theory and GR are als interesting. These are shown in Fig. 6.6 for an ideal degenerate neutro gas.

## 6.6 Bounds of the Mass of Nonrotating Neutron Stars

The value of the largest possible mass, $M_{\max}$, of a nonrotating neutron star is interesting for at least two reasons. First of all, it is important for the theory of stellar evolution. If $M_{\max}$ were fairly small ($\simeq 1.5\,M_\odot$) then black holes could be formed with high probability, as mentioned at the beginning of this chapter (see also Addendum 2). Its value also plays a decisive role in the observational identification of black holes. If one can find a very compact object which can be shown convincingly not to rotate very fast, and to have a mass which is definitely larger than $M_{\max}$, then one has a serious canditate for a black hole. Precisely such arguments have been used to identify the x-ray source Cygnus X-1 with a black hole, as will be discussed in Chap. 7.

In view of the large uncertainties of the equation of state for high densities, it is important to find reliable limits for $M_{\max}$. This is what we are going to do next (see also [28]). We shall assume that for densities less than some given density $\varrho_0$ (which may depend on our current knowledge), the equation of state is reasonably well known; for densities higher than $\varrho_0$, the equation of state will be required to satisfy only quite general conditions.

### 6.6.1 Basic Assumptions

We shall make the following minimal assumptions about the matter in the interior of a nonrotating neutron star:

(M1) The energy-momentum tensor is described in terms of the mass-energy density and an isotropic pressure $p$, which satisfies an equation of state $p(\varrho)$ [see (6.2.3), for example]. Stresses which may have arisen as a result of the slowing down of the star's rotation, are neglected.

(M2) The density is positive: $\varrho \geqq 0$.

(M3) Matter is microscopically stable: $\dfrac{dp}{d\varrho} \geqq 0$. Since the pressure is certainly positive at low density, we then have $p \geqq 0$.

(M4) The equation of state is known for $\varrho \leqq \varrho_0$. We may choose for $\varrho_0$ a value which is not significantly below nuclear densities.

Some authors also assume $dp/d\varrho \leqq 1$, since $(dp/d\varrho)^{1/2}$ is the phase velocity of sound waves. However, sound waves propagating through matter in neutron stars are subject to dispersion and absorption, and hence $(dp/d\varrho)^{1/2}$ is not the velocity at which a signal propagates. Thus causality does not require $dp/d\varrho \leqq 1$. On the other hand, most of the "realistic" equations of state do satisfy this inequality. Although we shall not require this "causality assumption", we shall occasionally mention the implications of this additional assumption.

We repeat the basic equations from Sect. 6.2, since these form the basis for the following derivations (again $G = c = 1$).

The metric is

$$g = e^{2a(r)}\, dt^2 - \frac{dr^2}{1 - 2m(r)/r} - r^2 (d\vartheta^2 + \sin^2\vartheta\, d\varphi^2)\,. \tag{6.6.2}$$

The structure equations are:

$$-dp/dr = (\varrho + p)\,\frac{m + 4\pi r^3 p}{r^2 (1 - 2m/r)} \tag{6.6.3a}$$

$$dm/dr = 4\pi r^2 \varrho \tag{6.6.3b}$$

$$da/dr = \frac{m + 4\pi r^3 p}{r^2 (1 - 2m/r)}\,. \tag{6.6.3c}$$

The boundary conditions are, if $\varrho_c$ is the central density,

$$p\,(r = 0) = p\,(\varrho_c) =: p_c \tag{6.6.4a}$$

$$m\,(0) = 0 \tag{6.6.4b}$$

$$e^{a(r)} = 1 - 2M/R\,, \tag{6.6.4c}$$

where $R$ is the stellar radius (the point at which the pressure vanishes), and $M = m(R)$ is the gravitational mass of the star.

Examination of (6.6.3a) leads one to expect that

$$2m(r)/r < 1\,. \tag{6.6.5}$$

The justification of this conclusion is as follows. Let $r_*$ be the first point at which, starting from the center of the star, $2m(r_*) = r_*$. In a neighborhood of $r_*$, the TOV equation (6.6.3a) becomes

$$\frac{1}{\varrho + p}\,\frac{dp}{dr} = \frac{1}{2(r_* - r)}\,\frac{1 + 8\pi r_*^2 p(r_*)}{1 - 8\pi r_*^2 \varrho(r_*)} + O(1)\,. \tag{*}$$

The left hand side of this equation is the logarithmic derivative of the relativistic enthalpy $\eta$, defined by

$$\eta := \frac{\varrho + p}{n}\,, \tag{6.6.6}$$

where $n$ is the baryon number density. This is an immediate consequence of the first law of thermodynamics

$$d\varrho/dn = (\varrho + p)/n\,. \tag{6.6.7}$$

Thus, if we integrate (*) and use the fact that the right hand side is negative for $r < r^*$, we obtain $\eta\,(r_*) = 0$. Examination of (6.6.6) shows that this is not possible for a realistic equation of state.

We can also conclude from our assumptions that the density does not increase with increasing radius. From (6.6.3a) we have, in fact

$$\frac{d\varrho}{dr} = \frac{d\varrho}{dp}\frac{dp}{dr} = -\left(\frac{dp}{d\varrho}\right)^{-1}(\varrho + p)\,\frac{m + 4\pi r^3 p}{r^2(1 - 2m/r)}\,.$$

According to (M2) and (M3), the quantities $\varrho$, $p$, and $dp/d\varrho$ are positive. The positive sign of the last factor is a consequence of (6.6.5) and

$$m(r) = \int_0^r \varrho\, 4\pi r^2\, dr\,. \tag{6.6.8}$$

We have thus shown that $d\varrho/dr \leqq 0$.

This allows us to divide the star into two regions.

The *envelope* is the part having $\varrho < \varrho_0$ ($r > r_0$), for which the equation of state is known. The interior region with $\varrho > \varrho_0$ and $r < r_0$ is called the *core*. In this region we assume only the general requirements (M1) – (M4).

The mass $M_0$ of the core is given by

$$M_0 = \int_0^{r_0} \varrho\, 4\pi r^2\, dr\,. \tag{6.6.9}$$

If $M_0$ is given, the structure equations (6.6.3a) and (6.6.3b) can be integrated outward from $r_0$. The corresponding boundary conditions are $p(r_0) = p(\varrho_0) =: p_0$, $m(r_0) = M_0$. The total mass $M$ of the star is

$$M = M_0 + M_{\mathrm{env}}(r_0, M_0)\,. \tag{6.6.10}$$

We may regard the mass of the envelope as a known function of $r_0$ and $M_0$, since the equation of state is known there. We may now obtain bounds for $M$ by following the strategy: (1) Determine the range of values ($M_0$, $r_0$) which is allowed by the assumptions (M1) – (M4). This range of possible core values will be known as the *allowed region* in the ($r_0$, $M_0$) plane. (2) Look for the maximum of the function $M$ defined by (6.6.10), in the allowed region of the variables $r_0$ and $M_0$.

### 6.6.2 Simple Bounds for Allowed Cores

One obtains a simple, but not optimal, bound for the allowed region as follows: For $r = r_0$, we find from (6.6.5)

$$M_0 < \tfrac{1}{2} r_0\,. \tag{6.6.11}$$

If the density does not increase with $r$, then

$$M_0 \geqq \frac{4\pi}{3} r_0^3\, \varrho_0\,. \tag{6.6.12}$$

These two inequalities imply the following bounds

$$M_0 < \frac{1}{2}\left(\frac{3}{8\pi\varrho_0}\right)^{1/2}$$
$$r_0 < \left(\frac{3}{8\pi\varrho_0}\right)^{1/2} . \qquad (6.6.13)$$

A simple numerical example shows that these are not completely uninteresting. If $\varrho_0 = 5\times 10^{14}\ \mathrm{g/cm^3}$, then (6.6.13) gives $M_0 \leqq 6\,M_\odot$ and $r_0 < 18$ km.

### 6.6.3 Allowed Core Region

In order to determine precisely the allowed region for the cores, we investigate the quantity

$$\zeta(r) = \mathrm{e}^{a(r)}, \qquad (6.6.14)$$

which is finite and positive everywhere. $\zeta(r)$ can vanish only at the center, in the limit in which the pressure approaches infinity. This is a consequence of (6.6.3c), which can be integrated inward from the surface, subject to the boundary condition (6.6.4c), without encountering a singularity.

One can derive a differential equation relating $\zeta(r)$ and $m(r)$ from the structure equations (6.6.3a–c):

$$(1-2m/r)^{1/2}\frac{1}{r}\frac{d}{dr}\left[\left(1-\frac{2m}{r}\right)^{1/2}\frac{1}{r}\frac{d\zeta}{dr}\right] = \frac{\zeta}{r}\frac{d}{dr}\left(\frac{m}{r^3}\right). \qquad (6.6.15)$$

(It is easy to verify that this equation is correct.) Since $\varrho$ does not increase with increasing $r$, the mean density does not either. Therefore the right hand side of (6.6.15) is not positive. If one introduces the new independent variable

$$\xi = \int_0^r dr\, r\,(1-2m/r)^{-1/2} \qquad (6.6.16)$$

the resulting inequality can be written in the simple form

$$d^2\zeta/d\xi^2 \leqq 0 . \qquad (6.6.17)$$

Using the mean value theorem, we conclude

$$d\zeta/d\xi \leqq \frac{\zeta(\xi)-\zeta(0)}{\xi} . \qquad (6.6.18)$$

This inequality is optimal, since equality holds for a star having constant density (the proof is an exercise). Since $\zeta(0) \geqq 0$, we have

$$\zeta^{-1}\, d\zeta/d\xi \leqq 1/\xi . \qquad (6.6.19)$$

Equality holds for a star having constant density when the pressure at the center diverges.

When it is rewritten in terms of $r$ and $a(r)$, (6.6.19) reads

$$(1-2m/r)^{1/2}\,1/r\,da/dr \leqq \left[\int_0^r dr\, r\left(1-\frac{2m}{r}\right)^{-1/2}\right]^{-1}. \tag{6.6.20}$$

We now estimate the right hand side in an optimal fashion. Since $m/r^3$ does not increase outwards, we have $m(r')/r' \geqq [m(r)/r](r'/r)^2$ for all $r' \leqq r$, and hence

$$\begin{aligned}\int_0^r dr'\, r'\left(1-\frac{2m(r')}{r'}\right)^{-1/2} &\geqq \int_0^r dr'\, r'\left(1-\frac{2m(r)}{r^3}r'^2\right)^{-1/2}\\ &= \frac{r^3}{2m(r)}\left[1-\left(1-\frac{2m(r)}{r}\right)^{1/2}\right].\end{aligned} \tag{6.6.21}$$

Again, equality holds for a star having uniform density. If we now use the structure equation (6.6.3c) for the left hand side of (6.6.20), we obtain

$$\frac{m+4\pi r^3 p}{r^3(1-2m/r)^{1/2}} \leqq \frac{2m(r)}{r^3}\left[1-\left(1-\frac{2m(r)}{r}\right)^{1/2}\right]^{-1}$$

and this gives the following bound on $m(r)/r$:

$$m(r)/r \leqq 2/9\,\{1-6\pi r^2 p(r)+[1+6\pi r^2 p(r)]^{1/2}\}. \tag{6.6.22}$$

Equality holds for a uniformly dense star with infinite central pressure.

The first interesting consequence of (6.6.22) is obtained for $r=R$ [where $p(R)=0$]:

$$2M/R \leqq 8/9. \tag{6.6.23}$$

Hence the red shift at the surface satisfies

$$z_{\text{surf}} \leqq 2. \tag{6.6.24}$$

If (6.6.22) is evaluated at the core boundary, one obtains

$$M_0 \leqq 2/9\, r_0\,[1-6\pi r_0^2 p_0+(1+6\pi r_0^2 p_0)^{1/2}]. \tag{6.6.25}$$

This represents the optimal improvement of (6.6.11). Together with (6.6.12), i.e. $M_0 \geqq 4\pi/3\, r_0^3\, \varrho_0$, which is already optimal, (6.6.25) determines the allowed region in the $(r_0, M_0)$-plane for cores which satisfy the assumptions (M1) – (M4). This region is shown in Fig. 6.7.

One finds the largest core mass for the stiffest equation of state, namely, for incompressible matter with constant density $\varrho$. Even then the mass of the core inside a radius $r_0$ is limited; otherwise the pressure becomes infinite at the center and equilibrium is lost.

As long as $\varrho_0$ is not significantly larger than nuclear matter density, then $p_0 \ll \varrho_0$ (see Sect. 6.5), and we have with high accuracy the bounds

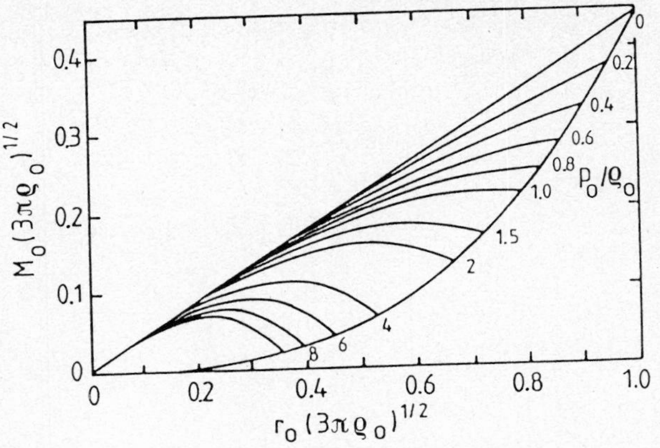


**Fig. 6.7.** Allowed region for cores in the $(r_0, M_0)$-plane. The *lower curve* is determined by (6.6.12) and the *upper curve* by (6.6.25) for $p_0 = 0$. The upper boundary (6.6.25) is shown for various values of the ratio $p_0/\varrho_0$

[to be compared with the cruder bound (6.6.13)]:

$$\begin{aligned} M_0 &< \frac{4}{9}\left(\frac{1}{3\pi\varrho_0}\right)^{1/2} \\ r_0 &< \left(\frac{1}{3\pi\varrho_0}\right)^{1/2}. \end{aligned} \tag{6.6.26}$$

## 6.6.4 Upper Limit for the Total Gravitational Mass

As an illustration, let us choose $\varrho_0 = 5.1 \times 10^{14}$ g/cm$^3$ and use an equation of state due to Baym, Bethe, Pethick and Sutherland (BBPS), given in [37]. We then have $p_0 = 7.4 \times 10^{33}$ dyn/cm$^2$, and thus $p_0/\varrho_0 = 0.016 \ll 1$. Recall that for nuclear matter, $\varrho_{nuc} = 2.8 \times 10^{14}$ g/cm$^3$. The allowed region corresponding to these values of $\varrho_0$ and $p_0$ is shown in Fig. 6.8. Superimposed are contours of constant total mass [by (6.6.10),

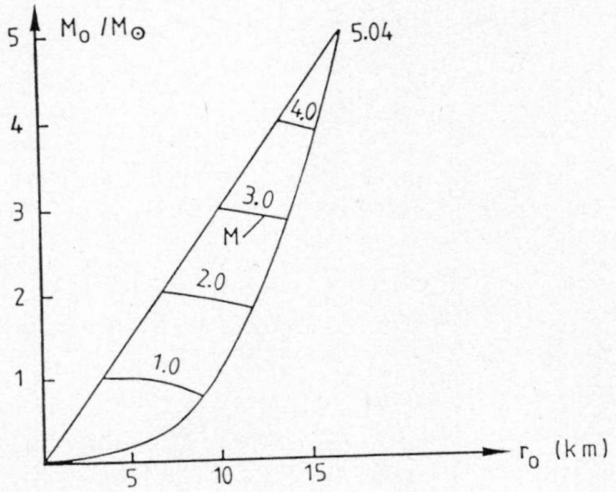


**Fig. 6.8.** The function $M(r_0, M_0)$ for the BBPS equation of state and $\varrho_0 = 5 \times 10^{14}$ g/cm$^3$. The allowed region is also shown. The maximum limit is $5\,M_\odot$ (from [28])

this is a function of $M_0$ and $r_0$]. The optimum upper bound is $5\,M_\odot$. If one also required $dp/d\varrho < 1$, then one would obtain $3\,M_\odot$, instead, which explains why one obtains an upper limit of $\approx 3\,M_\odot$ for "realistic" equations of state.

The calculation shows that the contribution of the envelope to the limiting mass is less than 1%. The limit is reached for the maximum value of the core mass. One can show (see [28]) that this is true as long as $p_0 \ll \varrho_0$.

We thus obtain, to a good approximation

$$M \leqq 4/9 \left(\frac{1}{3\pi\varrho_0}\right)^{1/2} \tag{6.6.27}$$

or

$$M \leqq 6.8 \left(\frac{\varrho_{\mathrm{nuc}}}{\varrho_0}\right)^{1/2} M_\odot . \tag{6.6.28}$$

The additional "causality assumption" $dp/d\varrho \leqq 1$ would have the effect of replacing the factor 6.8 by 4.0.

## 6.7 Cooling of Neutron Stars

### 6.7.1 Introduction

When the iron-nickel core ($M_{\mathrm{core}} \simeq 1.5\,M_\odot$) of a massive star becomes unstable and collapses to nuclear density, roughly 10% of its gravitational energy, or about $10^{53}$ erg is released. Most of this energy is emitted almost immediately in the form of neutrinos[5] (see [38] and references therein). Therefore, the internal temperature of a neutron star drops rapidly from about 10 MeV to about 1 MeV, and the matter is to all extents and purposes in its ground state. The Fermi energies of the neutrons, protons, and electrons are much larger. Neutrino emission dominates the further cooling of the neutron star, until the interior temperature falls to about $10^8$ K (with a corresponding surface temperature of about $10^6$ K). Only then does photoemission also begin to play an important role.

[5] At the beginning of the collapse, the neutrinos are produced mainly by electron capture. The mean free path of these neutrinos soon becomes so short that they remain trapped during the continued collapse and a highly degenerate neutrino gas is formed. (For details, see Sect. 6.9.) Within a few seconds, this diffuses out of the core and gives rise to a huge pulse of neutrinos. Part of the energy is converted to heat by the resulting shock wave, and the temperature rises to $\sim 10$ MeV. A large part of this energy is also radiated away in the form of neutrinos arising from pair emission processes, such as $e^- + e^+ \rightarrow \nu + \bar{\nu}$. (This pair annihilation process is considered in detail in Sect. 6.9.)

**Table 6.1.** Hot (putative) neutron stars in historical supernova remnants

| Name | Age [yr] | Distance [kpc] | $T$ [$10^6$ K] |
|---|---|---|---|
| Cas A | ~ 300 | 2.8 | < 1.5 |
| Kepler | 376 | 8.0 | < 2.1 |
| Tycho | 408 | 3.0 | < 1.8 |
| Crab | 926 | 2.0 | < 3.0 |
| SN 1006 | 974 | 1.0 | < 1.0 |

We shall see that the cooling curves (surface temperature as a function of time) depend on a number of interesting aspects of the physics of neutron stars. The equation of state at high densities is particularly important, as is the possible existence of a pion condensate in the central region of a neutron star. Magnetic fields and the possible existence of a superfluid state of the nucleons also play some role.

Observations of the Crab pulsar during lunar eclipses have shown that the effective surface temperature $T_s$ is less than $3.0 \times 10^6$ K. Since the age of the Crab pulsar is known, this upper limit provides an important restriction on models of neutron stars.

The Einstein observatory has made it possible to determine also upper limits for the thermal emission temperature of other suspected neutron stars contained in young supernova remnants. The limits given in Table 6.1 are taken from [39].

One might expect to find a neutron star near the center of the remnant of every historical supernova, but this is definitely the case only for the Crab nebula. The observations force us to conclude either that neutron stars are not formed in other supernova explosions[6] or that these have cooled off so fast that their present temperatures are below the given limits. We remark that as a result of interstellar x-ray absorption, these limits have an uncertainty of about 50%.

In order to determine the cooling curve, we need to know both the total internal energy of the star and the total luminosity as functions of the interior temperature. We must also know the relation between interior temperature and surface temperature.

### 6.7.2 Thermodynamic Properties of Neutron Stars

*A. Basic Equations*

In the nonrelativistic theory the equation for thermal equilibrium (energy balance) is

$$\partial L/\partial r = -4\pi r^2 n T \, \partial s/\partial t \, .$$

[6] A neutron star will not be observed as a pulsar if it does not radiate in our direction or if its magnetic field (or its rotation) is not large enough.

Here the total luminosity $L$ is the sum of the neutrino luminosity $L_\nu$ and the contribution $L_\gamma$ of radiation and heat conduction. $T$ is the local temperature, $n$ the baryon number density and $s$ denotes the specific entropy per baryon.

The relativistic generalization of this equation is

$$\partial(L\,e^{2\phi})/\partial r = -\frac{4\pi r^2}{(1-2Gm/r\,c^2)^{1/2}}\,n\,e^{\phi}\,T\,\partial s/\partial t\,. \tag{6.7.1}$$

Here we have denoted the metric coefficient of $dt^2$ by $e^{2\phi}$ [instead of $e^{2a}$ as in (6.6.2)]. One may find a simple derivation of (6.7.1) in Sect. 3.2.5 of [32]. The gradient of the neutrino luminosity is given by

$$\partial(L_\nu\,e^{2\phi})/\partial r = \frac{4\pi r^2}{(1-2Gm/r\,c^2)^{1/2}}\,n\,e^{2\phi}\,q_\nu\,, \tag{6.7.2}$$

where $q_\nu$ is the neutrino emissivity per baryon.

For thermal neutrino energies $\lesssim 1$ MeV, the mean free path of the neutrinos is considerably larger than the stellar radius [see Eq. (6.9.47)]. Therefore, the temperature gradient inside the star is determined by the energy flux $L_\gamma$ due to radiation and heat conduction. The well known nonrelativistic equation for thermal energy transport is (see any book on astrophysics, such as [40]):

$$\partial T/\partial r = \frac{-3\varkappa\varrho}{4acT^3}\,\frac{L_\gamma}{4\pi r^2}\,,$$

where $\varkappa$ is the total opacity: $\varkappa^{-1} = \varkappa_{\rm rad}^{-1} + \varkappa_{\rm cond}^{-1}$ and $a$ is the Stefan-Boltzmann constant. The relativistic generalizations is (see Sect. 3.2.7 of [32])

$$\partial(T\,e^{\phi})/\partial r = \frac{-3\varkappa\varrho}{4acT^3}\,\frac{L_\gamma\,e^{\phi}}{4\pi r^2(1-2Gm/r\,c^2)^{1/2}}\,. \tag{6.7.3}$$

The degenerate matter in the interior of a neutron star has a very high thermal conductivity and hence (6.7.3) implies $T\,e^{\phi} = \text{const}$, except for a thin envelope of a few meters thick. The constancy of $T\,e^{\phi}$ also follows simply from the expression for the gravitational red shift. The factor $e^{\phi}$ from general relativity is quite important, since the neutrino emissivities depend strongly on temperature (as $\sim T^n$ with $n = 6, 8$; see Sect. 6.7.4).

Once we know the opacity, (6.7.1–3) determine the relation between the surface temperature $T_{\rm s}$ and $T' = T\,e^{\phi}$ in the interior, provided we impose also the photospheric boundary conditions (see Sect. 3.3 of [32]):

$$\begin{aligned} P_{\rm s} &= \frac{2}{3}\,\frac{GM}{R^2\,e^{\phi_{\rm s}}}\,\frac{1}{\varkappa_{\rm s}(P_{\rm s},T_{\rm s})} \\ L_{\gamma,{\rm s}} &= 4\pi R^2\,\sigma\,T_{\rm s}^4\,. \end{aligned} \tag{6.7.4}$$

We remark that the hydrostatic structure equations almost completely decouple from the thermal equations (6.7.1 – 3), since the equation of state for $T = 0$ is applicable (except for the atmosphere).

## B. Relation Between Surface and Interior Temperature

Figure 6.9 shows a typical temperature profile for a neutron star with mass $M = 1.25\, M_\odot$, taken from [41]. The opacities and equation of state used are discussed there.

Figure 6.10 shows the relation between the surface temperature $T_s$ and the interior temperature at the core-envelope boundary for two equations of state and two different values of the magnetic field $B$. The effect of the magnetic field on the opacity has been taken into account very crudely.

## C. Specific Heats

In the absence of superfluidity, the internal energy of the neutrons dominates. According to the Landau theory of (normal) Fermi liquids, the specific heat of degenerate fermions is that of an ideal gas, provided the mass is replaced by the effective mass $m^*$ of the quasi-particles:

$$c\,(T) = \frac{\pi^2}{3} \frac{k_B^2 T}{n} \mathscr{D}(\varepsilon_F) . \tag{6.7.5}$$

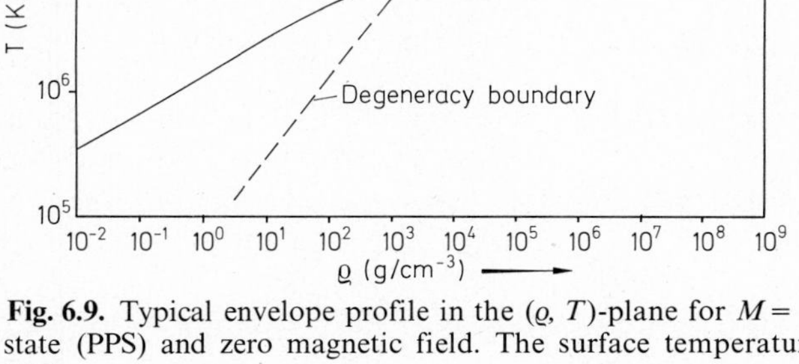

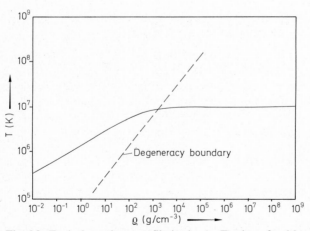

**Fig. 6.9.** Typical envelope profile in the $(\varrho, T)$-plane for $M = 1.25\, M_\odot$, a stiff equation of state (PPS) and zero magnetic field. The surface temperature is $10^{5.5}$ K (adapted from [41])

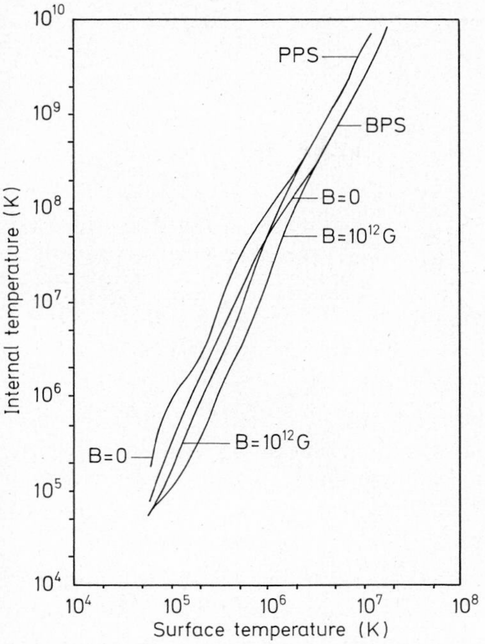


**Fig. 6.10.** Relationship between surface temperature and temperature at the boundary between core and envelope for $M = 1.25\,M_\odot$, for two equations of state (PPS) and (BPS) and for two values of the magnetic field $B = 0$, $10^{12}$ G. BPS is a fairly soft equation of state while PPS is quite stiff (adapted from [41])

Here, $\mathscr{D}(\varepsilon_F)$ is the density of states at the Fermi surface:

$$\mathscr{D}(\varepsilon_F) = 3\,n\,\frac{\varepsilon_F}{p_F^2} = \begin{cases} \dfrac{3\,n\,m^*}{p_F^2} & \text{(nonrelativistic)} \\[2ex] \dfrac{3\,n}{\varepsilon_F} & \text{(ultrarelativistic).} \end{cases} \tag{6.7.6}$$

For $\varrho \gg 10^6\,\mathrm{g/cm^3}$ the electrons are always ultrarelativistic, while the nucleons remain nonrelativistic. The Fermi momentum of the neutrons is determined by their number density, which in turn is given by the equation of state[7]. The Fermi momentum of the protons (for $\varrho > 10^{14}\,\mathrm{g/cm^3}$, such that the nuclei have dissolved) is determined by the condition of $\beta$-equilibrium, as discussed in Sect. 6.4.2.

Rough values are (compute these for an ideal mixture)

$$\begin{aligned} p_F(n) &\simeq (340\ \mathrm{MeV/c})\,(\varrho/\varrho_0)^{1/3} \\ p_F(p) = p_F(e) &\simeq (85\ \mathrm{MeV/c})\,(\varrho/\varrho_0)^{2/3}. \end{aligned} \tag{6.7.7}$$

---

[7] From thermodynamics we have

$$T\,ds = 0 = d(\varrho/n) + P\,d(1/n)$$

or

$$d\varrho/dn = (P+\varrho)/n\,, \quad \frac{d\ln\varrho}{d\ln n} = \frac{P+\varrho}{\varrho}\,.$$

If we know $\varrho(n)$, then $P(n)$ and $P(\varrho)$ are determined.

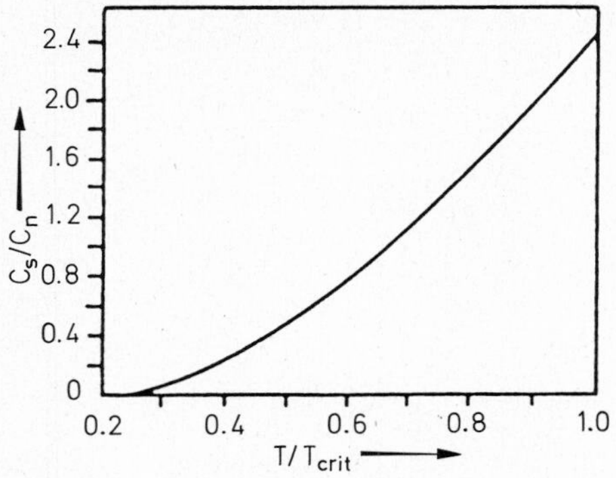


**Fig. 6.11.** Superfluid specific heat $c_s$, in units of the normal specific heat as a function of temperature (in units of the critical temperature)

In the following $\varrho_0$ always stands for nuclear matter density $\varrho_0 = 2.8 \times 10^{14}\,\mathrm{g/cm^3}$. The Landau parameters $m_n^*$ and $m_p^*$ can in principle be computed microscopically, but their numerical values are uncertain. In the numerical results of [41] the values $m_p^*/m_p = 1$ and

$$m_n^*/m_n = \mathrm{Min}\left[1, 0.885\left(\frac{\varrho}{\varrho_0}\right)^{-0.032}, \quad 0.815\left(\frac{\varrho}{\varrho_0}\right)^{-0.135}\right] \tag{6.7.8}$$

were used.

If the nucleons are superfluid, the specific heat below the transition temperature $T_c$ is modified. For $T \ll T_c$, it decreases as $\exp(-\Delta_0/kT)$, where $\Delta_0$ is the energy gap at $T = 0$. This modification, which is well known from superconductivity, is shown in Fig. 6.11.

Below $\varrho \sim 10^{14}\,\mathrm{g/cm^3}$ the ions can also contribute significantly to the internal energy. For the bulk of the crust, the melting temperature of the ion lattice is so high that the ions form a lattice very early. The specific heat is then

$$c_v\,(\mathrm{ions}) = 3\,k_B\,n_{\mathrm{ions}}\,D\,(\theta_D/T)\,, \tag{6.7.9}$$

where the Debeye function has the limiting values

$$D(x) = \begin{cases} 1, & x \ll 1 \\ \dfrac{4\pi^4}{5x^3}, & x \gg 1. \end{cases} \tag{6.7.10}$$

The Debeye temperature $\theta_D$ is not far from $T_p = \hbar\,\omega_p/k_B$, where $\omega_p$ is the ion plasma frequency. A more precise calculation gives

$$\theta_D \simeq 0.45\,T_p.$$

*D. Superfluidity*

As is the case in superconductivity and in nuclei, superfluidity of neutrons and of protons can occur as a result of pairing interactions. At fairly low densities the $^1S_0$ interaction is predominantly attractive and can lead to an s-wave pairing. At higher densities ($k_F > 1.6\ \mathrm{fm}^{-1}$), the $^1S_0$ interaction becomes repulsive, while the $^3P_2$ interaction becomes attractive, that p-wave superfluidity can occur.

It is very difficult to calculate reliably the energy gaps, since they depend exponentially on the strength of the interaction and effective masses. Figure 6.12 shows a typical result, together with the corresponding transition temperatures.

These results should be regarded as reasonable estimates. They show that the p-wave energy gaps are clearly smaller than the s-wave gaps. If the protons are superfluid, they are then obviously also superconducting. The results shown in Fig. 6.12 indicate that the Ginzburg-Landau parameter for the protons is much larger than $1/\sqrt{2}$ and hence that they form a type II superconductor (see the exercise below).

We have already remarked that the specific heat of nucleons differs from that given by (6.7.5) below the transition temperature. In Sect. 6.7.3 we shall see that the energy gaps in the excitation spectrum of the nucleons at the Fermi surfaces also suppress the internal neutrino processes.

Figure 6.13 shows the contributions to the specific heat of a neutron star for $\varrho = 10^{14}\ \mathrm{g/cm^3}$. The equations of state BPS and PPS are identical for $\varrho < \varrho_0$.

---

**Exercise:** Suppose that the protons are superfluid and thus superconducting. Estimate the Ginzburg-Landau parameter $\varkappa$ and show that the protons probably form a type II superconductor.

*Solution:* The Ginzburg-Landau parameter is the ratio

$$\varkappa = \lambda(T)/\xi(T),$$

where $\lambda$ is the penetration depth for magnetic fields and $\xi$ is the Ginzburg-Landau coherence length. We use the microscopic expressions for these length scales, which can be found, for example, in [171, Chap. 13].

One finds

$$\varkappa = 0.957 \frac{\lambda_L(0)}{\xi_0}$$

with

$$\xi_0 = \frac{\hbar v_F}{\pi \Delta_0} = 0.180 \frac{\hbar v_F}{k_B T_c}$$

and

$$\lambda_L(0) = \left( \frac{m c^2}{4\pi n e^2} \right)^{1/2}.$$

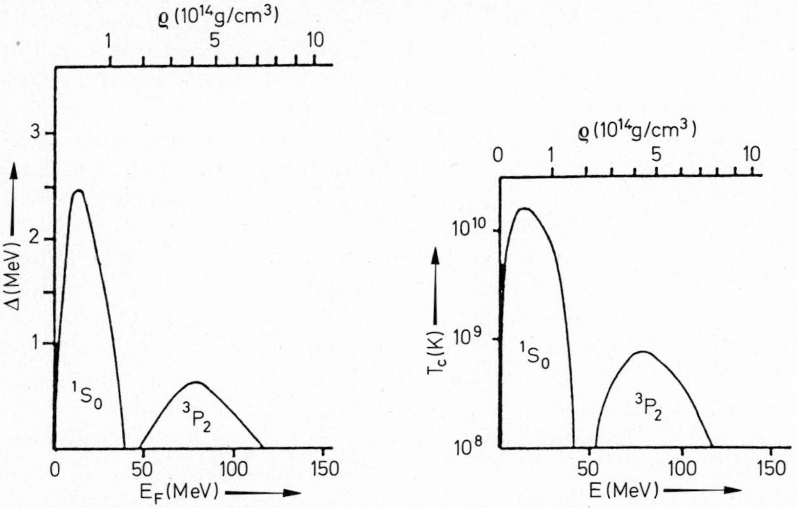


**Fig. 6.12.** Density dependence of the ${}^1S_0$ and ${}^3P_2$ energy gaps and transition temperatures (adapted from [42])

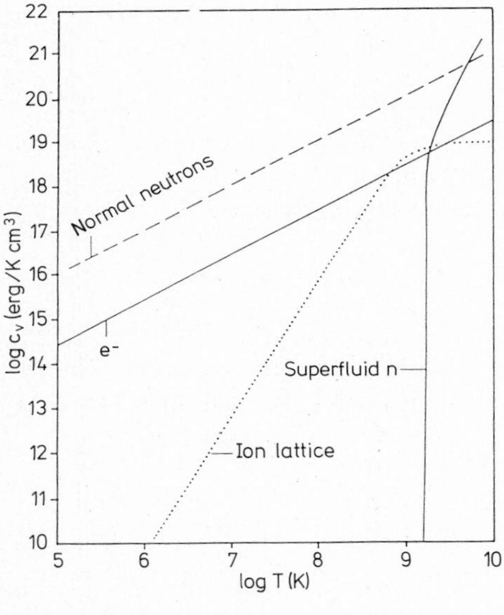


**Fig. 6.13.** Contributions to the specific heat for $\varrho = 10^{14}\,\mathrm{g/cm^3}$ (from [41])

Hence

$$\varkappa \simeq \left( \frac{8}{3\pi^2} \frac{e^2}{\hbar c} \frac{\hbar k_F}{m c} \right)^{1/2} \frac{\varepsilon_F}{\Delta_0} .$$

As an example, we take $k_F = 0.7 \times 10^{13}\,\mathrm{cm}^{-1}$, corresponding to a proton density $2 \times 10^{13}\,\mathrm{g/cm^3}$ and a Fermi energy $\varepsilon_F \approx 10$ MeV. Then $\varkappa > 1/\sqrt{2}$

if $\Delta_0 > 0.005\,\varepsilon_F$. This shows that the protons probably form a type II superconductor.

---

### 6.7.3 Neutrino Emissivities

In this section we discuss the most important neutrino emission processes.

*A. Electron-Ion Bremsstrahlung*

In the crust, the dominant neutrino process is neutrino bremsstrahlung in Coulomb scattering of electrons:

$$e^- + (Z, A) \rightarrow e^- + (Z, A) + \nu + \bar{\nu}. \qquad (6.7.11)$$

There is no particular problem in the calculation of this process, as long as one is satisfied with an accuracy within a factor 2. Of course, one must take into account screening corrections for the dense plasma in the crust of the neutron star. In [43] further many-body effects were also taken into account. The following result [44]

$$\varepsilon_{\text{ions}} = (2.1 \times 10^{20}\ \text{erg cm}^{-3}\,\text{s}^{-1}) \frac{Z^2}{A} \frac{\varrho}{\varrho_0} \left(\frac{T}{10^9\,\text{K}}\right)^6 \qquad (6.7.12)$$

is sufficiently accurate. The value of the factor $Z^2/A$ for the crust has probably been calculated most reliably in [45].

*B. Modified Urca Process and Nucleon-Nucleon Bremsstrahlung*

In the absence of a pion condensate, the dominant neutrino processes in the interior are:

– modified Urca Process $n + n \rightarrow n + p + e^- + \bar{\nu}_e$ (6.7.13)

– neutrino bremsstrahlung in nucleon-nucleon collisions (resulting from the neutral current interaction):

$$n + n \rightarrow n + n + \nu + \bar{\nu} \qquad (6.7.14)$$

$$n + p \rightarrow n + p + \nu + \bar{\nu}. \qquad (6.7.15)$$

Before discussing these processes in any detail, let us clarify why the normal Urca processes

$$n \rightarrow p + e^- + \bar{\nu}_e, \quad e^- + p \rightarrow n + \nu_e$$

do not play a role in degenerate matter.

Beta equilibrium requires that the chemical potentials satisfy $\mu_n = \mu_e + \mu_p$. Those neutrons which are not forbidden to decay by the

Pauli principle are within $kT$ of the Fermi surface. This also holds for the protons and for the electrons in the final state; the neutrino energy is then also $\sim kT$. From (6.7.7) we know that the Fermi momenta of the electrons and protons are small compared with the neutron Fermi momentum and thus all the final particles must have small momenta. This means that the Urca process is strongly suppressed by energy and momentum conservation. This is no longer the case when an additional neutron, which can absorb energy and momentum, takes part in the process, as in (6.7.13). A pion condensate would have the same effect.

The calcuations [46] of all three processes (6.7.13–15) are very similar. As an example, we consider neutron-neutron bremsstrahlung (6.7.14) in some detail.

We have to evaluate the Feynman diagrams in Fig. 6.14.

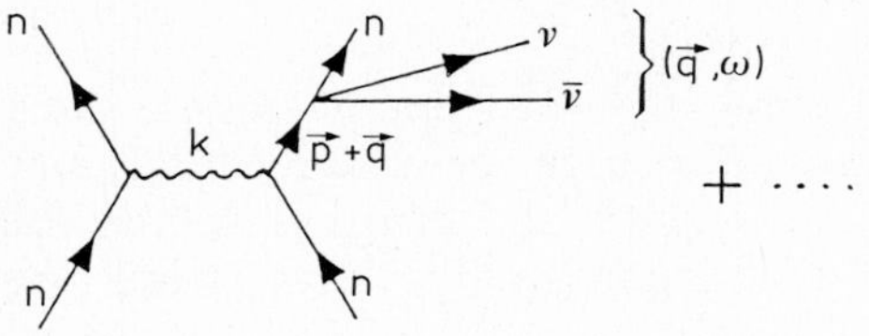


**Fig. 6.14.** Feyman diagrams for n-n bremsstrahlung. The *dots* indicate the diagrams in which the neutrino pair is attached to the other external lines

The amplitude for this process is a product of three factors: The T-matrix for nucleon-nucleon scattering (this can be taken on the mass shell, since off shell corrections are small in the vicinity of the Fermi energy), a nucleon propagator and a matrix element for neutrino emission.

For the weak vertex the standard $SU(2)\times U(1)$ gauge model gives the following effective four-fermion interaction for each neutrino type $(N=p, n)$:

$$\mathscr{L}_{\text{weak}} = \frac{G_{\text{F}}}{2\sqrt{2}} \bar{N}\gamma^{\alpha}(g_N - g'_N\gamma_5)\, N\, \bar{\nu}\gamma_{\alpha}(1-\gamma_5)\,\nu\,, \tag{6.7.16}$$

where

$$g_N = \begin{cases} -1 & (N=n) \\ 1-4\sin^2\theta_w & (N=p) \end{cases}$$

$$g'_N = \begin{cases} -g_A & (N=n) \\ \phantom{-}g_A & (N=p); \quad g_A \simeq 1.24\,. \end{cases} \tag{6.7.17}$$

We treat the nucleons nonrelativistically. The nucleon propagator in Fig. 6.14 is

$$G(\boldsymbol{p}\pm\boldsymbol{q}, E_{\boldsymbol{p}}\pm\omega) = \frac{1}{E_{\boldsymbol{p}}\pm\omega - E_{\vec{p}\pm\vec{q}}}\,, \tag{6.7.18}$$

where $+(-)$ denotes an outgoing (incoming) neutron. The neutrinos are thermal, i.e., $\omega \sim kT$. We expand $G$ in powers of $1/m_{\text{n}}$ and keep only the lowest order term

$$G(\boldsymbol{p}\pm\boldsymbol{q}, E_p\pm\omega)\simeq 1/\omega. \tag{6.7.19}$$

Recoil corrections to the neutrino luminosity are of order $(p_F/m_n)^2\sim 20\%$.

We must now consider the n – n interaction.

The long range part is dominated by one pion exchange (OPE). Since the nucleons involved are restricted to a narrow energy band in the vicinity of the Fermi surface, it is natural to describe the short range contribution in terms of nuclear Fermi liquid parameters.

Because of its long range and its tensor character, OPE makes the largest contribution to the neutrino luminosity. The long range is very important since the mean interparticle separation in neutron star matter is fairly large, about 2.2 fm. The tensor character turns out to be important because of the structure of the weak interaction (6.7.16), (6.7.17). One should also note that in the Landau limit $k\to 0$, the OPE interaction vanishes. Thus there is no obvious double counting in the procedure sketched here.

The Landau parameters for matter in neutron stars are not very well known. Fortunately this is not too important for our problem, since the OPE contribution is the deciding factor for the neutrino luminosity. Approximating the neutron propagator by (6.7.19) results in:

(i) The vector contributions to the weak interaction cancel, for the OPE contribution *and* for the Landau interaction. This is true for all three processes (6.7.13, 14, 15).

(ii) The Landau interaction also does not contribute to the axial part of the weak interaction for n – n bremsstrahlung. This is not true for the other processes considered.

Thus only the OPE contributes to n – n bremsstrahlung. The luminosity per unit volume is in general (see Fig. 6.14)

$$\varepsilon_\nu=\int\prod_{i=1}^{4}\frac{d^3p_i}{(2\pi)^3}\frac{d^3q_1}{(2\pi)^3}\frac{d^3q_2}{(2\pi)^3}(2\pi)^4\,\delta^4(P_f-P_i)\frac{1}{s}\left(\sum_{\text{spins}}|M|^2\right)\omega\cdot\mathscr{S}, \tag{6.7.20}$$

where $s$ is a symmetry factor (equal to the product of the number of identical outgoing particles and the number of identical incoming particles) and $\mathscr{S}$ is the product of statistical Fermi-Dirac factors.

The neutrons are very close to the Fermi surface and the neutrinos are thermal. Hence the phase space integral can be factorized into angular and energy integrals, to a good approximation. The entire temperature dependence is then contained in the following energy integral:

$$I_{\nu\bar\nu}=\int_0^\infty d\omega\,\omega^4\int\prod_{i=1}^{4}dE_i\,\delta(\omega-E_1-E_2+E_3+E_4)\mathscr{S}. \tag{6.7.21}$$

The power of $\omega$ is determined as follows:

- Lepton matrix element $\omega^2$
- Nucleon propagator $1/\omega^2$
- Phase space $\omega^3\, d\omega$
- Energy loss $\omega$.

After introducing dimensionless variables, one obtains

$$\begin{aligned} I_{\nu\bar{\nu}} &= (kT)^8 \int \prod_{i=1}^{4} dx_i \frac{1}{e^{x_i}+1}\, \delta\left(\sum_{i=1}^{4} x_i - y\right) \\ &= \frac{41}{60480}(2\pi)^8 (kT)^8 . \end{aligned} \tag{6.7.22}$$

Various corrections to this result ($\varrho$-exchange, suppression of the short range part of OPE, etc.) have been investigated in [46]. The following numerical results for the processes (6.7.13–15) were obtained:

$$\varepsilon_{\rm nn} = (7.8\times 10^{19}\ \mathrm{erg\ cm^{-3}\ s^{-1}}) \left(\frac{m_{\rm n}^*}{m_{\rm n}}\right)^4 \left(\frac{\varrho}{\varrho_0}\right)^{1/3} \left(\frac{T}{10^9\,\mathrm{K}}\right)^8$$

$$\varepsilon_{\rm np} = (7.5\times 10^{19}\ \mathrm{erg\ cm^{-3}\ s^{-1}}) \left(\frac{m_{\rm n}^*}{m_{\rm n}}\right)^2 \left(\frac{m_{\rm p}^*}{m_{\rm p}}\right)^2 \left(\frac{\varrho}{\varrho_0}\right)^{2/3} \left(\frac{T}{10^9\,\mathrm{K}}\right)^8$$

$$\varepsilon_{\rm Urca} = (2.7\times 10^{21}\ \mathrm{erg\ cm^{-3}\ s^{-1}}) \left(\frac{m_{\rm n}^*}{m_{\rm n}}\right)^3 \frac{m_{\rm p}^*}{m_{\rm p}} \left(\frac{\varrho}{\varrho_0}\right)^{2/3} \left(\frac{T}{10^9\,\mathrm{K}}\right)^8 .$$

From these expressions one sees that the modified Urca process dominates (for normal nucleons). Note also the rather strong dependence on the effective masses.

### *C. Neutrino Emission in the Presence of a Pion Condensate*

In the interior of a neutron star the difference between the neutron and proton chemical potentials is of the order

$$\mu_{\rm n} - \mu_{\rm p} = \mu_{\rm e} \approx 100\ \mathrm{MeV}, \tag{6.7.24}$$

which is comparable to the pion mass. One thus expects that the self energy of the pions due to interaction with the surrounding nucleons (repulsive s-wave, attractive p-wave) might be sufficient for the spontaneous production of pions. Since these are bosons, they could occupy the energetically most favorable state macroscopically. Such a state corresponds to a nonvanishing expectation value of the pion field. A global gauge symmetry would thus be spontaneously broken; in the normal ground state, without pions, this expectation value vanishes.

The results of numerous calculations indicate that condensation of the charged pion field begins at about twice the nuclear density.

However, it is possible that the critical density is considerably higher, or that condensation does not take place at all. Two good review articles ([47] and [48]) on the subject are available.

We now give a simple argument which shows that if a pion condensate is present, the cooling of a neutron star by neutrino radiation is speeded up considerably.

In the condensed state, the nucleons are coherent superpositions of protons and neutrons (i.e. they are rotated in isospin space). The neutron component of such a quasiparticle $u$ can $\beta$-decay, resulting in a proton state which has a nonvanishing overlap with the quasiparticle in the final state of the process

$$\mathrm{u} \to \mathrm{u}' + \mathrm{e}^- + \bar{\nu}_\mathrm{e} . \tag{6.7.25}$$

In this decay, the quasiparticle can absorb energy and momentum $(\mu_\pi, \boldsymbol{k})$ from the condensed pions and hence the process (6.7.25), unlike the Urca process, is not suppressed, provided $k_\mathrm{F}(u) \lesssim k/2$. This condition is satisfied, since the pion wave vector $k$ is typically of the order $\sim 400$ MeV/c.

One can see, without a detailed calculation, that the process (6.7.25) is a very important cooling mechanism. In order to show this, we compare (6.7.25) with the modified Urca process (i.e., we compare the Feynman diagrams of Fig. 6.15).

One expects $\varepsilon_\pi/\varepsilon_\mathrm{Urca} \sim n_\pi/n_\mathrm{n} \times$ ratio of phase space factors, where $n_\pi$ is a measure of the degree of pion condensation.

Technically, $n_\pi/n_\mathrm{n} \sim \theta^2$, where $\theta$ is the angle describing the chiral rotation from the normal to the condensed state. The spectator neutron in the modified Urca process gives rise to an additional phase space reduction factor $(kT/E_\mathrm{F})^2$ and we thus expect

$$\varepsilon_\pi/\varepsilon_\mathrm{Urca} \sim \theta^2 (E_\mathrm{F}/kT)^2 \sim \theta^2 \frac{10^6}{T_9^2}, \qquad T_9 := \frac{T}{10^9\,\mathrm{K}} . \tag{6.7.26}$$

This is confirmed by a detailed calculation. One finds [49]

$$\varepsilon_\pi \simeq (2.1 \times 10^{27}\ \mathrm{erg\ cm^{-3}\ s^{-1}})\, \theta^2\, T_9^6 . \tag{6.7.27}$$

This gives

$$\varepsilon_\pi/\varepsilon_\mathrm{Urca} \simeq 0.8 \times 10^6\, \theta^2/T_9^6 . \tag{6.7.28}$$

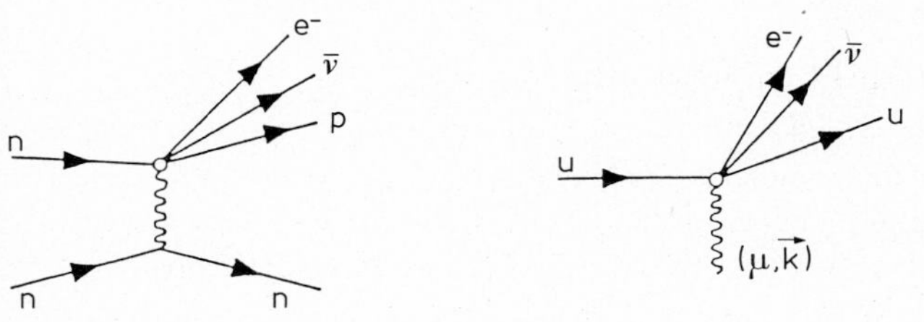


**Fig. 6.15**

*D. Neutrino Processes for Superfluid Nucleons*

The gaps in the excitation spectra of nucleons at the Fermi surfaces suppress the interior processes (6.7.13–15) since the number of thermal excitations is reduced. The reduction factor for (6.7.14) is $\exp(-2\Delta_n/kT)$ and the temperature dependence of the energy gap is given approximately by $\Delta(T) \simeq \Delta(0)\,(1-T/T_c)^{1/2}$.

In [50] it has been remarked that if the neutrons are superfluid, it is possible that two quasiparticle excitations can decay into neutrino pairs. In other words, broken "Cooper pairs" recombine and join the condensate. (In ordinary superconductors, this recombination is accompanied by emission of phonons and photons.) The calculation of the corresponding emissivity is a nice exercise in BCS theory [8]. This process can be more important than bremsstrahlung in the crust within a certain temperature range.

### 6.7.4 Cooling Curves

It is now possible to compute cooling curves. Let us start with a simple estimate. Consider a homogeneous star ($\varrho = \text{const}$) consisting of normal nucleons without a pion condensate. Then the modified Urca process dominates and the interior temperature varies according to (for the moment we ignore GR)

$$-dT/dt \simeq \frac{\varepsilon_{\text{Urca}}}{c_n}, \tag{6.7.29}$$

where $c_n$ is the specific heat of the normal neutrons. If we now take $m_p^*/m_p = 1$, $m_n^*/m_n = 0.8$, and use (6.7.5–7) and 6.7.23), we obtain from (6.7.29)

$$-dT_9/dt = 1.1\times 10^{-8}\, T_9^7 \left(\frac{\varrho}{\varrho_0}\right)^{1/3} \tag{6.7.30}$$

*independent of M.* From this differential equation, we obtain for the cooling time $\Delta t$ from the initial temperature $T(i)$ to a final temperature $T(f)$:

$$\Delta t(T(i) \to T(f)) = (1.58\times 10^7\,\text{s}) \left(\frac{\varrho_0}{\varrho}\right)^{1/3} [T_9^{-6}(f) - T_9^{-6}(i)]. \tag{6.7.31}$$

---

[8] The nonrelativistic neutron field is given by

$$\psi(\boldsymbol{x}) = \sum_{\boldsymbol{k}} (a_{\boldsymbol{k}\uparrow}\chi_\uparrow + a_{\boldsymbol{k}\downarrow}\chi_\downarrow)\, e^{i\boldsymbol{k}\cdot\boldsymbol{x}}.$$

The Bogoliubov transformation for the quasiparticles is

$$a_{\boldsymbol{k}\uparrow} = u_k\,\alpha_{\boldsymbol{k}} + v_k\,\beta^*_{-\boldsymbol{k}}, \quad a_{\boldsymbol{k}\downarrow} = u_k\,\beta_{-\boldsymbol{k}} - v_k\,\alpha^*_{\boldsymbol{k}}.$$

The Fermi matrix element (for example) is then given by

$$|\langle \text{G.S.}\,|\psi^*\psi|\underbrace{\boldsymbol{k},\boldsymbol{k}'}_{\text{quasiparticles}}\rangle|^2 = (u_k v_{k'} + u_{k'} v_k)^2.$$

As an example, take $T(i) = 10$ MeV and $\varrho/\varrho_0 = 1$. Then

$$\log \Delta t(s) = 7.2 - 6 \log T_9(f). \tag{6.7.32}$$

This result is shown in Fig. 6.16; the relation between interior and surface temperature given in Fig. 6.10 was also used. For the age of the Crab pulsar ($\Delta t = 2.9 \times 10^{10}$ s) one obtains $T_9(f) = 0.286$, which corresponds to a surface temperature $T_s \simeq 2 \times 10^6$ K; this is just the observed upper limit.

Corrections to the cooling curves due to general relativity have been studied in [51]. They can be computed analytically for a homogeneous model. The result for the temperature $T_\infty$, which would be measured by a distant observer, is also shown in Fig. 6.16.

Fig. 6.17 shows results [41] for a stellar model with $M = 1.25\,M_\odot$ based on the stiff PPS equation of state. For this calculation, one took $B = 0$ and assumed that the nucleons are not superfluid. For the first few thousand years, this curve is nearly identical to the lower curve in Fig. 6.16. The upper limits given in Table 6.1 are also indicated in Fig. 6.17. One sees that the theoretical curve is consistent with observation, except perhaps for the case of SN 1006, without assuming the presence of exotic forms of matter.

Figure 6.18 shows how much the results shown in Fig. 6.17 can change in the presence of magnetic fields and/or superfluidity.

Figure 6.19 shows the cooling curves for an equally massive star, but assuming the soft BPS equation of state. The different curves shown correspond to various assumptions as to magnetic field, superfluidity, and pion condensation. The presence of a pion condensate for $\varrho > 2\,\varrho_0$ was assumed in the lower curves. Since the central density is high

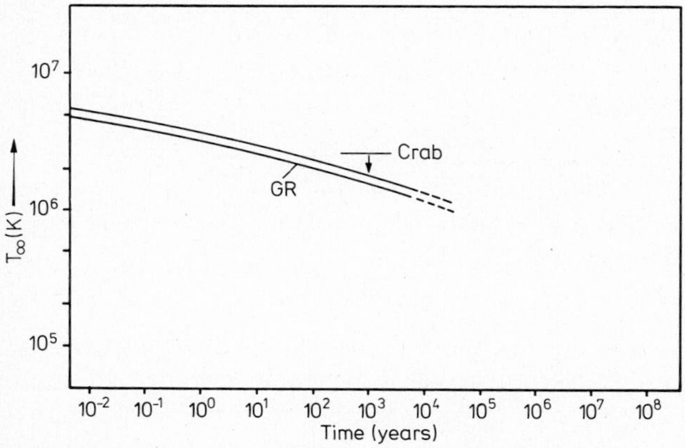


**Fig. 6.16.** Observed temperature $T_\infty$ as a function of time for a homogeneous stellar model with normal nucleons and no pion condensate. The *lower curve* includes general relativistic corrections

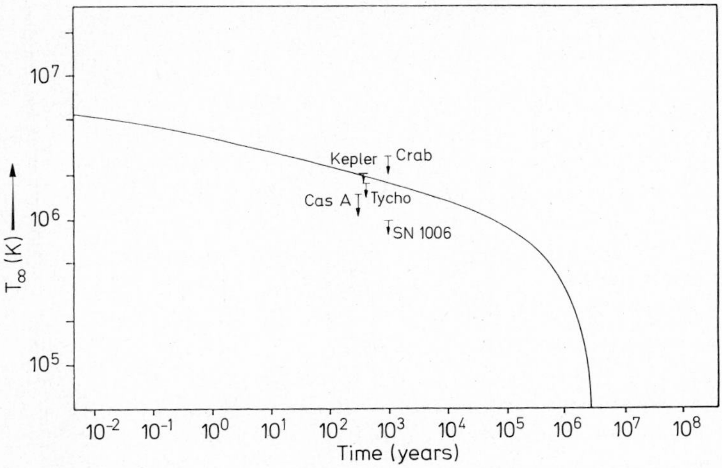


**Fig. 6.17.** Observed temperature $T_\infty$ as a function of time for a stellar model based on the PPS equation of state, using $M = 1.25\ M_\odot$, $B = 0$ and without superfluidity (adapted from [41]). The upper limits from Table 6.1 are also indicated

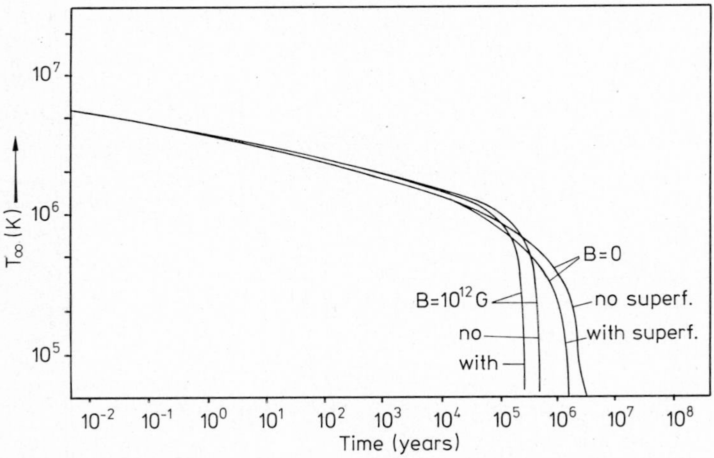


**Fig. 6.18.** Dependence of the results of Fig. 6.17 when magnetic fields and/or superfluidity are taken into account

($2.7 \times 10^{15}$ g/cm$^3$) for a soft equation of state, pion condensation, as expected, plays a dramatic role in this case. For a stiff equation of state, the central density is only $3.8 \times 10^{14}$ g/cm$^3$ and it is improbable that a pion condensate is present.

Superfluidity plays a more important role for the stiff equation of state, since more of the mass is in the relevant density range.

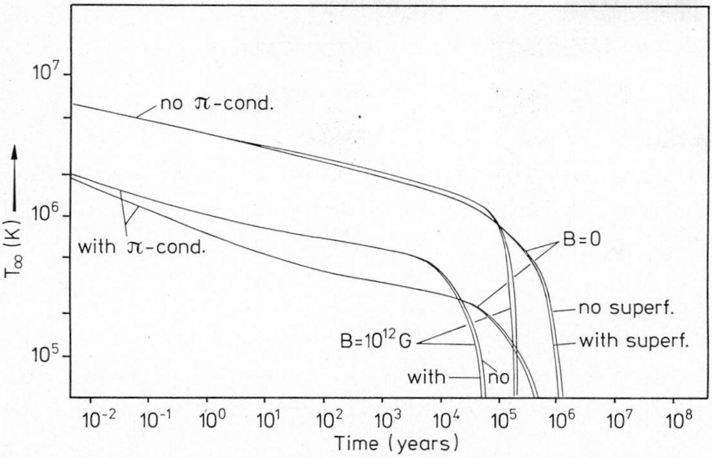


**Fig. 6.19.** Observed temperature $T_\infty$ as a function of time for a stellar model based on the BPS equation of state, using $M = 1.25\ M_\odot$, and various assumptions about pion condensation, magnetic fields and superfluidity (adapted from [41])

As we have previously emphasized, magnetic fields have been treated rather crudely, and the results must be regarded as qualitative.

The dependence on the total mass is weak, as one might have expected from the homogeneous model.

At the present time, the upper limits on the temperature (Table 6.1) are consistent with models for neutron stars without exotic forms of matter, such as pion condensates or quark phases. This might change when it becomes possible to reduce these limits significantly with future x-ray telescopes.

The cooling of neutron stars has been discussed so extensively because it provides an excellent example of how astrophysical problems require the simultaneous application of several different physical disciplines.

**Note:** Since this chapter has been written new studies [107–109] of neutron star cooling have been published.

# 6.8 Addendum 1: Ground State Energy of Macroscopic Matter

## 6.8.1 Stability of Matter with Negligible Self-Gravity

From experience we know that "ordinary" matter around us is *stable:* The total energy of a (finite) system has a lower bound which is extensive, i.e., proportional to the number of particles.

The coarse structure of real matter should be described correctly by nonrelativistic quantum mechanics based on the Hamiltonian

$$H_N = \sum_{j=1}^{N} -\frac{\hbar^2}{2m_j}\Delta_j + \sum_{i<j}\frac{e_i e_j}{|\boldsymbol{x}_i - \boldsymbol{x}_j|}, \tag{6.8.1}$$

whose potential energy describes the Coulomb interactions between electrons and nuclei. It is far from trivial that the ground-state energy of this Hamiltonian is extensive,

$$H_N \geqq -cN, \tag{6.8.2}$$

where $c$ is a constant independent of $N$. That the inequality (6.8.2) really holds, if the Pauli principle for the electrons is taken into account, was first proven in a remarkable analytic tour de force by *Dyson* and *Lenard* [78]. The many estimates of the proof led, however, to an extremely large value of $c\,(\sim 10^{14}\,\mathrm{Ry})$. Since then a much simpler proof was found by *Lieb* and *Thirring* [79], which was inspired by the Thomas-Fermi theory.

Before quoting their result, we will first discuss the problem on a heuristic level.

### A. Heuristic Discussion

For a neutral Coulomb system screening effects reduce the effective interaction essentially to one between nearest neighbors. Thus the potential energy is roughly (for bosons and fermions)

$$V \simeq -N\frac{e^2}{(R/N^{1/3})}, \tag{6.8.3}$$

where $R$ is the dimension of the system.

Let us leave the statistics of the "electrons" (mass $m$) open. The kinetic energy is given by $T \simeq N p^2/2m$, where $p$ is the average momentum of the "electrons". For bosons, the uncertainty relation implies $p \gtrsim \hbar/R$ whereas for fermions the Pauli principle allows at most one electron in a de Broglie cube $(\hbar/p)^3$ and thus

$$p \gtrsim N^{1/3}\,\hbar/R. \tag{6.8.4}$$

For the total energy of the system, we find therefore the inequalities

$$E = T + V \gtrsim \begin{cases} \dfrac{\hbar^2}{2m}\dfrac{N}{R^2} - e^2\dfrac{N^{4/3}}{R} & \text{(bosons)} \\[2ex] \dfrac{\hbar^2}{2m}\dfrac{N^{5/3}}{R^2} - e^2\dfrac{N^{4/3}}{R} & \text{(fermions)}. \end{cases} \tag{6.8.5}$$

The minimum of $E$ is reached for

$$R \simeq \begin{cases} \dfrac{\hbar^2}{m e^2} N^{-1/3} & \text{(bosons)} \\ \dfrac{\hbar^2}{m e^2} N^{1/3} & \text{(fermions)} \end{cases} \tag{6.8.6}$$

with the value

$$E_{\min} \simeq -\,\mathrm{Ry} \begin{cases} N^{5/3} & \text{(bosons)} \\ N & \text{(fermions)}\,. \end{cases} \tag{6.8.7}$$

For fermions we obtain indeed saturation of the Coulomb interaction. Whether for bosons the ground-state energy in fact goes as $N^{5/3}$ has not yet been shown rigorously. With the help of a clever trial wave function, *Dyson* [80] has found an *upper* limit proportional to $-N^{7/5}$. In any case, a hypothetical world with bosonic "electrons" would not be stable. Joining two macroscopic pieces of matter would lead to an energy release larger than $\sim (N/10^{23})$ megatons (1 megaton $\sim 10^{32}$ Ry). Clearly *Ehrenfest* [81] was underestimating the role of the Pauli principle for matter in bulk in his address to Pauli in 1931 on the occasion of the award of the Lorentz medal, when he said: "You must admit, Pauli, that if you would only partially repeal your prohibition, you could relieve many of our practical worries, for example the traffic problem on our streets."

The number density in the ground-state of macroscopic matter is, using (6.8.6) for fermions,

$$n \simeq \frac{N}{R^3} \simeq \left(\frac{m e^2}{\hbar^2}\right)^3 = a_0^{-3} = \frac{\alpha^3}{\lambda\!\!\!^{-3}_{\mathrm{e}}} \tag{6.8.8}$$

and the corresponding density is

$$\varrho \simeq \frac{\alpha^3}{\lambda\!\!\!^{-3}_{\mathrm{e}}} m_N \simeq 10\ \mathrm{g/cm^3}, \tag{6.8.9}$$

which is fine.

## B. Rigorous Results

It is most satisfactory that the following rigorous result can be proved (for detailed proofs see [82], [83], [84]):

**Theorem** (Dyson, Lenard, Lieb, Thirring): For the Hamilton operator (6.8.1) for $N$ electrons and $M$ nuclei with charges $Z_j$ one has the operator inequality

$$H \geqq -\,1.54\,(4\pi)^{2/3}\,\mathrm{Ry}\,N \left\{1 + \left[\sum_{j=1}^{M} Z_j^{7/3}\,N\right]^{1/2}\right\}^2 . \tag{6.8.10}$$

**Remarks**

1. The numerical factor in (6.8.10) is not optimal. The quality of the estimate is illustrated for neutral hydrogen, ($Z_j = 1$, $M = N$):

$$H > -22.24\, N \,\mathrm{Ry}.$$

2. The theorem does not presuppose charge neutrality.
3. Only the Fermi statistics of the electrons is used. The bound (6.8.10) holds independent of the statistics of the nuclei.
4. The danger of stability comes from the short distance behavior of the Coulomb potential. Smoothing it out at $r = 0$ can lead trivially to stability. If we replace, for instance, $1/r$ by

$$v(r) = \frac{1}{r}(1 - \mathrm{e}^{-\mu r}),$$

we can estimate the potential energy as follows.

Clearly we have

$$V = \sum_{i<j} e_i e_j v(|\boldsymbol{x}_i - \boldsymbol{x}_j|) = \tfrac{1}{2} \sum_{i,j} e_i e_j v(|\boldsymbol{x}_i - \boldsymbol{x}_j|) - \frac{N}{2} e^2 v(0),$$

if, for simplicity, all $|e_j| = e$. The double sum is, however, equal to

$$\int d^3p \,|\sum_j \mathrm{e}^{\mathrm{i}\boldsymbol{p}\cdot\boldsymbol{x}_j} e_j|^2 \,\tilde{v}(\boldsymbol{p}).$$

Since the Fourier transform $\tilde{v}(\boldsymbol{p})$ of $v(|\boldsymbol{x}|)$ is positive:

$$\tilde{v}(\boldsymbol{p}) = \frac{\mu^2}{\boldsymbol{p}^2(\boldsymbol{p}^2 + \mu^2)} > 0,$$

we find

$$V \geqq -\frac{N}{2} e^2 v(0),$$

which proves stability (even for bosons), since the kinetic energy is positive. For a realistic potential we would, however, expect $v(0) = O\,[\mathrm{MeV}]$, instead of $O\,[\mathrm{Ry}]$, which shows that this argument misses the point.

### 6.8.2 Nonsaturation of Gravitational Forces

Because of their purely attractive character, gravitational forces do not saturate [85].

#### *A. Heuristic Discussion*

We consider first $N$ particles of mass $m$. Let $p$ be the average momentum of a particle and $R$ the average distance between two particles (~ dimension of the system). For purely gravitational interactions and nonrelativistic situations we have

$$E \simeq N \frac{p^2}{2m} - \tfrac{1}{2} N^2 \frac{Gm^2}{R}. \qquad (6.8.11)$$

As in the Coulomb case,

$$R \gtrsim \begin{cases} \hbar/p & \text{(bosons)} \\ N^{1/3}\,\hbar/p & \text{(fermions)} \end{cases} \qquad (6.8.12)$$

and thus

$$E \gtrsim \begin{cases} N \dfrac{p^2}{2m} - N^2 \dfrac{Gm^2}{2\hbar} p & \text{(bosons)} \\ N \dfrac{p^2}{2m} - N^{5/3} \dfrac{Gm^2}{2\hbar} p & \text{(fermions)}. \end{cases} \qquad (6.8.13)$$

The minimum of the right hand side is attained for

$$p \simeq \begin{cases} N \dfrac{Gm^3}{2\hbar} & \text{(bosons)} \\ N^{2/3} \dfrac{Gm^3}{2\hbar} & \text{(fermions)} \end{cases} \qquad (6.8.14)$$

and the corresponding value for the ground-state energy is

$$E_{\min} \simeq -\frac{G^2 m^5}{8\hbar^2} \cdot \begin{cases} N^3 & \text{(bosons)} \\ N^{7/3} & \text{(fermions)}. \end{cases} \qquad (6.8.15)$$

This shows that the exclusion principle reduces the degree of instability, but is by far insufficient to guarantee the saturation of gravitational forces.

## *B. Rigorous Discussion*

The following theorem [85] is a rigorous version of the estimate (6.8.15) for fermions.

**Theorem:** For a system of $N$ identical fermions with mass $m$ and purely gravitational attraction, the ground-state energy $E_0(N)$ obeys

$$-AN(N-1)^{4/3} \frac{G^2 m^5}{\hbar^2} \leqq E_0(N) \leqq -BN^{1/3}(N-1)^2 \frac{G^2 m^5}{\hbar^2}, \qquad (6.8.16)$$

$A$ and $B$ being positive constants (non-optimal values are obtained in the proof).

*Proof:* a) *Lower bound:* We write the Hamiltonian in the following way:

$$H_N = \sum_{i=1}^{N} \left[ \sum_{j \neq i} \left( \frac{\boldsymbol{p}_j^2}{2(N-1)\,m} - \frac{G}{2} \frac{m^2}{|\boldsymbol{x}_i - \boldsymbol{x}_j|} \right) \right] \equiv \sum_{i=1}^{N} h_i. \qquad (6.8.17)$$

Each $h_i$ represents $N-1$ *independent* particles ($j \neq i$) in the "Coulomb" field of a fixed one (the $i$th particle). The ground-state of $h_i$ is obtained by distributing $(N-1)$ fermions over the first $(N-1)$ levels of this hydrogenic atom with energies

$$\varepsilon_n = -\frac{1}{n^2}\frac{1}{8}(N-1)\frac{G^2 m^5}{\hbar^2}, \quad n = 1, 2, \ldots \tag{6.8.18}$$

and degree of degeneracy $n^2$. The last level $\varepsilon_\nu$ to be completely filled is determined by

$$\sum_{n=1}^{\nu} n^2 \leqq N-1 < \sum_{n=1}^{\nu+1} n^2,$$

i.e.

$$\tfrac{1}{3}\nu(\nu+\tfrac{1}{2})(\nu+1) \leqq N-1 < \tfrac{1}{3}(\nu+1)(\nu+\tfrac{3}{2})(\nu+2). \tag{6.8.19}$$

Hence we obtain

$$\begin{aligned}\inf\langle h_i\rangle &\geqq \sum_{n=1}^{\nu+1} n^2\,\varepsilon_n = -(\nu+1)\tfrac{1}{8}(N-1)\frac{G^2 m^5}{\hbar^2}\\ &\geqq -\tfrac{1}{4}(N-1)^{4/3}\frac{G^2 m^5}{\hbar^2} \quad \text{for} \quad N \geqq 2,\end{aligned} \tag{6.8.20}$$

since (6.8.19) implies $\nu+1 \leqq 2(N-1)^{1/3}$ for $N \geqq 2$. Clearly

$$E_0(N) \geqq N\inf\langle h_i\rangle = -\tfrac{1}{4}N(N-1)^{4/3}\frac{G^2 m^5}{\hbar^2}. \tag{6.8.21}$$

This proves the lower bound with $A \leqq \frac{1}{4}$.

b) *Upper bound:* We use the variational principle with a Slater determinant as trial wave function:

$$\psi(\boldsymbol{x}_1, \ldots, \boldsymbol{x}_N) = (N!)^{-1/2}\det[\varphi_k(\lambda\,\boldsymbol{x}_l)], \tag{6.8.22}$$

where $\lambda$ is a scale parameter and $\{\varphi_k\}$ a set of localized wavefunctions of the form

$$\varphi_k(\boldsymbol{x}) = \phi(\boldsymbol{x}-\boldsymbol{a}_k). \tag{6.8.23}$$

The $\{\boldsymbol{a}_k\}$ are $N$ fixed points, chosen such that they satisfy

$$|\boldsymbol{a}_k - \boldsymbol{a}_l| \geqq 1 \quad \text{for} \quad k \neq l. \tag{6.8.24}$$

Furthermore, $\phi$ is a smooth function which vanishes outside a sphere of radius $\frac{1}{2}$.

From atomic physics, one knows that

$$\frac{(\psi, H\psi)}{(\psi, \psi)} = \lambda^2 N\frac{\hbar^2}{2m}\alpha - \lambda\tfrac{1}{2}Gm^2\sum_{k\neq l}\beta_{kl}, \tag{6.8.25}$$

where

$$\alpha \ = \int |\nabla \phi|^2 \, d^3x \tag{6.8.26}$$

$$\beta_{kl} = \int [|\psi_k(\boldsymbol{x})\, \psi_l(\boldsymbol{x}')|^2 - \overset{*}{\psi_k}(\boldsymbol{x})\, \psi_k(\boldsymbol{x}')\, \psi_l(\boldsymbol{x})\, \psi_l^*(\boldsymbol{x}')] \frac{d^3x\, d^3x'}{|\boldsymbol{x}-\boldsymbol{x}'|}\,.$$

The support property of $\phi$ implies that the exchange term in $\beta_{kl}$ vanishes.

One finds that $\alpha$ becomes maximal for $\phi \propto \sin(2\pi r)$, giving

$$\alpha_{\max} = 4\pi^2 \tag{6.8.27}$$

and

$$\beta_{kl} = \frac{1}{|\boldsymbol{a}_k - \boldsymbol{a}_l|}$$

because $\beta_{kl}$ is the Coulomb potential between two spherically symmetric charge distributions with total charge 1 around $\boldsymbol{a}_k$ and $\boldsymbol{a}_l$.

We now choose the $N$ points $\{\boldsymbol{a}_k\}$ on a cubic lattice with period 1 within a cube with side $(\mu - 1)$, $\mu$ being an integer, such that

$$(\mu - 1)^3 \leqq N \leqq \mu^3 . \tag{6.8.28}$$

Then $|\boldsymbol{a}_k - \boldsymbol{a}_l| \leqq \sqrt{3}\,(\mu - 1) \leqq \sqrt{3}\, N^{1/3}$ for all $k, l$.

Using all this in (6.8.25) gives

$$\frac{(\psi, H\psi)}{(\psi, \psi)} \leqq \lambda^2 \cdot 2\pi^2 N \frac{\hbar}{m} - \frac{\lambda}{2\sqrt{3}} N^{2/3} (N-1)\, G m^2; \tag{6.8.29}$$

compare this with (6.8.13). Minimizing with respect to $\lambda$ gives

$$E_0(N) \leqq \frac{1}{3 \cdot 2^5 \pi^2} N^{1/3} (N-1)^2 \frac{G^2 m^5}{\hbar^2}\,. \tag{6.8.30}$$

### 6.8.3 Newton vs. Coulomb

It is not difficult to extend the analysis to systems of particles where gravitational and electrostatic forces are operating simultaneously.

Since the Coulomb forces tend to establish local neutrality, we expect that the spatial distribution of the nuclei, and hence their momentum distribution, is much the same as those of the electrons. From (6.8.3) and (6.8.11) the energy of a system of $N$ identical fermions with mass $m$ and charge $-e$, and $N$ particles with mass $M$, charge $e$, and unspecified statistics is therefore estimated to be

$$E \simeq N \frac{p^2}{2m} - \tfrac{1}{2} N^2 \frac{GM^2}{R} - N \frac{e^2}{(R/N^{1/3})}\,. \tag{6.8.31}$$

Since the light particles satisfy the exclusion principle, we can use (6.8.4) and obtain

$$E \gtrsim \frac{N p^2}{2m} - N^{5/3} \frac{GM^2}{2\hbar} p - N \frac{e^2 p}{\hbar}\,. \tag{6.8.32}$$

The expression on the right hand side reaches the minimum for

$$p_0 = \frac{m e^2}{\hbar}\left(1 + N^{2/3}\frac{GM^2}{2e^2}\right) \tag{6.8.33}$$

giving

$$E_0 \simeq -N\,\mathrm{Ry}\left(1 + N^{2/3}\frac{GM^2}{2e^2}\right)^2 . \tag{6.8.34}$$

The dimension of the system belonging to (6.8.33) is $R_0 \simeq N^{1/3}\hbar/p_0$, or

$$R_0 \simeq R_{\mathrm{Coulomb}}\left(1 + N^{2/3}\frac{GM^2}{2e^2}\right)^{-1}, \tag{6.8.35}$$

where $R_{\mathrm{Coulomb}}$ is the value of $R_0$ for $G = 0$.

These equations show that for $N \gg N_c$, where

$$N_c = \left(\frac{GM^2}{e^2}\right)^{-3/2} \tag{6.8.36}$$

the system behaves like a collection of neutral identical fermions in gravitational interaction with the light mass $m$ as their inertial mass and the heavy mass $M$ as their gravitational mass.

These heuristic expectations follow also from a rigorous result [85]. Combining the two previous theorems, it is not difficult to prove the following.

**Theorem:** For a system consisting of $N$ identical fermions with mass $m$ and charge $-e$, and $N$ particles with mass $M$, charge $e$, and unspecified statistics, the ground-state energy obeys

$$-CN\frac{m e^4}{2\hbar^2}\left(1 + c N^{2/3}\frac{GM^2}{e^2}\right)^2$$

$$\leqq E_0(N) \leqq -DN\frac{m e^4}{2\hbar^2}\left(1 + dN^{2/3}\frac{GM^2}{e^2}\right)^2, \tag{6.8.37}$$

$C$, $c$, $D$, $d$ being positive constants.

### 6.8.4 Semirelativistic Systems

The discussion in the last section shows that for $N \gg N_c$, a system consisting of $N$ electrons and $N$ protons behaves as a system of $N$ identical fermions governed by an effective Hamiltonian

$$H_{\mathrm{eff}} = \sum_i \frac{1}{2m}\boldsymbol{p}_i^2 - \sum_{i<j}\frac{GM^2}{|\boldsymbol{x}_i - \boldsymbol{x}_j|} . \tag{6.8.38}$$

In the ground-state the average momentum (6.8.33) is

$$p_0 \simeq N^{2/3}\frac{GM^2 m}{2\hbar} . \tag{6.8.39}$$

For $N \gtrsim (GM^2/\hbar c)^{-3/2} = 1.86\, N_\odot$ this average momentum becomes relativistic ($p_0 \gtrsim m\,c$) for the electrons. Since the heavy particles remain nonrelativistic, it is reasonable to try to describe the system by the following semirelativistic effective Hamiltonian

$$H'_{\text{eff}} = \sum_i (\boldsymbol{p}_i^2 c^2 + m^2 c^4)^{1/2} - \sum_{i<j} \frac{GM^2}{|\boldsymbol{x}_i - \boldsymbol{x}_j|}. \tag{6.8.40}$$

Let us first estimate the ground-state energy heuristically. If $p$ denotes the average momentum, we find instead of (6.8.13)

$$E(N) \gtrsim N(p^2 c^2 + m^2 c^4)^{1/2} - N^{5/3} \frac{GM^2}{2\hbar}\, p. \tag{6.8.41}$$

For the ground-state

$$p_0 \simeq m\,c \left(\frac{N}{N_r}\right)^{2/3} [1-(N/N_r)^{4/3}]^{-1/2} \tag{6.8.42}$$

with

$$N_r = \left(\frac{GM^2}{2\hbar\, c}\right)^{-3/2}, \tag{6.8.43}$$

$$E_0(N) \simeq N\,m\,c^2 [1-(N/N_r)^{4/3}]^{1/2}. \tag{6.8.44}$$

This minimum exists, however, only for $N < N_r$. For $N > N_r$ the expression (6.8.41) is not bounded from below, since in the ultrarelativistic limit it becomes $N\left[1-\left(\frac{N}{N_r}\right)^{2/3}\right] p\,c$. The system collapses for $N > N_r$. This shows again that no white dwarfs can have a mass larger than $\simeq N_r M$.

The existence of the *Chandrasekhar limit* can also be shown rigorously.

**Theorem:** The effective Hamiltonian, interpreted as a quadratic form for a $N$-fermion system, is not bounded from below if

$$N > (\sqrt{12}\,\pi)^{3/2} N_r. \tag{6.8.45}$$

(This bound is not optimal!)

*Proof:* We establish again an upper bound with the trial wavefunction (6.8.22). This time we find instead of (6.8.29):

$$\frac{(\psi, H\psi)}{(\psi, \psi)} \leqq N(\phi, (\lambda^2 p^2 c^2 + m^2 c^4)^{1/2} \phi) - \frac{\lambda}{2\sqrt{3}} N^{2/3}(N-1)\, GM^2. \tag{6.8.46}$$

The first term on the right can be estimated with the Schwarz inequality

$$(\phi, \sqrt{p^2+m^2}\,\phi) \leqq \|\phi\| \cdot \|\sqrt{p^2+m^2}\,\phi\| = [(\phi, (p^2+m^2)\,\phi)]^{1/2}.$$

Hence

$$\frac{(\psi, H\psi)}{(\psi, \psi)} \leqq N[\lambda^2 c^2 \alpha\, \hbar^2 + m^2]^{1/2} - \frac{\lambda}{2\sqrt{3}} N^{2/3}(N-1)\, GM^2 \tag{6.8.47}$$

with

$$\alpha = \int |\nabla \phi|^2 \, d^3x . \tag{6.8.48}$$

Again we chose $\phi \propto \sin(2\pi r)\, \chi_{[0, 1/2]}(r)$. Then $\alpha = 4\pi^2$ and thus

$$\frac{(\psi, H\psi)}{(\psi, \psi)} \leqq \lambda \left\{ N\,[\hbar^2\, 4\pi^2\, c^2 + \lambda^{-2}\, m^2]^{1/2} - \frac{1}{2\sqrt{3}}\, N^{2/3}\, (N-1)\, GM^2 \right\} .$$

With increasing $\lambda$ this expression becomes arbitrarily negative for

$$N > (\sqrt{12}\;\pi)^{3/2}\, N_r .$$

The discussion in this section is based directly on the fundamental principles of physics (no phenomenological concepts have been used). It would be nice if some of the estimates, for example (6.8.45), could be improved.

On a phenomenological level, the structure of matter and the equation of state for white dwarf material are relatively well known. At these high densities, the atoms are completely ionized and one can apply well-known techniques to study this high density plasma. (For the equation of state and its application to the structure of white dwarfs see [86], [87].)

## 6.9. Addendum 2: Core Collapse Models of Type II Supernova Explosions

In this Addendum, we sketch the present status of model calculations for type II supernova explosions, which are believed to be induced by core collapse of relatively massive stars to neutron star densities.

### 6.9.1 Some Observational Facts

On August 31, 1885, Hartwig discovered a "nova" near the center of the Andromeda nebula (M 31), which was visible for about 18 months. In 1919, Lundmark estimated the distance of M 31 to be 650,000 light years and concluded from this that some stars could flare up to luminosities thousands of times larger than a normal nova. Another such event was observed in 1895 in NGC 5253 ("nova" Z Centauri), which appeared to be around 5 times brighter than the entire galaxy. The word super-novae was coined by Zwicky who began in 1934 organized searches for supernovae and discovered himself 122 extragalactic events. (The historical literature is quoted in the recent review article [89].)

Current estimates of the supernova rate are about 1 supernova per $(70 \pm 30)$ years in giant spirals like our own.

For a typical supernova outburst the optical luminosity at maximum is about $10^{10} L_{\odot}$. Two major classes of supernovae, called type I and type II are distinguished.

The spectra of type I supernovae show no strong hydrogen lines, which implies that the progenitor stars should be hydrogen-deficient stars. Type I events have been observed in elliptical galaxies as well as in spiral and irregular galaxies. They are not concentrated in spiral arms. Another characteristic of type I supernovae is the exponential tail in their light curves (see Fig. 6.20). Such a tail requires an additional energy source at late times. [Radioactive decay models ($^{56}Ni \rightarrow {}^{56}Co \rightarrow {}^{56}Fe$) have recently been quite successful in fitting the observed light curves.] The properties of type I events suggest that the progenitor stars are perhaps low mass objects of about 1 to 2 $M_{\odot}$ (white dwarfs or He-stars?). The origin of type I supernovae is still uncertain. (For a review, see [90].)

Type II supernovae appear only in the arms of spiral galaxies. They are generally less bright and form a less homogeneous class. The spectra show normal abundances, including hydrogen. The light curves

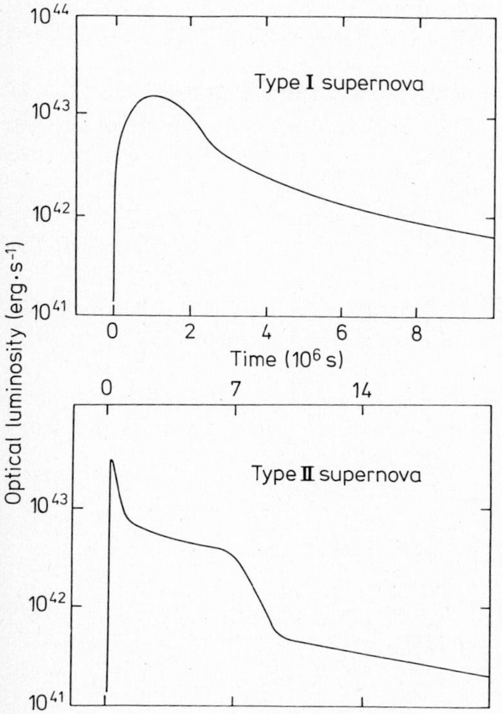


**Fig. 6.20.** Schematic light-curves of type I (upper curve) and type II supernovae. Type II events are characterized by a sharp maximum and a pronounced plateau after maximum light, which indicates the presence of an extended stellar envelope

**Table 6.2.** Some observed properties of supernovae

| Energetics | | Light curves and spectra | | |
|---|---|---|---|---|
| | | | Type I | Type II |
| Maximum luminosity | $10^{44}$ erg s$^{-1}$ | Near Maximum light for | 30 days | 10 days |
| Energy in visual light | $10^{48}-10^{50}$ erg | Plateau | no | yes, 100 days |
| Total energy output | $10^{51}-10^{52}$ erg | Duration | 2 years | 1 year |
| Expansion velocity | $10^4$ km s$^{-1}$ | H-lines | no | strong |
| Temperature near maximum light | 15 000 K | Abundances | Co, Fe? | solar |

(see Fig. 6.20) are concordant with models based on an explosion within an extended red giant envelope. These features suggest that type II supernovae are exploding massive stars ($M \gtrsim 8\ M_\odot$) with extended H-envelopes.

Some observed properties are listed in Table 6.2. For more extensive discussions, we refer to [89], [90].

We add some remarks about the energetics. The integrated energies of a supernova outburst range from $10^{48}$ to $10^{50}$ erg in visual light, which is only of the order of a few percent of the total energy of the ejected material. The observations indicate that about $10^{51}$ erg are totally released in a supernova. Such a large amount is only available in gravitational or nuclear energy release.

If thermonuclear reactions are responsible, then at least about 0.5 to $1\ M_\odot$ of He, C or O have to be fused explosively to iron group elements. This requirement can be met by white dwarfs or degenerate carbon cores of stars with masses of around 6 to $8\ M_\odot$.

Since the binding energy of a neutron star is around $2 \times 10^{53}$ erg, only about 1% of the gravitational energy release is needed for the explosion.

Thermonuclear disruption of accreting white dwarfs in close binaries is a much discussed possibility for type I supernovae [90], [91], [92].

We discuss here only models triggered by core collapse and therefore describe next briefly the evolution of relatively massive stars ($\gtrsim 10\ M_\odot$).

### 6.9.2 Presupernova Evolution of Massive Stars

These stars achieve progressively higher values of temperature and density in the center (see Fig. 6.21). The history of such a star consists of a continued contraction of its core, with halts during stages of nuclear energy generation. There is a succession of nuclear fuels which can be effective in producing such halts: these include hydrogen burning, helium burning, carbon burning, neon burning, oxygen burning and silicon burning. (The major burning phases and their time scales are indicated in Fig. 6.22.)

The last of these stages of nuclear energy generation produces the nuclear statistical equilibrium abundance peak centered around $^{56}$Fe. At this stage, all the available sources of nuclear energy have been extracted from matter. Only massive stars ($\gtrsim 10\, M_\odot$) go through this full sequence, whereby an onion structure is built up around the core of iron peak elements [93].

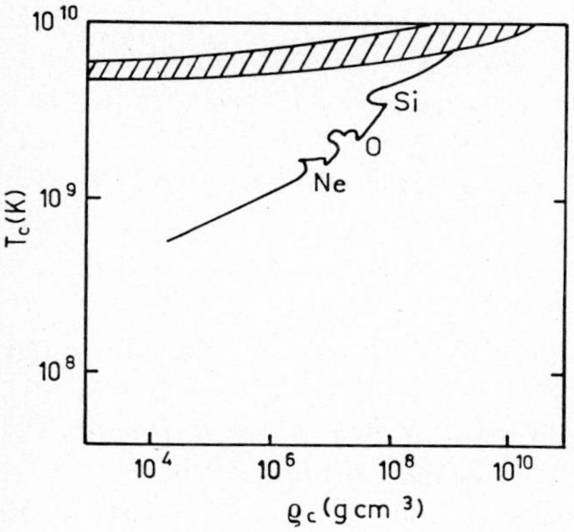


**Fig. 6.21.** Evolutionary track of a 25 $M_\odot$ star. When the central values reach the shaded region, the star becomes unstable due to photodisintegration into helium (from [30])

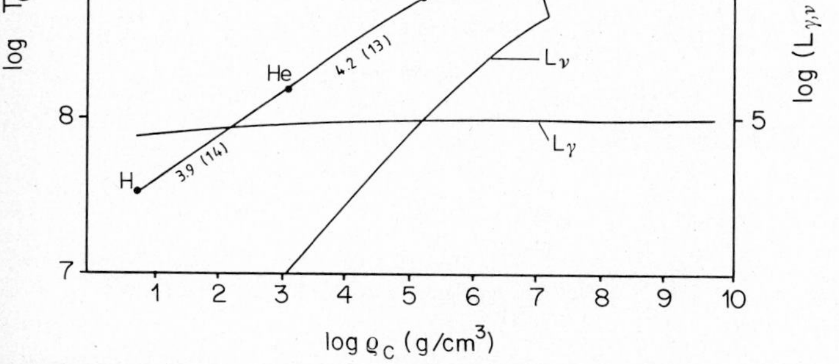

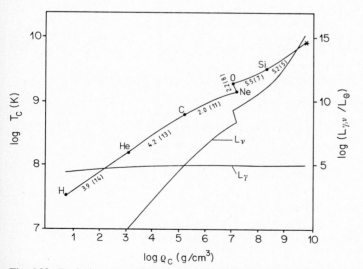

**Fig. 6.22.** Evolutionary track a 15 $M_\odot$ star (smoothed out); total optical ($L_\gamma$) and neutrino ($L_\nu$) luminosities (adapted from [30])

The evolution of these massive stars is relatively short (Fig. 6.22) and becomes rapid even by human standards once the central temperature reaches $10^9$ K. At this point, the neutrino luminosity greatly exceeds the photon luminosity (see Fig. 6.22).

The most important neutrino processes are the following pair emission reactions:

pair annihilation: $e^- + e^+ \to \nu + \bar{\nu}$ (6.9.1 a)

photoneutrino process: $\gamma + e^- \to e^- + \nu + \bar{\nu}$ (6.9.1 b)

plasmon neutrino process: plasmon $\to \nu + \bar{\nu}$. (6.9.1 c)

The importance of these reactions was noticed shortly after Feynman and Gell-Mann had postulated the existence of a universal current-current interaction. The direct electron-neutrino terms, contained in this interaction, were at the time purely hypothetical. We know now from laboratory experiments that processes like (6.9.1) do exist and are well described by the standard electroweak gauge model (see, e.g. [94], [95]).

---

**Exercises:**

1. Calculate the cross section for pair annihilation in the standard electro-weak gauge model.
2. Determine the energy loss per $cm^3$ for pair annihilation at $T \gtrsim 10^9$ K. Give analytic expressions for limiting cases (nonrelativistic, ultrarelativistic, nondegenerate, degenerate).
3. Estimate the total energy loss of a polytropic star model and compare this with the curve shown in Fig. 6.22.

*Partial solution:* Once the temperature approaches $T_0 = m_e c^2/k \simeq 6 \times 10^9$ K, electron-positron pairs are created in equilibrium with radiation: $e^- + e^+ \leftrightarrow 2\gamma$.

Under equilibrium conditions, we have for the chemical potentials:

$$\mu_{e^-} + \mu_{e^+} = \mu_\gamma = 0. \tag{6.9.2}$$

The electron and positron number densities, $n_\pm(p)$, are given by Fermi distributions

$$n_\pm(p) = \frac{2}{(2\pi)^3} \frac{1}{e^{(E(p) \pm \mu)/kT} + 1}; \tag{6.9.3}$$

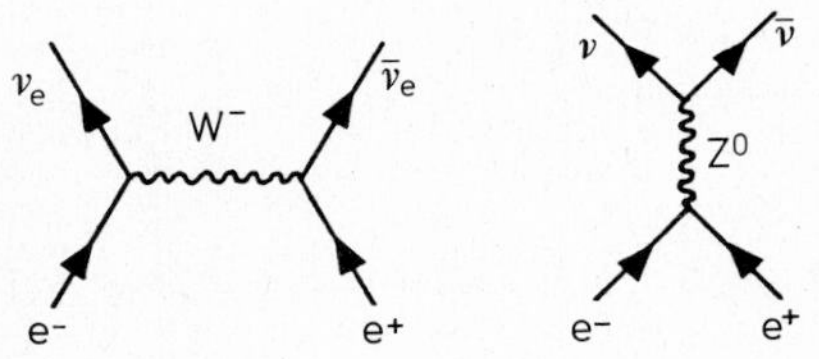


**Fig. 6.23.** Feynman diagrams for the process $e^- + e^+ \to \nu + \bar{\nu}$

the chemical potential $\mu$ is related to the lepton number density. Charged and neutral currents contribute to the pair annihilation (see Fig. 6.23).

The reaction rate for the production of a $\nu_e - \bar{\nu}_e$ pair is easily found to be ($c = \hbar = 1$):

$$v \cdot \sigma = \frac{G_F^2}{12\pi} \frac{1}{E_- \cdot E_+} \{(C_V^2 + C_A^2)[m^4 + 3m^2 p_- \cdot p_+ + 2(p_- \cdot p_+)^2] + 3(C_V^2 - C_A^2)[m^4 + m^2 p_- \cdot p_+]\}, \tag{6.9.4}$$

where in the standard model

$$C_V = \tfrac{1}{2} + 2\sin^2\theta_w, \quad C_A = \tfrac{1}{2}. \tag{6.9.5}$$

The rate of energy loss per unit volume is

$$\varepsilon_{\text{pair}} = \int (E_- + E_+)\, \sigma \cdot v\, n_-(p_-)\, n_+(p_+)\, d^3p_-\, d^3p_+ . \tag{6.9.6}$$

Unfortunately, this integral cannot be done analytically, except in limiting regions.

In the nonrelativistic case we get from (6.9.4)

$$v \cdot \sigma \simeq \frac{G_F^2 m^2}{\pi} C_V^2 \tag{6.9.7}$$

and thus

$$\varepsilon_{\text{pair}}^{\text{NR}} = C_V^2\, 2m \frac{G_F^2 m^2}{\pi} n_+ n_- . \tag{6.9.8}$$

In the nondegenerate limit we have

$$n_- n_+ = [n^{(0)}]^2$$

with

$$n^{(0)} = \frac{1}{\pi^2} \int_0^\infty dp\, p^2\, e^{-\sqrt{p^2+m^2}/kT} . \tag{6.9.9}$$

In the nonrelativistic limit this is

$$n^{(0)} = \frac{1}{\pi^2} (m\,kT)^{3/2} \sqrt{\frac{\pi}{2}}\, e^{-m/kT} .$$

Hence we find in the nonrelativistic and nondegenerate limit

$$\varepsilon_{\text{pair}}^{\text{NR, ND}} = C_V^2 \frac{G_F^2}{\pi^4} m^9 \left(\frac{kT}{m}\right)^3 e^{-2m/kT} = C_V^2 (4.9 \times 10^{18}\ \text{erg cm}^{-3}\,\text{s}^{-1})\, T_9^3 \exp\left(-\frac{11.86}{T_9}\right), \tag{6.9.10}$$

($T_9 = T/10^9$ K). In the extreme relativistic region we find from (6.9.4), after averaging over the angles,

$$\sigma \cdot v = (C_V^2 + C_A^2) \frac{8}{3} \frac{G_F^2}{12\pi} E_+ E_- . \tag{6.9.11}$$

Inserting this into (6.9.6) gives

$$\varepsilon_{\text{pair}}^{\text{ER}} = \frac{2}{9} (C_V^2 + C_A^2) \frac{G_F^2}{\pi^5} m^9 \left(\frac{kT}{m}\right)^9 F\left(\frac{\mu}{kT}\right), \tag{6.9.12}$$

where

$$F(y) = \int_0^\infty \frac{u^4\, du}{1+e^{u-y}} \int_0^\infty \frac{v^3\, dv}{1+e^{v+y}} + \int_0^\infty \frac{u^3\, du}{1+e^{u-y}} \int_0^\infty \frac{v^4\, dv}{1+e^{v+y}} . \tag{6.9.13}$$

In the nondegenerate case ($\mu/kT \simeq 0$) we have

$$F(0) = 2\,(3!)\,(4!) \sum_{n=1}^{\infty} \frac{(-1)^{n+1}}{n^5} \sum_{m=1}^{\infty} \frac{(-1)^{m+1}}{m^4} = 265.1 .$$

Hence

$$\varepsilon_{\text{pair}}^{\text{ER, ND}} = 58.9\, (C_V^2 + C_A^2) \frac{G_F^2}{\pi^5} m^9 \left(\frac{kT}{m}\right)^9 \tag{6.9.14}$$

$$= \frac{C_V^2 + C_A^2}{2} \cdot (4.1 \times 10^{24}\ \text{erg cm}^{-3}\,\text{s}^{-1})\, T_{10}^9 . \tag{6.9.15}$$

For comparison note that the average energy loss per unit volume of the sun is $\bar{\varepsilon}_\odot = 2.8$ erg cm$^{-3}$ s$^{-1}$.

If we include the $\mu$- and $\tau$-neutrinos, then we have to replace in (6.9.12), (6.9.14), and (6.9.15)

$$\tfrac{1}{2}(C_V^2 + C_A^2) \rightarrow \tfrac{1}{2}(C_V^2 + C_A^2) + [(C_V - 1)^2 + C_A^2] . \tag{6.9.16}$$

This factor, which multiplies the old $V-A$ result, is numerically equal to 0.84 for $\sin^2\theta_{\text{w}} = 0.23$.

It is instructive to compare the expressions (6.9.10) and (6.9.15) with the radiation energy density

$$aT^4 = 7.6 \times 10^{21}\, T_9^4\ \text{erg/cm}^3 ,$$

which dominates the thermal energy density for advanced stages of massive stars ($T_9 \gtrsim 1$). We obtain, for example,

$$\frac{aT^4}{\varepsilon_{\text{pair}}} = \begin{cases} 10\ \text{yr} & \text{for } T_9 = 1 \\ 10\ \text{min} & \text{for } T_9 = 5 . \end{cases}$$

---

The neutrino luminosity reaches $10^{15}\, L_\odot$ for a $15\, M_\odot$ star before the star becomes unstable. This is at least $10^4$ times the optical luminosity of a whole galaxy. This strong neutrino radiation accelerates the late phases very much. As a consequence, a lot of unburned material remains outside the central region. The star tends to become centrally condensed with the formation of strongly degenerate cores. (Without neutrino radiation, the mass of the central core would be larger and degeneracy would be reduced.)

Quasi hydrostatic evolution continues until the degenerate iron core, supported mainly by relativistic electron pressure, approaches the Chandrasekhar mass limit[9]

$$M_{Ch} = 5.76 \langle Y_e^2 \rangle M_\odot . \tag{6.9.17}$$

Once Si-burning starts, this core convergence happens in the course of a few days. The core is initially marginally stable and finally becomes unstable through photodisintegration of the iron-peak elements to $\alpha$-particles. The nuclear disintegration of iron to helium costs a lot of internal energy and reduces the adiabatic index $\Gamma_1$ below the critical value 4/3 (see the exercises below). The region where $\Gamma_1$ drops below 4/3 is shown in Fig. 6.21. We see from this figure that the cores of massive stars become unstable once the central density approaches $10^{10}$ g cm$^{-3}$.

For stars in the lower part of the considered mass range, electron capture, which also reduces pressure support within the core, is a further destabilizing factor. General relativity also has a destabilizing effect, because it increases the critical value $\Gamma_1$ above $\frac{4}{3}$ by $O(GM/Rc^2)$, where $M$ and $R$ are the mass and radius of the core (see the exercise on p. 293).

Although there are differences in detail, all the models computed up to this point give similar results. Thus, the dynamic collapse of the iron core is unavoidable. This phase will be discussed next.

---

**Exercises:**

1. Consider a star of uniform density. Show that the equilibrium is only stable if

$$\Gamma = -\frac{V}{P}\frac{\partial P}{\partial V} > \frac{4}{3},$$

where $V$ is the volume of the star and $P$ the pressure.

---

[9] The average number of electrons per nucleon, $Y_e$, becomes $Y_e \simeq 0.41 - 0.43$, corresponding to $M_{Ch} \simeq 1\,M_\odot$. Note, however, that finite temperature effects increase the critical mass, because the electron pressure is larger,

$$p = \frac{1}{4} n_e \mu_e \left(1 + \frac{2}{3}\frac{\pi^2 T^2}{\mu_e^2}\right). \tag{6.9.18}$$

Here $\mu_e$ is the chemical potential of the electrons

$$\mu_e = 11.1\,(\varrho_{10} Y_e)^{1/3}\ \mathrm{MeV}. \tag{6.9.19}$$

Verify these formulae as an exercise!

2. Generalize this and consider adiabatic radial perturbations of an inhomogeneous star. Compute the change of the total energy up to second order. Show that the first order variation vanishes due to the equilibrium conditions and that the second order is given by

$$\delta^2 W = \frac{1}{2} \int \left[ P\Gamma_1 \left( \frac{d\delta V}{dV} \right)^2 + \frac{4}{3} \frac{1}{V} \frac{dP}{dV} (\delta V)^2 \right] dV,$$

where

$$\Gamma_1 = \left( \frac{\partial \ln P}{\partial \ln \varrho} \right)_{\mathrm{ad}} .$$

Stability requires $\delta^2 W > 0$ for every Lagrangian variation $\delta V$ of $V = 4\pi r^3/3$. In particular, this must be true for a homologous variation $\delta V \propto V$. With a partial integration, we obtain the *necessary* condition

$$\int (\Gamma_1 - \tfrac{4}{3}) P\, dV > 0 . \qquad (6.9.20)$$

3. Estimate $\Gamma_1$ in the neighborhood of the iron-helium transition, using the value $\Delta\varepsilon = 2 \times 10^{18}$ erg/g for the transition energy, and show that $\Gamma_1$ becomes smaller than 4/3.
4. Study the equilibrium conditions for

$$\gamma + {}^{56}_{26}\mathrm{Fe} \rightleftarrows 13\,\alpha + 4\,\mathrm{n} ,$$

treating the nuclei and nucleons as ideal nondegenerate gases and approximating the nuclear partition sums by the ground-state contribution. Compare the result with Fig. 6.21.

---

### 6.9.3 The Physics of Stellar Collapse

*A. Dynamics of the Collapse*

For a number of reasons, which will be discussed in the next section, the stellar core collapses almost adiabatically and there is little change of the composition of matter until the very center of the star reaches nuclear density, where the complex nuclei finally break up. For these reasons, the hydrodynamics of the collapse is qualitatively quite simple. The pressure is dominated by ultrarelativistic electrons and hence the adiabatic index is close to $\frac{4}{3}$. But then the inner part of the core, given approximately by the Chandrasekhar mass, collapses homologously (velocity proportional to radius). This is shown in the following exercise.

---

**Exercise:** Consider a spherically symmetric collapse for a polytropic equation of state,

$$P = K\varrho^{\gamma} . \qquad (6.9.21)$$

Show that there are homologous solutions exactly for $\gamma=\frac{4}{3}$ and set up the differential equations for this case. Solve these (for one of them this has to be done numerically) and determine the mass $M_{\mathrm{hc}}$ of the homologously collapsing core, if for an initial static configuration the value $K$ is suddenly reduced (e.g., by electron capture).

*Solution:* We adopt a Lagrangian description, in which time $t$ and interior mass $m$ (total mass interior to a spherical shell) are regarded as independent variables. The instantaneous radius $r$ interior to which lies the mass $m$ is given implicitly by the equation

$$m=\int_0^r 4\pi r'^2 \varrho(r',t)\, dr' ,$$

where $\varrho(r,t)$ is the density at radial distance $r$ and time $t$. In differential form we have

$$\frac{\partial r}{\partial m}=\frac{1}{4\pi r^2 \varrho} . \tag{6.9.22}$$

Newton's equation, when only gravity and pressure gradient forces are present, is

$$\varrho \frac{\partial^2 r}{\partial t^2}=-\frac{1}{\varrho}\left(\frac{\partial P}{\partial r}\right)_t-\frac{\varrho G m}{r^2}$$

or, with (6.9.22),

$$\frac{\partial^2 r}{\partial t^2}=-4\pi r^2 \frac{\partial P}{\partial m}-\frac{G m}{r^2} . \tag{6.9.23}$$

In addition, we have the polytropic equation of state (6.9.21). Now we try a separation ansatz

$$r(t)=f(t)\, g(m) . \tag{6.9.24}$$

The normalization of $f(t)$ will be fixed later. Eq. (6.9.22) implies

$$\frac{dg}{dm}=\frac{1}{4\pi g^2 (f^3 \varrho)} , \tag{6.9.25}$$

which requires that $f^3\varrho$ is time-independent. Inserting (6.9.24) into the equation of motion (6.9.23) gives

$$f^2 \ddot{f} g=-4\pi g^2 \frac{\partial (f^4 P)}{\partial m}-\frac{G m}{g^2} .$$

We multiply this equation by $g^{-2}(\partial/\partial g)\, g^2$ and use (6.9.25) [note that $g(m)$ is a monotonic function]. This gives

$$3 f^2 \ddot{f}=-\frac{1}{g^2}\frac{\partial}{\partial g}\left(\frac{g^2}{f^3 \varrho}\frac{\partial (f^4 P)}{\partial g}\right)-4\pi G f^3 \varrho . \tag{6.9.26}$$

One sees immediately from this equation that the ansatz (6.9.24) is only possible for $\gamma = \frac{4}{3}$. For this case, we set

$$\varrho(g, t) = \varrho_c(t)\, \theta^3(g) . \tag{6.9.27}$$

Then $\varrho_c(t)\, f^3(t)$ is independent of $t$, and (6.9.26) gives

$$\begin{aligned} 3 f^2 \ddot{f} \left[\frac{1}{4\pi G \varrho_c f^3}\right] &= -\left[\frac{1}{\pi G} K \varrho_c^{-2/3} f^{-2}\right] \frac{1}{g^2} \frac{d}{dg}\left(g^2 \frac{d\theta}{dg}\right) - \theta^3 \\ &= -\lambda = \text{const} . \end{aligned} \tag{6.9.28}$$

Now we normalize $f$ such that the square bracket on the right hand side of (6.9.28) is equal to one, i.e.

$$f(t) = \left(\frac{K}{\pi G}\right)^{1/2} \varrho_c^{-1/3}(t) . \tag{6.9.29}$$

Then (6.9.28) leads to the following equations for $f$ and $\theta$:

$$\frac{3}{4}\left(\frac{\pi G}{K^3}\right)^{1/2} f^2 \ddot{f} = -\lambda \tag{6.9.30 a}$$

$$\frac{1}{g^2} \frac{d}{dg}\left(g^2 \frac{d\theta}{dg}\right) + \theta^3 = \lambda . \tag{6.9.30 b}$$

The boundary conditions for $\theta$ are obviously

$$\theta(0) = 1 , \qquad \theta'(0) = 0 . \tag{6.9.31}$$

If $R(t) = f(t)\, g(M)$ is the radius of the star, then the total mass $M$ is given by

$$M = \int_0^R 4\pi r^2 \varrho\, dr = \int_0^{g_s} 4\pi g^2 f^3 \varrho\, dg ,$$

where $g_s$ is the surface value of $g$, determined by $\theta(g_s) = 0$ [see (6.9.27)]. Hence the average density $\bar{\varrho}$ is given by [using (6.9.27)]:

$$\frac{\bar{\varrho}}{\varrho_c} = \frac{M}{\frac{4\pi}{3}(f g_s^3)\, \varrho_c} = \frac{1}{\frac{4\pi}{3} g_s^3} \int_0^{g_s} 4\pi g^2 \theta^3(g)\, dg$$

or, with (6.9.30 b),

$$\frac{\bar{\varrho}}{\varrho_c} = \lambda - \frac{3}{g_s} \frac{d\theta}{dg}(g_s) . \tag{6.9.32}$$

Let us consider the limiting case when the surface falls freely. From (6.9.23) we find that this is the case when

$$\frac{\ddot{f}}{f} = -\frac{4\pi}{3} G \bar{\varrho} .$$

But the left hand side is by (6.9.30a) and (6.9.29) equal to $-4\pi\lambda/3$. Thus, in this limiting case, we have $\bar{\varrho}/\varrho_c = \lambda$. Eq. (6.9.32) shows that the corresponding $\lambda$, which we denote by $\lambda_m$, satisfies

$$\theta(g_s) = 0, \quad \theta'(g_s) = 0 \quad \text{for} \quad \lambda_m. \tag{6.9.33}$$

Physical values are in the interval $[0, \lambda_m]$. Numerically, one finds with (6.9.30b) and (6.9.31)

$$\lambda_m = 0.00654. \tag{6.9.34}$$

The corresponding domains of variation for $g_s$ and $\bar{\varrho}/\varrho_c$ [obtained from (6.9.32)] are with increasing $\lambda$:

$$\begin{aligned} g_s&: \quad 6.897 \nearrow 9.887 \\ \frac{\bar{\varrho}}{\varrho_c}&: \quad 0.01846 \searrow 0.006544. \end{aligned} \tag{6.9.35}$$

The total mass is, using (6.9.29),

$$M = \frac{4\pi}{3} f^3 g_s^3 \bar{\varrho} = \frac{4\pi}{3} g_s^3 \frac{\bar{\varrho}}{\varrho_c} \left(\frac{K}{\pi G}\right)^{3/2} \tag{6.9.36}$$

and thus varies (for a given $K$) only by a factor 1.045. If one begins, for example, with a static core, then $\lambda = 0$. For a given core mass $M_0$ the value $K_0$ of $K$ is determined by (6.9.35) and (6.9.36). If for some reason (e.g., electron capture) $K_0$ is reduced to $K$, an inner part of the core will collapse homologously; its mass is

$$M_{hc} = 1.045 \left(\frac{K}{K_0}\right)^{3/2} M_0. \tag{6.9.37}$$

This shows that $M_{hc}$ is very close to the Chandrasekhar mass belonging to $K$. In particular $M_{hc} = M_0$ if $K$ is decreased by only about 3%.

The time dependence $f(t)$ of the collapse is easily obtained from (6.9.30a):

$$f(t) = (6\lambda)^{1/3} \left(\frac{K^3}{\pi G}\right)^{1/6} t^{2/3}, \tag{6.9.38}$$

if $t$ is chosen such that $f(0) = 0$.

(I thank D. Vaucher for the numerical computations; for a different treatment, see [96].)

---

The infall velocity of the homologous core is subsonic. Since no signals can travel upstream from the subsonic region to the material above the sonic point (infall velocity equal to sound velocity) to cause it to adjust to the homologous core, one expects that matter outside the sonic point moves essentially in free fall. This is confirmed by numerical calculations.

Once the density exceeds nuclear matter density, the nuclei are broken up and matter suddenly becomes very stiff ($\Gamma_1 \simeq \frac{5}{3}$). Because the core is homologous, this happens practically throughout it at once. As a result of the large infall kinetic energy, the collapsing core overshoots nuclear density, then bounces back, and produces a strong shock wave at the sonic point near $M_{\rm Ch}$. The strength of this shock, its propagation and damping through the outer parts of the core, and most especially whether such a shock wave is capable of producing mass ejection in the overlying mantle and envelope, will be discussed in Sect. 6.9.4. There we will also present the results of recent numerical calculations which give a strong explosion of the envelope for certain stars.

### *B. Neutrino Trapping*

At the beginning of the core collapse, the electron chemical potential $\mu_e$ is $\sim 6$ MeV. During the collapse, it increases with increasing density, thereby increasing the rate of electron capture reactions. Neutrinos produced in these processes escape until the core becomes sufficiently opaque that the time for a neutrino to diffuse out of the core becomes large compared with the dynamical time scale. Let us investigate when this happens.

Various processes contribute to the neutrino opacity. Scattering from free neutrons and from nuclei through neutral current interactions are the dominant sources.

The effective neutrino-hadron interaction in the standard model is (see, e.g., [94]):

$$\mathcal{L} = -\frac{G_{\rm F}}{\sqrt{2}}\,\bar{\nu}\,\gamma^\mu\,(1-\gamma_5)\,\nu\,J^Z_\mu\,, \tag{6.9.39}$$

where the neutral hadron current $J^Z_\mu$ has the following form

$$J^Z_\mu = (V^3_\mu - A^3_\mu) - 2\sin^2\theta_{\rm W}\,J^{\rm elm}_\mu\,. \tag{6.9.40}$$

For nucleons $V^3_\mu$ and $A^3_\mu$ are the isospin partners of the charged vector and axial currents which govern $\beta$-decay.

The axial term is not important for heavy nuclei because it gives spin-dependent contributions which are largely averaged out.

---

**Exercise 1:** Compute the differential cross section for MeV neutrinos on heavy nuclei.

*Solution:* For simplicity, we take $\sin^2\theta_{\rm W} = 0.25$, a value which is not far from the experimental one ($\simeq 0.23$). Neutrinos can then scatter only by neutrons, not by protons, as is obvious from (6.9.40), if we neglect the axial contribution. The current matrix element is

$$\langle f \,|\, J^Z_\mu \,|\, i\rangle = \tfrac{1}{2}\,(P_f + P_i)_\mu\, N\, F(q^2)\,, \tag{6.9.41}$$

where $N$ is the number of neutrons and $F(q^2)$ is the form factor of the neutrons, normalized such that $F(0) = 1$. The momentum transfer is sufficiently small that we have coherence: $F(q^2) \simeq F(0) = 1$. From this point, the calculation is straightforward and short. One finds ($\hbar = c = 1$):

$$\frac{d\sigma}{d\Omega} = \frac{G_F^2}{(2\pi)^2}\,\frac{N^2}{4}\,(1+\cos\theta)\,\nu^2\,,$$

where $\nu$ is the neutrino energy. We write this result as follows. Let

$$\sigma_0 = \frac{4\,G_F^2\, m_e^2\, \hbar^2}{\pi\, c^2} = 1.7\times 10^{-44}\ \mathrm{cm}^2. \tag{6.9.42}$$

Then

$$\frac{d\sigma^{\nu A}}{d\Omega} = \frac{1}{4\pi}\,\frac{N^2}{16}\,\sigma_0 \left(\frac{\nu}{m_e c^2}\right)^2 (1+\cos\theta)\,. \tag{6.9.43}$$

The corresponding cross section $\Sigma$ for momentum transfer is

$$\Sigma_{\nu A} = \int (1-\cos\theta)\,\frac{d\sigma}{d\Omega}\,d\Omega = \frac{1}{24}\,N^2\,\sigma_0 \left(\frac{\nu}{m_e c^2}\right)^2. \tag{6.9.44}$$

**Exercise 2:** Calculate the same quantity for neutrino-neutron scattering. (Here the axial contribution is important!) The answer is

$$\Sigma_{\nu n} = \frac{1}{24}\,(1 + 5\,g_A^2)\,\sigma_0 \left(\frac{\nu}{m_e c^2}\right)^2, \tag{6.9.45}$$

where $g_A$ is the axial coupling constant ($g_A \simeq -1.25$).

From the cross sections (6.9.44), (6.9.45) we obtain for the mean free path

$$\lambda_\nu = \frac{m_n}{\varrho\, \Sigma_{\nu n}} \left(X_n + \frac{\bar{N}^2}{A\,(1+5\,g_A^2)}\,X_A\right)^{-1}. \tag{6.9.46}$$

Here $X_A$ and $X_n$ are the mass fractions of heavy nuclei and "free" neutrons and $\bar{N}$ is the average number of neutrons in each heavy nucleus.

Numerically we obtain

$$\lambda_\nu \simeq (1.0\times 10^6\ \mathrm{cm})\,\varrho_{12}^{-1} \left(\frac{\nu}{10\ \mathrm{MeV}}\right)^{-2} \left(X_n + \frac{\bar{N}^2}{A\,(1+5\,g_A^2)}\,X_A\right)^{-1}. \tag{6.9.47}$$

If we take $X_A \simeq 1$, $\varrho_{12} = 1$, $\nu = 10$ MeV, we get for typical values of $A$ and $N$: $\lambda_\nu \simeq 1$ km.

The radius at which $\varrho_{12} = 1$ is much larger, namely about 40 km, and a typical dynamical time scale is $\simeq 1$ ms, because the free fall time is

$$\tau_{ff} = \left(\frac{8\pi}{3} G\varrho\right)^{-1/2} = (1.3 \times 10^{-3}\,\text{s})\, \varrho_{12}^{-1/2}. \tag{6.9.48}$$

The distance $\Lambda$ over which neutrinos diffuse in this time is[10] for $\varrho_{12} = 1$

$$\Lambda = (\tfrac{1}{3}\lambda_\nu c\, \tau_{ff})^{1/2} \simeq 10\,\text{km}. \tag{6.9.49}$$

From this, we conclude that for $\varrho > 10^{12}\,\text{g cm}^{-3}$ neutrinos will be *trapped* in the star for the duration of the collapse. The density of the neutrinos increases very fast and they become degenerate. The Pauli principle forbids further electron capture and matter stays close to beta equilibrium. For this reason, little entropy is generated (the electromagnetic and strong interactions are always in equlilibrium) and the collapse is essentially adiabatic, shortly after neutrino trapping, with a fixed lepton fraction $Y_l$. This is confirmed by detailed numerical calculations which will be discussed later. It turns out that the entropy per nucleon starts from a value $s \simeq 1.0\, k_B$ (make a rough ideal gas calculation) and does not increase much. This is the reason why the nuclei are not broken up. Qualitatively, this can be seen as follows.

The entropy per nucleon of a perfect Fermi gas is $s = k_B \dfrac{\pi^2}{2}(T/T_F)$, where $T_F$ is the Fermi temperature. For nuclei $T_F \simeq 35$ MeV and thus $s = 1.5\, k_B$ corresponds to a temperature of only about 10 MeV. At such low temperatures, thermal dissociation of the nuclei does not take place until the nuclei fuse into one stellar sized nucleus above nuclear density.

---

### 6.9.4 Numerical Studies

Many authors have studied the gravitational collapse in detail, in order to find out whether $\sim 1\%$ of the gravitational energy released by the collapsing core can be transmitted to the weakly bound matter further out such that the outer part is blown off. All workers in this field agree that a strong shock forms when the inner core bounces at about nuclear density. Its detailed properties, propagation and damping are still controversial and a matter of intense research. The current status of model calculations has recently been presented by *W. Hillebrandt* [97]. Here is a short summary.

---

[10] Elementary kinetic theory gives for the diffusion constant $D = \frac{1}{3} c\, \lambda_\nu$. The fundamental mode of the diffusion equation for a homogeneous sphere of radius $R$ is proportional to $\exp\left(-\dfrac{\pi^2 D}{R^2} t\right)$ and thus the diffusion time is $\tau_{diff} \simeq \dfrac{R^2}{\pi^2 D}$.

Most important and crucial for detailed numerical studies is the equation of state. This is because the deviation of the adiabatic index from 4/3 determines the size of the homologously collapsing part of the core and thus where the shock front forms and how much energy is put into the shock. It would be out of place to discuss these complicated matters here and we must refer to [97] (and references therein).

The amount of mass within the homologous core is also a function of $Y_e$. We have seen that the value of this quantity depends on $e^-$-capture processes at densities from a few times $10^{10}$ up to $10^{12}$ g cm$^{-3}$. At such densities and moderate entropies [$(1-2)\,k_B$/nucleon], the capture on free protons is the dominant contribution, because shell-blocking effects reduce the capture rates of the most abundant nuclei. This is fortunate because the elementary processes $e^- + p \leftrightarrow n + \nu_e$ are easy to calculate.

It turns out that $Y_e$ is typically decreased from 0.44 to about 0.35.

Of less importance, at least for the collapse phase, is an adequate treatment of the neutrino transport. This process plays some role in the damping of the outgoing shock wave. (The propagating shock gains energy from neutrinos diffusing out of the core up to the shock front, but loses energy via neutrino pair production and electron capture.) The main damping is, however, due to photodisintegration of heavy nuclei as the shock propagates through the outer part of the stellar core. A typical shock energy of about $10^{52}$ erg would be used up by about 0.5 $M_\odot$. Some additional energy is, however, added to the shock from the kinetic energy of the infalling matter. On the other hand, the mass of the homologous core will be typically 0.8 $M_\odot$. For a 1.5 $M_\odot$ core, there is thus a danger that the dissociation of nuclei causes so much damping that the shock finally cannot overcome the ram pressure of the infalling matter and changes into a standing accretion shock.

After these general remarks, we turn to a brief description of recent results [97].

Numerical simulations have led to the present general agreement that models with Fe-Ni cores more massive than about 1.5 $M_\odot$ do not explode. Such massive cores are expected to form at the end of thermonuclear burning of stars with main sequence masses above 12 to 15 $M_\odot$. [It is, however, not excluded that the predicted core masses and (or) entropies are too large.]

These "negative" results led Hillebrandt and others to perform collapse computations starting from a 10 $M_\odot$ star model that has been evolved numerically from the main sequence to the onset of core collapse by Weaver and Woosley. This model consists of a 1.38 $M_\odot$ "Fe"-core in nuclear statistical equilibrium surrounded by a Ne-O-shell of 0.13 $M_\odot$ and a He-H-envelope expanding with $100-300$ km s$^{-1}$ out to $10^{16}$ cm when the core began to collapse. This shell was ejected through several Ne-flashes under semidegenerate conditions. Further properties of the model at the onset of core-collapse are: central density $\varrho_c = 5 \times 10^9\,\mathrm{g\,cm^{-3}}$, central temperature $T_c = 7.6 \times 10^9\,\mathrm{K}$, central electron concentration $Y_e^c = 0.43$ and an entropy per nucleon of $0.8\,k_B$. As the core mass of this model was larger than the Chandrasekhar mass, it was gravitationally unstable.

An additional important feature of the model is an enormous drop in density by about 10 orders of magnitude at the edge of the core over a tenth of a solar mass.

In Hillebrandt's computation [98], the collapse phase of this model turned out to be quite similar to that obtained for more massive stellar cores. The lepton concentration at neutrino trapping was found to be 0.4, slightly larger than in previous models, corresponding to a Chandrasekhar mass of about $0.9\,M_\odot$ which should be approximately equal to the homologously collapsing part of the core. Numerically, the shock formed at $0.8\,M_\odot$ when the core bounced at a density of about $4 \times 10^{14}\,\mathrm{g\,cm^{-3}}$, and the energy originally put into the shock was about $8 \times 10^{51}$ erg (see Fig. 6.24). This energy is sufficient to dissociate $0.5\,M_\odot$ of heavy elements into nucleons. Indeed, the shock was able to propagate out to $1.3\,M_\odot$, but was significantly damped on its way. At this point, the strong density decrease speeded up the shock and matter reached escape velocity approximately 30 ms after the core bounce. As

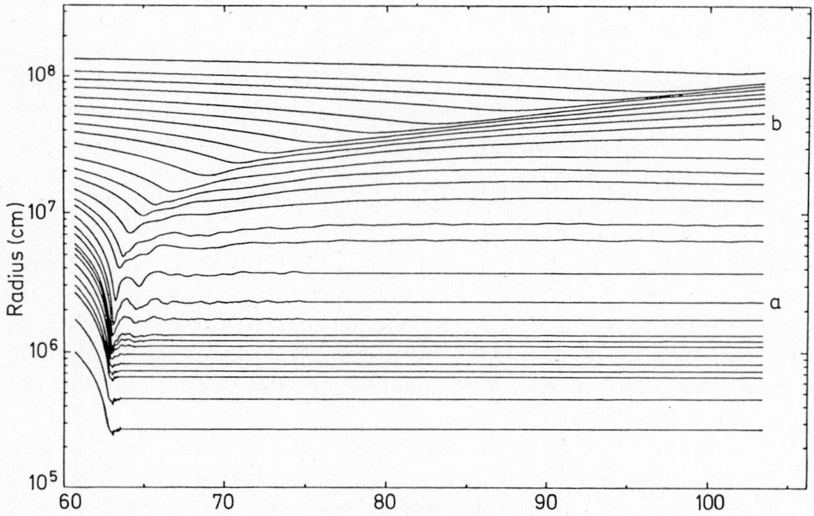


**Fig. 6.24.** Radius versus time for selected mass zones of a $10\,M_\odot$ star (from [98]). Only the inner $1.5\,M_\odot$ are plotted. The time is measured in ms from the beginning of the computation when the central density was $2.7 \times 10^{10}\,\mathrm{g\,cm^{-3}}$. The curve labeled *a* gives the boundary of the unshocked inner core of $0.75\,M_\odot$, the curve labeled *b* the boundary of the original Fe-core ($M \simeq 1.36\,M_\odot$)

a result, $0.06\,M_\odot$ from the core were ejected with velocities of about 25,000 km s$^{-1}$, leaving behind a neutron star of $1.44\,M_\odot$. The total energy in the explosion was found to be $5 \times 10^{50}$ erg, which is a bit low for a type II supernova. Since the He-H-envelope is very extended, this should, however, be enough to induce a very bright event, which may even have some similarity with the Crab supernova.

This is the first successful core-collapse supernova model which is based on a realistic stellar model and realistic microphysics input data. It might, however, not be characteristic for the average type II event, because it does not lead to the heavy element enrichment which is needed to explain the heavy element content of the galaxy. Further information is contained in Figs. 6.25 and 6.26.

The previous discussion shows that the final outcome of core collapse, namely whether a neutron star forms in a type II supernova explosion or the collapse continues to a black hole, depends strongly on the initial pre-supernova stellar structure.

It remains to be seen which mass range of models can lead to an explosion. More realistic collapse models should probably also include rotation (and magnetic fields). First steps in this direction have already been made [112].

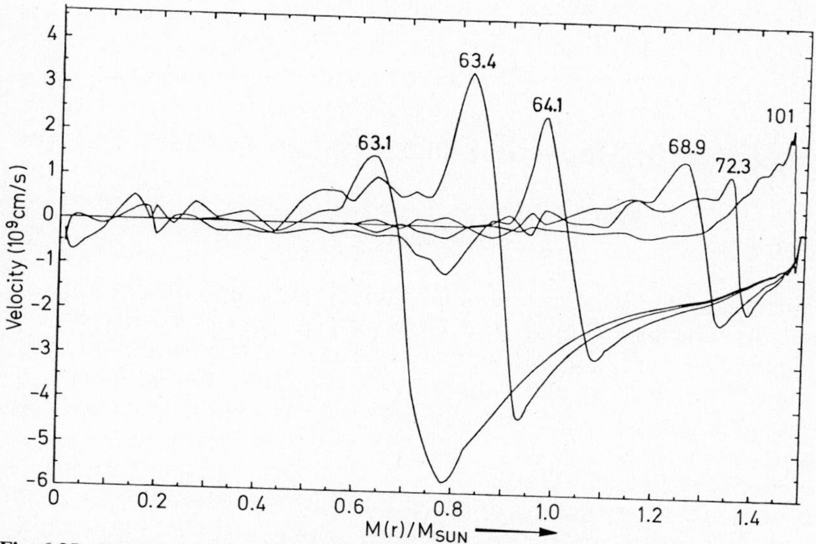


**Fig. 6.25.** Snap-shots of velocity profiles after core-bounce (from [98]). Inward velocities are negative. The curves are labeled with time in ms as in Fig. 6.24. One sees the *damping of the shock due to nuclear* dissociations and the speeding up near the edge of the core. The outermost layers of the original core reach escape velocity approximately 30 ms after core bounce

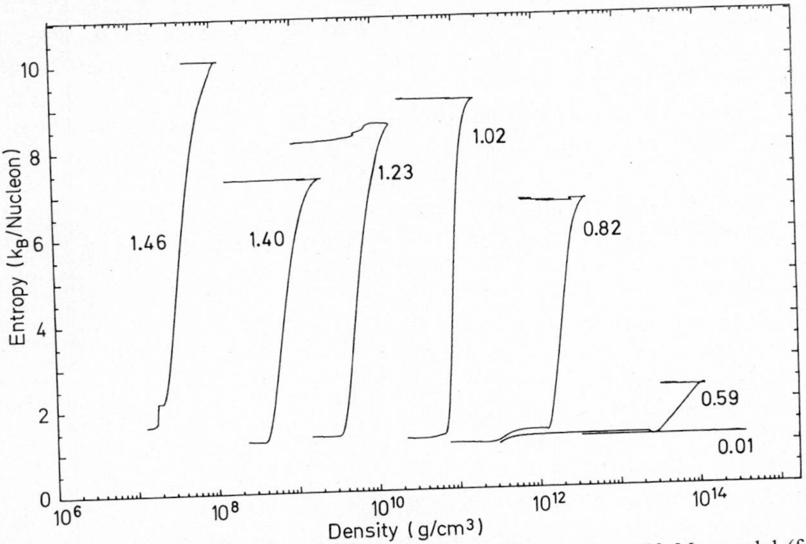


**Fig. 6.26.** Entropy versus density for selected mass zones of the 10 $M_\odot$ model (from [98]). Curves are labeled with the mass $M(r)$ in units of $M_\odot$. The steep increase in entropy reflects the passage of the shock front. The time evolution of the mass zones begins at the lower left end of each curve 60 ms after the start of the computations. The decrease in maximum entropy with increasing mass shows the effect of nuclear dissociation. The decrease of entropy after maximum at 1.23 $M_\odot$ is caused by neutrino losses from this zone

## 6.10 Addendum 3: Magnetic Fields of Neutron Stars, Pulsars

### 6.10.1 Introduction

The high conductivity of the stellar interior ensures conservation of magnetic flux during collapse (see Exercise 1 below). For this reason, the magnetic field strength will increase quadratically with the decrease of the linear dimensions, and consequently magnetic fields as high as $10^{12}$ G are not unreasonable for neutron stars. We shall see that such high fields are required if the observed slowdown of pulsars is the result of electromagnetic radiation.

The most accurate method for determining the magnetic fields of neutron stars in cyclotron line spectroscopy which was first successfully applied by Trümper and coworkers [57] for the x-ray pular Her X-1. (x-ray pulsars will be discussed in Chap. 8.) In 1976, they discovered a line feature in the hard x-ray spectrum at 58 keV for this source (see Fig. 6.27). This line corresponds most probably to the transition energy

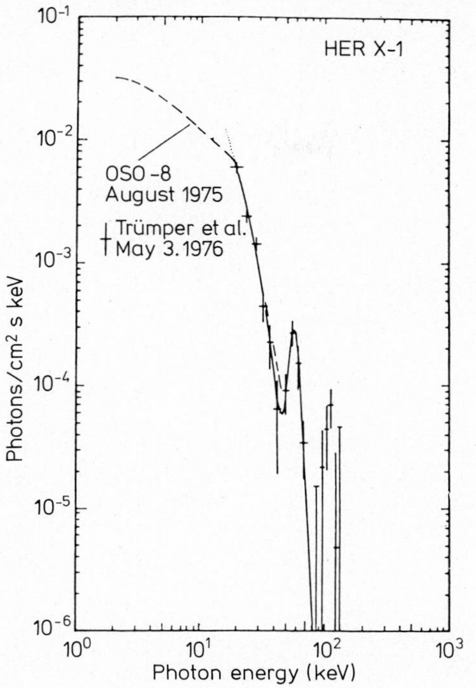


**Fig. 6.27.** The hard x-ray spectrum of Her X-1

$\hbar\,\omega_B$ between the ground state and the first excited Landau level. Depending on the interpretation of the spectral structure as an emission or an absorption line, the field strength must be 4 or 5 times $10^{12}$ G. These results have been confirmed with the HEAO-1 satellite and for two other sources cyclotron structures have been observed [99], which correspond again to magnetic fields of about $10^{12}$ G.

---

**Exercises:**

1 a) Use Ohm's law

$$\boldsymbol{J} = \sigma\left(\boldsymbol{E} + \frac{1}{c}\,\boldsymbol{v} \times \boldsymbol{B}\right), \tag{6.10.1}$$

where $\boldsymbol{v}$ is the (center of mass) fluid velocity of the plasma, and the Maxwell equations

$$\operatorname{div}\boldsymbol{B} = 0\,, \qquad \operatorname{curl}\boldsymbol{E} + \frac{1}{c}\,\partial_t \boldsymbol{B} = 0 \tag{6.10.2}$$

$$\operatorname{curl}\boldsymbol{B} = \frac{4\pi}{c}\,\boldsymbol{J} \qquad \text{(neglect displacement current)} \tag{6.10.3}$$

to derive the following equations

$$\partial_t \boldsymbol{B} - \operatorname{curl}(\boldsymbol{v} \times \boldsymbol{B}) = -c \operatorname{curl}(\boldsymbol{J}/\sigma) \tag{6.10.4}$$

$$= -\frac{c^2}{4\pi\sigma} \Delta \boldsymbol{B} \quad (\text{if } \sigma \text{ is constant}) . \tag{6.10.5}$$

1 b) Estimate from Eq. (6.10.5) the diffusion time scale for decay of the magnetic field of a neutron star, using calculated values for $\sigma$ from [61].

1 c) Make use of the following mathematical theorem (for a derivation see Exercise 2): For any vector field $\boldsymbol{u}$ we have

$$\frac{d}{dt} \int_{S_t} \boldsymbol{u} \cdot d\boldsymbol{S} = \int_{S_t} [\partial_t \boldsymbol{u} - \operatorname{curl}(\boldsymbol{v} \times \boldsymbol{u}) + \boldsymbol{v} \operatorname{div} \boldsymbol{u}] \cdot d\boldsymbol{S} , \tag{6.10.6}$$

where $S_t$ is a surface, bounded by a closed loop $\partial S_t$, which is moving with the fluid, to deduce that the magnetic flux $\phi(t)$ through $S_t$ changes according to

$$\frac{d\phi}{dt} = -c \int_{\partial S_t} \frac{1}{\sigma} \boldsymbol{J} \cdot d\boldsymbol{s} . \tag{6.10.7}$$

Thus in the limit $\sigma \to \infty$ the flux lines are frozen in.

1 d) Include the displacement current in (6.10.3) and generalize (6.10.5) to

$$\partial_t \boldsymbol{B} - \operatorname{curl}(\boldsymbol{v} \times \boldsymbol{B}) = \frac{1}{4\pi\sigma} (\partial_t^2 \boldsymbol{B} - c^2 \Delta \boldsymbol{B}) . \tag{6.10.8}$$

Write the left hand side as

$$\partial_t \boldsymbol{B} - \operatorname{curl}(\boldsymbol{v} \times \boldsymbol{B}) = D_t \boldsymbol{B} - (\underset{\sim}{\omega} + \underset{\sim}{\sigma} - \tfrac{2}{3} \theta \underset{\sim}{1}) \cdot \boldsymbol{B} , \tag{6.10.9}$$

where

$$D_t = \partial_t + \nabla_{\boldsymbol{v}} \tag{6.10.10}$$

$$\theta = \operatorname{div} \boldsymbol{v}$$

$$\sigma_{ik} = \tfrac{1}{2} (v_{i,k} + v_{k,i}) - \tfrac{2}{3} \delta_{ik} \theta$$

$$\omega_{ik} = \tfrac{1}{2} (v_{i,k} - v_{k,i}) . \tag{6.10.11}$$

For $\sigma \to \infty$ we obtain

$$D_t \boldsymbol{B} = (\underset{\sim}{\omega} + \underset{\sim}{\sigma} - \tfrac{2}{3} \theta \underset{\sim}{1}) \cdot \boldsymbol{B} . \tag{6.10.12}$$

However, in regions of plasma, where the field has a strong gradient, the right hand side of (6.10.8) cannot be neglected. This term may lead to topological changes in the magnetic field structure referred to as *magnetic field line reconnection.* At the interface between adjacent cells of a chaotic field the gradients are so strong that the frozen-in behavior of the field is destroyed. The phenomenon of rapid magnetic field dissipation (solar flares, etc.), often accompanied by bursts, is due to this mechanism. (For a detailed discussion we refer to [100].)

2) Let $C \subset M$ be a $k$-dimensional submanifold of $M$ and $\alpha$ be a (time-dependent) $k$-form on $M$ (with compact support). Let $\phi_t$ denote the flow of a (time-dependent) vector field $\boldsymbol{v}$. By the change-of-variable formula and the definition of the Lie derivative we have

$$\frac{d}{dt}\int_{\phi_t(C)} \alpha = \int_{\phi_t(C)} (\partial_t \alpha + L_v \alpha)\,. \qquad (6.10.13)$$

Derive from this Eq. (6.10.6) of Exercise 1.

*Solution of Exercise 2:* We consider $M = \mathbb{R}^3$ as a Riemannian manifold, and choose for $C$ a hypersurface $S$, $S_t = \phi_t(S)$. The two-form $\alpha$ can be represented as $\alpha = i_{\boldsymbol{u}}\eta$, where $\eta$ is the volume form of $\mathbb{R}^3$. It is easy to see that

$$\int_S \alpha = \int_S \boldsymbol{u} \cdot d\boldsymbol{S}\,; \qquad (6.10.14)$$

see Eq. (4.44) of Part I. Now

$$L_v \alpha = L_v i_{\boldsymbol{u}} \eta = i_v d i_{\boldsymbol{u}} \eta + d i_v i_{\boldsymbol{u}} \eta\,.$$

But $d i_{\boldsymbol{u}} \eta = (\operatorname{div} \boldsymbol{u})\,\eta$, $i_v i_{\boldsymbol{u}} \eta = (\boldsymbol{u} \times \boldsymbol{v})^\flat$ and thus $d i_v i_{\boldsymbol{u}} \eta = i_{\operatorname{curl}(\boldsymbol{u} \times \boldsymbol{v})}\,\eta$ [since for any vector field $\boldsymbol{w}$ we have $d(\boldsymbol{w})^\flat = i_{\operatorname{curl} \boldsymbol{w}}\,\eta$].

Thus

$$\partial_t \alpha + L_v \alpha = i_{\boldsymbol{w}}\,\eta \qquad (6.10.15)$$

with

$$\boldsymbol{w} = \partial_t \boldsymbol{u} + \boldsymbol{v} \operatorname{div} \boldsymbol{u} + \operatorname{curl}(\boldsymbol{u} \times \boldsymbol{v})\,. \qquad (6.10.16)$$

This proves Eq. (6.10.6).

3) Consider ideal relativistic magnetohydrodynamics. Define electric and magnetic one-forms by

$$E = i_u F\,, \qquad B = i_u {*F}\,, \qquad (6.10.17)$$

where $u$ is the four-velocity field.

a) Show that (6.10.17) implies

$$F = u^\flat \wedge E - *(u^\flat \wedge B)\,. \qquad (6.10.18)$$

The conductivity part of the four-current is equal to $\sigma E$ (why?). Thus for $\sigma \to \infty$ we have $i_u F = 0$ and hence

$$L_u F = i_u dF + d i_u F = 0\,. \qquad (6.10.19)$$

Furthermore,

$$*F = u^\flat \wedge B\,.$$

b) Write the homogeneous Maxwell equations $dF = 0$ as $\delta * F = 0$ and translate this to

$$(u^\alpha B^\beta - u^\beta B^\alpha)_{;\alpha} = 0\,. \qquad (6.10.20)$$

Conclude from this equation

$$\nabla_u B = \nabla_{B\#} u^\flat - \theta \cdot B - (B, a)\, u^\flat , \tag{6.10.21}$$

where $\theta = \operatorname{div} u$, $a = \nabla_u u$.
Equation (6.10.19) implies the flux conservation

$$\int_C F = \int_{\phi_t(C)} F \tag{6.10.22}$$

for a two-dimensional submanifold $C \subset \mathbb{R}^4$, if $\phi_t$ denotes the flow of $u$.

c) Show that the energy-momentum tensor of the electromagnetic field for infinite conductivity is

$$T^{\text{mag}}_{\alpha\beta} = \frac{1}{8\pi}\left[(h_{\alpha\beta} - u_\alpha u_\beta)\, B_\varrho B^\varrho - 2 B_\alpha B_\beta\right] , \tag{6.10.23}$$

where

$$h_{\alpha\beta} = g_{\alpha\beta} - u_\alpha u_\beta . \tag{6.10.24}$$

---

### 6.10.2 Magnetic Dipole Radiation

The very high stability of the basic periodicities of radio pulsars and the short periods (in particular of the Crab pulsar for which $P \simeq 0.33$ ms) led quickly to the general acceptance of the rotating neutron star model of pulsars. (Rapidly spinning neutron stars were already considered by *Pacini* in 1967, before the discovery of pulsars, and first proposed as a pulsar model by *Gold* in 1968.)

High angular velocities $\Omega$ of neutron stars are very natural, since the angular momentum of an imploding stellar core is nearly conserved during collapse. This implies that $\Omega$ increases as $R^{-2}$, where $R$ is the radius of the collapsing core. The initial radius of the core is typically $\sim 5 \times 10^8$ cm, while the final radius is $\sim 10^6$ cm. This means that the period of rotation of the core will decrease by a factor $10^5 - 10^6$ during collapse. Therefore, its final rotation period may approach the neutron star limiting period $\sim 0.5$ ms ($\Omega_{\max} \simeq (GM/R^3)^{1/2} \simeq 10^4\, \text{s}^{-1}$).

The shortest pulsar period known at present is $P = 1.56$ ms, corresponding to an equatorial velocity of $\simeq 0.13\, c$.

The period of a rotating neutron star with a strong magnetic field will increase mainly due to magnetic dipole radiation. The energy radiated by a time-varying dipole moment is (in vacuum)

$$\frac{dE}{dt} = -\frac{2}{3c^3} |\ddot{\boldsymbol{\mu}}|^2 = -\frac{2}{3c^3} \mu_\perp^2 \Omega^4 , \tag{6.10.25}$$

where $\mu_\perp$ is the component of the magnetic moment vector $\boldsymbol{\mu}$ perpendicular to the rotation axis: $\mu_\perp = \mu \sin\theta$, $\theta = \measuredangle(\boldsymbol{\mu}, \boldsymbol{\Omega})$.

As an example, choose $\Omega = 10$ rad/s, $\theta = \frac{\pi}{4}$, $R = 10^6$ cm and for the polar field $B_p = 2\,\mu/R^3 = 10^{12}$ G. Then $-dE/dt \simeq 3 \times 10^{33}$ erg/s, which is comparable to the energy loss required to explain the observed slowdown of most pulsars.

Let us assume that the rotational energy is entirely changing due to (6.10.25). Then we obtain the differential equation

$$\dot{\Omega} = -\frac{2\mu^2 \sin^2\theta}{3c^3 I}\,\Omega^3\,, \tag{6.10.26}$$

where $I$ is the moment of inertia of the neutron star. The solution is

$$\Omega/\Omega_0 = \left[1 + \frac{t}{\tau_m/2}\right]^{-1/2}, \tag{6.10.27}$$

where $\Omega_0 = \Omega\ (t = 0)$ and

$$\tau_m = \frac{3c^3 I}{2\mu^2 \Omega_0^2 \sin^2\theta}\,. \tag{6.10.28}$$

If $t$ is set equal to zero at the present time, then the age of the pulsar, which is defined to be the time at which $\Omega/\Omega_0$ was very large, becomes

$$t_{age} = \tfrac{1}{2}\,\tau_m\,. \tag{6.10.29}$$

On the other hand, we can observe the present period derivative $\tau_0 = -(\Omega/\dot{\Omega})_0$. According to (6.10.26) this time agrees with $\tau_m$,

$$\tau_0 = \tau_m\,. \tag{6.10.30}$$

For the Crab pulsar $\frac{1}{2}\tau_0 = 1.2 \times 10^3$ yr, which is not very far from its true age (the birth year is 1054).

There are, however, a number of factors (deformation of the magnetic field, etc.) that change the Eq. (6.10.26). It is common practice to describe the slow down by an equation of the form

$$\dot{\Omega} = -K\,\Omega^n\,, \tag{6.10.31}$$

where $K$ is a positive constant and $n$ is known as the *braking index*. Since $n = (\Omega\,\ddot{\Omega}/\dot{\Omega}^2)_0$, one can, in principle, determine $n$. For the Crab pulsar, one finds $n = 2.5$.

With (6.2.30) and (6.2.28) we can determine the magnetic field. The average surface field is

$$\langle B_s \rangle = \left(\frac{3c^3 I}{2\tau_0 \Omega_0^2 R^6}\right)^{1/2}. \tag{6.10.32}$$

For $M = 1.4\,M_\odot$, $I = 1.4 \times 10^{45}$ g cm$^2$ (taken from neutron star models; show that this is a reasonable number), one finds for $\langle B_s \rangle$ a few times $10^{12}$ G.

Note that the energy loss of the Crab pulsar is very large

$$-\dot{E} = I\,\Omega_0\,\dot{\Omega}_0 = 6.5 \times 10^{38}\ \mathrm{erg/s}\,. \tag{6.10.33}$$

The energy in the pulsed radiation is only a tiny fraction ($\lesssim 10^{-9}$) of this.

---

**Exercise:** Investigate the influence of the gravitational energy radiation on the pulsar age. Derive for this the following rate of energy loss for a rotating neutron star:

$$\frac{dE^{\mathrm{grav}}}{dt} = -\frac{32\,G}{5\,c^5}\,I^2\,\varepsilon^2\,\Omega^6\,, \tag{6.10.34}$$

where $\varepsilon$ is the ellipticity in the equatorial plane. How large would $\varepsilon$ have to be in order that magnetic dipole radiation (6.10.25) and gravitational quadrupole radiation give the correct age for the Crab pulsar?

---

### 6.10.3 Synchrotron Radiation from the Crab Nebula

Most of the energy (6.10.33) is pumped into the Crab nebula and is partly radiated there as synchrotron radiation. (A large fraction of the pulsar energy loss is required to maintain the expansion of the nebula.) The observed spectrum extends from low radio frequencies to the hard $\gamma$-ray region. (This is also the case for the pulsed radiation from the pulsar.) The total luminosity derived from the nebular spectrum is $\approx 10^{38}$ erg/s, of which about 12 percent is emitted at radio frequencies. The total energy output from the nebula is, within uncertainties, comparable to the rotational energy loss of the pulsar.

The synchrotron spectrum has its maximum at the frequency $\nu_{\mathrm{m}} = 0.29\,\nu_{\mathrm{c}}$, where $\nu_{\mathrm{c}}$ is the critical frequency (see, for example, [101]). Numerically

$$\nu_{\mathrm{m}} = 0.07\,\frac{e\,B_\perp}{m\,c}\left(\frac{\varepsilon}{m\,c^2}\right)^2, \tag{6.10.35}$$

where $B_\perp$ is the component of the magnetic field perpendicular to the electron velocity and $\varepsilon$ is the electron energy. For typical values $B_\perp = 5\times 10^{-4}$ G we obtain in the $\gamma$-ray region: $\varepsilon \simeq 2\times 10^6$ GeV for $\nu = 10^{22}$ Hz.

The relativistic Larmor formula tells us that

$$-\dot{\varepsilon} = \frac{2}{3}\,\frac{e^4\,B_\perp^2}{m^2\,c^3}\left(\frac{\varepsilon}{m\,c^2} - 1\right). \tag{6.10.36}$$

This implies that the lifetime of the ultrarelativistic electrons is

$$\begin{aligned} t_{1/2}[\mathrm{s}] &= \frac{5.1 \times 10^8}{(B_\perp[G])^2}\,\frac{m\,c^2}{\varepsilon_0} \\ &= 6.1 \times 10^{11}\,(B_\perp[G])^{-3/2}\,(\nu_{\mathrm{m}}[\mathrm{Hz}])^{-1/2}\,. \end{aligned} \tag{6.10.37}$$

As an example, we take $\nu_m = 10^{20}$ Hz; then $t_{1/2} \simeq 10$ weeks. This requires a continuous injection or acceleration of relativistic electrons. Before the discovery of the Crab pulsar, this represented a major problem in understanding the physics of the Crab Nebula. In 1967, *Pacini* anticipated that a rotating magnetized neutron star could be the source of the nebular energy output. We know now that the pulsar is efficiently accelerating (directly and indirectly) particles to ultrarelativistic energies. (Since the x-ray source at high energies has a diameter of about one light year, the radiating electrons cannot all be accelerated at the pulsar.)

### 6.10.4 The Pulsar Magnetosphere

The induced electric fields of a rapidly rotating magnetized neutron star are so strong that the region surrounding the star cannot be a vacuum, but must contain a substantial space charge. This was first pointed out by *Goldreich* and *Julian* [102]. We repeat here their argument.

Consider the simple case of an aligned dipole field, where $\boldsymbol{\mu}$ is parallel to $\boldsymbol{\Omega}$, and assume that the star is surrounded by vacuum. Then for a given magnetic field, we can calculate the electric field. Let us assume that $\boldsymbol{B}$ is a dipole field. In polar coordinates:

$$\boldsymbol{B} = \frac{2\mu}{r^3}(\cos\vartheta, \tfrac{1}{2}\sin\vartheta, 0) . \tag{6.10.38}$$

The magnitude of $\boldsymbol{B}$ at the poles is thus

$$B_0 = \frac{2\mu}{R^3}, \tag{6.10.39}$$

where $R$ is the stellar radius. Since the conductivity of neutron star matter is extremely high, the interior electric field just below the surface is

$$\boldsymbol{E}^{\text{int}} = -\frac{1}{c}(\boldsymbol{\Omega} \times \boldsymbol{x}) \times \boldsymbol{B} = \frac{B_0 R^3 \Omega}{c\, r^2} \cdot (\tfrac{1}{2}\sin^2\vartheta, -\sin\vartheta\cos\vartheta, 0) . \tag{6.10.40}$$

The exterior field $\boldsymbol{E}^{\text{ext}}$ must be a potential field: $\boldsymbol{E}^{\text{ext}} = -\operatorname{grad}\phi$. We show now that a quadrupole ansatz

$$\phi = \text{const} \cdot \frac{P_2(\cos\vartheta)}{r^3} = C\,\frac{3\cos^2\vartheta - 1}{r^3}$$

leads to the unique solution with the correct boundary condition. The $\vartheta$-component of the exterior field is

$$E_\vartheta^{\text{ext}} = -\frac{1}{r}\frac{\partial\phi}{\partial\vartheta} = 6\,C\,\frac{\sin\vartheta\cos\vartheta}{r^4} .$$

Since the tangential component of $\boldsymbol{E}$ is continuous, we must require

$$C = -B_0 R^5 \Omega / 6c$$

and hence we have

$$\phi = -\frac{B_0 R^5 \Omega}{6c} \frac{3\cos^2\vartheta - 1}{r^3} \tag{6.10.41}$$

and thus

$$E_r^{\text{ext}} = -\frac{B_0 R^5 \Omega}{2c} \frac{3\cos^2\vartheta - 1}{r^3}$$

$$E_\vartheta^{\text{ext}} = -\frac{B_0 R^5 \Omega}{c r^4} \sin\vartheta \cos\vartheta, \qquad E_\varphi^{\text{ext}} = 0\,. \tag{6.10.42}$$

The normal component of the electric field is discontinuous, corresponding to a surface charge density

$$\varrho_{\mathrm{s}} = \frac{1}{4\pi}(E_r^{\text{ext}} - E_r^{\text{int}}) = -\frac{B_0 R \Omega}{4\pi c} \cos^2\vartheta\,. \tag{6.10.43}$$

The Lorentz invariant scalar product $\boldsymbol{E}^{\text{ext}} \cdot \boldsymbol{B}$ does not vanish. This quantity gives a measure of the force which a co-rotating charged particle feels in the direction of the magnetic field. From (6.10.38) and (6.10.42) we find

$$\boldsymbol{E}^{\text{ext}} \cdot \boldsymbol{B} = -\frac{B_0 R \Omega}{c} \frac{R^7}{r^7} \cos^3\vartheta\,, \tag{6.10.44}$$

while the corresponding quantity vanishes inside the neutron star. Thus inside a thin transition layer at the surface of the star, there is a non-vanishing $\boldsymbol{E} \cdot \boldsymbol{B}$ of magnitude [we take $\frac{1}{2}$ of (6.10.44)]

$$\boldsymbol{E} \cdot \boldsymbol{B} = -\frac{B_0^2 R \Omega}{2c} \cos^3\vartheta \tag{6.10.45}$$

leading to an acceleration of a particle in the direction of the magnetic field:

$$a = \frac{e\,\boldsymbol{E} \cdot \boldsymbol{B}}{m\,|\boldsymbol{B}|}\,.$$

Let us compare this with the gravitational acceleration $g = GM/R^2$,

$$\frac{a}{g} = -\frac{e B_0 R^3 \Omega}{m G c M} f(\vartheta)\,, \tag{6.10.46}$$

where

$$f(\vartheta) = \cos^3\vartheta\,(3\cos^2\vartheta + 1)^{-1/2}\,.$$

For typical values the ratio (6.10.46) is huge, $\sim 10^{11}$.

The electric fields parallel to $\boldsymbol{B}$ are very strong. From (6.10.45)

$$E_{\parallel} \approx \frac{\Omega R}{c} B_0 \approx 6 \times 10^{10} B_{12} P^{-1} [\mathrm{V\,cm^{-1}}] , \tag{6.10.47}$$

if $P$ is measured in seconds, and $B_{12} = B/10^{12}$ G. Fields of this magnitude give rise to field emission and charge will flow from the star to fill the surrounding region. In the plasma filled magnetosphere we have

$$\boldsymbol{E} = -\frac{1}{c}(\boldsymbol{\Omega} \times \boldsymbol{x}) \times \boldsymbol{B} \tag{6.10.48}$$

corresponding to a space charge density

$$\varrho = \frac{1}{4\pi} \operatorname{div} \boldsymbol{E} = -\frac{1}{2\pi c} \boldsymbol{\Omega} \cdot \boldsymbol{B} . \tag{6.10.49}$$

Numerically, this corresponds to a charge number density $n_e = 7 \times 10^{-2} B_{\parallel} P^{-1}\,\mathrm{cm^{-3}}$ ($B_{\parallel}$ [G] and $P$ [s]).

Since $\boldsymbol{E} \cdot \boldsymbol{B} = 0$, the magnetic field lines become equipotentials and the strong magnetic fields force the charged particles to corotate with the star in regions where magnetic field lines form closed loops ($\boldsymbol{E} \times \boldsymbol{B}/B^2$ drift). Corotation can, however, not be maintained beyond the light cylinder of radius

$$R_{\mathrm{L}} = \frac{c}{\Omega} \approx 5 \times 10^9 P\,[\mathrm{s}]\,\mathrm{cm} , \tag{6.10.50}$$

where the tangential velocity equals to the velocity of light.

There are also regions with open field lines where $\boldsymbol{E} \cdot \boldsymbol{B} \neq 0$ and along which charges may suffer large acceleration.

So far things are clear. But now the problems begin. It is very difficult to model the pulsar magnetosphere. Even for the aligned pulsar, it has not yet been possible to construct a self-consistent description of the currents and fields surrounding the star. This is largely the reason why no generally accepted model for pulsar emission exists, which would allow us to understand even the gross features of the large body of pulsar data [103].

It seems likely that the radio emission is generated not far above the polar caps. A substantial fraction of the voltage arising from unipolar induction across the polar cap is probably available for accelerating charges along open field lines close to the stellar surface. The spiral motion of electrons in this region generates $\gamma$-radiation. This curvature radiation in turn initiates electron-positron pair production in the intense magnetic field. (This process is treated, for example, in [132].) Repetition generates a cascade, positrons moving in one direction and electrons in the other. The result may be a sustained electron-positron discharge. There are models of how this could eventually lead to charge-bunching and finally to the emission of radio waves by coherent curvature radiation at perhaps $10^2-10^3$ km from the surface.

If the spin and magnetic axes are not aligned, one gets a "lighthouse" effect. The collimated radiation above the polar caps will appear as a pulsed radio signal to an observer in the cone swept out by the radio beam as it precesses about the spin axis.

Beside these "polar cap" models, there are the "light cylinder" models in which the conical beam is tangential to the light cylinder and perpendicular to the rotation axis.

It must be admitted, however, that we do not yet really know how radio pulsars work. (It is instructive to read the report of the panel discussion "From Whence The Pulses" in the Proceedings of the IAU Symposium on Pulsars in 1980 [104].)

The current status and the problems involved are described in great detail in a recent review article [105], which contains also an extensive bibliography of the pulsar literature.

### 6.10.5 Matter in Strong Magnetic Fields

In the strong magnetic fields of neutron stars, atoms become distorted into needle-like objects, changing dramatically the chemistry and solid-state physics of matter.

The influence of the magnetic field becomes very important once its strength is such that $\hbar\,\omega_{\mathrm{B}} = \hbar\, e\, B/m_{\mathrm{e}}\, c$ is larger than the Rydberg-energy $\alpha^2\, m_{\mathrm{e}}\, c^2/2$, i.e. for

$$B > \tfrac{1}{2}\,\alpha^2\, m_{\mathrm{e}}^2\, c^3/e\,\hbar \approx 10^9\ \mathrm{G}\,, \qquad (6.10.51)$$

A considerable literature has already been devoted to the states of atoms, molecular chains, and condensed matter in strong fields. (For a list of references, see [105].) As an illustration, we discuss here the ground state of hydrogen.

---

**Exercise:** Estimate with a variational calculation the ground-state energy of an isolated hydrogen atom in a magnetic field $B \sim 10^{12}$ G.

*Solution:* For such strong fields the Coulomb interaction is only a perturbation of the magnetic interaction in the transversal direction. It is, therefore, reasonable to choose for the ground state the $m = 0$ Landau orbital, except that $\exp(\mathrm{i}\, k\, z)$ is replaced by a real function $f(z)$ chosen to minimize the total energy.

The ground state of an electron in a homogeneous field is degenerate. The wave functions in cylindrical coordinates $(\varrho, \varphi, z)$ are

$$\psi = \mathrm{const} \cdot R_m(\varrho)\, \mathrm{e}^{-\mathrm{i}m\varphi}\, \mathrm{e}^{\mathrm{i}p_z \cdot z/\hbar}\,,$$

where

$$R_m(\varrho) = 2^{m/2}\,(2\pi\, m!\, \hat{\varrho}^2)^{-1/2}\, \mathrm{e}^{-\varrho^2/4\hat{\varrho}^2} \left(\frac{\varrho^2}{4\,\hat{\varrho}^2}\right)^{m/2} \quad (m = 0, 1, 2 \ldots) \qquad (6.10.52)$$

with

$$\hat{\varrho}=\left(\frac{\hbar c}{eB}\right)^{1/2}=\frac{2.6\times 10^{-10}}{B_{12}^{1/2}}\,\mathrm{cm}\,. \tag{6.10.53}$$

The condition (6.10.51) can also be expressed as $\hat{\varrho} \ll a_0$ (Bohr radius).

As a simple trial function we choose

$$\psi = \mathrm{const}\; R_0(\varrho)\cdot f(z)$$
$$R_0(\varrho) = \mathrm{const}\;\mathrm{e}^{-\varrho^2/4\hat{\varrho}^2},\quad f(z) = \alpha^{1/2}\,\mathrm{e}^{-\alpha|z|}, \tag{6.10.54}$$

where $\alpha$ is a variational parameter. Notice that this wavefunction is an eigenfunction of $[(p_x - eA_x)^2 + (p_y - eA_y)^2]/2m$. Hence the energy in the state $\psi$ relative to the energy without a Coulomb field is

$$E(\psi)=\left(\psi,\left[-\frac{\hbar^2}{2m_\mathrm{e}}\frac{\partial^2}{\partial z^2}-\frac{Ze^2}{\sqrt{\varrho^2+z^2}}\right]\psi\right)/(\psi,\psi)\,. \tag{6.10.55}$$

We have to minimize this expression in $\alpha$. To do this, we must first evaluate (6.10.55). The kinetic energy is easy:

$$E_\mathrm{kin}=\frac{\hbar^2}{2m_\mathrm{e}}\alpha^2\,. \tag{6.10.56}$$

For the potential energy we use [see (29.3.55) in [106]]:

$$\frac{1}{\sqrt{\varrho^2+z^2}}=\int_0^\infty dk\,\mathrm{e}^{-k|z|}J_0(k\varrho)$$

and find

$$E_\mathrm{pot}=-Ze^2\int_0^\infty dk\,\frac{2\alpha}{2\alpha+k}\int_0^\infty\frac{d\varrho\,\varrho}{\hat{\varrho}^2}\,\mathrm{e}^{-\varrho^2/2\hat{\varrho}^2}J_0(k\varrho)\,.$$

With the formula {(11.4.29) in [106]}:

$$\int_0^\infty \mathrm{e}^{-a^2t^2}t^{\nu+1}J_\nu(bt)\,dt=\frac{b^\nu}{(2a^2)^{\nu+1}}\,\mathrm{e}^{-b^2/4a^2},$$

we get for $\nu = 0$, $a^2 = 1/2\,\hat{\varrho}^2$, $b = k$:

$$E_\mathrm{pot}=-Ze^2\int_0^\infty dk\,\mathrm{e}^{-(k\hat{\varrho})^2/2}\,\frac{2\alpha}{2\alpha+k}$$

or

$$E_\mathrm{pot}=-Ze^2\,2\alpha f(\sqrt{2}\,\alpha\hat{\varrho}), \tag{6.10.57}$$

where $f$ is defined as

$$f(x)=\int_0^\infty\frac{\mathrm{e}^{-y^2}}{y+x}\,dy\,. \tag{6.10.58}$$

**Table 6.3.** Ground-state energy of the hydrogen atom in strong magnetic fields

| $B_{12}$ | $\alpha\, a_0$ | $E_{tot}$ [eV] |
|---|---|---|
| 2 | $\simeq 2$ | $\simeq -183$ |
| 4 | $\simeq 2.5$ | $\simeq -230$ |

This function is fortunately tabulated in Sect. 27.6 of [106]. If one plots the sum (6.10.56) and (6.10.57) as a function of $\alpha$ one finds the minimum for the values listed in Table 6.3. These agree with more accurate variational calculations.

---

# Chapter 7. Rotating Black Holes

## Introduction

All stars rotate more or less rapidly. When a horizon is formed during gravitational collapse, a Schwarzschild black hole is thus never produced. One expects, however, that the horizon will quickly settle down to a stationary state as a result of the emission of gravitational waves. The geometry of the stationary black hole is of course no longer spherically symmetric.

It is remarkable that we know all stationary black holes. Surprisingly, they are fully characterized by just three parameters, namely the mass, angular momentum and electric charge of the hole. These quantities can all be determined, in principle, by a distant observer.

Thus when matter disappears behind a horizon, an exterior observer sees almost nothing of its individual properties. One can no longer say for example how many baryons formed the black hole[1]. A huge amount of information is thus lost. The mass, angular momentum, and charge completely determine the external field, which is known analytically (Kerr-Newman solution). This led J. A. Wheeler to make the remark "A black hole has no hair", and our previous statement is thus known as the "no-hair-theorem". The proof is very difficult and was completed only in the course of a number of years, with important contributions by various authors (Israel, Carter, Hawking, Robinson). The proof is now also complete for the charged case [110]. For a thorough discussion and a list of references, see Chap. 6 of [21].

Unfortunately, there is still no physically natural route leading to the Kerr-Newman solution. Originally, it was found more or less by accident [52] and its physical meaning was not recognized until later.

---

[1] For this, it is essential that baryon number B does not correspond to an unbroken gauge symmetry, since it would otherwise act as the source of long-range external field, like the electric charge. In that case, the massless gauge boson corresponding to B would have to couple, however, exceedingly weakly to matter, in order to avoid inconcistency with the Eötvos-Dicke experiment. (This has been pointed out by Lee and Yang.)
*Exercise:* Show this.

Newman and co-workers showed in 1965 that the Kerr-Newman solution can be inferred from the Schwarzschild solution (the Reissner-Nordström solution for charged holes) by means of a complex coordinate transformation [53, 54]. One also obtains the Kerr-Newman family in a systematic study of the algebraically degenerate solutions of the Einstein-Maxwell equations [55]. The most natural path leading to the Kerr solution (zero charge) is described in Sect. 7.4 of [21]. The amount of analytical work is, however, considerable.

Obviously, the Schwarzschild solution is contained in the Kerr family (zero charge). It is the only static solution. One thus has as a corollary to the "no hair" theorem that every static uncharged black hole is a Schwarzschild black hole and thus automatically spherically symmetric. Historically, this theorem by Israel was the starting point for the developments which led to the "no hair" theorem.

## 7.1 Analytic Form of the Kerr-Newman Family ($G=c=1$)

We shall give the solution in terms of the so-called Boyer-Lindquist coordinates $(t, r, \vartheta, \varphi)$. It contains three parameters which we shall interpret below. We shall often make use of the abbreviations

$$\begin{aligned} \Delta &= r^2 - 2Mr + a^2 + Q^2, \\ \varrho^2 &= r^2 + a^2 \cos^2 \vartheta . \end{aligned} \tag{7.1.1}$$

The metric is given by

$$g = \frac{\Delta}{\varrho^2}[dt - a\sin^2\vartheta\, d\varphi]^2 - \frac{\sin^2\vartheta}{\varrho^2}[(r^2+a^2)\, d\varphi - a\, dt]^2 - \frac{\varrho^2}{\Delta}dr^2 - \varrho^2 d\vartheta^2 \tag{7.1.2}$$

and the electromagnetic field by

$$\begin{aligned} F = {} & Q\varrho^{-4}(r^2 - a^2\cos^2\vartheta)\, dr \wedge (dt - a\sin^2\vartheta\, d\varphi) \\ & + 2Q\varrho^{-4} a r \cos\vartheta \sin\vartheta\, d\vartheta \wedge [(r^2+a^2)\, d\varphi - a\, dt]. \end{aligned} \tag{7.1.3}$$

As special cases we have

$Q = a = 0$: Schwarzschild solution
$a = 0$: Reissner-Nordström solution
$Q = 0$: Kerr solution.

## 7.2 Asymptotic Field and $g$-Factor of a Black Hole

The parameter $M$, $a$ and $Q$ are most easily interpreted from the asymptotic form of the fields. The leading terms in an expansion of the metric in powers of $1/r$ give

$$g=\left[1-\frac{2M}{r}+O\left(\frac{1}{r^2}\right)\right]dt^2+\left[\frac{4aM}{r}\sin^2\vartheta+O\left(\frac{1}{r^2}\right)\right]dt\,d\varphi$$
$$-\left[1+O\left(\frac{1}{r}\right)\right][dr^2+r^2(d\vartheta^2+\sin^2\vartheta\,d\varphi^2)]. \tag{7.2.1}$$

If we now introduce the "Cartesian" coordinates

$$x=r\sin\vartheta\cos\varphi,\quad y=r\sin\vartheta\sin\varphi,\quad z=r\cos\vartheta,$$

then $g$ has the general form (4.4.11). In Sect. 4.4, it was shown, by performing the flux integrals for the energy and angular momentum, that $M$ is the total mass and that

$$S=aM \tag{7.2.2}$$

is the total angular momentum (which can be determined using a gyroscope).

The asymptotic form of the electric and magnetic field components in the $r$, $\vartheta$, and $\varphi$ directions is given by

$$\left.\begin{aligned}
E_{\hat{r}}&=E_r=F_{rt}=\frac{Q}{r^2}+O\left(\frac{1}{r^3}\right)\\
E_{\hat{\vartheta}}&=\frac{E_\vartheta}{r}=\frac{1}{r}F_{\vartheta t}=O\left(\frac{1}{r^4}\right)\\
E_{\hat{\varphi}}&=\frac{E_\varphi}{r\sin\vartheta}=\frac{F_{\varphi t}}{r\sin\vartheta}=0
\end{aligned}\right\} \tag{7.2.3}$$

$$\left.\begin{aligned}
B_{\hat{r}}&=F_{\hat{\vartheta}\hat{\varphi}}=\frac{F_{\vartheta\varphi}}{r^2\sin\vartheta}=\frac{2Qa}{r^3}\cos\vartheta+O\left(\frac{1}{r^4}\right)\\
B_{\hat{\vartheta}}&=F_{\hat{\varphi}\hat{r}}=\frac{F_{\varphi r}}{r\sin\vartheta}=\frac{Qa}{r^3}\sin\vartheta+O\left(\frac{1}{r^4}\right)\\
B_{\hat{\varphi}}&=F_{\hat{r}\hat{\vartheta}}=\frac{F_{r\vartheta}}{r}=0.
\end{aligned}\right\} \tag{7.2.4}$$

Asymptotically the electric field is radial and the corresponding Gaussian flux integral[2] is $4\pi Q$. This shows that $Q$ is the *charge* of the black hole.

---

[2] In general,

$$Q=\int_\Sigma *J=-\frac{1}{4\pi}\int_\Sigma d*F=-\frac{1}{4\pi}\oint *F,$$

where $\Sigma$ is a spacelike hypersurface. The last integral extends over a two-dimensional surface "at infinity".

Asymptotically, the magnetic field is a dipole field with dipole moment

$$\mu = Q\,a = \frac{Q}{M}\,S =: g\left(\frac{Q}{2M}\,S\right) \tag{7.2.5}$$

and one thus obtains the completely unexpected result:

$$g = 2, \tag{7.2.6}$$

exactly as for a Dirac electron!

## 7.3 Symmetries of $g$

The metric coefficients are independent of $t$ and $\varphi$. Hence

$$K = \partial/\partial t \quad \text{and} \quad \tilde{K} = \partial/\partial\varphi \tag{7.3.1}$$

are Killing fields ($L_K\,g = L_{\tilde{K}}\,g = 0$).

The scalar products of the Killing fields are

$$\begin{aligned}
(K, K) &= g_{tt} = 1 - \frac{2Mr - Q^2}{\varrho^2} \\
(K, \tilde{K}) &= g_{t\varphi} = \frac{(2Mr - Q^2)\,a\sin^2\vartheta}{\varrho^2} \\
(\tilde{K}, \tilde{K}) &= g_{\varphi\varphi} = -\frac{[(r^2+a^2)^2 - \Delta a^2\sin^2\vartheta]}{\varrho^2}\,\sin^2\vartheta\,.
\end{aligned} \tag{7.3.2}$$

These scalar products have an intrinsic meaning and thus we see that the Boyer-Lindquist coordinates are extremely well adapted. They provide a natural generalization of the Schwarzschild coordinates.

## 7.4 Static Limit and Stationary Observers

An observer moving along a world line with constant $r$, $\vartheta$ and uniform angular velocity sees an unchanging space-time geometry, and is thus a stationary observer. His angular velocity, measured at infinity, is

$$\Omega = \frac{d\varphi}{dt} = \frac{\dot{\varphi}}{\dot{t}} = \frac{u^\varphi}{u^t}, \tag{7.4.1}$$

where $u$ is the four-velocity. The four-velocity of a stationary observer is proportional to a Killing field:

$$u = u^t\left(\frac{\partial}{\partial t} + \Omega\,\frac{\partial}{\partial\varphi}\right) = \frac{K + \Omega\tilde{K}}{\|K + \Omega\tilde{K}\|}\,. \tag{7.4.2}$$

Obviously $K + \Omega \tilde{K}$ must be timelike:

$$g_{tt} + 2\Omega\, g_{t\varphi} + \Omega^2\, g_{\varphi\varphi} > 0\,.$$

The left hand side vanishes for

$$\Omega = \frac{-g_{t\varphi} \pm \sqrt{g_{t\varphi}^2 - g_{tt}\, g_{\varphi\varphi}}}{g_{\varphi\varphi}}\,.$$

Let $\omega = -g_{t\varphi}/g_{\varphi\varphi}$; then

$$\Omega_{\min} = \omega - \sqrt{\omega^2 - g_{tt}/g_{\varphi\varphi}} \tag{7.4.3}$$

$$\Omega_{\max} = \omega + \sqrt{\omega^2 - g_{tt}/g_{\varphi\varphi}}\,. \tag{7.4.4}$$

Note that

$$\omega = \frac{a\,(2Mr - Q^2)}{(r^2 + a^2)^2 - \Delta a^2 \sin^2 \vartheta}\,. \tag{7.4.5}$$

For an interpretation of $\omega$ consider stationary observers who are nonrotating with respect to local freely falling test particles that have been dropped in radially from infinity. Since the angular momentum of such test particles vanishes, we have for these special stationary observers (so-called "Bardeen observers") $(u, \tilde{K}) = 0$, hence $(K + \Omega \tilde{K}, \tilde{K}) = 0$, i.e., $\Omega = \omega$. We assume $a > 0$. Obviously $\Omega_{\min} = 0$ if and only if $g_{tt} = 0$, i.e., for

$$(K, K) = 0\colon\ r = r_o(\vartheta) := M + \sqrt{M^2 - Q^2 - a^2 \cos^2 \vartheta}\,. \tag{7.4.6}$$

We assume that $M^2 > Q^2 + a^2$; otherwise there exists a naked singularity (see below).

An observer is said to be *static* (relative to the "fixed stars") if $\Omega = 0$, so that $u$ is proportional to $K$. Static observers can exist only outside the *static limit* $\{r = r_o(\vartheta)\}$. At the static limit, $K$ becomes lightlike. An observer would then have to move at the speed of light in order to remain at rest with respect to the fixed stars.

We now consider the redshift which an asymptotic observer measures for light from a source at rest ($u = K/\|K\|$) outside of the static limit.

We have from (1.9.19)

$$\nu_e/\nu_o = \frac{(K, K)_o^{1/2}}{(K, K)_e^{1/2}} \simeq \frac{1}{(K, K)_e^{1/2}} \to \infty \quad \text{at the static limit.} \tag{7.4.7}$$

## 7.5 Horizon and Ergosphere

For $\omega^2 = g_{tt}/g_{\varphi\varphi}$, $\Omega_{\min}$ and $\Omega_{\max}$ coincide. Then

$$\Omega_{\min} = \Omega_{\max} =: \Omega_{\mathrm{H}} = \omega\,.$$

Since by definition $\omega^2 = g_{t\varphi}^2/g_{\varphi\varphi}^2$, we have for this case $g_{\varphi\varphi}g_{tt} = g_{t\varphi}^2$ (

$$-\frac{(\Delta - a^2\sin^2\vartheta)}{\varrho^2}\,\frac{[(r^2+a^2)^2 - \Delta a^2\sin^2\vartheta]\sin^2\vartheta}{\varrho^2} = \frac{a^2\sin^4\vartheta}{\varrho^4}(\Delta - r^2 - a^2)$$

This is equivalent to

$$\Delta = 0, \qquad (7.5.$$

which means that

$$r = r_+ := M + \sqrt{M^2 - a^2 - Q^2}\,. \qquad (7.5.$$

In Sect. 7.8 we shall see that this surface is a *horizon*, if $M^2 > a^2 + Q$ The region between the static limit and this horizon is the so-calle *ergosphere* (for reasons which will be clarified later). The static lim and the horizon come together at the poles (see Fig. 7.1).

Inside the ergosphere, $K$ is spacelike. "All the king's horses and a the king's men" cannot prevent an observer from rotating about th black hole. The ergosphere disappears when $a = Q = 0$ (Schwarzschi solution).

According to our previous discussion, the angular velocity $\Omega_{\rm H}$ at th horizon is given by

$$\Omega_{\rm H} = \omega|_{\Delta=0} = \frac{a(2Mr_+ - Q^2)}{(r_+^2 + a^2)^2}$$

or, using (7.5.2),

$$\Omega_{\rm H} = \frac{a}{r_+^2 + a^2}\,. \qquad (7.5.$$

Thus $\Omega_{\rm H}$ is constant on the horizon; the black hole as a whole rotat like a rigid body.

The static limit is timelike (its normal vectors are spacelike), exce at the poles. Thus one can pass through this surface in *both* directions.

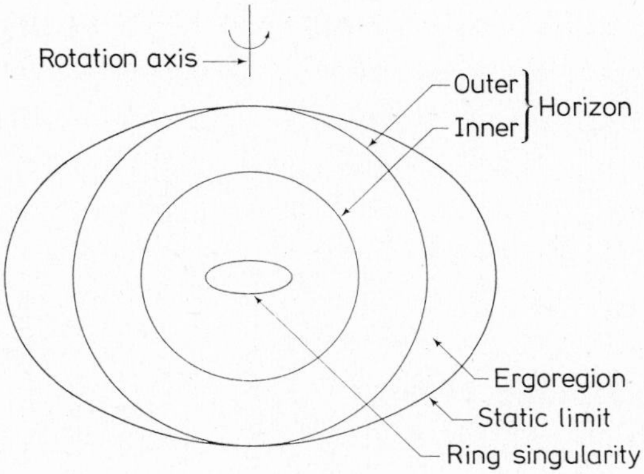


**Fig. 7.1.** Cross section throu the axis of rotation of a Ke Newman solution

## 7.6 Coordinate Singularity at the Horizon, Kerr Coordinates

For $\Delta = 0$ the Kerr metric expressed in terms of the Boyer-Lindquist coordinates appears singular. However, this is merely a coordinate "singularity" as can be seen by transforming to the so-called Kerr coordinates. These new coordinates are generalizations of the Eddington-Finkelstein coordinates and are defined by

$$\begin{aligned} d\tilde{V} &= dt + (r^2 + a^2)\, dr/\Delta \\ d\tilde{\varphi} &= d\varphi + a\, dr/\Delta\,. \end{aligned} \tag{7.6.1}$$

Note that the exterior differentials of the right hand sides vanish. The metric can be written in terms of the new coordinates $(\tilde{V}, r, \vartheta, \tilde{\varphi})$ as follows:

$$\begin{aligned} g = {} & [1 - \varrho^{-2}(2Mr - Q^2)]\, d\tilde{V}^2 - 2dr\, d\tilde{V} - \varrho^2\, d\vartheta^2 \\ & - \varrho^{-2}[(r^2 + a^2)^2 - \Delta a^2 \sin^2\vartheta] \sin^2\vartheta\, d\tilde{\varphi}^2 + 2a \sin^2\vartheta\, d\tilde{\varphi}\, dr \\ & + 2a\varrho^{-2}(2Mr - Q^2) \sin^2\vartheta\, d\tilde{\varphi}\, d\tilde{V}. \end{aligned} \tag{7.6.2}$$

This expression is regular at the horizon. In place of $\tilde{V}$ one often uses $\tilde{t} = \tilde{V} - r$. In terms of the Kerr coordinates, the Killing fields $K$ and $\tilde{K}$ are given by

$$K = \left(\frac{\partial}{\partial \tilde{t}}\right)_{r,\vartheta,\tilde{\varphi}}, \quad \tilde{K} = \left(\frac{\partial}{\partial \tilde{\varphi}}\right)_{\tilde{t},r,\vartheta}. \tag{7.6.3}$$

## 7.7 Singularities of the Kerr-Newman Metric

The Kerr-Newman metric has a true singularity which lies inside the horizon for $M^2 > a^2 + Q^2$. It has a rather complicated structure, which is described in detail in Sect. 5.6 of the book by *Hawking* and *Ellis* [15]. The maximal analytic continuation of the Kerr solution is also discussed there. These aspects have no relevance for astrophysics and hence will not be discussed further here.

## 7.8 Structure of the Light Cones

Visualizing space-time geometry is made easier by considering the structure of the light cones. We examine this more closely in the equatorial plane, as indicated in Fig. 7.2. Each point in this plane represents an integral curve of the Killing field $K$. The wave fronts of

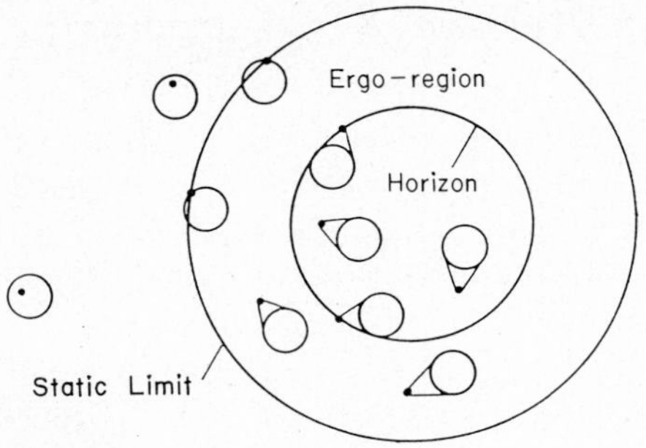


**Fig. 7.2.** Structure of light cones in the equational plane of a Kerr-Newman black hole

light signals which have formed shortly after being emitted from the marked points are shown in Fig. 7.2.

We note the following facts:

(i) Since $K$ is timelike outside the static limit, the points of emission are inside the wave fronts.

(ii) At the static limit $K$ becomes lightlike, and the point of emission lies thus on the wave front.

(iii) Inside the ergosphere $K$ is spacelike and hence the emitting points are outside the wave fronts.

(iv) For $r = r_+$, $K$ is still spacelike, but the wave fronts arising from a point of emission on this surface lie entirely inside the surface, except for touching points *(Exercise)*. This shows that the surface $(r = r_+)$ is indeed a horizon. Show also that the horizon is a lightlike surface which is invariant with respect to $K$ and $\tilde{K}$.

## 7.9 Penrose Mechanism

Since $K$ is spacelike inside the ergosphere, it is possible in principle to extract energy form a black hole, thereby reducing its angular velocity and thus also the size of the ergosphere.

For example, imagine a piece of matter which falls from a large distance into the ergosphere, as indicated in Fig. 7.3, where it breaks up

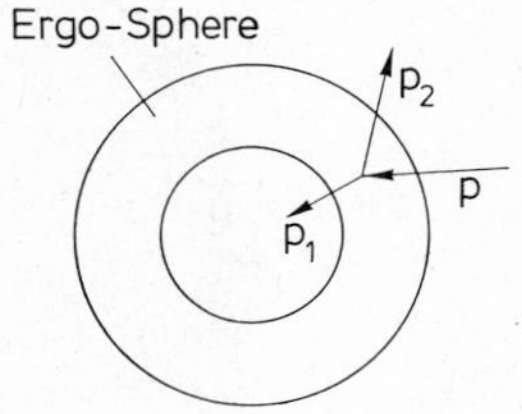


**Fig. 7.3.** Penrose mechanism

into two fragments in such a way that $E_1 := (p_1, K) < 0$. Here $p$, $p_1$ and $p_2$ are the corresponding four momenta, and $p = p_1 + p_2$. It is important that $(p, K)$ is constant along a geodesic, since

$$\nabla_u (u, K) = (\nabla_u u, K) + (u, \nabla_u K) = u^\mu u^\nu K_{\mu;\nu}$$
$$= \tfrac{1}{2} u^\mu u^\nu (K_{\mu;\nu} + K_{\nu;\mu}) = 0.$$

The scalar product $(p, K)$ is also the asymptotic energy of the particle. Since $E = (p, K) = (p_1, K) + (p_2, K)$, we see that $E_2 = (p_2, K) > E$. Hence the second fragment can leave the ergosphere and carry away more energy than the incoming object had. This process could provide simultaneous and permanent solutions to our energy and waste disposal problems.

## 7.10 The Second Law of Black Hole Dynamics

The most general form of the second law has been proved by Hawking. It states:

**Theorem:** In any (classical) interaction of matter and radiation with black holes, the total surface area of the boundaries of these holes (as formed by their horizons) can never decrease.

The proof can be found in the book of *Hawking* and *Ellis* [15]. The key idea is to use the following fact which was found previously by Penrose: Light rays generating a horizon never intersect. If one would assume that null geodesic generators of a horizon begin to converge and the energy density is always non-negative, then the focusing effect of the gravitational field would lead to the intersection of light rays. One must therefore conclude that the light rays generating the horizon cannot converge, so that the surface area of a horizon cannot decrease.

The limitation to classical interactions means that we do not consider changes in the quantum theory of matter due to the presence of the strong external gravitational fields of black holes. For macroscopic black holes, this is completely justified. If "miniholes" were to exist, quantum effects, such as spontaneous radiation, would become important.

As a special example, the second law implies that if two black holes collide and coalesce to a single black hole, then the surface area of the resulting black hole is larger than the sum of the surface of the event horizons of the two original black holes (see Fig. 7.4).

**Applications:** For a Kerr-Newman black hole, one easily finds for the surface area

$$A = 4\pi (r_+^2 + a^2) = 4\pi [(M + \sqrt{M^2 - a^2 - Q^2})^2 + a^2]. \qquad (7.10.1)$$

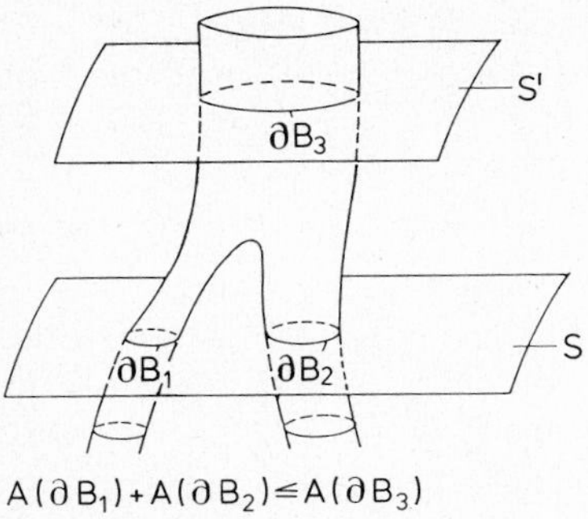


**Fig. 7.4.** Coalescence of two black holes as an illustration of the second law

---

**Exercise:** Prove (7.10.1).

---

We now set

$$A = 16\pi M_{\text{irr}}^2 . \tag{7.10.2}$$

We can then write (7.10.1) in the interesting form

$$M^2 = \underbrace{\left(M_{\text{irr}} + \frac{Q^2}{4M_{\text{irr}}}\right)^2}_{\text{“electromagnetic self-energy”}} + \underbrace{\frac{S^2}{4M_{\text{irr}}}}_{\text{“rotational energy”}} \tag{7.10.3}$$

According to the second law, the "irreducible mass" $M_{\text{irr}}$ cannot decrease when an isolated black hole interacts with matter and radiation. The maximum energy which can be extracted when $Q = 0$ is

$$\Delta M = M - M_{\text{irr}}; \quad \frac{\Delta M}{M} = 1 - \frac{1}{\sqrt{2}}[1 + \sqrt{1 - a^2/M^2}]^{1/2}. \tag{7.10.4}$$

Since $a^2 \leqq M^2$, this means that

$$\Delta M/M \leqq 1 - 1/\sqrt{2} . \tag{7.10.5}$$

We now consider two Kerr holes which collide and coalesce to a single hole, as shown in Fig. 7.4. The final hole is assumed to be stationary. The second law implies the inequality

$$M(M + \sqrt{M^2 - a^2}) \geqq M_1(M_1 + \sqrt{M_1^2 - a_1^2}) + M_2(M_2 + \sqrt{M_2^2 - a_2^2}) .$$

Hence, we have for the efficiency

$$\varepsilon := \frac{M_1 + M_2 - M}{M_1 + M_2} < \frac{1}{2} . \tag{7.10.6}$$

As a special case, take $M_1 = M_2 = \frac{1}{2}\mathcal{M}$ and $a_1 = a_2 = a = 0$. Then $M^2 \geqq 2(\frac{1}{2}\mathcal{M})^2$, or $M > \mathcal{M}/\sqrt{2}$ and hence

$$\varepsilon \leqq 1 - 1/\sqrt{2} = 0.293. \tag{7.10.7}$$

In principle, a lot of energy can be released.

One must also investigate the stability of the Kerr solution. The linearized perturbation theory of this solution has been extensively developed by *Chandrasekhar* in Chap. 7 of [21], and in [111].

---

**Exercise:** Generalize the Kerr-Newman solution to non-Abelian gauge theories.

---

## 7.11 Remarks on the Realistic Collapse

It is not possible to say much about the realistic collapse. Computer simulations for the nonspherical case are in progress [112], but it will be some time before these give a reliable picture and permit us to answer questions such as:

- When, if ever, are black holes formed in stellar core collapse?
- How much mass is driven off from the outer layers of a collapsing star?
- How much gravitational radiation and neutrino radiation is emitted?
- How do the answers to these questions depend on the mass and angular momentum of the star before the supernova explosion took place?
- What are the effects of magnetic fields?

At present, one can make the following qualitative remarks about the realistic collapse:

(i) The formation of trapped surfaces[3] is a stable phenomenon, because the stability of the Cauchy development (see Sect. 7.5 of [15]) shows that a trapped surface will be formed whenever the deviation from spherical symmetry is not too large.

A general theorem of *Hawking-Penrose* [15] then implies that a singularity must occur. According to the (unproven) “cosmic censorship hypothesis”, this will be hidden behind an event horizon. If this is the case, then not only the development of singularities, but also the occurrence of horizons is a stable phenomenon.

(ii) Trapped surfaces, and thus also singularities and horizons are presumably formed whenever an isolated mass $M$ becomes enclosed

[3] This is a spacelike 2-surface such that *both* the outgoing and the ingoing families of future-directed null geodesics orthogonal to it are *converging*. This means that if one imagined that the 2-surface emitted an instantaneous flash of light, both the outgoing and ingoing wave fronts would be decreasing in area. Thus, the outgoing wavefront would shrink to zero in a finite affine distance trapping within it all the material of the star (because it cannot travel outwards faster than light).

The reader is invited to illustrate this concept in the Schwarzschild-Kruskal manifold.

within a two-dimensional surface $S$ of "diameter" $d \sim 2\pi GM/c^2$, i.e., any two points of $S$ can be joined by a curve in $S$ whose length is smaller than $d$. The trapped surface then lies inside the horizon.
(iii) Calculations based on perturbation theory indicate that a black hole quickly (in $10^{-3} \cdot M/M_{\odot}$ s) approaches a stationary state after it is formed. The emission and partial reabsorption of gravitational waves is primarily responsible for this. The black hole is then a member of the Kerr-Newman family and thus characterized by its mass, angular momentum and charge (the latter is likely to be negligible in all realistic cases). These are the only properties of its previous history which are retained. A reliable calculation of the gravitational radiation emitted during this relaxation process does not exist.

In summary: We have acquired a crude picture of the realistic collapse by induction from the spherically symmetric case, a mixture of rigorous mathematical results, physical arguments, extrapolation from perturbation calculations and blind faith. It remains for the future to refine this picture in a reliable manner.

If a black hole is surrounded by matter, originating for example from a normal star which forms a close binary system with the black hole, then the matter will be sucked in by the black hole and heated up so strongly that it becomes a source of intense x-ray radiation. This possibility leads us to the next chapter, in which the reasons which strongly indicate the existence of a black hole in the x-ray source Cyg X-1 are discussed.

# Chapter 8. Binary X-Ray Sources

"I am prepared to say unequivocally that the beginning of x-ray astronomy, opening up a new window into the universe and revealing the existence of several new classes of astronomical objects, was the *most important* single scientific fruit of the whole space program. The newly discovered x-ray sources gave an entirely fresh picture of the universe, dominated by violent events, explosions, shocks and rapidly varying dynamical processes. X-ray observations finally demolished the ancient Aristotelian view of the celestical universe as a serene region populated by perfect objects moving in eternal peace and quietness. The old quiescent universe of Aristotle, which has survived essentially intact the intellectual revolutions associated with the names of Copernicus, Newton and Einstein, disappeared forever as soon as the x-ray telescopes went to work."

F. J. Dyson

## 8.1 Brief History of X-Ray Astronomy

X-rays from space are completely absorbed by the Earth's atmosphere and thus x-ray astronomy could not develop before it became possible to carry instruments aloft by balloons and little sounding rockets which were able to observe the x-ray sky for only a few minutes before they fell back down. As a first step, x-ray detectors were directed at the Sun and in 1948, the first solar x-rays were detected with a rocket-borne instrument.

In July 1962, the first point source outside the solar system was detected, using considerably improved instrumentation. This source is in the constellation Scorpio and thus was named Sco X-1. A brilliant achievement of this era was the measurement of the angular size of the x-ray source in the Crab Nebula in 1964 using the moon as an occultating disk. The sample of galactic x-ray sources was soon increased to about 30 and M 87 was identified as the first extragalactic x-ray source. In addition, a strong, apparently diffuse, background radiation was already discovered in 1962.

However, the "x-ray window" was really opened when NASA launched the famous x-ray satellite UHURU from the coast of Kenya on December 12, 1970. Uhuru is the Swahili word for "freedom" and was chosen because the launch took place on Kenya's independence day. Within less than two years, about one hundred galactic and fifty extragalactic x-ray sources were discovered with this orbiting x-ray observatory in the spectral range 2–20 keV. Some of these sources show very regular short x-ray pulsations (*x-ray pulsars*). One of the best known examples is Hercules X-1 with a period of 1.24 s. This period is, however, not strictly constant, but varies with a period of 1.70017 days; this is due to the fact that Her X-1 is a member of a close binary system. We now know of many such systems, for which the optical partner has also been unambiguously identified.

The completely irregular, rapidly fluctuating source Cyg X-1, which may contain a black hole (see Sect. 8.5) was also discovered with UHURU.

UHURU observations also showed that the space between galaxies in clusters of galaxies contains hot gas with a temperature of 10 million to 100 million K.

In 1975 and 1976 astronomers discovered a new class of x-ray sources, the so-called bursters, using the satellites ANS, SAS-3, OSO 7, 8 and others. One observed brief outbursts of x-rays from sources near the center of our galaxy or in globular clusters. They repeat themselves at irregular intervals which can lie between a few hours and several days. Typically the maximum intensity is reached in a few seconds and then falls back to its original value after about a minute. In this brief period some $10^{39}$ ergs of x-ray energy are emitted; this is comparable to the energy radiated by the Sun in about two weeks.

The detectors which were carried by these satellites did not have a very high sensitivity and thus only the strongest sources could be observed. An x-ray telescope having a sensitivity comparable to that of optical and radio telescopes in the corresponding spectral range became available with the launch of the "Einstein Observatory" in November, 1978. In addition, the telescope's resolution of four seconds of arc is 1000 times higher than the resolution of the x-ray detectors used previously. Thus, in just fifteen years, a development took place which is comparable to the progress achieved in optical astronomy from Galilei's telescope of 1610 to the five meter Hale reflector on Mount Palomar. This instrument is described in a pleasant Scientific American article by *R. Giacconi* [56]. The short lifetime of the "Einstein Observatory" ended already after two years. It allowed us for the first time to penetrate close to the central core of the mysterious engines which drive the most violent objects (active galaxies and quasars) in the Universe.

## 8.2 Mechanics of Binary Systems

We shall see that not only x-ray pulsars, but also bursters and Cyg X-1 are components of close binary systems. Therefore, it is necessary to make a few preliminary remarks about the mechanics of such systems.

Suppose for simplicity that the two components of the binary system move about each other in a circular orbit. The orbital period $P$ and the angular frequency $\Omega$ are given by Kepler's third law:

$$\Omega^2 = \left(\frac{2\pi}{P}\right)^2 = G\,\frac{M_1 + M_2}{a^3}\,, \tag{8.2.1}$$

where $M_1$ and $M_2$ are the masses of the two stars and $a$ is the distance between their centers of mass. We now consider a single gas particle in the field of the two masses (restricted three body problem). Its motion is most conveniently described in a co-rotating system:

$$\dot{\boldsymbol{v}} = -\nabla\psi - 2\boldsymbol{\Omega}\times\boldsymbol{v}\,. \tag{8.2.2}$$

Here $\boldsymbol{v}$ is the velocity and $\psi$ is the sum of Newtonian and gravitational potentials:

$$\psi(\boldsymbol{x}) = -\frac{GM_1}{r_1} - \frac{GM_2}{r_2} - \frac{1}{2}(\boldsymbol{\Omega}\times\boldsymbol{x})^2\,. \tag{8.2.3}$$

The equilibrium positions in the co-rotating system are thus precisely the critical points of $\psi$. There are five of these, of which three are collinear with $M_1$ and $M_2$ (*Euler* 1767) and two are equilateral (*Lagrange* 1773), if the units are suitably chosen. Moreover, the three collinear equilibrium points are unstable (*Plummer* 1901). The equilateral equilibrium points can be stable if the mass ratio is sufficiently small ("troyan asteroids").

After taking the inner product of (8.2.2) with $\boldsymbol{v}$, one finds the *Jacobi integral*

$$\tfrac{1}{2}\boldsymbol{v}^2 + \psi = \text{const}\,. \tag{8.2.4}$$

Obviously $\psi$ cannot increase above the value of this integral as the particle moves. For this reason, the structure of the equipotential surfaces of $\psi$ is extremely important. These are shown in Fig. 8.1, along

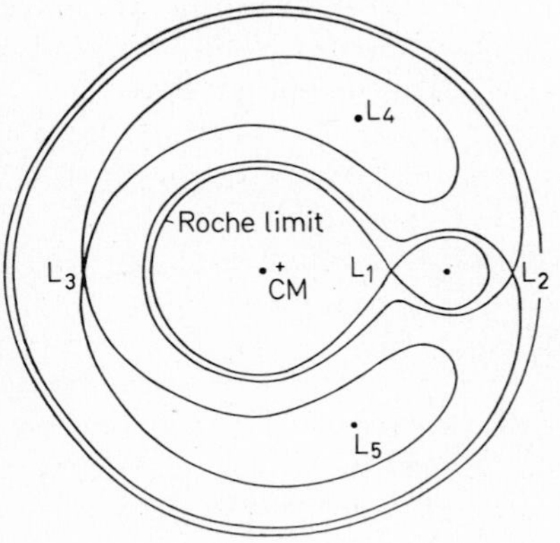


**Fig. 8.1.** Equipotential lines in the orbital plane of a binary star system, equilibrium points and Roche limit for the restricted three body problem. $L_1$, $L_2$ and $L_3$ are unstable equilibrium positions

with the five equilibrium positions. The equipotential surface passing through the "inner Lagrange point" $L_1$ is particularly important. Inside this critical Roche surface the equipotential surfaces, which enclose the two centers of mass, are disjoint; outside, this is not the case. The Roche limit thus determines the maximum volume (Roche volume) of a star. If this is exceeded, a portion of the outer layer of the star will flow over to the other star.

For a gas, the Navier-Stokes equation in the co-rotating system reads

$$D_t \boldsymbol{v} = -\nabla\psi - 2\boldsymbol{\Omega} \times \boldsymbol{v} - \frac{1}{\varrho}\nabla p + \text{friction terms}, \tag{8.2.5}$$

where $D_t \boldsymbol{v}$ denotes the hydrodynamic derivative of the velocity field. For a stationary situation ($\boldsymbol{v} = 0$ in the co-rotating system), one obtains, as expected,

$$\nabla P = -\varrho\nabla\psi\,. \tag{8.2.6}$$

Hence the surfaces of constant pressure, and in particular, the surface of the star, are equipotential surfaces of $\psi$.

We now imagine a situation in which the more massive of the two stars enters an evolutionary phase (red giant phase) during which it fills up the Roche limit. If this happens, matter flows through the inner Lagrange saddle point into the Roche volume of its partner. Part of it will escape and the rest will be accreted by the secondary. This mechanism will play an important role in the following.

---

**Exercise:** a) Show that (8.2.2) is the Euler-Lagrange equation of the Lagrange function

$$L = \tfrac{1}{2}\dot{\boldsymbol{x}}^2 + (\boldsymbol{\Omega} \times \boldsymbol{x}) \cdot \boldsymbol{x} - \psi\,.$$

Derive the Hamiltonian of the corresponding autonomous Hamiltonian system and show that it is just the Jacobi integral.
b) Show that the equilibrium positions of the corresponding Hamiltonian vector field are given by $\dot{\boldsymbol{x}} = 0$, $\nabla\psi = 0$. Determine the five equilibrium points and prove that the collinear positions are unstable. Demonstrate finally that the two Lagrangian points are elliptic, if

$$\frac{M_1}{M_1 + M_2} < \frac{1}{2} - \frac{\sqrt{69}}{18} = 0.03852\ldots.$$

(Note that this does not prove stability, which is a much more difficult problem that has been solved not long ago by V. I. Arnold.)

For the solution of this exercise, see [4], Sect. 10.2.

---

## 8.3 X-Ray Pulsars

We now know of about fifty compact galactic x-ray sources with an x-ray luminosity $L_x > 10^{34}$ erg/s, with star-like optical counterparts. About twenty of these are x-ray pulsars and have been identified unambiguously as binary star systems. (Observational details are described in [113].) In almost all cases, the companion star is a bright O or B star having a mass of about ten to twenty solar masses. Her X-1 is an exception to this rule, since the mass of its companion is only about $2 M_\odot$.

We believe that we have a definite qualitative interpretation of the x-ray pulsars. The companion of the normal star is a neutron star which sucks up gas which is flowing over. The accreted[1] matter falls eventually onto the surface of the neutron star, releasing about 10% of its rest energy. Neutron stars frequently have a strong magnetic field $B \sim 10^{12}$ G, since the magnetic flux is conserved when highly conductive stellar material collapses to a neutron star. This gives rise to an amplification factor for B of the order of $10^{10}$. (The evidence for such high fields is discussed in Sect. 6.10.) Such a strong field causes the plasma to move to the magnetic poles, where the falling matter gives rise to two "hot spots" of intense radiation (Fig. 8.2). The magnetic field axis does not in general coincide with the axis of rotation, and hence the hot spots rotate with the star. The direction of the emitted x-rays also rotates, similar to the beam emitted by a lighthouse. If the rotating beam sweeps by the Earth, the star appears as an x-ray pulsar[2]. In many cases, the x-ray emission is periodically eclipsed by the optical companion, with the orbital period of the binary system.

The observed x-ray luminosities are in the range $L_x \simeq 10^{36} - 10^{38}$ erg/s. Thermal radiation of this magnitude from a source with such small dimensions requires a temperature

$$kT = k\left(\frac{L_x}{\sigma A}\right)^{1/4} = 10\left(\frac{L_x}{10^{38}\ \mathrm{erg\ s^{-1}}}\right)^{1/4}\left(\frac{A}{\mathrm{km}^2}\right)^{-1/4}\ \mathrm{keV}, \tag{8.3.3}$$

---

[1] accrescere: increase by accumulation.

[2] The extension of the magnetosphere can be estimated as follows. At the Alfvén radius $r_A$, the magnetic pressure $B^2/8\pi$ is comparable to the momentum flux density $\varrho v_f^2$, where $v_f$ is the velocity of a freely falling body. From

$$B^2/8\pi \simeq \varrho v_f^2, \quad \tfrac{1}{2} v_f^2 = GM/r_A, \quad 4\pi r^2 \varrho v_f = \dot{M}, \tag{8.3.1}$$

one obtains for a typical accretion rate $\dot{M} \simeq 10^{17}$ g/s, $M \simeq M_\odot$, and a magnetic field strength at the pole of $B = 5 \times 10^{12}$ G an Alfvén radius $r_A \simeq 100\, R \simeq 10^8$ cm.

Since the integral curves of a dipole field satisfy $\sin^2 \vartheta / r = \mathrm{const}$, we expect for the surface area of the radiating polar cap

$$A \simeq \pi (R \sin \vartheta_0)^2 \quad \text{and} \quad r_A \sin^2 \vartheta_0 = R, \tag{8.3.2}$$

which means that $A \simeq \pi R^2 (R/r_A)^2 \simeq 1\ \mathrm{km}^2$.

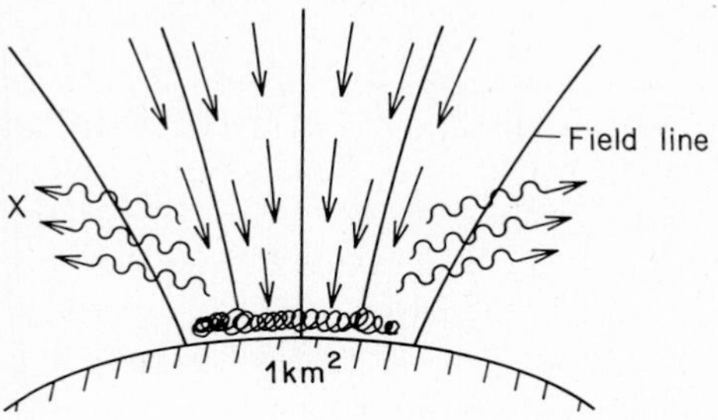


**Fig. 8.2.** Infalling matter at the polar cap. The x-ray luminosity for an accretion rate $\dot{M}$ is $L_X = \dot{M}\,GM/R \simeq 0.1\,\dot{M}c^2 = 10^{37}$ erg s$^{-1}$ $\dot{M}_{17}$. The corresponding radiation temperature is in the keV region

which implies that the x-rays must have energies of a few keV, in agreement with observation.

The mass-flow rates required to produce the observed x-ray luminosity can be sustained in a close binary system quite readily. Possible mechanisms are:

*(i) Stellar Wind:* The wind from O or B supergiants have an intensity of about $10^{-6}\,M_\odot$/year and this can easily lead to an accretion rate of $10^{-9}\,M_\odot$/year onto the compact companion. Since about 10% of the rest energy can be converted to radiation, this is sufficient to account for the observed x-ray luminosity.

The orbiting collapsed star is an obstacle in the supersonic stellar wind, and hence a bow-shaped shockfront is formed around it. Some of the material flowing through this shock front is decelerated enough to be captured by the compact star.

*(ii) Roche Lobe Overflow:* The normal companion can expand to the Roche limit if it is a blue or red supergiant. In this case, a considerable mass transfer of $(10^{-8} - 10^{-3})\,M_\odot$/year can occur. However, since column densities larger than about 1 g/cm$^2$ are opaque to keV x-rays, it is possible that this Roche overflow extinguishes the x-ray source. This type of mass transfer probably produces an x-ray source only if the normal component has a mass not larger than a few solar masses. (It certainly plays a crucial role in bursts sources; see Sect. 8.4.)

Figure 8.3 shows schematically these two canonical pictures of x-ray sources.

*(iii) X-Ray Heating:* A transfer of mass can also be induced if the x-rays heat up the outer layers of the optical companion. Such a self-sustaining mechanism may be operating in Her X-1.

The details of how accretion takes place, the conversion of potential energy to radiation and the transport of the radiation through the hot plasma which is present above the poles are extremely complicated problems which do not yet have a satisfactory solution. We refer the interested reader to the review by *Boerner* [58]. The spherical and disk accretion process will be discussed in great detail in Chap. 9.

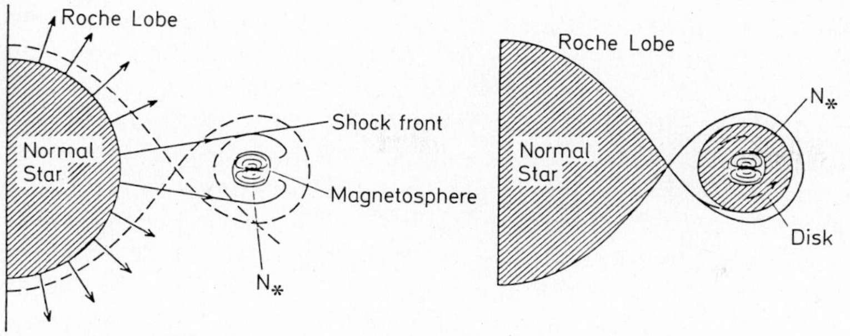


**Fig. 8.3.** Schematic picture of wind and disk accretion

In the case of five x-ray pulsars, one has succeded in determining the mass of the neutron star. Although the uncertainties are quite large, one is quite sure that they lie in the range $1-2\,M_\odot$ and that they certainly do not exceed $2.5\,M_\odot$ [59]. The methods used for the mass determination will be discussed in more detail in connection with Cyg X-1.

We recall that the mass of the binary radio pulsar PSR 1913+16 is known quite accurately (see Sect. 5.6).

## 8.4 Bursters

The bursters mentioned in Sect. 8.1 represent a completely different class of x-ray sources. We shall give the main reasons why the following interpretation is believed to be correct, and refer to [60] and [114] for details.

We are again dealing with compact binary systems of which one member is a neutron star. In this case, however, the normal optical companion is a rather low-mass star with perhaps $0.5-1\,M_\odot$. If this evolves to the red giant stage and fills its critical Roche volume, gas can flow over through the inner Langrange point to the neutron star and form an accretion disk, because the captured plasma has enough angular momentum relative to the compact partner. The critical volume can be attained by low-mass stars, since the orbital periods, at least in the cases when they could be determined, are (with few exceptions) only a few hours. The critical radius is given by the following approximate expression (Paczynski)

$$R_{\mathrm{cr}} = 0.46\left(\frac{M_1}{M_1+M_2}\right)^{1/3} a \quad \text{for} \quad 0 < \frac{M_1}{M_2} < 0.8\,. \tag{8.4.1}$$

Estimate $R_{\mathrm{cr}}$ for periods of a few hours.

Only in a few cases (x-ray transient sources), it has been possible to observe also absorption lines in the optical spectrum during quiescent periods, which are characteristic for a low mass star. This indicates that the dominant portion of the optical spectrum is emitted from the hot accretion disk, producing the emission lines and that the luminosity of the optical member is low. In this connection, it is important to recall that the x-ray sources which do not pulsate and which do not show eclipses are almost always found close to the center of the galaxy (galactic bulge sources) or in globular clusters, i.e., in rather old stellar populations. For this reason, one would expect that the neutron star's magnetic field has at least partially decayed [3] or that the dipole axis is aligned with the axis of rotation. This would explain the absence of pulsations. For a relatively weak magnetic field, matter falling onto the neutron star is distributed more or less uniformly over the surface of the neutron star and the gravitational energy released is emitted in the form of approximately constant x-radiation. Eclipses are not observed because the accretion disk is optically thick for x-rays and the tiny star thus lies in the x-ray shadow of the accretion disk.

The accreted matter consists mainly of H and He and thus fusion reactions can take place. In fact, it is fairly certain that most of the x-ray outbursts observed over and above the constant background (with one known exception) are due to thermonuclear explosions. Let us now explain this in more detail.

Underneath a layer of hydrogen about 1 m thick, the density has risen to about $10^{4-6}\,\mathrm{g/cm^3}$. (*Exercise:* Verify this by a simple estimation.) If the core of the neutron star is sufficiently hot ($\sim$ a few times $10^8$ K), the hydrogen can ignite at a depth of one meter. As detailed calculations show, an approximately equally thick layer of helium is produced just below it. At a still lower depth, helium becomes unstable with respect to fusion into carbon. Thus at least two thin shells are burning, as indicated in Fig. 8.4. As Schwarzschild and Härm discovered in a different connection, these shells are unstable against thermonuclear reactions [4]. The existence and strength of this instability are immediate consequences of the very strong temperature dependence of the thermonuclear reaction rates. For the situation under discussion, it is augmented by the partial degeneracy of the burning matter, so that the pressure does not respond as strongly to the rising temperature.

The $p-p$ chains are not sufficiently temperature sensitive to induce a runaway in the hydrogen-burning layer. The CNO cycle would be

---

[3] In the magnetohydrodynamic approximation, the characteristic diffusion time of the magnetic field is $\tau \simeq 4\,\pi\,\sigma\,R^2/c^2$. If one now inserts typical values for the conductivity $\sigma$, see [61], one obtains extremely long times. However, the field of very old neutron stars has probably partially decayed.

[4] An analytic discussion of this type of instability has been given by *Giannone* and *Weigert* [115].

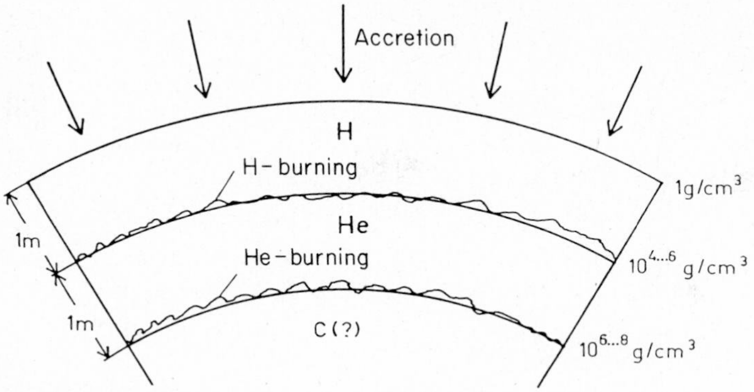


**Fig. 8.4.** Outer layers of an accreting weakly magnetized neutron star

sufficiently temperature dependent, but is saturated at high reaction rates by the long ($\sim$ 100 s) lifetimes of the $\beta$-unstable nuclei $^{13}$N, $^{14}$O, $^{15}$O and $^{17}$F which participate in the cycle.

On the other hand, the helium burning layer is unstable for a wide range of conditions, as has been confirmed by several detailed calculations [114]. After about $10^{21}$ g of matter have accumulated on the surface, a helium flash is produced in which almost all of the combustible matter is used up. Most of the energy ($\lesssim 10^{39}$ erg) is transported to the photosphere and radiated away in the form of x-rays. The calculated properties of such x-ray bursts (rise time, maximum luminosity, decay time, etc.) are quite similar to those observed [60], [114]. For a typical accretion rate of $10^{17}$ g/s, the interval between bursts is about three hours. In the fusion of helium to carbon, the nuclear energy is only about 1 MeV per nucleon. The time average of the energy in the bursts should, therefore, be roughly one percent of the energy released in the continuous flow of x-rays. This is indeed observed (with exceptions).

This thermonuclear flash model has been confirmed by further observations. During the cooling off phase of a burst, the spectrum is nearly that of a black body. The emitting surface area can then be determined from the luminosity and the approximate distance of the x-ray source. In all cases [62], one obtains radii of about ten kilometers after the first 15 seconds. (During the initial phase the radius is about 100 km.)

In addition, it has been shown that an outburst in the optical region occurs nearly simultaneously with the x-ray burst. Its maximum is, however, delayed by a few seconds. The interpretation is obvious: the x-ray burst heats up the accretion disk and a delayed optical "echo" results. The data show that the radius of the accretion disk is close to $10^6$ km.

In conclusion, we mention one (up to now) unique source, namely the Rapid Burster MXB 1730-335. During its active periods (which are separated by about six months) it displays x-ray bursts of the type already described every 3–4 hours. In addition, it emits in rapid succession, another kind of x-ray bursts, often a few thousand in one day. They are probably due to some sort of instability in the accretion flow. However, no one can really claim to have understood this machine-gun fire.

## 8.5 Cyg X-1: A Black Hole Candidate

Cyg X-1 is also a binary x-ray source. We shall now discuss the reasons why we believe that a black hole might be present in this compact x-ray source.

First of all, let us describe the observed properties of the system [116], [117]. As previously mentioned, we are dealing with a single-line spectroscopic binary system. The optical partner is a blue supergiant of ninth magnitude ($V = 8.87$, $B - V = 0.81$, $U - B = -0.30$), spectral type O 9.7 ab and is catalogued as HDE-226868. The orbital period is 5.6 days and the x-ray luminosity[5] $L_x$ is about $10^4 L_\odot$. The x-ray source fluctuates irregularly over all time intervals between 20 s and the shortest resolved time of about 0.01 s. No periodicity is observed and also no clear signal for x-ray eclipses has been detected. It is also worth noting that the x-ray spectrum is unusually hard, corresponding to a temperature of $3 \times 10^8$ K, with roughly equal amounts of energy in the bands $1 < E < 10$ keV and $10 \text{ keV} < E < 100$ keV.

The only reasonable explanation of these observations is that the x-radiation is emitted by gas falling onto a highly compact object, which can only be a neutron star or a black hole. The first possibility would be excluded if we can show convincingly that the compact object is too massive to be a neutron star. We give now the arguments which show that this is most probably the case.

For this, it is necessary to discuss some mechanical facts. With the notation in Fig. 8.5, we define the *mass function* of the system by

$$F(M_1, M_2, i) = M_1^3 \sin^3 i / (M_1 + M_2)^2 . \tag{8.5.1}$$

[5] Compare this with the *Eddington limit*, at which the rate of momentum transfer from the radiation to the gas is equal to the gravitational attraction. With obvious notation, we have

$$\sigma_T L_{\text{Edd}} / 4\pi r^2 c = G M m_p / r^2$$

which implies

$$L_{\text{Edd}} = 4\pi G M m_p c / \sigma_T = 1.3 \times 10^{38} \,[\text{erg/s}]\, M / M_\odot .$$

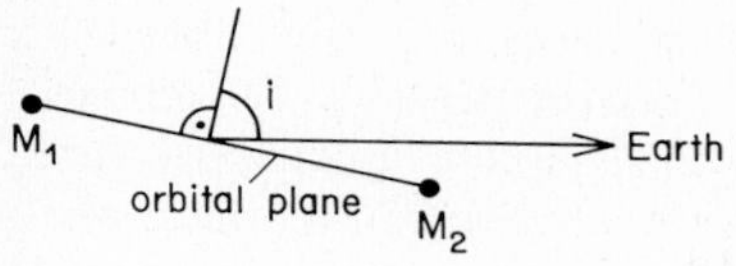


**Fig. 8.5.** Angle of inclination $i$

Equation (8.5.1) can be rewritten in the form

$$M_x = F_x\,(1+q)^2/\sin^3 i\,, \tag{8.5.2}$$

where $q = M_{opt}/M_x$ ($M_x$ is the mass of the x-ray star) and $F_x = F(M_x, M_{opt}, i)$.

From Kepler's third law, one obtains immediately

$$F(M_1, M_2, i) = \frac{4\pi^2}{G P^2}(a_2 \sin i)^3\,. \tag{8.5.3}$$

Here $a_1$ and $a_2$ denote the semimajor axes of the two orbits. If one now calculates the velocity of $M_2$ in the direction of observation, one finds for the arithmetic mean, $v_2$, of the maximum and minimum values

$$v_2 = \frac{2\pi}{P}\frac{a_2 \sin i}{\sqrt{1-e^2}}\,, \tag{8.5.4}$$

where $e$ is the eccentricity. Inserting this in (8.5.3) gives

$$F(M_1, M_2, i) = \frac{P}{2\pi G}(1-e^2)^{3/2} v_2^3\,. \tag{8.5.5}$$

Except for the eccentricity, the right-hand side contains only directly measurable quantities. The observed value of $v_2$ is 75.6 km/s. In order to determine $e$ one needs to know the precise form of the velocity curve. The component in the direction of observation is

$$(v_2)_{obs} = \frac{M_1}{[(M_1+M_2)\,a\,(1-e^2)]^{1/2}}\sin i\,[e\cos\omega + \cos(v+\omega)]\,, \tag{8.5.6}$$

where $v$ is the true anomaly and $\omega$ is the argument of the periastron (angular distance from the node).

---

**Exercise:** Derive Eqs. (8.5.4) and (8.5.6).

---

A fit to the observations [118] gives the rather small value $e \lesssim 0.02$ for Cyg X-1 and thus from (8.5.5)

$$F_x = (0.252 \pm 0.010)\,M_\odot\,. \tag{8.5.7}$$

Using this value for the mass function, it would be easy to establish a lower bound for $M_x$ from (8.5.2) if $M_{opt}$ were known. However, it is dangerous to infer the mass from the spectral type in the case of binary stars. We show now that it is possible to infer a (conservative) lower limit from the fact that prominent x-ray eclipses are not observed, without making any assumptions about the mass of the optical star. The following argument is due to *Paczynski* [119]. From geometric consideration, the radius $R_2$ of the visible star must satisfy the inequality $R_2 \leqq a \cos i$, where $a = a_1 + a_2$. Using (8.5.4), we can write this inequality in the form

$$x \leqq (1+q)\frac{\cos i}{\sin i}, \tag{8.5.8}$$

where $x$ is defined by

$$x := \frac{2\pi R_2}{P v_2}.$$

We next recall that if $D$ is the distance from the source, the Stefan-Boltzmann law implies

$$4\pi D^2 L_{\text{obs}} = 4\pi\sigma R_2^2 T_e^4.$$

The observed values give from the apparent magnitude with bolometric and absorption corrections

$$R_2/R_\odot \simeq 9.5\left[\frac{D}{1\,\text{kpc}}\right] \tag{8.5.9}$$

and thus

$$x = 1.14\left[\frac{D}{1\,\text{kpc}}\right]. \tag{8.5.10}$$

Inequality (8.5.8) is equivalent to

$$\sin^2 i \leqq \frac{(1+q)^2}{x^2 + (1+q)^2}.$$

If we now insert this into (8.5.2), we obtain

$$M_x = (1+q)^2 F_x/\sin^3 i \geqq \frac{F_x}{1+q}[x^2 + (1+q)^2]^{3/2}. \tag{8.5.11}$$

The right-hand side has a *minimum* with respect to $q$ for $(1+q)^2 = \frac{1}{2}x^2$ and hence, together with (8.5.7) and (8.5.10),

$$M_x \geqq \frac{3\sqrt{3}}{2} F_x x^2 = 3.4\, M_\odot \left[\frac{D}{2\,\text{kpc}}\right]^2. \tag{8.5.12}$$

From studies [120] of the absorption with distance for a large sample of stars in the direction of HDE-226868, the distance of Cyg X-1 is known to be approximately 2.5 kpc and certainly larger than 2 kpc. In this way, we arrive at the reliable lower limit $M_x \gtrsim 3.4\, M_\odot$.

One finds a tighter bound by assuming in addition that the optical star is confined to its Roche lobe:

$$R_2 \leqq R_{cr} . \tag{8.5.13}$$

In the mass range of interest, one can use the formula (Paczynski)

$$R_{cr} = a\left[0.38 + 0.2 \log\left(\frac{M_1}{M_2}\right)\right], \tag{8.5.14}$$

which is valid for $0.3 < M_1/M_2 < 20$.

From the definition of $x$ and (8.5.4) (for $e \approx 0$) the inequality (8.5.13) can be written as

$$x \leqq (1+q)\frac{0.38 + 0.2 \log q}{\sin i} . \tag{8.5.15}$$

The problem is now to find the minimum of the expression (8.5.2) for $M_x$ with respect to $q$ and $\sin i$, subject to the constraints (8.5.8) and (8.5.15).

---

**Exercise:** Show that the minimum occurs when equality holds in (8.5.8) and (8.5.15). Derive the numerical inequalities:

$$M_x \geqq \begin{cases} 6.5\, M_\odot & \text{for} \quad D = 2.0 \text{ kpc} \\ 9.5\, M_\odot & \text{for} \quad D = 2.5 \text{ kpc} . \end{cases} \tag{8.5.16}$$

---

More precise determinations make use of the observed optical light variations. This is a rather complicated matter and the interested reader is referred to the original literature (quoted in [59]). Even so, the determination of the mass involves rather large uncertainties. As an example, we mention that the optical velocity curve, which is inferred from the absorption lines, is subject to all sorts of disturbances, including x-ray heating of the optical companion's photosphere, tidal deformations of its surface, emission and absorption from gas streaming between the two stars. In addition, erratic variations of unknown origin occur.

The following result is given in [59]:

$$M_x = (9 - 15)\, M_\odot .$$

If this is correct, then, according to the discussion of Sect. 6.6, the compact object in Cyg X-1 cannot be a neutron star.

In conclusion, we can say that the mass $M_x$ is probably too large for a neutron star, but that not all doubts have been removed. (It is somewhat disappointing that the situation has not improved since a number of years.)

Two other sources, namely LMC X-3 and Cir X-1 are candidates for containing a black hole. However, the evidence is not as strong as it is for Cyg X-1.

## 8.6 Evolution of Binary Systems

We conclude with a sketch of a possible evolutionary scenario which may lead to the formation of binary x-ray sources.

We start off with a normal close binary system consisting of upper main-sequence stars. As a numerical example, we take their masses to be $20\,M_\odot$ and $8\,M_\odot$, and an orbital period of 4.5 days.

After about 6 million years, the more massive of the two stars has burned up all the hydrogen in the interior (massive stars have a more intense but shorter life). Core contraction sets in and the hydrogen continues to burn in a shell. The outer layers expand considerably in the process. If the star were isolated, it would become a red giant. However, in the binary system, the outer layers fill the Roche volume, and begin to deposit mass onto its companion. In the short time interval of about 30,000 years, nearly all of the hydrogen envelope of the primary streams over to the less massive star and a helium star having a mass of about $5\,M_\odot$ remains. The original less massive star has now become a very massive main sequence star with a mass of $23\,M_\odot$. As a result of angular momentum conservation, the orbital period has increased to about 11 days.

The further evolution of the helium star proceeds rather rapidly. After about half a million years, the core becomes unstable and collapses. A portion of the outer shell may be driven off in a supernova explosion[6]. The remnant is a neutron star (radio pulsar) or a black hole.

At that time, the secondary is still unevolved. At some point, however, it will also leave the main sequence and after perhaps four million years it becomes a blue supergiant with a strong stellar wind. Accretion of some of this matter by the compact companion turns the latter into a strong x-ray source. This stage lasts only about 40,000 years. Then the supergiant expands to fill its Roche lobe and it is possible that the x-ray source is thereby extinguished by the excessive accretion rate. The compact star can only accept a small fraction of the matter streaming over to it. The rest will be lost from the system. At the end of the process, a binary system, consisting of a helium star with about $6\,M_\odot$ and a neutron star, remains. Its period is about 0.2 days.

Finally, the helium star also becomes unstable. If the instability leads to a strong explosion, the system is probably disrupted. It may, however, be possible in some cases that the binary system survives. Then a system such as the binary pulsar PSR 1913+16 might come into being.

---

[6] The system will probably not be disrupted because the secondary has a much larger mass.

The observed phenomena in binary star systems are exceedingly diverse. Nova outbursts are also known to occur only in binary systems with periods less than a day, in which one component is a hot white dwarf. The white dwarf accretes hydrogen-rich material, which is heated by nearly adiabatic compression to high temperatures until hydrogen burning sets in. The hydrogen burning shell can (as for bursters) become thermally unstable due to the strong temperature dependence of the CNO-reactions and a nova event may result. It is possible that some type I supernova explosions also occur in binary systems in which one member is a white dwarf. The strange object SS 433 shows again that our imagination is often not sufficient to forsee even qualitatively the phenomena which occur in binary systems.

This all too brief discussion should still have made clear that astronomy is in an extremely fruitful phase of development. More surprises are ahead of us.

# Chapter 9. Accretion onto Black Holes and Neutron Stars [1]

We have seen in the last chapter that gas accretion by compact objects in close binaries from wind or Roche-lobe overflow drives these systems into strong x-ray sources. For an understanding of sources like Cyg X-1, it is important to construct models for the flow of gas onto a black hole and the properties of the radiation which is produced in this process.

Such models may also be relevant on a galactic scale. We know from observations that quasars and active galactic nuclei have enormous power outputs produced in very small volumes. A typical quasar can produce a hundred to a thousand times the luminosity of a normal galaxy, i.e. up to $10^{47}$ erg/s, or even more. The rapid fluctuations of optical and x-ray luminosity, which are sometimes as short as 1 d, show that the primary source must be very compact.

Many people have proposed that the energy source of quasars and other active galactic nuclei involves accretion onto a supermassive ($\sim 10^8 M_\odot$) black hole, because this is an efficient way to convert rest mass into radiation. It is quite plausible that massive collapsed objects are formed in the cores of some galaxies. A supermassive black hole may easily be fed by sufficient material from its surroundings (a few solar masses per year) to generate in a small volume the enormous luminosities observed from quasars [2]. For some recent reviews of this subject, see e.g. [124–127].

Realistic models for gas accretion are very difficult to construct and one must start with highly idealized situations. Great simplifications occur for radial-infall models and for thin accretion-disk configurations. Even then, a self-consistent solution of the hydrodynamic and energy transport equations seems at present to be outside of our possibilities.

---

[1] This chapter was written in collaboration with M. Camenzind.

[2] A useful measure is the Eddington luminosity $L_{\mathrm{Edd}} \sim 10^{46} (M/10^8 M_\odot)$ erg s$^{-1}$. If the radiation efficiency of the infalling gas is 0.1, then an accretion rate $\dot{M}$ with $0.1 \dot{M} c^2 \sim L_{\mathrm{Edd}}$ would correspond to $M/\dot{M} \sim 5 \times 10^7$ yr. This is in rough agreement with other independent estimates of quasar lifetimes.

Another useful number is: $(1\,\mathrm{d}) \cdot c \sim 100\,(M/10^8 M_\odot)$ Schwarzschild radii.

Spherical accretion onto a black hole becomes simple if the flow is nearly adiabatic. This idealization is not entirely academic, because it can be justified for a stellar mass black hole accreting gas in a typical interstellar environment. We shall develop the relativistic theory of adiabatic spherical accretion in Sect. 9.1.

In Sect. 9.2 we shall study thin accretion-disk models and their stability. In view of the great uncertainties (viscosity, etc.), we develop mainly the nonrelativistic theory. The basic equations and some immediate consequences of the relativistic theory will also be derived. In an Appendix, we give a short summary of nonrelativistic and relativistic hydrodynamics of viscous fluids.

We hope that this chapter will help the reader to find his way through the enormous literature on the subject.

## 9.1 Spherically Symmetric Accretion onto a Black Hole

In this section, we consider a Schwarzschild black hole at rest in an infinitely extended medium (plasma) and study the steady spherically symmetric accretion flow.

Even this idealized problem is difficult if one wants to find self-consistent flows which satisfy both the hydrodynamic equations and the transport equation for radiation. Under certain conditions, the flow is, however, nearly adiabatic. We discuss first this simpler problem which is a good starting point for the study of more realistic models.

The treatment will be fully relativistic since general relativity modifies the flow close to the horizon. Moreover, the relativistic generalization of Bondi's theory [128] for polytropic equations of state is no real complication. (The nonrelativistic theory is presented in detail in [129].)

### 9.1.1 Adiabatic Flow

We use a hydrodynamic description and postpone the discussion of its validity.

In the relativistic theory, we must distinguish between the rest-mass density, $\varrho_0$, and the total density of mass energy $\varrho$, which appears in the energy-momentum tensor for the (ideal fluid),

$$T^{\mu\nu} = (\varrho + P)\, U^\mu U^\nu - P\, g^{\mu\nu}\,. \tag{9.1.1}$$

For a chemically homogeneous fluid $\varrho_0 = n\, m$, where $n$ is the number density of baryons and $m$ is the mean rest mass of a baryon in the fluid.

The conservation of baryon number in differential form is ($\eta$ is the volume form):

$$L_U(\varrho_0 \eta) = 0 \quad \text{or} \quad \nabla \cdot (\varrho_0 U) = 0 \; ; \tag{9.1.2}$$

see the exercises at the end of Sect. 4.7 of Part I.

Next we derive the relativistic Bernoulli equation. Let $\xi$ be a Killing field whose flow leaves $U$, $\varrho$, and $P$ invariant,

$$L_\xi U = L_\xi \varrho = L_\xi P = 0 \, . \tag{9.1.3}$$

Then $\nabla \cdot T = 0$ and the Killing equation for $\xi$ imply (see Sect. 2.4)

$$(T^{\mu\nu} \xi_\nu)_{;\mu} = 0 \, . \tag{9.1.4}$$

In particular, for the expression (9.1.1) we obtain

$$[(\varrho + P) \, U^\mu \, \xi \cdot U]_{;\mu} - (P \, \xi^\mu)_{;\mu} = 0 \, .$$

The second term vanishes because $L_\xi (P \, \eta) = 0$. Thus

$$L_U[(\varrho + P)(\xi \cdot U)] + (\varrho + P) \, \xi \cdot U \operatorname{div} U = 0 \, .$$

Now (9.1.2) implies

$$(L_U \varrho_0) \, \eta + \varrho_0 \operatorname{div} U \, \eta = 0 \, ,$$

i.e.

$$L_U \varrho_0 = - \varrho_0 \operatorname{div} U \, . \tag{9.1.5}$$

Thus we find

$$L_U\left(\frac{\varrho + P}{\varrho_0} \, \xi \cdot U\right) = 0 \, , \tag{9.1.6}$$

which is the relativistic *Bernoulli equation.* In this equation $(\varrho + P)/\varrho_0$ is the relativistic enthalpy.

We now apply these results for our accretion problem in the Schwarzschild metric

$$g = e^{2\phi} \, dt^2 - [e^{-2\phi} \, dr^2 + r^2 (d\vartheta^2 + \sin^2 \vartheta \, d\varphi^2)],$$
$$e^{2\phi} = 1 - 2GM/r \, . \tag{9.1.7}$$

For stationary flows, the Killing field $\xi = \partial/\partial t$ satisfies the conditions (9.1.3).

The continuity equation (9.1.2), or

$$\partial_\mu (\sqrt{-g} \, \varrho_0 \, U^\mu) = 0$$

reduces to

$$\partial_r (r^2 \varrho_0 \, u) = 0 \, , \tag{9.1.8}$$

where $u \equiv - U^r$.

The Bernoulli equation (9.1.6) takes the form

$$\partial_r \left( \frac{\varrho + P}{\varrho_0} \Gamma \right) = 0 \,, \tag{9.1.9}$$

where

$$\Gamma = \xi \cdot U = g_{tt} \, U^t = \mathrm{e}^{2\phi} \, U^t \,.$$

Since

$$1 = (U, U) = \mathrm{e}^{2\phi} (U^t)^2 - \mathrm{e}^{-2\phi} (U^r)^2 \,,$$

we have

$$\Gamma = [\mathrm{e}^{2\phi} + (U^r)^2]^{1/2} = \left( 1 - \frac{2GM}{r} + u^2 \right)^{1/2} . \tag{9.1.10}$$

Integrating (9.1.8) and (9.1.9) gives

$$4\pi r^2 \varrho_0 u = \dot{M} \quad \text{(independent of } r\text{)}, \tag{9.1.11}$$

$$\frac{\varrho + P}{\varrho_0} \Gamma = \frac{\varrho_\infty + P_\infty}{\varrho_\infty} \,. \tag{9.1.12}$$

In (9.1.12) we have used the boundary condition $u_\infty = 0$ far away where the density and the pressure are $\varrho_\infty$ and $P_\infty$. The quantity $\dot{M}$ is the mass accretion rate.

For the discussion of these basic equations, we show next that the radial flow profiles $u(r)$ satisfy the following Pfaffian equation:

$$\left( a^2 - \frac{u^2}{\Gamma^2} \right) \frac{du}{u} + \left[ 2a^2 - \frac{GM}{r\Gamma^2} \right] \frac{dr}{r} = 0 \,. \tag{9.1.13}$$

This equation can be derived as follows.

From the basic equations (9.1.8) and (9.1.9) we obtain

$$\frac{d\varrho_0}{\varrho_0} + \frac{du}{u} + 2\frac{dr}{r} = 0 \tag{9.1.14}$$

$$\frac{d(\varrho + P)}{\varrho + P} + \frac{d\Gamma}{\Gamma} - \frac{d\varrho_0}{\varrho_0} = 0 \,. \tag{9.1.15}$$

Eliminating $d\varrho_0/\varrho_0$ gives

$$\frac{d(\varrho + P)}{\varrho + P} + \left[ 1 + \frac{u^2}{\Gamma^2} \right] \frac{du}{u} + \left[ 2 + \frac{GM}{r\Gamma^2} \right] \frac{dr}{r} = 0 \,. \tag{9.1.16}$$

For the first term we can write

$$\frac{d(\varrho + P)}{\varrho + P} = \left( \frac{\partial \ln(\varrho + P)}{\partial \ln \varrho_0} \right)_{\mathrm{ad}} d\ln \varrho_0 = (a^2 + 1) \left( -\frac{2\,dr}{r} - \frac{du}{u} \right), \tag{9.1.17}$$

where $a$ is the velocity of sound. Here we have used (9.1.14) and

$$\left(\frac{\partial \ln(\varrho+P)}{\partial \ln \varrho_0}\right)_{\mathrm{ad}} = \frac{\varrho_0}{\varrho+P}\left(\frac{\partial(\varrho+P)}{\partial \varrho_0}\right)_{\mathrm{ad}}$$

$$= \frac{\varrho_0}{\varrho+P}\left[\left(\frac{\partial \varrho}{\partial \varrho_0}\right)_{\mathrm{ad}} + \left(\frac{\partial P}{\partial \varrho_0}\right)_{\mathrm{ad}}\right] = \frac{\varrho_0}{\varrho+P}(1+a^2)\left(\frac{\partial \varrho}{\partial \varrho_0}\right)_{\mathrm{ad}} = 1+a^2 .$$

Inserting (9.1.17) into (9.1.16) gives Eq. (9.1.13).

Notice that the fluid velocity, measured by a stationary observer, is equal to $u/\Gamma$.

Suppose now that we have additional information which determines the thermodynamics of the fluid. Together with the asymptotic conditions and Eq. (9.1.11) we can then consider the velocity of sound as a given function of $r$, $u$ and the parameter $\dot{M}$. Instead of $\dot{M}$ we use also the dimensionless parameter $\lambda$, defined by

$$\dot{M} = 4\pi\lambda\left(\frac{GM}{a_\infty^2}\right)^2 \varrho_\infty a_\infty . \tag{9.1.18}$$

Thus (9.1.13) determines for each $\lambda$ a unique curve in the $(r, u)$-plane for which $u_\infty = 0$.

Consider first a value of $\lambda$ for which the Pfaffian form in (9.1.13) has no critical point. Then an accretion solution always remains subsonic, because if it would become sonic ($a = u/\Gamma$) at some point, then $du/dr$ would diverge. If we exclude such a singular behavior, the velocity goes to zero at the horizon, because (9.1.10) implies that $u/\Gamma = 1$ or $0$ for $r = r_g \equiv 2GM$. Clearly, only the first possibility is physically reasonable for an accretion flow. In that case, the fluid velocity, measured by stationary observers, approaches the velocity of light at the horizon.

Our discussion shows that this is only possible if $\lambda$ is chosen such that the Pfaffian form in (9.1.13) has a critical point, where

$$a_c^2 = u_c^2/\Gamma_c^2 , \qquad 2a_c^2 = \frac{GM}{r_c \Gamma_c^2} . \tag{9.1.19}$$

Together with (9.1.10) we find from these relations

$$u_c^2 = \frac{1}{4}\frac{r_g}{r_c} , \qquad a_c^2 = \frac{u_c^2}{1-3u_c^2} . \tag{9.1.20}$$

---

**Exercise:** Show that the radial component of the relativistic Euler equation is

$$u\frac{du}{dr} + \frac{\Gamma^2}{\varrho+P}\frac{dP}{dr} = -\frac{GM}{r^2} . \tag{9.1.21}$$

*Hint:* Use the connection forms relative to the usual orthonormal tetrad (Sect. 9.3.1).

---

We now study the special case of a polytropic equation of state

$$P = K\varrho_0^\gamma . \tag{9.1.22}$$

Let $\varrho = \varrho_0 + \varepsilon$ ($\varepsilon$ is the internal energy density). From the thermodynamic relation

$$P(\varrho_0, s) = \varrho_0 \left(\frac{\partial \varrho}{\partial \varrho_0}\right)_s - \varrho \tag{9.1.23}$$

and (9.1.22) we find

$$\left(\frac{\partial(\varepsilon/\varrho_0)}{\partial \varrho_0}\right)_s = K\varrho_0^{\gamma-2}$$

or integrated

$$\varepsilon = \frac{K}{\gamma-1}\varrho_0^\gamma = \frac{P}{\gamma-1} . \tag{9.1.24}$$

---

**Exercise:** It will turn out that close to the horizon the electrons become relativistic. Thus the thermal and caloric equations of state for an ideal mixture of protons and electrons are ($n_e = n_p \equiv n$):

$$P = 2nkT , \quad \varepsilon = (3+\tfrac{3}{2})\, nkT .$$

Show that this implies a polytropic equation of state for adiabatic changes, where the index $\gamma = 13/9$.

---

The velocity of sound is

$$a^2 = \left(\frac{\partial P}{\partial \varrho}\right)_s = \left(\frac{\partial P}{\partial \varrho_0}\right)_s \left(\frac{\partial \varrho}{\partial \varrho_0}\right)^{-1} = \left(\frac{\partial P}{\partial \varrho_0}\right)_s \frac{\varrho_0}{\varrho + P} = \frac{\gamma P}{\varrho + P} . \tag{9.1.25}$$

From (9.1.24), i.e. $\varrho + P = \varrho_0 + \dfrac{\gamma}{\gamma-1}P$, we obtain also

$$a^2 = \frac{\gamma(P/\varrho_0)}{1 + \dfrac{\gamma}{\gamma-1}(P/\varrho_0)} \tag{9.1.26}$$

and from this

$$\gamma K \varrho_0^{\gamma-1} = \frac{a^2}{1-(a^2/\gamma-1)} . \tag{9.1.27}$$

The Bernoulli equation (9.1.12) becomes, using (9.1.10) and (9.1.24)

$$\left(1 + \frac{\gamma K}{\gamma-1}\varrho_0^{\gamma-1}\right)^2 \left(1 - \frac{r_g}{r} + u^2\right) = \left(1 + \frac{\gamma K}{\gamma-1}\varrho_\infty^{\gamma-1}\right)^2 . \tag{9.1.28}$$

This equation, together with (9.1.11), determines the velocity profile. For the physically relevant solution, we must choose $\dot{M}$ (or $\lambda$) such that (9.1.20) holds.

At the critical point, Eq. (9.1.28) gives with the aid of (9.1.27) and (9.1.20)

$$\left(1-\frac{a_c^2}{\gamma-1}\right)^2(1+3a_c^2)=\left(1-\frac{a_\infty^2}{\gamma-1}\right)^2. \qquad (9.1.29)$$

In practice, the critical point is sufficiently far away from the horizon that we can use there nonrelativistic approximations. (The situation at infinity is assumed to be nonrelativistic.) Expanding (9.1.29) gives

$$\begin{aligned} a_c &= a_\infty\left(\frac{2}{5-3\gamma}\right)^{1/2} \quad \text{for} \quad \gamma \neq \tfrac{5}{3} \\ a_c &= \left(\tfrac{2}{3}\,a_\infty\right)^{1/2} \quad \text{for} \quad \gamma = \tfrac{5}{3}\,. \end{aligned} \qquad (9.1.30)$$

(For $\gamma=\frac{5}{3}$ one has to go to second order!) From the nonrelativistic approximation of (9.1.20), i.e.

$$u_c^2 \simeq a_c^2 \simeq \frac{1}{4}\frac{r_g}{r_c} \qquad (9.1.31)$$

we obtain

$$r_c = \begin{cases} \dfrac{5-3\gamma}{4}\dfrac{GM}{a_\infty^2} & \text{for} \quad \gamma \neq \tfrac{5}{3} \\[2ex] \dfrac{3}{4}\dfrac{GM}{a_\infty} & \text{for} \quad \gamma = \tfrac{5}{3}\,. \end{cases} \qquad (9.1.32)$$

If we use this and $a_c^2/a_\infty^2 \simeq (\varrho_{0c}/\varrho_\infty)^{\gamma-1}$ [see (9.2.26)] in

$$\dot{M} = 4\pi\, r_c^2\, u_c\, \varrho_{0c}$$

then we find for the critical value of $\lambda$ in (9.1.18)

$$\lambda_c \simeq \frac{1}{4}\left(\frac{5-3\gamma}{2}\right)^{\frac{3\gamma-5}{2(\gamma-1)}}. \qquad (9.1.33)$$

We have thus determined the accretion rate uniquely. [The approximate expression (9.1.33) agrees with that of the nonrelativistic theory.] The corresponding transsonic solution can be determined numerically from (9.1.28) and (9.1.11). An example is given in Fig. 9.1.

For a more realistic equation of state for the plasma (ideal mixture, treating the electrons relativistically), the accretion rate is almost the same, but otherwise the flow is quantitatively quite different [131].

Numerically, (9.1.18) and (9.1.32) give for $\gamma=\frac{5}{3}$

$$\dot{M} \simeq 10^{-15}\left(\frac{M}{M_\odot}\right)^2\left(\frac{\varrho_\infty}{10^{-24}\,\mathrm{g\,cm^{-3}}}\right)\left(\frac{a_\infty}{10\ \mathrm{km/s}}\right)^{-3} M_\odot\,\mathrm{yr}^{-1}\,. \qquad (9.1.34)$$

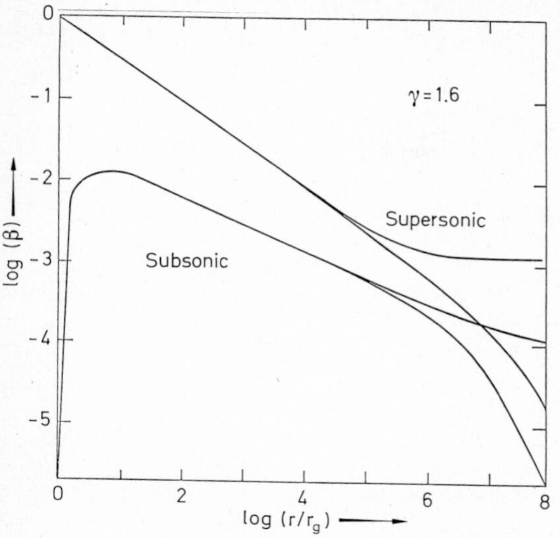


**Fig. 9.1.** Accretion flow for $\gamma = 1.6$ and boundary conditions $T_\infty = 10^4$ K, $\varrho_\infty = 10^{-22}$ g cm$^{-3}$ (from [130])

**Exercise:** Eliminate $\varrho_0$ in (9.1.28) with (9.1.11) and show that $z = u^{2/3}$ satisfies a fifth-order polynomial equation, whose coefficients depend on $\lambda$. Solve this equation for various values of $\lambda$ on a computer and determine the critical value of $\lambda$. Compare the result with the approximation (9.1.33).

### 9.1.2 Thermal Bremsstrahlung from the Accreting Gas

In order to estimate the radiation from the accreting gas, we need approximate expressions for the density and temperature profiles at small $r$ ($r \ll r_c$), where the temperature becomes high. Well inside the critical point the fluid velocity $u$ is much larger than the velocity of sound and we obtain from (9.1.28) and (9.1.27) for $\gamma \neq \frac{5}{3}$

$$u^2 \simeq \frac{r_g}{r} \quad (\gamma \neq \tfrac{5}{3}) . \tag{9.1.35}$$

In other words, the deceleration due to the gas pressure becomes negligible and the flow velocity approaches the free-fall speed.

The case $\gamma = \frac{5}{3}$ is special since $u$ remains comparable to $a$. Let us evaluate for this case $u$ and $\varrho_0$ at the horizon (indexed by $h$). For $r = r_g$, Eq. (9.1.28) gives

$$u_h\left(1+\frac{\gamma K}{\gamma-1}\,\varrho_{0h}^{\gamma-1}\right) \simeq 1 , \tag{9.1.36}$$

since $a_\infty \sim a^2$, by (9.1.30), and is thus of second order. Now (9.1.18) and (9.1.11) give

$$4\pi r^2 u\, \varrho_0 = 4\pi \lambda_c (GM)^2 \varrho_\infty a_\infty^{-3} . \tag{9.1.37}$$

If we use this for $r = r_g$ to eliminate $\varrho_0$ in (9.1.36), we obtain, together with $a_\infty^2 \approx \gamma K \varrho_\infty^{\gamma-1}$, at the horizon:

$$u_h \left[ 1 + \frac{1}{\gamma - 1} (a_\infty)^{-(3\gamma-5)} \left( \frac{\lambda_c}{4 u_h} \right)^{\gamma-1} \right] \simeq 1 .$$

For $\gamma \neq \frac{5}{3}$ this implies, as expected, $u_h = 1$. For $\gamma = \frac{5}{3}$ we find ($\lambda_c \simeq \frac{1}{4}$)

$$u_h + \frac{3}{2^{11/3}} u_h^{2/3} \simeq 1 , \qquad (\gamma = \tfrac{5}{3})$$

giving

$$u_h \simeq 0.78 , \qquad (\gamma = \tfrac{5}{3}) \tag{9.1.38}$$

and thus from (9.1.37)

$$\varrho_{0h} = \frac{\lambda_c \varrho_\infty}{4 u_h a_\infty^3} \simeq \frac{1}{16 u_h} \frac{\varrho_\infty}{a_\infty^3} , \qquad (\gamma = \tfrac{5}{3}) . \tag{9.1.39}$$

We use (9.1.37) also for $\gamma \neq \frac{5}{3}$ to obtain the density profile. Inserting (9.1.35) gives

$$\frac{\varrho_0}{\varrho_\infty} \simeq \frac{\lambda_c}{\sqrt{2}} \left( \frac{GM}{r a_\infty^2} \right)^{3/2} , \qquad (\gamma \neq \tfrac{5}{3}) . \tag{9.1.40}$$

Let us now assume an ideal gas law:

$$P = \frac{\varrho_0}{m} k T . \tag{9.1.41}$$

For an adiabatic compression, we have from (9.1.40)

$$\frac{T}{T_\infty} = \left( \frac{\varrho_0}{\varrho_\infty} \right)^{\gamma-1} \simeq \left( \frac{\lambda_c}{\sqrt{2}} \right)^{\gamma-1} \left( \frac{GM}{r a_\infty^2} \right)^{3(\gamma-1)/2} , \qquad (\gamma \neq \tfrac{5}{3}) . \tag{9.1.42}$$

The limiting values of the density and temperature are

$$\frac{\varrho_{0h}}{\varrho_\infty} \simeq \frac{\lambda_c}{4} \frac{1}{a_\infty^3} , \qquad \frac{T_h}{T_\infty} \simeq \left[ \frac{\lambda_c}{4} \frac{1}{a_\infty^3} \right]^{\gamma-1} , \qquad (\gamma \neq \tfrac{5}{3}) . \tag{9.1.43}$$

For the special case $\gamma = \frac{5}{3}$ we obtain

$$\frac{\varrho_{0h}}{\varrho_\infty} \simeq 0.08 \frac{1}{a_\infty^3} , \qquad \frac{T_h}{T_\infty} \simeq 0.19 \frac{1}{a_\infty^2} , \qquad (\gamma = \tfrac{5}{3}) . \tag{9.1.44}$$

Now the asymptotic speed of sound is

$$a_\infty = \left( \frac{5}{3} \frac{k T_\infty}{m} \right)^{1/2} \simeq 11.7 \left( \frac{T_\infty}{10^4\,\mathrm{K}} \right)^{1/2} \mathrm{km/s} . \tag{9.1.45}$$

Thus, for $\gamma = \frac{5}{3}$ the temperature at the horizon is independent of $T_\infty$:

$$T_h \simeq 0.11 \frac{m c^2}{k} = 1.2 \times 10^{12}\,\mathrm{K}\,, \qquad (\gamma = \tfrac{5}{3})\,. \tag{9.1.46}$$

A more accurate calculation [131] gives $T_h = 6.075 \times 10^{11}$ K for $T_\infty = 10^4$ K. However, a more realistic equation of state (as in the exercise on page 393) gives a lower temperature by about an order of magnitude [131]:

$$T_h = 7.6 \times 10^{10}\,\mathrm{K} \tag{9.1.47 a}$$

(practically independent of $T_\infty$) and

$$\frac{\varrho_{0h}}{\varrho_\infty} = 3.9 \times 10^{11} \left(\frac{T_\infty}{10^4\,\mathrm{K}}\right)^{-1}; \tag{9.1.47 b}$$

the $T_\infty$-dependence is approximate. We use these values to estimate the total luminosity due to thermal bremsstrahlung (free-free emission).

---

**Exercises:** The cross sections for bremsstrahlung in electron-ion and electron-electron collisions are derived in great detail in Vol. 4 of Landau-Lifschitz, § 92, 93 [132]. Use the results to derive the following formulae for the thermal bremsstrahlung emissivity.

1. Write the non-relativistic cross section (L.L.92.15) (i.e. formula (92.15) in [132]) in the form ($v$ is the electron velocity):

$$\frac{d\sigma}{d\nu} = \frac{16\pi}{3\sqrt{3}} \frac{\alpha c^2}{v^2} \frac{Z^2 r_0^2}{\nu} G(\nu, v)\,, \qquad \left(r_0 = \frac{e^2}{m_e c^2}\right) \tag{9.1.48}$$

and deduce the following spectral emissivity for a fully ionized plasma

$$\varepsilon_\nu = \frac{2^5 \pi e^6}{3 m_e c^3} \left(\frac{2\pi}{3 k m_e T}\right)^{1/2} Z^2 n_i n_e \, \mathrm{e}^{-h\nu/kT} \bar{G}(\nu, T)\,. \tag{9.1.49}$$

Here $n_e$, $n_i$ are the electron and ion number densities and $\bar{G}(\nu, T)$ is the average Gaunt factor

$$\bar{G}(\nu, T) = \int_0^\infty G\left(\nu, v = \sqrt{\frac{2(y k T + h\nu)}{m_e}}\right) \mathrm{e}^{-y}\, dy\,. \tag{9.1.50}$$

Show that in the Born approximation (L.L.92.16)

$$\bar{G}(\nu, T) = \left(\frac{3}{\pi} \frac{kT}{h\nu}\right)^{1/2}. \tag{9.1.51}$$

Derive (L.L.92.16) also from scratch.

Numerically (9.1.49) is in cgs-units:

$$\varepsilon_\nu = (6.8 \times 10^{-38}\,\mathrm{erg\,s^{-1}\,cm^{-3}\,Hz^{-1}})\, Z^2 n_e n_i T^{-1/2} \mathrm{e}^{-h\nu/kT} \bar{G}(\nu, T)\,. \tag{9.1.52}$$

Integrate $\varepsilon_\nu$ over $\nu$ to find

$$\varepsilon = \left(\frac{2\pi k T}{3m}\right)^{1/2} \frac{2^5 \pi e^6}{3 h m c^3} Z^2 n_e n_i \bar{\bar{G}}(T)$$
$$= (1.4 \times 10^{-27} \text{ erg s}^{-1} \text{ cm}^{-3})\, T^{1/2} n_e n_i Z^2 \bar{\bar{G}}(T)\,, \tag{9.1.53}$$

where

$$\bar{\bar{G}}(T) = \int_0^\infty \bar{G}(\nu = x\, kT/h, T)\, e^{-x}\, dx\,. \tag{9.1.54}$$

Numerical values for $\bar{G}$ and $\bar{\bar{G}}$ can be found in [133].
2. The ultrarelativistic cross section for electron-ion bremsstrahlung is given in Eq. (L.L.93.17) and the integral over the spectrum in (L.L.92.24):

$$\int h\nu \frac{d\sigma}{d\nu} d\nu = 4 Z^2 \alpha r_0^2 E \left(\ln \frac{2E}{m_e c^2} - \frac{1}{3}\right). \tag{9.1.55}$$

Here $E$ denotes the electron energy. (To logarithmic accuracy this can be derived quickly with the method of virtual quanta; see [132], § 99.)

Deduce from this the emissivity

$$\varepsilon_{ei} = 12 \alpha Z^2 r_0^2 n_e n_i c\, kT \left[\frac{3}{2} + \ln\left(\frac{2kT}{m_e c^2}\right) - \gamma\right], \tag{9.1.56}$$

where $\gamma = 0.577$ is the Euler constant.

Show that the ultrarelativistic expression (L.L.97.12) for electron-electron bremsstrahlung leads to [134]

$$\varepsilon_{ee} = 24 \alpha r_0^2 n_e^2 c\, kT \left[\frac{5}{4} + \ln\left(\frac{2kT}{m_e c^2}\right) - \gamma\right]. \tag{9.1.57}$$

---

Inserting (9.1.47) into the formulae (9.1.56), (9.1.57) gives for the emission rate per unit volume close to the horizon

$$\varepsilon_{ff}|_{\text{horizon}} = (8.7 \times 10^3 \text{ erg s}^{-1} \text{ cm}^{-3}) \left(\frac{n_\infty}{\text{cm}^{-3}}\right)^2 \left(\frac{T_\infty}{10^4 \text{ K}}\right)^{-2}. \tag{9.1.58}$$

The luminosity gradient in the optically thin case is given by [see (6.7.2)]:

$$\frac{d}{dr}(L_r e^{2\phi}) = 4\pi r^2 e^{\phi} \varepsilon_{ff}\,. \tag{9.1.59}$$

The total emitted luminosity of the plasma is thus

$$L_{ff} = \int_{r_g}^\infty \varepsilon_{ff} e^{\phi} 4\pi r^2\, dr\,. \tag{9.1.60}$$

An order of magnitude estimate is

$$L_{\mathrm{ff}} \sim \frac{4\pi}{3} r_g^3 \varepsilon_{\mathrm{ff}}|_{\mathrm{horizon}} \sim 9\times 10^{20} \left(\frac{n_\infty}{\mathrm{cm}^{-3}}\right)^2 \left(\frac{T_\infty}{10^4\,\mathrm{K}}\right)^{-2} \left(\frac{M}{M_\odot}\right)^3 \mathrm{erg/s}. \tag{9.1.61}$$

**Exercises:** For stellar mass black holes and typical interstellar conditions show that:

1. The adiabatic compressional heating is much larger than the emission due to thermal bremsstrahlung. (Hence the adiabatic approximation is justified.)
2. The optical depth due to the dominant Thomson scattering is much less than unity. [This justifies the use of (9.1.60).]

The luminosity (9.1.61) is very low for stellar mass black holes. Note, however, that (9.1.61) would become very large for supermassive black holes. The efficiency for the conversion of rest-mass energy into radiation is from (9.1.61), (9.1.34), and (9.1.45):

$$\varepsilon \equiv \frac{L_{\mathrm{ff}}}{\dot{M} c^2} \sim 10^{-11} \left(\frac{n_\infty}{\mathrm{cm}^{-3}}\right) \left(\frac{T_\infty}{10^4\,\mathrm{K}}\right)^{-1/2} \left(\frac{M}{M_\odot}\right). \tag{9.1.62}$$

This is extremely low for normal conditions, but would become interesting for $M \sim 10^9 M_\odot$.

The spectrum for bremsstrahlung is rather flat below $kT_h$ and falls off exponentially above this energy [see (9.1.49) and (9.1.51)]. Thus, much of the emission would be in the form of hard x-rays and $\gamma$-rays.

If tangled magnetic fields are present at the "accretion radius" $r_A = GM/a_\infty^2$, one expects that the plasma becomes magneto-turbulent and that synchrotron radiation will be important. (For a rough estimate, based on equipartition, see [129]; the problem has recently been studied in great detail in [135].)

The presence of (weak) magnetic fields would also justify the use of the hydrodynamic description, because the Larmor radius

$$\varrho_L = \frac{m_p c v}{eB} \simeq (1\times 10^8\,\mathrm{cm}) \left(\frac{B}{10^{-6}\,\mathrm{G}}\right)^{-1} \left(\frac{T}{10^4\,\mathrm{K}}\right)^{1/2} \tag{9.1.63}$$

is small compared to the scales over which the flow varies substantially. Otherwise, the average distance over which a proton is appreciably deflected by Coulomb scattering is not very much less than the accretion radius. This is shown in the next exercise.

**Exercise:** Consider the accretion of a hydrogen plasma without magnetic fields. The deflection of a particle is dominated by the cumulative effect of many small-angle scatterings.

Show that the mean-square deflection angle for a proton which penetrates a depth $L$ is

$$\langle(\Delta\theta)^2\rangle = \frac{8\pi n L e^4}{m^2 v^4} \ln\left(\frac{b_{\max}}{b_{\min}}\right). \tag{9.1.64}$$

Take for the maximal impact parameter $b_{\max}$ the Debye length,

$$b_{\max} = \lambda_{\mathrm{D}} = \left(\frac{k\,T}{4\pi\,n\,e^2}\right)^{1/2}, \tag{9.1.65}$$

and for the minimum impact parameter $b_{\min}$ the value which corresponds to a 90° scattering,

$$b_{\min} = \frac{e^2}{m\,v^2} \simeq \frac{e^2}{3k\,T}. \tag{9.1.66}$$

Thus

$$\langle(\Delta\theta)^2\rangle = L\,\frac{8\pi\,n\,e^4}{m^2\,v^4}\,\ln\Lambda, \tag{9.1.67}$$

where

$$\Lambda = \frac{3}{2}\left(\frac{k^3\,T^3}{\pi\,n}\right)^{1/2}\frac{1}{e^3}. \tag{9.1.68}$$

Setting $\langle(\Delta\theta)^2\rangle = 1$ gives the length on which a proton scatters, by random walk, through an angle of about 90°:

$$L_{90°} \simeq \frac{9}{8\pi}\,\frac{(k\,T)^2}{n\,e^4\ln\Lambda}. \tag{9.1.69}$$

Numerically

$$\ln\Lambda \simeq 23 + \tfrac{3}{2}\ln\left(\frac{T}{10^4\,\mathrm{K}}\right) - \tfrac{1}{2}\ln\left(\frac{n_{\mathrm{e}}}{\mathrm{cm}^{-3}}\right). \tag{9.1.70}$$

Compute the ratio

$$\frac{L_{90°}}{r_{\mathrm{A}}} \simeq 0.1\left(\frac{M}{M_\odot}\right)^{-1}\left(\frac{\varrho_\infty}{10^{-24}\,\mathrm{g/cm^3}}\right)^{-1}\left(\frac{T_\infty}{10^4\,\mathrm{K}}\right)^3. \tag{9.1.71}$$

This shows that the hydrodynamic description is not really justified (without magnetic fields).

---

Finally, we refer to some further recent studies [135–141] of spherically symmetric accretion, which include dissipation and the transport of radiation through the plasma. The problem then becomes much more complicated, since the transport of the photons cannot be treated everywhere in the diffusion approximation. (Some authors use Monte Carlo simulations for this process.)

## 9.2 Disk Accretion onto Black Holes and Neutron Stars

### 9.2.1 Introduction

Under many circumstances the matter accreting onto a compact object will have significant angular momentum and will therefore form a disk.

The gas elements in the disk lose angular momentum, due to friction between adjacent layers and spiral inwards. Part of the released gravitational energy increases the kinetic energy of rotation and the other part is converted into thermal energy which is radiated from the disk surface. Thus, viscosity converts gravitational potential energy in an efficient manner into radiation.

The behavior of rotating gas masses and the formation of disks have been studied already long ago – before the interest in accretion driven x-ray sources – in connection with the evolution of the early solar nebula. An important example is *von Weizsäcker's* paper of 1948 [142], entitled "The rotation of cosmic gas masses", in which he derived some of the basic equations which will be important in this section. He suggested also that turbulent viscosity should be the dominant dissipation process. Later, in 1952, this problem was taken up again by *Lüst* [143].

Early studies of accretion disks to explain x-ray binaries and active galactic nuclei are described in [144–147].

Fairly direct evidence for the presence of disk accretion is available in many cataclysmic variables. We have also seen that there is good evidence for disk accretion in x-ray bursters.

### 9.2.2. Basic Equations for Thin Accretion Disks (Non-Relativistic Theory)

In this section we set up the basic equations which govern the structure of thin accretion disks. We use nonrelativistic hydrodynamics. In view of the uncertainties as to the nature and magnitude of the viscosity, general relativistic corrections are at the moment of secondary importance, except for the region close to the black hole (see Sect. 9.2.6).

For the stability analysis of disk models, it is necessary to keep the time dependence in the equations.

Notations and the basic equations of hydrodynamics are summarized in an Appendix.

#### *a) Equation of State*

Since we are concerned with fully ionized disk plasmas, the pressure is the sum of the gas and radiation pressures,

$$P = P_{\mathrm{g}} + P_{\mathrm{r}} = \frac{k}{\mu\, m_{\mathrm{H}}}\, \varrho\, T + \frac{a}{3}\, T^4. \qquad (9.2.1)$$

The internal specific energy is

$$\varepsilon = c_v\, T + \frac{a\, T^4}{\varrho}.$$

Let $\gamma = c_p/c_v$, $P_g = \beta P$. Using $c_p = c_v + k/\mu m_H$ we get

$$\varepsilon = \frac{3P_r}{\varrho} + c_v \frac{P_g}{\varrho(c_p - c_v)} = \frac{P}{\varrho}\left[\frac{\beta}{\gamma - 1} + 3(1-\beta)\right].$$

Thus

$$\varepsilon = A\frac{P}{\varrho}, \quad A = \frac{\beta}{\gamma - 1} + 3(1-\beta). \tag{9.2.2}$$

To describe the axisymmetric structure of an accretion disk, we employ cylindrical coordinates $(r, \varphi, z)$ with the $z$-axis chosen as the axis of rotation ($z = 0$ is the central plane of the disk). The mean velocity field $\boldsymbol{v}$ is a small perturbation of the Keplerian circular motion $(0, (GM/r)^{1/2}, 0)$.

### b) Mass Conservation

The continuity equation is (dropping the $\varphi$-derivatives)

$$\partial_t \varrho + \frac{1}{r}\partial_r(r\varrho v_r) + \partial_z(\varrho v_z) = 0. \tag{9.2.3}$$

Let us integrate this equation over the $z$-direction

$$\partial_t \int \varrho\, dz + \frac{1}{r}\partial_r\left(\int r\varrho v_r\, dz\right) = 0. \tag{9.2.4}$$

A useful variable is the surface density

$$S(r, t) = \int \varrho\, dz. \tag{9.2.5}$$

For thin disks (height much smaller than radial variable), we can neglect the vertical variation of $v_r$ in the second integral of (9.2.4) and thus obtain

$$\partial_t S + \frac{1}{r}\partial_r(r S v_r) = 0 \tag{9.2.6}$$

or, introducing the accretion rate

$$\dot{M}(r, t) = -2\pi r v_r S \tag{9.2.7}$$

we also have

$$\partial_t S = \frac{1}{2\pi r}\partial_r \dot{M}. \tag{9.2.6'}$$

Equation (9.2.6) can also be obtained by applying the mass conservation to an annulus of gas with inner radius $r$ and with radial extent $\delta r$. Similarly, the consequences of energy and angular momentum conservation can be deduced.

**Exercise:** Consider orthogonal coordinates $x^i$ in $\mathbb{R}^3$ in which the metric has the form

$$g = \sum_i g_i^2 (dx^i)^2.$$

Show that the connection forms $\omega_{ij}$ relative to the orthonormal tetrad $\theta_1 = g_1\, dx^1, \theta_2 = g_2\, dx^2, \theta_3 = g_3\, dx^3$ are

$$\omega_{ij} = \frac{1}{g_i g_j} [g_{i,j}\, \theta_i - g_{j,i}\, \theta_j] \quad \text{(no sums)}.$$

Use this to compute the covariant derivative

$$(\nabla_v v)_i = \sum_k \left[\frac{1}{g_k} v_k v_{i,k} + \frac{1}{g_i g_k} (g_{i,k} v_i v_k - g_{k,i} v_k^2)\right].$$

Specialize this to cylindrical coordinates.

*c) Angular Momentum Conservation*

Next we consider the $\varphi$-component of the momentum equation (A.11) of the Appendix. With the solution of the previous exercise, one finds

$$\varrho\left(D_t v_\varphi + \frac{v_r v_\varphi}{r}\right) = \frac{1}{r} \partial_r (r\, t_{\varphi r}) + \partial_z t_{\varphi z} + \frac{1}{r} t_{r\varphi}. \tag{9.2.8}$$

If we multiply this equation with $r$ and use

$$D_t = \partial_t + v_r \partial_r + v_z \partial_z \tag{9.2.9}$$

for $\varphi$-independent functions, we obtain

$$\varrho\, D_t (r\, v_\varphi) = \frac{1}{r} \partial_r (r^2 t_{r\varphi}) + \partial_z (r\, t_{z\varphi}). \tag{9.2.10}$$

Now we multiply the continuity equation (9.2.3) with $r\, v_\varphi$ and add the resulting equation to (9.2.10). This gives

$$\partial_t (\varrho\, r\, v_\varphi) + \frac{1}{r} \partial_r (r\, v_r\, \varrho\, r\, v_\varphi) + \partial_z (v_z\, \varrho\, r\, v_\varphi) = \frac{1}{r} \partial_r (r^2 t_{r\varphi}) + \partial_z (r\, t_{z\varphi}). \tag{9.2.11}$$

Integrating over $z$ gives the exact equation

$$\partial_t \int \varrho\, r\, v_\varphi\, dz + \frac{1}{r} \partial_r \int v_r\, \varrho\, r^2 v_\varphi\, dz = \frac{1}{r} \partial_r (W_{r\varphi} r^2), \tag{9.2.12}$$

where

$$W_{r\varphi} = \int t_{r\varphi}\, dz. \tag{9.2.13}$$

Equation (9.2.12) follows also from angular momentum conservation, applied to the "control volume" mentioned before.

For thin disks (9.2.12) is approximately

$$\partial_t (S\, r\, v_\varphi) + \frac{1}{r}\, \partial_r (v_r\, S\, r^2\, v_\varphi) = \frac{1}{r}\, \partial_r (r^2\, W_{r\varphi})\,. \tag{9.2.14}$$

With the aid of (9.2.6) this can also be written as

$$S\,[\partial_t (r\, v_\varphi) + v_r\, \partial_r (r\, v_\varphi)] = \frac{1}{r}\, \partial_r (r^2\, W_{r\varphi})\,. \tag{9.2.14'}$$

Note that $r\, v_\varphi$ is the specific angular momentum. Eq. (9.2.11) has the form of a conservation equation for angular momentum.

The component $t_{r\varphi}$ has the form

$$t_{r\varphi} = \eta\, r\, \partial_r \left(\frac{v_\varphi}{r}\right), \tag{9.2.15}$$

where $\eta$ is the dynamic viscosity coefficient. (Phenomenological expressions for $\eta$ will be discussed later.)

### *d) Radial Momentum Conservation*

The radial component of the momentum equation is

$$\varrho \left(D_t\, v_r - \frac{v_\varphi^2}{r}\right) = -\, \varrho\, \partial_r\, \phi - \partial_r\, P + \frac{1}{r}\, \partial_r (r\, t_{rr}) + \partial_z\, t_{rz} - \frac{1}{r}\, t_{\varphi\varphi}\,. \tag{9.2.16}$$

Except for $t_{r\varphi}$, all viscous stresses can be neglected. Neglecting also the $z$-component of $\boldsymbol{v}$, we have to sufficient accuracy

$$\varrho (\partial_t\, v_r + v_r\, \partial_r\, v_r) = \varrho \left(\frac{v_\varphi^2}{r} - \partial_r \phi\right) - \partial_r P \tag{9.2.17}$$

and after integration over $z$

$$S (\partial_t\, v_r + v_r\, \partial_r\, v_r) = S \left(\frac{v_\varphi^2}{r} - \frac{GM}{r^2}\right) - \partial_r\, W\,, \tag{9.2.18}$$

where

$$W = \int P\, dz\,. \tag{9.2.19}$$

### *e) Energy Conservation*

Our starting point is Eq. (A.25) of the Appendix. The dominant part of $\underset{\sim}{t} \cdot \boldsymbol{v}$ is the radial component of magnitude $t_{r\varphi}\, v_\varphi$ and thus

$$\mathrm{div}\, (\underset{\sim}{t} \cdot \boldsymbol{v}) \simeq \frac{1}{r}\, \partial_r (r\, t_{r\varphi}\, v_\varphi)\,.$$

Ignoring again the $z$-component of the velocity and in addition the radial component of the energy flux vector, we have

$$\varrho(\partial_t + v_r\,\partial_r)\left(\tfrac{1}{2}v_r^2 + \tfrac{1}{2}v_\varphi^2 + h + \phi\right) = \partial_t P + \frac{1}{r}\,\partial_r(r\,t_{r\varphi}\,v_\varphi) - \partial_z F\,, \tag{9.2.20}$$

where $F$ is the vertical energy flux density ($F = q_z$). We use (9.2.2) and integrate over $z$. In the thin disk approximation we obtain

$$S(\partial_t + v_r\,\partial_r)\left[\tfrac{1}{2}v_r^2 + \tfrac{1}{2}v_\varphi^2 + (A+1)\frac{W}{S} + \phi\right] = \partial_t W + \frac{1}{r}\,\partial_r(r\,W_{r\varphi}\,v_\varphi) - Q^-\,, \tag{9.2.21}$$

where $Q^-$ is the energy flux per unit area emitted at the disk surface: $Q^- = 2F$ (surface).

We also note that the dissipation function (A.20) is

$$\Upsilon = 2t_{r\varphi}\,\theta_{r\varphi} = t_{r\varphi}\,r\,\partial_r\left(\frac{v_\varphi}{r}\right) \tag{9.2.22}$$

and thus

$$Q^+ \equiv \int \Upsilon\,dz = W_{r\varphi}\,r\,\partial_r\left(\frac{v_\varphi}{r}\right) \tag{9.2.23}$$

is the energy produced per unit area.

### *f) Equations for the Vertical Structure*

The $z$-component of the momentum equation contains only the small components of the viscosity tensor. Ignoring these we have

$$\varrho\,D_t\,v_z = -\varrho\,\partial_z\phi - \partial_z P\,. \tag{9.2.24}$$

We assume that motions in the $z$-direction are subsonic. Then, as we show later, the left hand side is of order $(z/r)^2$. Neglecting such terms for thin disks, we obtain

$$\partial_z P = -\varrho\,\frac{GM}{r^2}\,\frac{z}{r}\,, \tag{9.2.25}$$

i.e., we have hydrostatic equilibrium in the $z$-direction. (This is not surprising since there is no net motion in the vertical direction.) From (9.2.25) we see that the disk half-thickness $z_0$ is roughly given by $z_0/r \approx c_s/v_\varphi$, where $c_s$ is the speed of sound. The thin disk requirement thus says that the circular flow velocity is highly supersonic. This poses restrictions on the disk interior temperature, which are only satisfied if the gas cools sufficiently fast.

We also see that the time it takes for a sound wave to travel the distance $z_0$ is $z_0/c_s \sim \Omega^{-1}$, where $\Omega$ is the Keplerian angular velocity.

Usually it is assumed that the energy dissipated into heat is radiated on the spot in the vertical direction. Then we have [see (9.2.22)]

$$\partial_z F = \Upsilon = t_{r\varphi}\, r\, \partial_r \left(\frac{v_\varphi}{r}\right). \tag{9.2.26}$$

In addition, we need an energy transport equation in the $z$-direction, which will be written down later.

Some of the equations are simplified if one approximates $v_\varphi$ by the circular Keplerian velocity,

$$v_\varphi \simeq \Omega\, r, \quad \Omega = \left(\frac{GM}{r^3}\right)^{1/2}. \tag{9.2.27}$$

Then $\partial_t v_\varphi = 0$ and we obtain from (9.2.14′)

$$\frac{\dot{M}\,\Omega\, r}{2} = -\,2\pi\, \partial_r (W_{r\varphi} r^2) \tag{9.2.28}$$

with

$$W_{r\varphi} = r \frac{d\Omega}{dr} \int \eta\, dz \tag{9.2.29}$$

and thus

$$\dot{M}\,\Omega\, r^2 + 2\pi\, r^3 \frac{d\Omega}{dr} \int \eta\, dz = \dot{I}\,, \tag{9.2.30}$$

where $\dot{I}$ is independent of $r$.

The Keplerian approximation follows from (9.2.18) if inertia and pressure gradient terms are neglected. Corrections are of order $(z_0/r)^2$ (see later).

For (9.2.26) we find

$$\partial_z F = \frac{9}{4} \eta \frac{GM}{r^3} \tag{9.2.31}$$

and (9.2.23) becomes

$$Q^+ = \frac{9}{4} \frac{GM}{r^3} \int \eta\, dz\,. \tag{9.2.32}$$

### 9.2.3 Steady Keplerian Disks

We summarize first the basic equations for steady disks in the Keplerian approximation $\Omega = (GM/r^3)^{1/2}$.

*a) Radial Structure Equations*

The continuity equation gives

$$\dot{M} = -\,2\pi\, r\, S\, v_r = \text{const}\,. \tag{9.2.33}$$

The angular momentum conservation (9.2.28) implies

$$\dot{M} r^2 \Omega = -2\pi r^2 W_{r\varphi} + \dot{I}. \tag{9.2.30'}$$

Here the constant $\dot{I}$ is the net inward flux of angular momentum, whose value is usually assumed to be of order $\dot{M} r_0^2 \Omega(r_0)$ ($r_0$ is the inner edge of the disk). Then we have for the specific angular momentum $l(r) = r^2 \Omega(r)$:

$$\dot{M}[l(r) - l(r_0)] = -2\pi r^2 W_{r\varphi}. \tag{9.2.34}$$

The torque $2\pi r^2 W_{r\varphi}$ is, on the other hand, determined by (9.2.29), i.e.,

$$W_{r\varphi} = -\tfrac{3}{2}\Omega \int \eta \, dz. \tag{9.2.35}$$

In the energy equation (9.2.21) we neglect derivatives of $W$, $S$, and $v_r$:

$$\frac{d}{dr}\left[\dot{M}\left(\tfrac{1}{2} v_\varphi^2 - \frac{GM}{r}\right) + 2\pi r^2 W_{r\varphi}\Omega\right] = 2\pi r Q^- .$$

With (9.2.34) this becomes

$$Q^- = \frac{3}{4\pi} \dot{M} \frac{GM}{r^3} [1 - (r_0/r)^{1/2}]. \tag{9.2.36}$$

Note that $Q^-$ is independent of $\eta$!

This equation also follows from $Q^- = Q^+$, (9.2.23), and (9.2.34).

### b) Vertical Structure Equations

For the vertical structure, we have the following equations

$$\frac{dP}{dz} = -\varrho \frac{GM}{r^2} \frac{z}{r} \tag{9.2.37}$$

$$\frac{dF}{dz} = \frac{9}{4} \frac{GM}{r^3} \eta \tag{9.2.38}$$

$$P = \frac{k}{\mu m_{\mathrm{H}}} \varrho T + \frac{a}{3} T^4 \tag{9.2.39}$$

$$\frac{dT}{dz} = \begin{cases} \dfrac{-3\varkappa \varrho F}{4 a c T^3} & \text{for} \quad \nabla_{\mathrm{rad}} \leqq \nabla_{\mathrm{ad}} \\[2ex] \varrho \dfrac{GM}{r^2} \dfrac{z}{r} \dfrac{T}{P} \nabla_{\mathrm{conv}} & \text{for} \quad \nabla_{\mathrm{rad}} > \nabla_{\mathrm{ad}} . \end{cases} \tag{9.2.40}$$

The last equation is the energy transport equation, which is well-known from the theory of stellar structure. The upper line has to be used if radiation transport dominates and the disk is optically thick. (This equation was already used in Sect. 6.7.) The lower line of (9.2.40) gives roughly the energy transport if convection dominates. (The notation is standard; see, e.g. [148].)

The opacity is dominated by electron scattering and free-free transitions. In the next exercise it will be shown that these contributions to the total opacity $\varkappa$ are

$$\frac{1}{\varkappa} \simeq \frac{1}{\varkappa_{es}} + \frac{1}{\varkappa_{ff}},$$

with

$$\varkappa_{es} = 0.40\ \mathrm{cm^2/g}$$

$$\varkappa_{ff} = (0.645 \times 10^{23}\ \mathrm{cm^2/g}) \frac{Z^2}{A} \bar{\bar{G}} \left( \frac{\varrho_0}{\mathrm{g/cm^3}} \right) T_{\mathrm{K}}^{-7/2}. \tag{9.2.41}$$

---

**Exercise:** Use Kirchhoff's law for the relation between the emission and absorption coefficients to determine from (9.1.49) the Rosseland-mean (9.2.41) of the absorption coefficient due to free-free transitions. Derive also the expression (9.2.41) for Thomson scattering.

---

If the disk is optically thin, the energy in the disk is mainly lost by free-free emission (9.1.53). Thus

$$Q^- = \int \varepsilon_{ff}\, dz \quad (\tau < 1) \tag{9.2.42}$$

$$\varepsilon_{ff} = (1.4 \times 10^{-27}\ \mathrm{erg\ s^{-1}\ cm^{-3}})\ T^{1/2}\, n_e\, n_i\, Z^2\, \bar{\bar{G}}(T). \tag{9.2.43}$$

Comptonization may also be important; see [147].

### *c) Efficiency, Surface Temperature Distribution, and Spectral Distribution*

Let us discuss now some conclusions, which are independent of the viscosity law.

From (9.2.36) we obtain immediately the total luminosity of the disk (from both sides):

$$L_D = 2\pi \int_{r_0}^{\infty} Q^-\, r\, dr = \frac{1}{2} \frac{GM}{r_0 c^2} \dot{M} c^2. \tag{9.2.44}$$

In other words, the efficiency for Kepler disks is

$$\varepsilon = \frac{1}{2} \frac{GM}{r_0 c^2}. \tag{9.2.45}$$

Thus, half of the potential energy is radiated away. The other half is in the form of kinetic energy of the matter just outside the boundary layer. For neutron stars or white dwarfs, this part of the energy will be radiated away when the plasma falls onto the stellar surface.

In those parts where the disk radiates as a black-body, the effective surface temperature is determined by the Stefan-Boltzmann law,

$$F\,(\mathrm{surface}) = \tfrac{1}{2}\, Q^- = \sigma\, T_{\mathrm{eff}}^4. \tag{9.2.46}$$

Specific disk models will show in which parts this is likely to be the case.

Equation (9.2.36) gives for $T_{\rm eff}$ as a function of $r$

$$T_{\rm eff}(r) = \left\{\frac{3GM\dot{M}}{8\pi r^3}\frac{}{\sigma}[1-(r_0/r)^{1/2}]\right\}^{1/4} \tag{9.2.47}$$

or

$$T_{\rm eff} = T_* \left(\frac{r}{r_0}\right)^{-3/4} [1-(r_0/r)^{1/2}]^{1/4} \tag{9.2.48}$$

with

$$T_* = \left(\frac{3GM}{8\pi r_0^3}\frac{\dot{M}}{\sigma}\right)^{1/4} = 1.4\times 10^7\,{\rm K}\left(\frac{3\,r_{\rm g}}{r_0}\right)^{3/4}\left(\frac{M}{M_\odot}\right)^{-1/2}\dot{M}_{17}^{1/4}, \tag{9.2.49}$$

where $\dot{M}_{17} = \dot{M}/10^{17}\,{\rm g\,s^{-1}}$.

The maximum temperature is reached for $r = \frac{49}{36}\,r_0$.

The temperature distribution gives us also the spectral distribution of the disk. At each radius the spectrum emitted by an elemental area of the disk is the black-body spectrum

$$B_\nu(T_{\rm eff}) = \frac{2h\nu^3}{c^2}(e^{h\nu/kT_{\rm eff}}-1)^{-1}. \tag{9.2.50}$$

Thus, the spectrum given by the disk as a whole is

$$S_\nu \propto \int_{r_0}^{r_{\rm out}} B_\nu(T_{\rm eff}(r))\,2\pi r\,dr.$$

For frequencies $\nu \gg kT_*/h$ the spectrum drops exponentially. In this regime, we just see the exponential tail of the hottest inner part of the disk. For $\nu \ll kT_*/h$ the radiation comes predominantly from radii $r \gg r_0$, so that we can use the approximation $T_{\rm eff} = (r/r_0)^{-3/4}\,T_*$. Setting $x = h\nu/kT_{\rm eff}(r)$, $x_{\rm out} = h\nu/kT_{\rm out}$, and $T_{\rm out} = T_{\rm eff}(r_{\rm out})$ gives

$$S_\nu \propto \nu^{1/3}\int_0^{x_{\rm out}} \frac{x^{5/3}}{e^x-1}\,dx, \qquad (h\nu \ll kT_*). \tag{9.2.51}$$

We distinguish two different regimes. For $\nu \ll kT_{\rm out}/h$, the Rayleigh-Jeans tail of the coolest disk elements dominates and we find $S_\nu \propto \nu^2$. On the other hand, for $kT_{\rm out}/h \ll \nu \ll kT_*/h$ we may take $x_{\rm out} \gg 1$ and find $S_\nu \propto \nu^{1/3}$. The resulting spectrum is shown in Fig. 9.2.

For supermassive objects, it is instructive to write (9.2.49) as

$$T_* = 1.4\times 10^5\,{\rm K}\left(\frac{M}{10^8\,M_\odot}\right)^{-1/2}\dot{M}_{25}^{1/4}\left(\frac{3\,r_{\rm g}}{r_0}\right)^{3/4}. \tag{9.2.52}$$

The total luminosity (9.2.44) is for $r_0 = 3\,r_{\rm g}$:

$$L_D \sim 10^{45}\,{\rm erg\,s^{-1}}\,\dot{M}_{25}. \tag{9.2.53}$$

Such disks would radiate mainly in the UV spectral region.

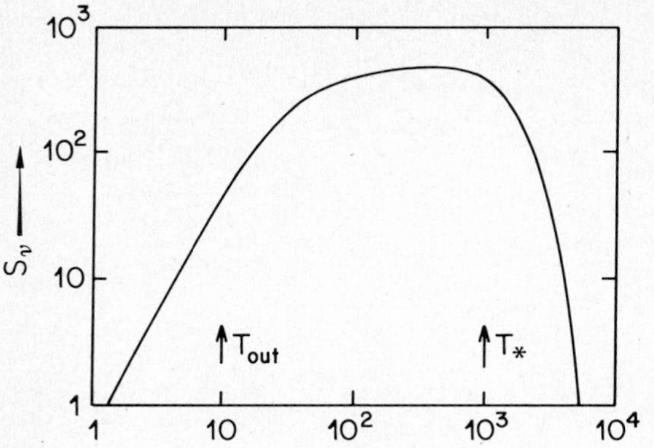


**Fig. 9.2.** The integrated spectrum of a steady accretion disk that radiates a local black body spectrum at each point (arbitrary units). The frequencies corresponding to $T_{out}$, the temperature of the outermost disk radius, and to $T_*$, the characteristic temperature of the inner disk, are marked

### 9.2.4 Standard Disks

*a) Viscosity*

Detailed disk models can only be constructed if we know the magnitude of the viscosity. Unfortunately, we can at best make order of magnitude estimates, since we are dealing with a highly supersonic strongly shearing fluid that is radiative, and has a large Reynolds number.

The sources of viscosity might be small-scale turbulence and transfer of angular momentum by magnetic stresses. Random magnetic fields are amplified by the differential rotation and turbulence of the disk. At the interface between adjacent cells of the resulting chaotic field, the gradients become so strong that magnetic field line reconnection occurs (see the exercise on p. 349).

Clearly, we are not able to handle these complicated phenomena in a quantitative manner. A reasonable parametrization has, however, been given by Shakura and Sunyaev [147] which we discuss next.

The contribution of the turbulent part of the term $\eta\,\Delta\boldsymbol{v}$ to the pressure in the Navier-Stokes equation will be of the order $\eta\, v_{\rm turb}/l_{\rm turb}$, where $v_{\rm turb}$ is the speed of the turbulent motions relative to the average motion, and $l_{\rm turb}$ is the characteristic size of the (largest) turbulent cells. For fully developed turbulence, this viscous force will be comparable to the inertial force $\varrho\, v_{\rm turb}^2$ and thus

$$\eta_{\rm turb} \sim \varrho\, v_{\rm turb} \cdot l_{\rm turb}\,. \tag{9.2.54}$$

The turbulent speed $v_{\rm turb}$ will be less than the sound speed $c_{\rm s}$; otherwise shocks would develop and quickly convert the turbulent energy into heat. Furthermore, we must require $l_{\rm turb} \lesssim z_0$. Consequently, the turbulent part of (9.2.15) will be bounded by

$$-t_{\varphi r}^{\text{turb}} \simeq \eta_{\text{turb}}\,\Omega \lesssim \varrho\, c_{\text{s}}\, z_0\, \Omega \simeq \varrho\, c_{\text{s}}^2 . \tag{9.2.55}$$

We expect the same inequality for the magnetic shear stress, because

$$-t_{\varphi r}^{\text{mag}} < P^{\text{mag}} \left(= \frac{B^2}{8\pi}\right) \lesssim P^{\text{therm}} \simeq \varrho\, c_{\text{s}}^2 . \tag{9.2.56}$$

Hence Shakura and Sunyaev propose

$$t_{r\varphi} = -\,\alpha\, P , \tag{9.2.57}$$

where $\alpha$ is a free parameter, satisfying $\alpha \lesssim 1$. Models which are based on this parametrization are called "$\alpha$-disks". They are clearly too primitive to be more than of qualitative value.

Comparison of (9.2.57) with (9.2.15) gives

$$\eta = \frac{2}{3}\,\alpha\, P \left(\frac{GM}{r^3}\right)^{-1/2} . \tag{9.2.57'}$$

*b) Vertical Structure*

Once a formula for the viscosity is given, we can determine with Eqs. (9.2.37–40) the local structure of the disk. Before we discuss results of numerical integrations, we derive rough analytic expressions.

We assume, following [149], a polytropic equation of state for fixed $r$,

$$P(z) = K\,\varrho(z)^{1+1/N} . \tag{9.2.58}$$

Equation (9.2.37) can then immediately be solved with the result

$$K(1+N)\,\varrho^{1/N} = \frac{1}{2}\,\frac{GM}{r}\left[\left(\frac{z_0}{r}\right)^2 - \left(\frac{z}{r}\right)^2\right] . \tag{9.2.59}$$

For the values in the central plane (indexed by c) we obtain

$$P_{\text{c}}/\varrho_{\text{c}} = \frac{1}{1+N}\,\frac{1}{2}\,\frac{GM}{r}\left(\frac{z_0}{r}\right)^2 . \tag{9.2.60}$$

Furthermore

$$\begin{aligned} S &= 2\varrho_{\text{c}}\, z_0\, I(N) \\ W &= 2 P_{\text{c}}\, z_0\, I(N+1) , \end{aligned} \tag{9.2.61}$$

where

$$I(N) = \frac{(2^N N!)^2}{(2N+1)!} . \tag{9.2.62}$$

From (9.2.57) we have generally

$$W_{r\varphi} = -\,\alpha\, W . \tag{9.2.63}$$

Next we calculate $Q^-$. In the optically thin case, we can use (9.2.42). If we write (9.2.43) as $\varepsilon_{\rm ff} = \varepsilon_0\, \varrho^2 T^{1/2}$, then

$$Q^- = 2\,\varepsilon_0\, \varrho_c^2\, T_c^{1/2} \int_0^{z_0} \left(\frac{\varrho}{\varrho_c}\right)^2 \left(\frac{T}{T_c}\right)^{1/2} dz\,.$$

With the aid of (9.2.59), the equation of state (without the radiation term in the optically thin case), and (9.2.58) we find

$$Q^- = 2\,\varepsilon_0\, J(2N)\, \varrho_c^2\, T_c^{1/2}\, z_0 \quad \text{(optically thin)}, \tag{9.2.64}$$

where

$$J(N) = \frac{(2N+2)!}{2^{2N+2}\,[(N+1)!]^2}\,. \tag{9.2.65}$$

For optically thick parts of the disk, we use the first equation of (9.2.40) (ignoring convection),

$$Q^- = -2\left(\frac{4\,a\,c\,T^3}{3\,\varkappa\,\varrho}\,\frac{dT}{dz}\right)_{\rm surface}. \tag{9.2.66}$$

Let us evaluate this under the assumption that either $P_r$, or $P_g$ dominates in the central disk.

If the radiation pressure dominates, we obtain from the equation of state and the hydrostatic equilibrium in the $z$-direction immediately $dT/dz$ and then

$$Q^- = 2\,\frac{c}{\varkappa}\,\frac{GM}{r^2}\,\frac{z_0}{r} \quad \text{(optically thick, } P_{\rm rc} > P_{\rm gc}\text{)}\,. \tag{9.2.67}$$

For the opposite case ($P_{\rm gc} > P_{\rm rc}$) we use an opacity law of the form

$$\varkappa = \varkappa_0\, \varrho^n\, T^{-s}\,. \tag{9.2.68}$$

The equation of state and (9.2.58) imply

$$T = T_c\left[1 - \left(\frac{z}{z_0}\right)^2\right].$$

Together with (9.2.59) we find

$$Q^- = \left(\frac{16\,a\,c\,T^4}{3\,\varkappa\,\varrho}\right)_c \frac{1}{z_0}\left[1 - \left(\frac{z}{z_0}\right)^2\right]^{3+s-N(1+n)}_{z\to z_0}.$$

The limit $z \to z_0$ exists only if $N = (3+s)/(1+n)$ and then

$$Q^- = \left(\frac{16\,a\,c\,T^4}{3\,\varkappa\,\varrho}\right)_c \frac{1}{z_0} \quad \text{(optically thick, } P_{\rm gc} > P_{\rm rc}\text{)}\,. \tag{9.2.69}$$

If $\varkappa$ is dominated by electron scattering, then [see (9.2.41)] $N = 3$.

*c) Radial Structure*

The vertical structure described in the previous paragraph depends on the parameters $z_0$ and $\varrho_c$ which are functions of the radius. The radial equations in Sect. 9.2.3 (a), together with the "α-law" (9.2.63) allow us to determine these functions (for given $M$, $\dot{M}$).

From (9.2.34) and (9.2.63) we have

$$W(r) = \frac{\dot{M}}{2\pi r^2 \alpha}[l(r) - l(r_0)] \tag{9.2.70}$$

with $l(r) = \sqrt{GMr}$.

On the left hand side, we insert (9.2.61) to obtain

$$P_c(r) = \frac{1}{I(N+1)} \frac{\dot{M}}{4\pi r^2 \alpha}\left(\frac{z_0}{r}\right)^{-1}\left(\frac{GM}{r}\right)^{1/2}\left[1 - \left(\frac{r_0}{r}\right)^{1/2}\right]. \tag{9.2.71}$$

This equation contains still the parameter $z_0$. Using the relation (9.2.60) between $\varrho_c$ and $P_c$ gives

$$\varrho_c(r) = \frac{2(N+1)}{I(N+1)} \frac{\dot{M}}{4\pi r^2 \alpha}\left(\frac{GM}{r}\right)^{-1/2}\left(\frac{z_0}{r}\right)^{-3}[1 - (r_0/r)^{1/2}]. \tag{9.2.72}$$

The equation of state allows us now to express $T_c$ as a function of $r$ and $z_0$:

$$T_c(r) = \begin{cases} \dfrac{1}{2(N+1)} \dfrac{\mu m_H}{k} \dfrac{GM}{r}\left(\dfrac{z_0}{r}\right)^2 & \text{for} \quad P_g > P_r \\ \left[\dfrac{3}{a} P_c(r)\right]^{1/4} & \text{for} \quad P_r > P_g. \end{cases} \tag{9.2.73}$$

Next we use the continuity equation (9.2.33), (9.2.61) and (9.2.72) to determine also $v_r$ as a function of $r$ and $z_0$:

$$-v_r(r) = \alpha \frac{I(N+1)}{2(N+1)\,I(N)}\left(\frac{GM}{r}\right)^{1/2}\left(\frac{z_0}{r}\right)^2 [1 - (r_0/r)^{1/2}]^{-1}. \tag{9.2.74}$$

This equation shows that indeed $v_r/v_\varphi = O\left(\left(\frac{z_0}{r}\right)^2\right)$. The previous relations imply also that $\frac{dW}{dr}\Big/\frac{GMS}{r^2} = O\left(\left(\frac{z_0}{r}\right)^2\right)$. A posteriori, this justifies that we have neglected such terms at several occasions.

Finally, if we use the expression (9.2.36) for $Q^-$ and compare it with those of the previous paragraph [Eqs. (9.2.67), (9.2.69), and (9.2.64)] the half-thickness $z_0$ of the disk is also determined. For the *optically thick* case we have:

$$\frac{3}{4\pi}\dot{M}\frac{GM}{r^3}[1-(r_0/r)^{1/2}] = \begin{cases} \left(\dfrac{16\,a\,c\,T^4}{3\,\varkappa\,\varrho}\right)_{\rm c}\dfrac{1}{z_0} & \text{for} \quad P_{\rm gc} > P_{\rm rc} \\ 2\,\dfrac{c}{\varkappa}\,\dfrac{GM}{r^2}\,\dfrac{z_0}{r} & \text{for} \quad P_{\rm rc} > P_{\rm gc}\,. \end{cases} \tag{9.2.75 a}$$

If the disk is optically thin, the relation is

$$\frac{3}{4\pi}\dot{M}\frac{GM}{r^3}[1-(r_0/r)^{1/2}] = 2\,\varepsilon_0\,J\,(2\,N)\,\varrho_{\rm c}^2\,T_{\rm c}^{1/2}\,z_0\,. \tag{9.2.75 b}$$

The five algebraic equations (9.2.71–75), together with the opacity law (9.2.41), can easily be solved in terms of $r$, $M$, and $\dot{M}$.

We distinguish three different regions of the disk, whose properties are given in Tables 9.1–9.3.

We use the notation:

$$\tilde{I} = \tfrac{3}{2}\,I\,(N+1)\,, \qquad \mathscr{I} = 1-(r_0/r)^{1/2}\,. \tag{9.2.76}$$

**Remarks:**

(i) Let us introduce a critical accretion rate by

$$\dot{M}_{\rm cr} = \frac{L_{\rm Edd}}{\varepsilon\,c^2} = (1.6\times10^{18}\ {\rm g\ s^{-1}})\left(\frac{M}{M_\odot}\right)\left(\frac{r_0}{3\,r_{\rm g}}\right), \tag{9.2.77}$$

where $L_{\rm Edd}$ is the Eddington luminosity (see p. 382) and $\varepsilon$ the efficiency (9.2.45). Since for the inner region $z_0/r_0 = 3\,(\dot{M}/\dot{M}_{\rm cr})\,\mathscr{I}$, we see that the

**Table 9.1.** Outer region: $P_{\rm g} > P_{\rm r}$, $\varkappa \approx \varkappa_{\rm ff}$

$$\frac{z_0}{r} = 7.7\times10^{-3}\,\frac{(N+1)^{19/40}}{\tilde{I}^{1/10}}\,\alpha^{-1/10}\,(2\,\mu)^{-3/8}\left(\frac{3\,r_{\rm g}}{r_0}\right)^{-1/8}\left(\frac{M}{M_\odot}\right)^{-1/4}\dot{M}_{17}^{3/20}\left(\frac{r}{r_0}\right)^{1/8}\mathscr{I}^{3/20}$$

$$T_{\rm c} = (2.7\times10^7\ {\rm K})\,\frac{(N+1)^{-1/20}}{\tilde{I}^{1/5}}\,\alpha^{-1/5}\,(2\,\mu)^{1/4}\left(\frac{3\,r_{\rm g}}{r_0}\right)^{3/4}\left(\frac{M}{M_\odot}\right)^{-1/2}\dot{M}_{17}^{3/10}\left(\frac{r}{r_0}\right)^{-3/4}\mathscr{I}^{3/10}$$

$$\varrho_{\rm c} = (5.3\ {\rm g\ cm^{-3}})\,\frac{(N+1)^{-17/40}}{\tilde{I}^{7/10}}\,\alpha^{-7/10}\,(2\,\mu)^{9/8}\left(\frac{3\,r_{\rm g}}{r_0}\right)^{15/8}\left(\frac{M}{M_\odot}\right)^{-5/4}\dot{M}_{17}^{11/20}\left(\frac{r}{r_0}\right)^{-15/8}\mathscr{I}^{11/20}$$

Optical depth:

$$\tau_{\rm ff} = 122\,\frac{(N+1)^{-1/5}}{\tilde{I}^{4/5}}\,\alpha^{-4/5}\,(2\,\mu)\,\dot{M}_{17}^{1/5}\,\mathscr{I}^{1/5} > 1$$

$$\tau_{\rm es} = 1.5\times10^4\,\frac{(N+1)^{1/20}}{\tilde{I}^{4/5}}\,\alpha^{-4/5}\,(2\,\mu)^{3/4}\left(\frac{3\,r_{\rm g}}{r_0}\right)^{3/4}\left(\frac{M}{M_\odot}\right)^{-1/2}\dot{M}_{17}^{7/10}\left(\frac{r}{r_0}\right)^{-3/4}\mathscr{I}^{7/10}$$

Inner boundary ($\tau_{\rm ff} = \tau_{\rm es}$) at:

$$\frac{r}{r_0} = 6.0\times10^2\,(N+1)^{1/3}\,(2\,\mu)^{-1/3}\left(\frac{3\,r_{\rm g}}{r_0}\right)\left(\frac{M}{M_\odot}\right)^{-2/3}\dot{M}_{17}^{2/3}$$

**Table 9.2.** Middle region: $P_g > P_r$, $\varkappa \approx \varkappa_{es}$

$$\frac{z_0}{r} = 1.2 \times 10^{-2} \frac{(N+1)^{1/2}}{\tilde{I}^{1/10}} \alpha^{-1/10} (2\mu)^{-2/5} \left(\frac{3r_g}{r_0}\right)^{-1/20} \left(\frac{M}{M_\odot}\right)^{-3/10} \dot{M}_{17}^{1/5} \left(\frac{r}{r_0}\right)^{1/20} \mathscr{I}^{1/5}$$

$$T_c = (6.5 \times 10^7\,\mathrm{K})\, \tilde{I}^{-1/5} \alpha^{-1/5} (2\mu)^{1/5} \left(\frac{3r_g}{r_0}\right)^{9/10} \left(\frac{M}{M_\odot}\right)^{-3/5} \dot{M}_{17}^{2/5} \left(\frac{r}{r_0}\right)^{-9/10} \mathscr{I}^{2/5}$$

$$\varrho_c = (1.4\,\mathrm{g\,cm^{-3}}) \frac{(N+1)^{-1/2}}{\tilde{I}^{7/10}} \alpha^{-7/10} (2\mu)^{6/5} \left(\frac{3r_g}{r_0}\right)^{33/20} \left(\frac{M}{M_\odot}\right)^{-11/10} \dot{M}_{17}^{2/5} \left(\frac{r}{r_0}\right)^{-33/20} \mathscr{I}^{2/5}$$

$$\left(\frac{P_g}{P_r}\right)_c = 0.3 \frac{(N+1)^{-1/2}}{\tilde{I}^{1/10}} \alpha^{-1/10} (2\mu)^{-2/5} \left(\frac{3r_g}{r_0}\right)^{-21/20} \left(\frac{M}{M_\odot}\right)^{7/10} \dot{M}_{17}^{-4/5} \left(\frac{r}{r_0}\right)^{21/20} \mathscr{I}^{-4/5}$$

Inner boundary ($P_g = P_r$) at:

$$\frac{r}{r_0} \mathscr{I}(r)^{-16/21} = 3.2 \frac{(N+1)^{10/21}}{\tilde{I}^{-2/21}} \alpha^{2/21} (2\mu)^{8/21} \left(\frac{3r_g}{r_0}\right) \left(\frac{M}{M_\odot}\right)^{-2/3} \dot{M}_{17}^{16/21}$$

**Table 9.3.** Inner region: $P_r > P_g$, $\varkappa \approx \varkappa_{es}$

$$\frac{z_0}{r_0} = 0.2 \left(\frac{3r_g}{r_0}\right) \left(\frac{M}{M_\odot}\right)^{-1} \dot{M}_{17}\, \mathscr{I}$$

$$T_c = (2.5 \times 10^7\,\mathrm{K})\, \tilde{I}^{-14} \alpha^{-1/4} \left(\frac{3r_g}{r_0}\right)^{3/8} \left(\frac{M}{M_\odot}\right)^{-1/4} \left(\frac{r}{r_0}\right)^{-3/8}$$

$$\varrho_c = (3.5 \times 10^{-4}\,\mathrm{g\,cm^{-3}}) \frac{(N+1)}{\tilde{I}} \alpha^{-1} \left(\frac{3r_g}{r_0}\right)^{-3/2} \left(\frac{M}{M_\odot}\right) \dot{M}_{17}^{-2} \left(\frac{r}{r_0}\right)^{3/2} \mathscr{I}^{-2}$$

Optical depth [a]:

$$\tau_{es} = 1.6 \frac{(N+1)}{\tilde{I}} \alpha^{-1} \left(\frac{3r_g}{r_0}\right)^{-1/2} \left(\frac{\dot{M}}{\dot{M}_{cr}}\right)^{-1} \left(\frac{r}{r_0}\right)^{3/2} \mathscr{I}^{-1}$$

Ratio of radial to azimuthal velocity:

$$-v_r/v_\varphi = 3\alpha \frac{\tilde{I}}{(N+1)\, I(N)} \left(\frac{\dot{M}}{\dot{M}_{cr}}\right)^2 \left(\frac{r_0}{r}\right)^2 \mathscr{I}$$

[a] The critical accretion rate is defined in Eq. (9.2.77)

disk remains geometrically thin only if

$$\dot{M} \ll \tfrac{1}{3} \dot{M}_{cr}. \tag{9.2.78}$$

(ii) For $r \to r_0$ the solution for the inner region becomes unphysical: $v_r \to 0$, $\varrho_c \to \infty$.
(iii) $T_c$ is independent of $\dot{M}$ in the inner region.
(iv) From our solution, we can determine the viscosity as a function of $r$, $\alpha$, $M$, and $\dot{M}$. For instance, in the inner zone, we find from $Q^+ = Q^-$

and the expressions (9.2.32) for $Q^+$ and (9.2.67) for $Q^-$:

$$\frac{9}{4}\frac{GM}{r^3}\int \eta\, dz = 2\frac{c}{\varkappa}\frac{GM}{r^2}\frac{z_0}{r}$$

and thus

$$\eta = \frac{4}{9}\frac{c\, m_p}{\sigma_T} = 3.5 \times 10^{10}\ \mathrm{erg\ s\ cm^{-3}}. \tag{9.2.79}$$

Surprisingly, this value is independent of all parameters of the accretion. It is instructive to compare it with the radiative viscosity

$$\eta_{rad} = \frac{4}{15}\frac{a T^4}{\sigma_T n_e c}, \tag{9.2.80}$$

For a derivation see, e.g., [150].

---

**Exercise:** Determine the maximum value of the radiative viscosity (9.2.80) in the inner region and show that it is always much smaller than (9.2.79).

---

(v) This standard inner disk model cannot explain the observed x-ray spectrum for Cyg X-1 [151–152] nor the stochastic time variability of the x-ray flux [153]. A more promising idea relies on the fact that large magnetic fields generated in the inner disk can heat up a corona which is cooled down by the emission of hard x-rays [154].
(vi) Whenever a neutron star is strongly magnetized with surface magnetic fields $B_* \approx 10^{12}$ G, the inner disk cannot reach the neutron star surface. For certain aspects of the interaction of an accretion disk with a magnetosphere see [58, 155, 156].

---

**Exercise:** Determine the inner edge $r_A$ of a disk around a strongly magnetized neutron star by using pressure balance between the magnetic pressure of a dipole field and the central pressure in the disk.

---

*d) Numerical Solutions*

Recently, the vertical structure equations (9.2.37–40) have been solved numerically [157], using accurate opacity tables and the following two parametrizations for the viscosity.

The first parametrization is similar to (9.2.57′), but includes a pressure scale height correction:

$$\eta = \varepsilon \left(\frac{GM}{2r^3}\right)^{-1/2} P \left[1 + \frac{2GM\varrho}{r^3 P} z^2\right]^{-1/2}. \tag{9.2.81}$$

In the central plane this reduces to (9.2.57′) with $\alpha = (2/\sqrt{2})\,\varepsilon$.

The following second parametrization is used to estimate an effective magnetic viscosity

$$\eta_{\mathrm{m}} = \varepsilon \left( \frac{GM}{2r^3} \right)^{-1/2} \beta P \left[ 1 + \frac{2GM\varrho}{r^3 P} z^2 \right]^{-1/2} \tag{9.2.82}$$

with

$$\beta = \frac{P_{\mathrm{g}}}{P_{\mathrm{g}} + P_{\mathrm{r}}} . \tag{9.2.83}$$

Together with the radial structure equations in Sect. 9.2.3 (a), the structure of the disk is now uniquely determined for given values of $M$ and $\dot{M}$.

Figure 9.3 shows the half-thickness of the disk. (A grey atmosphere is added to the surface of the disk; $z_0$ is the height of the photosurface where $\tau = 2/3$.) Three different regions are found: an inner region dominated by radiation pressure, a middle region dominated by gas pressure and an outer convective region.

It turns out that the inner region is also convective (the radiative temperature gradient is higher than the adiabatic), but convection is very inefficient. Note that $z_0/r$ increases toward the center. At some point, the thin disk approximation breaks down.

In the middle region, $z_0/r$ increases monotonically with $r$. The surface temperature finally drops below the ionization temperature of the gas and the disk becomes convective. The height $z_0$ decreases there slightly and then becomes constant. The computations end when the disk becomes optically thin.

Except when otherwise indicated, the value of $\varepsilon$ was taken as 1/30 and $M = 1\,M_\odot$.

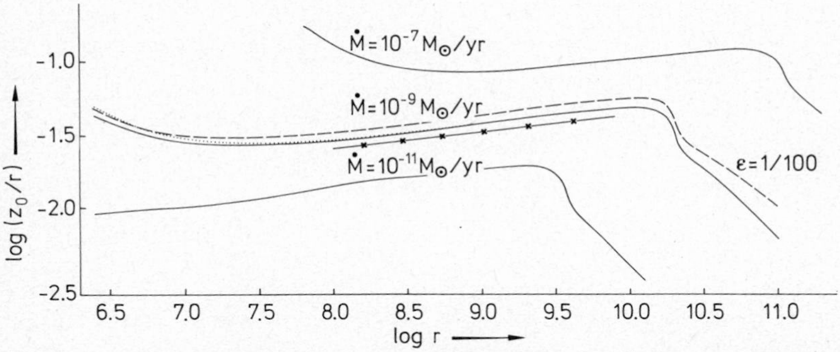


**Fig. 9.3.** Thickness of the disk for different accretion rates $\dot{M}$, given as $z_0/r$, $z_0$ = height of photosphere, $r$ distance from central object. *Dashed line* for the viscosity parameter $\varepsilon = 1/100$ (instead of 1/30). *Dotted line,* viscosity proportional to the gas pressure. *Crossed line,* central object $1.4\,M_\odot$ (instead of $1\,M_\odot$). In all figures cgs-units and $T$[K] except where noted (from [157])

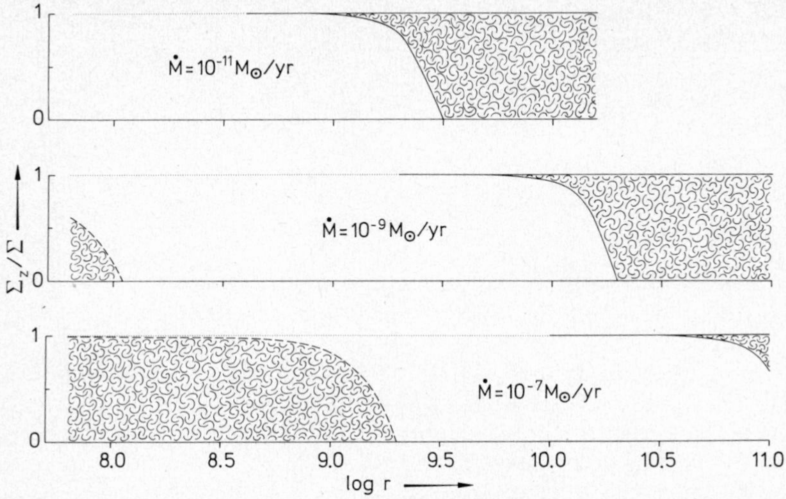


**Fig. 9.4.** Mass fraction $\Sigma_z/\Sigma$ of the convective regions for different accretion rates (from [157])

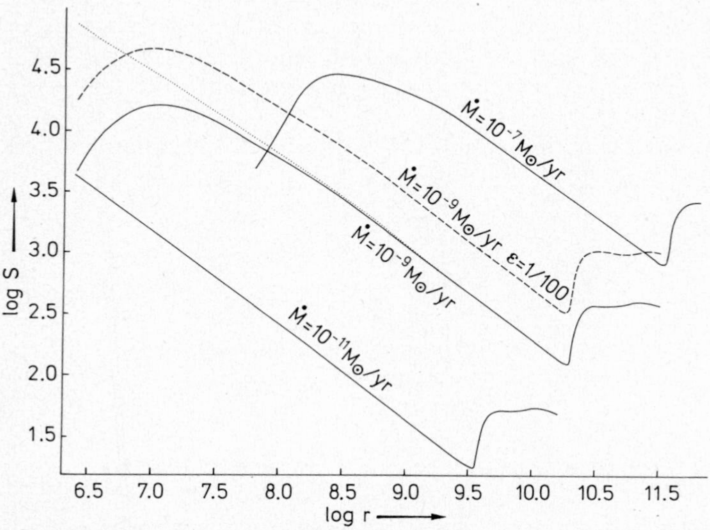


**Fig. 9.5.** Surface density $S$ versus distance $r$ from central object for different accretion rates. *Dashed* and *dotted lines* see Fig. 9.3 (from [157])

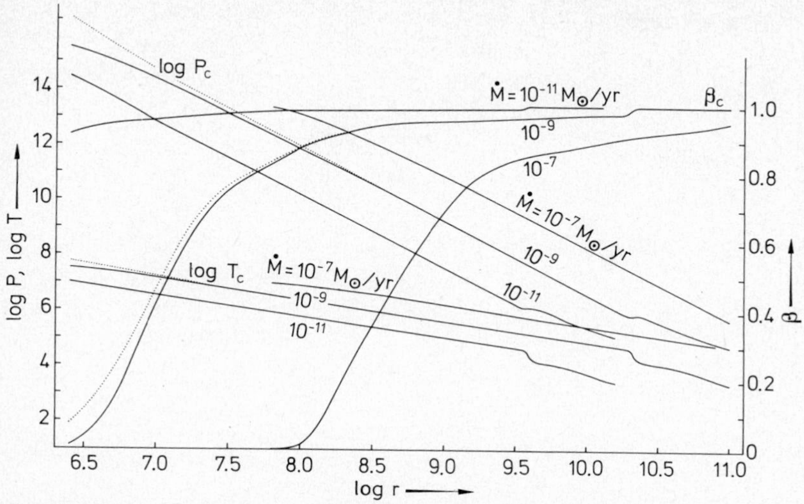


**Fig. 9.6.** Values of pressure, temperature and $\beta$ in the central plane versus distance $r$ from central object. *Dotted lines*, viscosity proportional to gas pressure instead of total pressure (from [157])

Figures 9.4–9.6 show the radial distributions of other quantities of interest. For details of the vertical disk structure we refer to [157]. Let us only mention that in the innermost radiation dominated region a density inversion appears.

### 9.2.5 Stability Analysis of Thin Accretion Disks

We investigate now, following the work of *Shakura* and *Sunyaev* [158], the stability of the steady disks described in Sect. 9.2.4. Only stable models have a chance to be physically relevant. It will turn out that possible instabilities depend strongly on the assumed viscosity.

We consider only axially symmetric perturbations of wavelength $\Lambda$, satisfying $z_0 \ll \Lambda \ll r$, and which change little on the dynamical time scale $\Omega^{-1}$. (Remember that $\Omega^{-1}$ is also roughly the time it takes for a sound wave to cross the disk in the transverse direction.)

The basic time-dependent equations have been derived in Sect. 9.2.2. For a linear stability analysis, we have to linearize these equations around the (analytic or numerical) equilibrium solutions.

For the type of perturbations, which we want to consider, we can still use in the vertical direction the hydrostatic equation (neglecting terms of order $(z_0/r)^2$). Furthermore, $v_\varphi$ is still Keplerian, up to terms of order $(z_0/r)^2$, $z_0^2/r\Lambda$.

**Exercise:** Prove these statements from the basic equations (9.2.6), (9.2.14′), (9.2.18), and (9.2.23).

*a) Dynamic Equation*

In the Keplerian approximation for $v_\varphi$ we get from (9.2.14′), since $v_\varphi$ is time independent,

$$r\,S\,v_r = \frac{\partial_r(r^2\,W_{r\varphi})}{\partial_r(r\,v_\varphi)}\,. \tag{9.2.84}$$

According to (9.2.29) we have, if $\nu = \eta/\varrho$ denotes the kinematic viscosity,

$$W_{r\varphi} = r\,\frac{d\Omega}{dr}\int \nu\,\varrho\,dz = r\,\frac{d\Omega}{dr}\,\nu\,S\,. \tag{9.2.85}$$

Hence, we obtain from (9.2.84)

$$r\,v_r\,S = -\frac{3}{r\,\Omega}\,\partial_r[\nu\,r^2\,\Omega\,S]\,. \tag{9.2.86}$$

Applying on this equation the operator $r^{-1}\,\partial_r$ and using the continuity equation (9.2.6) gives the following interesting diffusion type equation for $S$:

$$\partial_t S = \frac{3}{r}\,\partial_r\left\{\frac{1}{r\,\Omega}\,\partial_r[\nu\,S\,r^2\,\Omega]\right\}\,. \tag{9.2.87}$$

*b) Thermal Equation*

Instead of (9.2.21) we use another form of the energy equation, which we derive now.

The starting point is Eq. (A.21), which we write as follows

$$D_t(\varepsilon\,\varrho) - (\varepsilon\,\varrho + P)\,\frac{D_t\,\varrho}{\varrho} = \Upsilon - \operatorname{div}\boldsymbol{q}$$

or, using the continuity equation

$$\partial_t(\varepsilon\,\varrho) + \operatorname{div}[(\varepsilon\,\varrho + P)\,\boldsymbol{v}] - \nabla_v P = \Upsilon - \operatorname{div}\boldsymbol{q}\,.$$

Thus

$$\partial_t(\varrho\,\varepsilon) + \frac{1}{r}\,\partial_r[r\,v_r\,(\varrho\,\varepsilon + P)] + \partial_z[v_z\,(\varrho\,\varepsilon + P)]$$
$$- v_r\,\partial_r P - v_z\,\partial_z P = \Upsilon - \operatorname{div}\boldsymbol{q}\,.$$

Integrating over $z$ gives

$$\partial_t\int\varrho\,\varepsilon\,dz + \frac{1}{r}\,\partial_r\left(\int r\,v_r\,(\varrho\,\varepsilon + P)\,dz\right) - \int v_r\,\partial_r P\,dz - \int v_z\,\partial_z P\,dz$$
$$= Q^+ - Q^-\,. \tag{9.2.88}$$

This equation is so far exact. Now by (9.2.2)

$$\varrho\,\varepsilon = AP\,, \qquad A = \frac{\beta}{\gamma-1} + 3\,(1-\beta)\,.$$

In the thin disk approximation and using the hydrostatic equation for $\partial_z P$, we obtain

$$\partial_t(AW) + \frac{1}{r}\,\partial_r[r\,v_r\,(A+1)\,W] - v_r\,\partial_r W + \Omega^2 \int \varrho\, v_z\, z\, dz$$
$$= Q^+ - Q^-\,. \tag{9.2.89}$$

From (9.2.32) we have also

$$Q^+ = \tfrac{9}{4}\,\nu\, S\,\Omega^2\,. \tag{9.2.90}$$

For the α-model, (9.2.57′) gives

$$\nu = \frac{2\alpha}{3}\,\frac{W}{S}\,\frac{1}{\Omega}\,. \tag{9.2.91}$$

The expression (9.2.66) for $Q^-$ is also valid in the perturbed situation. We write this equation as [see also (9.2.69)]:

$$Q^- = e_s \frac{8}{3}\left(\frac{a\,c\,T^4}{\varkappa}\right)_c \frac{1}{S}\,, \tag{9.2.92}$$

where $e_s$ is a structure factor, which depends on the detailed vertical structure of the disk.

From now on, we choose for simplicity (as in [158]) a constant density in the $z$-direction and assume that the perturbations in the $z$-direction preserve this property.

Then the hydrostatic equation in the $z$-direction gives

$$P(z) = P_c\left[1 - \left(\frac{z}{z_0}\right)^2\right]\,, \qquad P_c = \frac{1}{4}\,S\,\Omega^2 z_0 \tag{9.2.93}$$

and thus the average pressure is

$$P = \tfrac{1}{6}\,S\Omega^2 z_0\,. \tag{9.2.94}$$

We also have

$$W = \tfrac{4}{3}\,P_c\, z_0 = \tfrac{1}{3}\,S\Omega^2 z_0^2 \tag{9.2.95}$$

and thus from (9.2.91) for the α-models

$$\nu = \tfrac{2}{9}\,\alpha\Omega^2 z_0^2\,. \tag{9.2.96}$$

Furthermore, since $v_z = \dfrac{z}{z_0}\,\partial_t z_0$,

$$\int \varrho\, v_z\, z\, dz = \tfrac{1}{3}\,S\, z_0\, \partial_t z_0\,.$$

Inserting these expressions into (9.2.89) gives

$$\tfrac{1}{3}\,\partial_t\,[A\,S\,\Omega^2 z_0^2] + \frac{1}{3}\,\frac{1}{r}\,\partial_r[r\,v_r(A+1)\,S\,\Omega^2 z_0^2] - \tfrac{1}{3}\,v_r\,\partial_r\,(S\,\Omega^2 z_0^2) + \tfrac{1}{3}\,S\,\Omega^2 z_0\,\partial_t\,z_0 = Q^+ - Q^- . \qquad (9.2.97)$$

In the second term on the left we use (9.2.86) to get the second basic equation:

$$\tfrac{1}{3}\,\partial_t\,[A\,S\,\Omega^2 z_0^2] - \frac{1}{r}\,\partial_r\left\{(A+1)\,\frac{\Omega\,z_0^2}{r}\,\partial_r\,[\nu\,\Omega\,r^2\,S]\right\} - \tfrac{1}{3}\,v_r\,\partial_r\,(S\,\Omega^2 z_0^2) + \tfrac{1}{3}\,S\,\Omega^2 z_0\,\partial_t\,z_0 = Q^+ - Q^- . \qquad (9.2.98)$$

*c) Linearization*

Equations (9.2.87) and (9.2.98) have to be linearized now around the equilibrium.

Let us introduce the following notations for the changes of $S$ and $z_0$ from their equilibrium values

$$\frac{\delta S}{S} = u\,, \qquad \frac{\delta z_0}{z_0} = h\,, \qquad |u|, |h| \ll 1\,. \qquad (9.2.99)$$

We set

$$\frac{\delta \nu}{\nu} = n\,u + m\,h + \ldots\,. \qquad (9.2.100)$$

If we use the α-law (9.2.96), then

$$n = 0\,, \qquad m = 2\,. \qquad (9.2.101)$$

Inserting (9.2.99) and (9.2.100) gives for the linearization of (9.2.87)

$$S\,\partial_t u = 3\,\nu\,S\,\partial_r^2\,[(n+1)\,u + m\,h]. \qquad (9.2.102)$$

The linearization of (9.2.98) is a bit more complicated.

Let

$$\frac{\delta Q^-}{Q^-} = l\,u + k\,h + \ldots\,, \qquad (9.2.103)$$

where the expansion coefficients depend on the opacity $\varkappa$.

One must also include variations of $A$. For definiteness we choose $\gamma = \frac{5}{3}$. Then

$$A = \tfrac{3}{2}\,(1+\bar{\beta})\,, \qquad \bar{\beta} \equiv 1 - \beta\,. \qquad (9.2.104)$$

(Note that $\beta$ in [158] is our $\bar{\beta}$.) First-order changes of $\bar{\beta}$ are obtained from the equation of state,

$$P = \bar{\beta}\,P + \frac{k}{\mu\,m_{\rm H}}\,\frac{S}{2\,z_0}\left(\frac{3\,\bar{\beta}\,P}{a}\right)^{1/4}$$

and from the expression (9.9.94) for $P$. Computing the variations of these two equations gives

$$\frac{\delta\bar{\beta}}{\bar{\beta}}=\frac{1-\bar{\beta}}{1+3\bar{\beta}}(7h-u)\,. \tag{9.2.105}$$

Now the linearization of (9.2.98) is straightforward. Using the equilibrium conditions (in particular $Q^+=Q^-$), one finds for the α-model, if only the dominant terms of order $(z_0/\Lambda)^2$ are kept:

$$\begin{aligned}&3\,(1+3\,\bar{\beta}+4\,\bar{\beta}^2)\,\partial_t u+(8+51\,\bar{\beta}-3\,\bar{\beta}^2)\,\partial_t h\\&\quad-3\,(1+3\,\bar{\beta})\,\alpha\,\Omega\,[(n+1-l)\,u+(m-k)\,h]\\&=\tfrac{2}{3}\,\alpha\,\Omega\,z_0^2\,(5+18\,\bar{\beta}+9\,\bar{\beta}^2)\,\partial_r^2\,[(n+1)\,u+m\,h]\,.\end{aligned} \tag{9.2.106}$$

---

**Exercise:** Derive Eq. (9.2.106).

---

If $\varkappa\approx\varkappa_{\mathrm{es}}$ we have from (9.2.92) and (9.2.93)

$$Q^-=e_{\mathrm{s}}\frac{8\,m_{\mathrm{p}}\,c}{\sigma_{\mathrm{T}}}\frac{\bar{\beta}\,P_{\mathrm{c}}}{S}=2\,e_{\mathrm{s}}\frac{m_{\mathrm{p}}\,c}{\sigma_{\mathrm{T}}}\,\bar{\beta}\,z_0\,\Omega^2\,. \tag{9.2.107}$$

Thus

$$\frac{\delta Q^-}{Q^-}=\frac{\delta\bar{\beta}}{\bar{\beta}}+h\,.$$

Using (9.2.105) this gives

$$k=\frac{8-4\,\bar{\beta}}{1+3\,\bar{\beta}},\qquad l=\frac{\bar{\beta}-1}{1+3\,\bar{\beta}}\,. \tag{9.2.108}$$

All perturbations are fully described by Eqs. (9.2.102) and (9.2.106), since all quantities of interest, in particular $\dot{M}$, can be expressed in terms of $u$ and $h$.

We consider harmonic perturbations

$$u\,(r,t)=u\,(r)\,\mathrm{e}^{\omega t}\,,\qquad h\,(r,t)=h\,(r)\,\mathrm{e}^{\omega t}$$

and write a single equation for the combination

$$\psi=(n+1)\,u+m\,h\,. \tag{9.2.109}$$

This amplitude describes the viscous perturbations, as can be seen from (9.2.90), (9.2.99), and (9.2.100).

From (9.2.102) we get for the α-law

$$u=\frac{2}{3}\,\alpha\,\frac{\Omega}{\omega}\,z_0^2\,\partial_r^2\,\psi\,. \tag{9.2.110}$$

Using (9.2.109), (9.2.110) in (9.2.106) gives

$$\omega\,\frac{A\,\omega-3\,(1+3\,\bar{\beta})\,\alpha\,\Omega\,(m-k)}{B\,\omega-3\,(1+3\,\bar{\beta})\,\alpha\,\Omega\,[m\,l-k\,(n+1)]}\,\psi=\frac{2}{3}\,\alpha\,\Omega\,z_0^2\,\partial_r^2\,\psi\,, \tag{9.2.111}$$

where

$$A(\bar{\beta}) = 8 + 51\bar{\beta} - 3\bar{\beta}^2$$
$$B(\bar{\beta}) = (n+1)A(\bar{\beta}) + m(2 + 9\bar{\beta} - 3\bar{\beta}^2)\,. \qquad (9.2.112)$$

For solutions proportional to $\sin(r/\Lambda)$ we obtain the dispersion relation

$$A\left(\frac{\omega}{3\alpha\Omega}\right)^2 + \left[2B\left(\frac{z_0}{3\Lambda}\right)^2 - (1+3\bar{\beta})(m-k)\right]\frac{\omega}{3\alpha\Omega} - 2\left(\frac{z_0}{3\Lambda}\right)^2(1+3\bar{\beta})[m\,l - (n+1)k] = 0\,. \qquad (9.2.113)$$

This agrees with Eq. (4.14) of [158] for the special values (9.2.101) and (9.2.108). For these we obtain

$$\frac{\omega}{\alpha\Omega} = \frac{3}{A}\left\{-\left[B\left(\frac{z_0}{3\Lambda}\right)^2 + (3-5\bar{\beta})\right] \pm\left[\left(B\left(\frac{z_0}{3\Lambda}\right)^2 + (3-5\bar{\beta})\right)^2 - 4A(5-3\bar{\beta})\left(\frac{z_0}{3\Lambda}\right)^2\right]^{1/2}\right\} \qquad (9.2.114)$$

with

$$A(\bar{\beta}) = 8 + 51\bar{\beta} - 3\bar{\beta}^2 > 0\,, \qquad B(\bar{\beta}) = 3(4 + 23\bar{\beta} - 3\bar{\beta}^2) > 0\,. \qquad (9.2.115)$$

Obviously, $\mathrm{Re}\,\{\omega\} < 0$, if $3 - 5\bar{\beta} > 0$. We thus have *stability* for

$$\bar{\beta} < \tfrac{3}{5}\,,$$

which is the case if the plasma pressure dominates.

When the radiation pressure dominates ($\bar{\beta} > \frac{3}{5}$) there is an unstable mode for $\Lambda/z_0$ larger than some value which depends on $\bar{\beta}$. (For $\bar{\beta} = 1$

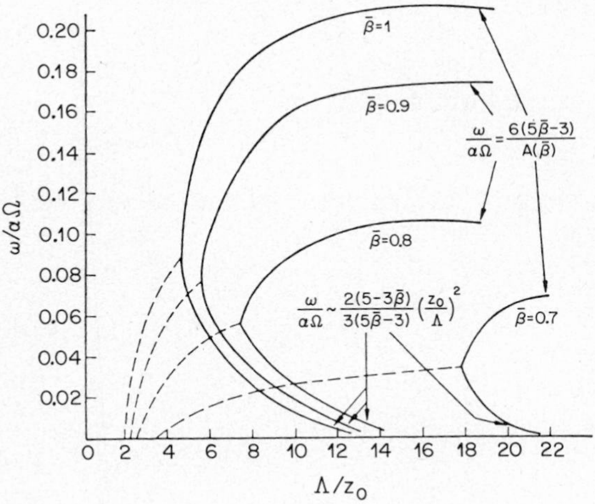


**Fig. 9.7.** Instability growth rate when radiation pressure dominates ($\bar{\beta} > \frac{3}{5}$) as a function of wavelength $\Lambda$. *Broken lines* denote travelling waves, *continuous lines* denote standing waves. Upper branches correspond to thermal instabilities and lower branches to dynamic instabilities (from [158])

this minimum value is equal to 2.) This unstable mode bifurcates at a higher value of $\Lambda/z_0$, as shown in Fig. 9.7.

The two branches correspond to physically quite distinct instabilities. For the lower branch, the perturbed viscosity becomes for long wavelengths very small ($Q^+ \approx Q^-$). On the other branch, the perturbation of the surface density becomes small compared to the viscous forces, disk thickness and other quantities. The growth of these perturbations is due to a thermal instability ($Q^+ \neq Q^-$). For a detailed discussion, we refer to [158].

If one uses a modified $\alpha$-law for which the viscosity is taken proportional to the gas pressure, instead of the total pressure [as in the model (9.2.82) for magnetic friction], then one finds [157] a dispersion relation which shows no instability. (Both modes in Fig. 9.7 disappear.) This illustrates that it is dangerous to draw definite conclusions from the previous analysis.

---

**Exercise:** Derive the values of the quantities $k$ and $l$ for the outer disk, $\varkappa = \varkappa_{\text{ff}}$, and use the dispersion relation (9.2.113) to show that the outer disk is always stable.

---

### 9.2.6 Relativistic Keplerian Disks

The previous analysis can also be carried out in the framework of GR.

GR is only relevant for the motion of the plasma in the immediate vicinity of the horizon of a black hole. Most of the observational properties of a disk, such as energy production and surface brightness, are contained in the radial structure equations (see Sect. 9.2.3). Since these equations follow directly from the conservation laws, it is not difficult to generalize them.

Relativistic effects would also slightly modify the vertical structure of thin disks, since the acceleration of gravity in the hydrostatic equilibrium should be replaced by the appropriate component of the Riemann tensor. This correction is, however, smaller than the uncertainties of the nonrelativistic theory (viscosity, etc.).

In order to illustrate the techniques used in a general relativistic treatment of accretion disks, we derive in detail the generalization of the luminosity distribution $Q^-(r)$, given in Eq. (9.2.36).

References to relevant recent work on inner disks will be given at the end of this section.

#### *a) Basic Equations*

Relativistic hydrodynamics of viscous fluids is briefly developed in the Appendix to this chapter, which the reader is encouraged to study first (at least for notation).

The fundamental equations governing the motion of a (single component) fluid in a given gravitational field are: the conservation of the particle current $N^\mu = n\,U^\mu$,

$$\operatorname{div}(n\,U) = 0\,, \tag{9.2.116}$$

the "conservation" of $T$, which splits into the relativistic Navier-Stokes equation ($\perp$ denotes the projection orthogonal to $U$),

$$(\nabla \cdot T)_\perp = 0 \tag{9.2.117}$$

and the energy equation

$$U \cdot (\nabla \cdot T) = 0\,. \tag{9.2.118}$$

According to (B.39) the energy-momentum tensor is

$$T = (\varrho + P)\,U \otimes U - P\,g + t + U \otimes q + q \otimes U\,; \tag{9.2.119}$$

magnetic fields are ignored. The energy equation (9.2.118) is given explicitly in (B.37).

We shall always ignore the self-gravity of the accreting gas as well as the increase in mass of the central object.

The metric is assumed to be (asymptotically) stationary and axisymmetric (e.g. the Kerr metric). Then $g$ has the form

$$g = g_{tt}\,dt^2 + 2\,g_{t\varphi}\,dt\,d\varphi + g_{\varphi\varphi}\,d\varphi^2 + g_{rr}\,dr^2 + g_{\vartheta\vartheta}\,d\vartheta^2\,, \tag{9.2.120}$$

where the metric coefficients are independent of $t$ and $\varphi$. (This is quite plausible. The reader may be challenged to give a rigorous proof.)

This space-time has two Killing fields, $k = \partial_t$ and $m = \partial_\varphi$ (which are commuting, $[k, m] = 0$). These two Killing fields give rise to two conserved currents

$$P^\alpha = T^\alpha_\beta\,k^\beta\,, \qquad J^\alpha = -\,T^\alpha_\beta\,m^\beta\,, \tag{9.2.121}$$

$$P^\alpha_{;\alpha} = J^\alpha_{;\alpha} = 0\,. \tag{9.2.122}$$

For later use, we write down the explicit expressions for these currents:

$$P^\alpha = (\mu\,U_t + q_t/n)\,N^\alpha - P\,k^\alpha + t^\alpha_t + U_t\,q^\alpha \tag{9.2.123}$$

$$J^\alpha = -\,(J - q_\varphi/n)\,N^\alpha - P\,m^\alpha + t^\alpha_\varphi + U_\varphi\,q^\alpha\,. \tag{9.2.124}$$

Here we have used the following notation:

$$\mu = \frac{\varrho + P}{n} \qquad \text{(relativistic enthalpy)}$$

$$\begin{aligned} U_t &= (U, k)\,, & U_\varphi &= (U, m) \\ t^\alpha_t &= t^\alpha_\beta\,k^\beta\,, & t^\alpha_\varphi &= t^\alpha_\beta\,m^\beta \\ q_t &= q_\alpha\,k^\alpha\,, & q_\varphi &= q_\alpha\,m^\alpha\,, \\ J &= -\,\mu\,U_\varphi\,. \end{aligned} \tag{9.2.125}$$

We introduce also the following quantities

$$\Omega = U^\varphi / U^t \quad \text{(angular velocity)}$$
$$l = - U_\varphi / U_t \quad \text{(specific angular momentum)}\,. \tag{9.2.126}$$

---

**Exercise:** A toroidal flow field $U = (U^t, 0, 0, U^\varphi)$ is completely specified by either $\Omega$ or $l$. Derive the following relations

$$U^t = (g_{tt} + 2\Omega\, g_{t\varphi} + \Omega^2\, g_{\varphi\varphi})^{-1/2} \tag{9.2.127}$$
$$U_t = \varrho\,(- l^2\, g_{tt} - 2\,l\, g_{t\varphi} - g_{\varphi\varphi})^{-1/2}\,, \tag{9.2.128}$$

where

$$\varrho^2 = g_{t\varphi}^2 - g_{tt}\, g_{\varphi\varphi}\,; \tag{9.2.129}$$
$$\Omega = - (l\, g_{tt} + g_{t\varphi})\,(l\, g_{t\varphi} + g_{\varphi\varphi})^{-1} \tag{9.2.130}$$
$$l = - (g_{t\varphi} + \Omega\, g_{\varphi\varphi})\,(g_{tt} + \Omega\, g_{t\varphi})^{-1} \tag{9.2.131}$$
$$U^t\, U_t = (1 - \Omega\, l)^{-1}\,. \tag{9.2.132}$$

---

*b) Dissipation Function for Disks*

Let $U = (U^t, U^A, U^\varphi)$, $A = r, \vartheta$. We assume that $U$ is almost toroidal, $|U^A| \ll |U^\varphi|$.

The energy conservation law (B.37) can be written in the form

$$n \nabla_U (\varrho/n) + \operatorname{div} q = \sigma_{\alpha\beta}\, t^{\alpha\beta} + \theta\,(\tfrac{1}{3}\, t^\alpha_\alpha - P) + q \cdot a \tag{9.2.133}$$

as is easily seen from (B.5) and (B.21). The right-hand side represents the rate at which energy is generated by viscous heating and compression; the last term, $a \cdot q$, is a special relativistic correction associated with the inertia of the heat flow $q$. We drop this term because the gas is in nearly geodesic orbits ($a \approx 0$). We drop also the terms proportional to $\theta$ because they are much smaller than the energy generated by frictional heating. The same is true for the first term on the left-hand side, which is the energy going into internal forms.

Thus, to sufficient accuracy, we have

$$\operatorname{div} q = \Upsilon\,, \tag{9.2.134}$$

where

$$\Upsilon = \sigma_{\alpha\beta}\, t^{\alpha\beta} \tag{9.2.135}$$

is the relativistic dissipation function which we are going to compute next. From now on, we specialize to toroidal flow fields.

As a first step, we derive the following expression

$$\Upsilon = - \frac{2\,\sigma_{A\varphi}\, t^A_\varphi}{\varrho^2\,(U^t)^2}\,. \tag{9.2.136}$$

In the expansion

$$\sigma_{\alpha\beta}\, t^{\alpha\beta} = 2\,(\sigma_{At}\, t^{At} + \sigma_{A\varphi}\, t^{A\varphi}) \tag{9.2.137}$$

we use the relations

$$\sigma_{At} = -\,\Omega\,\sigma_{A\varphi}, \qquad t_{At} = -\,\Omega\, t_{A\varphi}, \tag{9.2.138}$$

which follow from $U^\beta \sigma_{\alpha\beta} = U^\beta t_{\alpha\beta} = 0$. Since

$$\begin{aligned} t^{A\varphi} &= g^{AB}\,(g^{\varphi\varphi}\, t_{B\varphi} + g^{t\varphi}\, t_{Bt}) \\ &= g^{AB}\,(g^{\varphi\varphi} - \Omega\, g^{t\varphi})\, t_{B\varphi} = t^A_\varphi\,(g^{\varphi\varphi} - \Omega\, g^{t\varphi}) \end{aligned}$$

and similarly

$$t^{At} = t^A_\varphi\,(g^{t\varphi} - \Omega\, g^{tt})$$

we obtain from (9.2.137)

$$\tfrac{1}{2}\,\Upsilon = \sigma_{A\varphi}\, t^A_\varphi\,(g^{\varphi\varphi} - 2\Omega\, g^{t\varphi} + \Omega^2 g^{tt})\,.$$

But

$$g^{tt} = -\,g_{\varphi\varphi}/\varrho^2\,, \qquad g^{t\varphi} = g_{t\varphi}/\varrho^2\,, \qquad g^{\varphi\varphi} = -\,g_{tt}/\varrho^2\,.$$

If we insert these coefficients and use (9.2.127), the formula (9.2.136) follows.

Next we derive the following expression for $\sigma_{A\varphi}$, which is needed in (9.2.136),

$$\sigma_{A\varphi} = -\,\tfrac{1}{2}\,\varrho^2\,(U^t)^3\,\Omega_{;A}\,. \tag{9.2.139}$$

This gives for the dissipation function

$$\Upsilon = U^t\,\Omega_{,A}\, t^A_\varphi\,. \tag{9.2.140}$$

*Derivation of Eq. (9.2.139):* From

$$\sigma_{A\varphi} = g_{\varphi\alpha}\,\sigma^\alpha_A = g_{\varphi\varphi}\,\sigma^\varphi_A + g_{\varphi t}\,\sigma^t_A$$

and

$$\sigma^\varphi_A = \tfrac{1}{2}\, U^\varphi_{;A}\,, \tag{9.2.141}$$

(since $h^\alpha_A = g^\alpha_A$, $U_A = 0$), we have

$$\begin{aligned} \sigma_{A\varphi} &= \tfrac{1}{2}\, g_{\varphi\varphi}\, U^\varphi_{;A} + \tfrac{1}{2}\, g_{t\varphi}\, U^t_{;A} \\ &= \tfrac{1}{2}\, g_{\varphi\varphi}\,(\Omega\, U^t)_{;A} + \tfrac{1}{2}\, g_{t\varphi}\, U^t_{;A} \\ &= \tfrac{1}{2}\, g_{\varphi\varphi}\,\Omega_{,A}\, U^t + \tfrac{1}{2}\, U^t_{;A}\,(\Omega\, g_{\varphi\varphi} + g_{t\varphi})\,. \end{aligned} \tag{9.2.142}$$

Similarly

$$\sigma_{At} = \tfrac{1}{2}\, g_{t\varphi}\,\Omega_{,A}\, U^t + \tfrac{1}{2}\, U^t_{;A}\,(\Omega\, g_{t\varphi} + g_{tt})\,.$$

On the other hand, we know from (9.2.138) that $\sigma_{At} = -\,\Omega\,\sigma_{A\varphi}$. This allows us to solve the last equation for $U^t_{;A}$. If we insert the result in

(9.2.142), we find with the help of (9.2.127), (9.2.129) the formula (9.2.139).

We now integrate Eq. (9.2.134) over the four-dimensional domain which is traced out by the disk volume in the time interval $[t, t+\Delta t]$. Since the situation is stationary, we obtain with the divergence theorem for the disk luminosity

$$L_D = \int_{\text{disk}} \Upsilon\, i_k\, \eta\,, \tag{9.2.143}$$

since $dt \wedge i_k\, \eta = \eta$.

We specialize now to thin Keplerian disks ($\vartheta \approx \frac{\pi}{2}$) for which $U$ depends practically only on $r$. Then (9.2.140) gives

$$\Upsilon = U^t \Omega_{,r}\, t^r_\varphi \tag{9.2.144}$$

and thus

$$\int_{\text{disk}} \Upsilon\, i_k\, \eta = \int_{r_{\text{in}}}^{r_{\text{out}}} dr\, U^t \Omega_{,r} \int t^r_\varphi \sqrt{-g}\, d\vartheta\, d\varphi\,.$$

We write this as

$$L_D = \int_{r_{\text{in}}}^{r_{\text{out}}} 2\pi r\, Q^+(r)\, dr \tag{9.2.145}$$

with

$$Q^+(r) = \frac{1}{2\pi r}\, U^t \Omega_{,r}\, T_r\,, \tag{9.2.146}$$

where

$$T_r = \int_{r=\text{const}} t^r_\varphi \sqrt{-g}\, d\vartheta\, d\varphi \tag{9.2.147}$$

is the torque exerted by the shear flow onto the cylinder at radius $r$.

## c) Torque Equation for Keplerian Disks

We are left with computing $T_r$. This can be achieved with the conservation law for angular momentum. We integrate $J^\alpha_{;\alpha} = 0$ over the four-dimensional domain traced out in the time interval $[t, t+\Delta t]$ by the infinitesimal annulus between $r$ and $r+\Delta r$.

With Gauss' theorem we get, after dividing by $\Delta r \cdot \Delta t$,

$$\partial_r \left(2\pi \int J^r \sqrt{-g}\, dz\right) + 2\pi r\, 2 J^z \sqrt{-g}\,\big|_{\text{surface}} = 0\,. \tag{9.2.148}$$

Now from (9.2.124) we have (for $q^r \approx 0$)

$$\begin{aligned} J^r &= -J\, n\, U^r + t^r_\varphi \\ J^z &= U_\varphi\, q^z = -l\, U_t\, q^z\,. \end{aligned} \tag{9.2.149}$$

Thus

$$\partial_r [\dot{M} \tilde{J} + T_r] = - U_\varphi 2\pi r (2 q^z \sqrt{-g})|_{\text{surface}}, \tag{9.2.150}$$

where $\tilde{J} = J/m_{\rm H}$, and

$$\dot{M} = -2\pi \int m_{\rm H}\, n\, U^r \sqrt{-g}\, dz \tag{9.2.151}$$

is the accretion rate.

For thin disks, we assume again that the energy dissipated into heat is radiated on the spot in the vertical direction. Then

$$2 q^z \sqrt{-g}\,|_{\text{surface}} = Q^+ . \tag{9.2.152}$$

In the approximation $\mu \simeq m_{\rm H}$ we have also $U_\varphi = -\tilde{J} = -l\, U_t$, and so we get from (9.2.150)

$$\partial_r [\dot{M} \tilde{J} + T_r] = (2\pi r Q^+) \tilde{J} . \tag{9.2.153}$$

*d) Energy Production of Thin Keplerian Disks*

From (9.2.146) and (9.2.153) we can now determine $Q^+$.

Let

$$\tilde{F} = \frac{2\pi r Q^+}{\dot{M}}, \qquad \tilde{T} = \frac{T_r}{\dot{M}}. \tag{9.2.154}$$

Then (9.2.146) and (9.2.153) read, if we use also (9.2.132),

$$\partial_r [\tilde{J} + \tilde{T}] = \tilde{F} \tilde{J} \tag{9.2.155}$$

$$\tilde{T} = \frac{\tilde{F}}{U^t \Omega_{,r}} = \frac{U_t - \Omega \tilde{J}}{\Omega_{,r}} \tilde{F} . \tag{9.2.156}$$

Due to (9.2.156), Eq. (9.2.155) is a differential equation for $\tilde{F}$, with the solution:

$$\frac{(U_t - \Omega \tilde{J})^2}{-\Omega_{,r}} \tilde{F} = \int_{r_{\rm in}}^{r} (U_t - \Omega \tilde{J}) \tilde{J}_{,r}\, dr , \tag{9.2.157}$$

which satisfies the inner boundary condition $\tilde{F}(r_{\rm in}) = 0$. The correctness of (9.2.157) is proved in the next exercise.

---

**Exercise:** Prove the variation law

$$dE = \Omega\, dJ = 0 \tag{9.2.158}$$

for stationary and axisymmetric geodesic flows, where $E = m_{\rm H} U_t$, $J = -m_{\rm H} U_\varphi$. Use this relation to show that (9.2.157) solves (9.2.155).

*Proof of (9.2.158):* Introduce the "vorticity form" $d\Pi$, where

$$\Pi = m_{\rm H} U_\alpha dx^\alpha . \tag{9.2.159}$$

Clearly

$$E = i_k \Pi\,, \qquad J = -\,i_m \Pi\,. \tag{9.2.160}$$

Since the flow is stationary and axisymmetric, we have

$$L_k \Pi = L_m \Pi = 0 \tag{9.2.161}$$

and thus, if $\phi U = k + \Omega\, m$,

$$L_{\phi U} \Pi = L_{\Omega m} \Pi = \Omega\, L_m \Pi + d\Omega \wedge i_m \Pi = -\,J\, d\Omega\,.$$

On the other hand

$$\begin{aligned} L_{\phi U} \Pi &= i_{\phi U}\, d\Pi + d i_{k+\Omega m} \Pi \\ &= i_{\phi U}\, d\Pi + d\,(E - \Omega\, J)\,. \end{aligned}$$

The last two equations give

$$dE - \Omega\, dJ + i_{\phi U}\, d\Pi = 0\,. \tag{9.2.162}$$

For geodesic flows $i_{\phi U}\, d\Pi = 0$, because

$$i_U\, d\,(U_\alpha\, dx^\alpha) = U^\beta\,(U_{\alpha;\beta} - U_{\beta;\alpha})\, dx^\alpha = \nabla_U U = 0\,.$$

---

The solution (9.2.157) for the surface brightness of a relativistic thin Keplerian disk was found for the first time by *Page* and *Thorne* [159]. With partial integrations and using again (9.2.158), one finds the following alternative expressions:

$$\tilde{F} = -\,\frac{\Omega_{,r}}{(E - \Omega\, J)^2}\left(E J - E_{\text{in}}\, J_{\text{in}} - 2\int_{r_{\text{in}}}^{r} J\, E_{,r}\, dr\right) \tag{9.2.163}$$

$$\tilde{F} = -\,\frac{\Omega_{,r}}{(E - \Omega\, J)^2}\left(-\,E J + E_{\text{in}}\, J_{\text{in}} + 2\int_{r_{\text{in}}}^{r} E\, J_{,r}\, dr\right). \tag{9.2.164}$$

The final expression for the energy production in a Schwarzschild background now follows by inserting the values of $U_t$, $U_\varphi$ for circular orbits. These are immediately obtained from (3.2.7, 8):

$$U_t = \frac{1 - 2\, M/r}{(1 - 3\, M/r)^{1/2}}\,, \quad U_\varphi = \frac{(M\, r)^{1/2}}{(1 - 3\, M/r)^{1/2}}\,, \quad (G = 1)\,. \tag{9.2.165}$$

The integration can be done explicitly and one finds ($Q^- = Q^+$, $x = r/M$):

$$Q^- = \frac{3\dot{M}}{4\pi\, M^2}\,\frac{1}{(x-3)\, x^{5/2}} \cdot \left\{\sqrt{x} - \sqrt{6} + \frac{\sqrt{3}}{3}\ln\left[\frac{(\sqrt{x} + \sqrt{3}\,)(\sqrt{6} - \sqrt{3}\,)}{(\sqrt{x} - \sqrt{3}\,)(\sqrt{6} + \sqrt{3}\,)}\right]\right\}. \tag{9.2.166}$$

$Q^-$ satisfies the correct inner boundary condition $Q^-(x = 6) = 0$, and approaches for $x \gg 6$ the Newtonian expression (9.2.36). It attains its maximal value at $x_{\max} = 9.55$, i.e., most of the energy is produced in the region $6M < r \lesssim 30\,M$.

In the derivation of the energy production (9.2.166) we never used an explicit expression for the viscosity, because the angular momentum distribution is assumed to be known (almost free motion in the external field).

*Concluding Remarks*

We saw in Sect. 9.2.4 that the inner disk no longer remains geometrically thin if the accretion rate is larger than the critical value (9.2.77). Relativistic studies of inner disks (accretion tori) for this case can be found in [160–162]; see also [163].

Accretion tori around rotating holes may generate extended magnetospheres, since the strong differential rotation and the turbulent motion in the inner disk may fulfill the conditions for the operation of a plasma dynamo. The interaction of a magnetosphere of a disk with a rapidly rotating hole is the main subject of black hole electrodynamics. For aspects of this field of research see [164, 165].

## Appendix: Nonrelativistic and General Relativistic Hydrodynamics of Viscous Fluids

### A. Nonrelativistic Theory

We develop here briefly the principal equations of fluid dynamics from a phenomenological (continuum) viewpoint. For a detailed treatment we refer to [166–168].

Let $\boldsymbol{v}(\boldsymbol{x}, t)$ be the velocity field and $\varrho(\boldsymbol{x}, t)$ the matter density. The material (or substantial) derivative of a function $f$ is defined by

$$D_t f = \partial_t f + L_{\boldsymbol{v}} f. \tag{A.1}$$

We decompose the velocity-gradient tensor (in Euclidean coordinates) as

$$v_{i,k} = \theta_{ik} + \omega_{ik}, \tag{A.2}$$

where

$$\theta_{ik} = \tfrac{1}{2}(v_{i,k} + v_{k,i}) \tag{A.3}$$

is the rate-of-deformation tensor and

$$\omega_{ik} = \tfrac{1}{2}(v_{i,k} - v_{k,i}) \tag{A.4}$$

is the spin-tensor.

Denote by $\phi_t(\boldsymbol{x}) = \phi(\boldsymbol{x}, t)$ the trajectory of a fluid particle that is at position $\boldsymbol{x}$ at time $t = 0$.

The conservation of mass says that for a nice domain $D \subset \mathbb{R}^3$

$$\int_{\phi_t(D)} \varrho\,\eta = \int_D \varrho\,\eta, \tag{A.5}$$

where $\eta$ is the volume form of $\mathbb{R}^3$ (as a three-dimensional Riemannian manifold).

Using the change-of-variable formula and the definition of the Lie derivative, (A.5) for any $D$ is equivalent to

$$\frac{\partial \varrho}{\partial t}\eta + L_{\boldsymbol{v}}(\varrho\,\eta) = 0.$$

But

$$L_{\boldsymbol{v}}(\varrho\,\eta) = (L_{\boldsymbol{v}}\,\varrho)\,\eta + \varrho\,L_{\boldsymbol{v}}\,\eta = (L_{\boldsymbol{v}}\,\varrho + \varrho\,\mathrm{div}\,\boldsymbol{v})\,\eta$$

and thus we have the continuity equation

$$D_t\,\varrho + \varrho\,\mathrm{div}\,\boldsymbol{v} = 0 \tag{A.6}$$

or, equivalently,

$$\partial_t\,\varrho + \mathrm{div}\,(\varrho\,\boldsymbol{v}) = 0. \tag{A.6'}$$

As a corollary, we obtain the *transport theorem:*

$$\frac{d}{dt}\int_{\phi_t(D)} \varrho f\,dV = \int_{\phi_t(D)} \varrho\,D_t f\,dV. \tag{A.7}$$

Indeed, as before we first get

$$\frac{d}{dt}\int_{\phi_t(D)} \varrho f\,dV = \int_{\phi_t(D)} [\partial_t(\varrho f) + L_{\boldsymbol{v}}(\varrho f) + \varrho f\,\mathrm{div}\,\boldsymbol{v}]\,dV.$$

But the bracket is, with (A.6), equal to $\varrho\,D_t f$.

Next we formulate the balance of momentum in integral form. We consider again a comoving fluid element in $\phi_t(D)$. The forces which act on it are of two types. The first kind are external, or body forces, such as gravity or a magnetic field, which exert a force per unit volume on the continuum. The second kind of force consists of a surface force, which represents the action of the rest of the continuum through the surface of a fluid element. These stress forces are represented by the last term of the following momentum balance equation ($\boldsymbol{G}$ is the body force density):

$$\frac{d}{dt}\int_{\phi_t(D)} \varrho\,\boldsymbol{v}\,dV = \int_{\phi_t(D)} \varrho\,\boldsymbol{G}\,dV + \int_{\partial\phi_t(D)} \boldsymbol{T}(\boldsymbol{n})\,dS, \tag{A.8}$$

where $\boldsymbol{n}$ is the outward unit normal. One can show that the Cauchy traction vector $\boldsymbol{T}(\boldsymbol{n})$ depends linearly on $\boldsymbol{n}$ (Cauchy Lemma):

$$T_i(\boldsymbol{n}) = T_{ik}\,n_k. \tag{A.9}$$

With the transport theorem (A.7) and Gauss' theorem we find from (A.8)

$$\varrho\, D_t\, v_i = \varrho\, G_i + T_{ik,k}. \tag{A.10}$$

This holds in Cartesian coordinates, for which $L_v\, v_i = (\nabla_v \boldsymbol{v})_i$. Thus the invariant form of (A.10) reads

$$\partial_t\, \boldsymbol{v} + \nabla_v\, \boldsymbol{v} = \boldsymbol{G} + \frac{1}{\varrho} \operatorname{div} \underset{\sim}{T}. \tag{A.11}$$

One can show easily that these equations of motion are compatible with angular momentum conservation for the fluid element in $\phi_t(D)$ if and only if $T_{ik}$ is symmetric. (We exclude strongly polar media.)

We decompose $T_{ik}$ into an isotropic pressure term and a viscous part $t_{ik}$ which is due to velocity gradients

$$T_{ik} = -P\, \delta_{ik} + t_{ik}. \tag{A.12}$$

Since $\omega_{ik}$ represents a rigid rotation, the viscous-stress tensor $t_{ik}$ will be a linear function of $\theta_{ik}$. If we consider only isotropic media, we have the following decomposition into irreducible parts:

$$t_{ik} = 2\, \eta\, \sigma_{ik} + \zeta\, \theta\, \delta_{ik}, \tag{A.13}$$

where

$$\sigma_{ik} = \theta_{ik} - \tfrac{1}{3}\, \delta_{ik}\, \theta \tag{A.14}$$

is trace-free and

$$\theta = \theta_{kk} = \operatorname{div} \boldsymbol{v}. \tag{A.15}$$

In the stress law (A.13) $\eta$ is the shear viscosity and $\zeta$ the bulk viscosity.

Finally we consider various equivalent formulations of energy conservation.

The rate of energy increase for a material volume $\phi_t(D)$ is equal to the rate at which energy is transferred to the volume via work and heat:

$$\frac{d}{dt} \int_{\phi_t(D)} \varrho(\varepsilon + \tfrac{1}{2}\, v^2)\, dV = \int_{\phi_t(D)} \varrho\, \boldsymbol{G} \cdot \boldsymbol{v}\, dV + \int_{\partial\phi_t(D)} \boldsymbol{T}(\boldsymbol{n}) \cdot \boldsymbol{v}\, dS - \int_{\partial\phi_t(D)} \boldsymbol{q} \cdot \boldsymbol{n}\, dS. \tag{A.16}$$

Here, $\varepsilon$ is the specific internal energy and $\boldsymbol{q}$ in the last term is the heat flux.

Using again the transport theorem, the differential formulation of (A.16) reads, with Gauss' theorem,

$$\varrho\, D_t(\tfrac{1}{2}\, \boldsymbol{v}^2 + \varepsilon) = \varrho\, \boldsymbol{G} \cdot \boldsymbol{v} + \operatorname{div}(\underset{\sim}{T} \cdot \boldsymbol{v}) - \operatorname{div} \boldsymbol{q}. \tag{A.17}$$

For another form of this energy equation we rewrite the second term on the right-hand side with the help of the equation of motion (A.10) as follows

$$\begin{aligned}\operatorname{div}(\underset{\sim}{T}\cdot \boldsymbol{v}) &= \partial_k (v_i T_{ik}) = v_{i,k} T_{ik} + v_i T_{ik,k} \\ &= v_{i,k} T_{ik} + \varrho v_i D_t v_i - \varrho G_i v_i \\ &= \tfrac{1}{2}\varrho D_t \boldsymbol{v}^2 + v_{i,k} T_{ik} - \varrho G_i v_i .\end{aligned}$$

Using this in (A.17) gives

$$\varrho D_t \varepsilon = T_{ik}\theta_{ik} - \operatorname{div}\boldsymbol{q} \tag{A.18}$$

or, with the decomposition (A.12)

$$\varrho D_t \varepsilon = -P \operatorname{div}\boldsymbol{v} - \operatorname{div}\boldsymbol{q} + \Upsilon , \tag{A.19}$$

where the dissipation function $\Upsilon$ is given by

$$\Upsilon = \operatorname{Tr}(\underset{\sim}{t}\,\underset{\sim}{\theta}) = 2\eta \operatorname{Tr}\underset{\sim}{\sigma}^2 + \zeta\,\theta^2 \geqq 0 . \tag{A.20}$$

This represents the part of the viscous work going into the deformation of a fluid particle.

With (A.6) we can also write (A.19) in the form

$$\varrho\left[D_t\varepsilon + P D_t\left(\frac{1}{\varrho}\right)\right] = -\operatorname{div}\boldsymbol{q} + \Upsilon . \tag{A.21}$$

We now introduce the Gibbs equation

$$T\,ds = d\varepsilon + P\,d(1/\varrho) , \tag{A.22}$$

which allows us to write (A.21) as

$$T\varrho D_t s = -\operatorname{div}\boldsymbol{q} + \Upsilon . \tag{A.23}$$

We next derive still another alternative form of the energy equation. We start from (A.17) and write this time

$$\begin{aligned}\operatorname{div}(\underset{\sim}{T}\cdot \boldsymbol{v}) &= -(P v_k)_{,k} + (t_{ik} v_k)_{,i} \\ &= -v_i P_{,i} + \frac{P}{\varrho} D_t\left(\frac{1}{\varrho}\right) + (t_{ik} v_k)_{,i} .\end{aligned}$$

After a few manipulations we obtain from (A.17)

$$\varrho D_t\left(\varepsilon + \frac{P}{\varrho} + \frac{1}{2}\boldsymbol{v}^2\right) = \partial_t P + (t_{ik} v_k)_{,i} + \varrho\,\boldsymbol{G}\cdot\boldsymbol{v} - \operatorname{div}\boldsymbol{q} . \tag{A.24}$$

If furthermore, $\boldsymbol{G} = -\operatorname{grad}\phi$, and $\phi$ is stationary, then

$$\varrho D_t(\tfrac{1}{2}\boldsymbol{v}^2 + h + \phi) = \partial_t P + (t_{ik} v_k)_{,i} - \operatorname{div}\boldsymbol{q} . \tag{A.25}$$

Here, $h = \varepsilon + P/\varrho$ is the specific enthalpy. Equation (A.25) contains all the various equations which are called Bernoulli's equation. For example, if the flow is steady and inviscid ($\underset{\sim}{t} = 0$, $\boldsymbol{q} = 0$) then (A.25) implies that $\frac{1}{2}\boldsymbol{v}^2 + h + \phi$ is constant on any given streamline.

Finally we write down the constitutive relation between heat flux and temperature gradient,

$$\boldsymbol{q} = -\chi \operatorname{grad} T, \tag{A.26}$$

which is known as the Fourier heat conduction law.

## B. Relativistic Theory

We consider again a fluid that consists of a single component and leave out electromagnetic fields. Then the primary variables are the particle current $N^\mu = n U^\mu$ ($U^\mu$ is the four-velocity of the particle transport), the energy-momentum tensor $T^{\mu\nu}$ and the entropy flux $S^\mu$. $T^{\mu\nu}$ and $N^\mu$ are conserved,

$$T^{\mu\nu}_{;\nu} = 0, \quad N^\mu_{;\mu} = 0 \tag{B.1}$$

and the second law of thermodynamics requires

$$S^\mu_{;\mu} \geqq 0. \tag{B.2}$$

### *a) Equilibrium*

For equilibrium states we have $S^\mu = n\, s\, U^\mu$, where $s$ is the entropy per particle. Besides the Euler relation ($\mu$ is the chemical potential),

$$n\, s = \frac{1}{T}(\varrho + P) - \frac{\mu}{T} n \tag{B.3}$$

we have the fundamental thermodynamic equation of Gibbs

$$T\, ds = d(\varrho/n) + P\, d(1/n) \tag{B.4}$$

or

$$T\, n\, ds = d\varrho - \frac{\varrho + P}{n} dn. \tag{B.5}$$

From (B.3) we get

$$S^\mu = -\frac{\mu}{T} N^\mu + \frac{1}{T}(\varrho + P)\, U^\mu. \tag{B.6}$$

Since in the rest system, the matter-energy flux must vanish, we have

$$U_\lambda T^{\lambda\mu} = \varrho\, U^\mu \tag{B.7}$$

and thus we find the following expression for the entropy vector in equilibrium

$$S^\mu = -\frac{\mu}{T} N^\mu + \frac{1}{T} U_\lambda T^{\lambda\mu} + \frac{1}{T} P\, U^\mu. \tag{B.8}$$

Combining (B.3) with (B.5) gives

$$T\,d(n\,s) = d\varrho - \mu\,dn \tag{B.9}$$

and hence, if $U^\mu$ is kept fixed, we find with (B.7)

$$dS^\mu = \frac{1}{T}\,U_\lambda\,dT^{\lambda\mu} - \frac{\mu}{T}\,dN^\mu. \tag{B.10}$$

*b) Small Departures from Equilibrium*

We assume now that (B.10) remains valid for virtual variations from equilibrium states to arbitrary neighboring states. This means that the differentials are unchanged and that no extra differentials (of variables that vanish in equilibrium) enter. This postulate ("release of variations") can be justified in kinetic theories.

By addition of (B.8) and (B.10), we then obtain for an arbitrary state $(N^\mu, T^{\lambda\mu}, S^\mu)$ near some equilibrium state $(\mu/T, U^\mu/T, P, \ldots)$

$$S^\mu = -\frac{\mu}{T}\,N^\mu + \frac{1}{T}\,U_\lambda\,T^{\lambda\mu} + \frac{1}{T}\,P\,U^\mu - Q^\mu, \tag{B.11}$$

where $Q^\mu$ is of second order in the deviations from equilibrium. If one wants to arrive at hyperbolic equations, this term has to be taken seriously. (For a review and references, see [22], contribution by *W. Israel* and *J. M. Stewart.*) We are, however, not interested in transient effects implied by this contribution and continue with the conventional theory which assumes $Q^\mu = 0$.

In (B.11) $(\mu, U/T, P)$ are parameters of an equilibrium state which have to be fitted to the actual state. There is some arbitrariness in this fitting procedure. We assume with Eckart, that $N^\mu$ is unchanged $(N^\mu = n\,U^\mu)$ and that the energy-mass density is the same. With these assumptions, we have

$$T^{\mu\nu} = (\varrho + P)\,U^\mu\,U^\nu - P\,g^{\mu\nu} + \Delta T^{\mu\nu} \tag{B.12}$$

with

$$\Delta T^{\mu\nu}\,U_\mu\,U_\nu = 0 \tag{B.13}$$

and from (B.11), with the aid of (B.3)

$$S^\mu = n\,s\,U^\mu + \frac{1}{T}\,U_\lambda\,\Delta T^{\lambda\mu}. \tag{B.14}$$

*c) Relativistic Fourier Law and Navier-Stokes Equation*

The relativistic Fourier law and the Navier-Stokes equation are now a consequence of (B.1), (B.2), (B.5), and (B.12–14).

With the help of the projection-tensor

$$h_{\mu\nu} = g_{\mu\nu} - U_\mu\,U_\nu \tag{B.15}$$

on the space orthogonal to $U^\mu$, we can decompose

$$\Delta T^{\mu\nu} = t^{\mu\nu} + U^\mu q^\nu + U^\nu q^\mu \tag{B.16}$$

with

$$U_\alpha q^\alpha = 0, \quad U_\alpha t^{\alpha\beta} = t^{\alpha\beta} U_\beta = 0. \tag{B.17}$$

Inserting this into (B.14) gives

$$S^\mu = n\, s\, U^\mu + \frac{1}{T} q^\mu \tag{B.18}$$

which shows that $q^\mu$ is the heat flux.

Now we compute the entropy production. From (B.14) and (B.1)

$$S^\mu_{;\mu} = (n\, s\, U^\mu)_{;\mu} + \left(\frac{1}{T} U_\lambda \Delta T^{\lambda\mu}\right)_{;\mu} = n \nabla_U s - \frac{1}{T^2} T_{,\mu} U_\lambda \Delta T^{\lambda\mu}$$
$$+ \frac{1}{T}(U_\lambda \Delta T^{\lambda\mu}_{;\mu} + U_{\lambda;\mu} \Delta T^{\lambda\mu}). \tag{B.19}$$

The third term on the right can be obtained from

$$0 = U_\lambda T^{\lambda\mu}_{;\mu} = U_\lambda [(\varrho + P)\, U^\lambda U^\mu - P\, g^{\lambda\mu}]_{;\mu} + U_\lambda \Delta T^{\lambda\mu}_{;\mu},$$

which gives

$$\nabla_U \varrho + (\varrho + P)\, U^\mu_{;\mu} = -\, U_\lambda \Delta T^{\lambda\mu}_{;\mu}. \tag{B.20}$$

Using also the continuity equation (B.1)

$$\nabla \cdot (U\, n) = 0, \quad \text{i.e.,} \quad \operatorname{div} U = -\frac{1}{n} \nabla_U n \tag{B.21}$$

we can write (B.20) as

$$\left\langle d\varrho - \frac{\varrho + P}{n}\, dn,\, U \right\rangle = -\, U_\lambda \Delta T^{\lambda\mu}_{;\mu}$$

or, with (B.5),

$$U_\lambda\, \Delta T^{\lambda\mu}_{;\mu} = -\, T\, n\, \nabla_U s. \tag{B.22}$$

If we use this in (B.19), the first and third term cancel,

$$S^\mu_{;\mu} = -\frac{1}{T^2} T_{,\mu} U_\lambda \Delta T^{\lambda\mu} + \frac{1}{T} U_{\lambda;\mu} \Delta T^{\lambda\mu}. \tag{B.23}$$

Thus $S^\mu_{;\mu} = 0$ if $\Delta T^{\lambda\mu} = 0$, which is fine. Here we insert the decomposition (B.16) and the following identity

$$U_{\lambda;\mu} = \sigma_{\lambda\mu} + \omega_{\lambda\mu} + \tfrac{1}{3}\, \theta\, h_{\lambda\mu} + a_\lambda U_\mu, \tag{B.24}$$

where $a = \nabla_U U$, and

$$\omega_{\alpha\beta} = h^{\mu}_{\alpha}\, h^{\nu}_{\beta}\, U_{[\mu;\nu]} \quad \text{(vorticity tensor)} \tag{B.25}$$

$$\theta_{\alpha\beta} = h^{\mu}_{\alpha}\, h^{\nu}_{\beta}\, U_{(\mu;\nu)} \quad \text{(expansion tensor)} \tag{B.26}$$

$$\theta = h^{\alpha\beta}\,\theta_{\alpha\beta} = U^{\alpha}_{;\alpha} \quad \text{(volume expansion)} \tag{B.27}$$

$$\sigma_{\alpha\beta} = \theta_{\alpha\beta} - \tfrac{1}{3}\, h_{\alpha\beta}\,\theta \quad \text{(shear tensor)}. \tag{B.28}$$

---

**Exercise:** Verify (B.24).

---

We find immediately

$$S^{\mu}_{;\mu} = -\frac{1}{T^2}\,(T_{,\mu} - T\, a_{\mu})\, q^{\mu} + \frac{1}{T}\,\theta_{\lambda\mu}\, t^{\lambda\mu} \tag{B.29}$$

or, with the further decomposition

$$t_{\lambda\mu} = \hat{t}_{\lambda\mu} + \tfrac{1}{3}\, t\, h_{\lambda\mu}, \tag{B.30}$$

where $t = t^{\lambda}_{\lambda}$ and $\hat{t}^{\lambda\mu}$ is thus trace-free,

$$S^{\mu}_{;\mu} = -\frac{1}{T^2}\, q^{\mu}\, h^{\lambda}_{\mu}\,(T_{,\lambda} - T\, a_{\lambda}) + \frac{1}{T}\,(\sigma_{\mu\lambda}\,\hat{t}^{\mu\lambda} + \tfrac{1}{9}\,\theta\, t\, h_{\mu\lambda}\, h^{\mu\lambda}). \tag{B.31}$$

Now we impose (B.2). Since the individual terms in (B.31) are independent, each term must be separately non-negative for all fluid configurations. This is only possible if

$$q^{\mu} = \chi\, h^{\mu\lambda}(T_{,\lambda} - T\, a_{\lambda}), \quad \chi \geqq 0 \tag{B.32}$$

$$\hat{t}^{\mu\lambda} = 2\,\eta\,\sigma^{\mu\lambda}, \quad \eta \geqq 0 \tag{B.33}$$

$$t = 3\,\zeta\,\theta, \quad \zeta \geqq 0. \tag{B.34}$$

Equation (B.32) is the relativistic generalization of Fourier's law (A.26). The other equations give us

$$t_{\mu\nu} = 2\eta\,\sigma_{\mu\nu} + \zeta\,\theta\, h_{\mu\nu}, \tag{B.35}$$

which generalizes (A.13). The total energy-momentum tensor $T^{\mu\nu}$ is given by (B.12), (B.16), and (B.35).

We can now write (B.31) as

$$S^{\mu}_{;\mu} = -\frac{1}{\chi\, T^2}\, q_{\mu}\, q^{\mu} + \frac{2\eta}{T}\,\sigma_{\mu\nu}\,\sigma^{\mu\nu} + \frac{1}{T}\,\zeta\,\theta^2 \geqq 0. \tag{B.36}$$

Next we give a convenient form for the energy balance equation. Writing (B.22) in the form

$$\begin{aligned} T\, n\, \nabla_U s &= -(U_{\lambda}\,\Delta T^{\lambda\mu})_{;\mu} + U_{\lambda;\mu}\,\Delta T^{\lambda\mu} \\ &= -\, q^{\mu}_{;\mu} + U_{\lambda;\mu}(t^{\lambda\mu} + U^{\lambda} q^{\mu} + U^{\mu} q^{\lambda}) \\ &= U_{\lambda;\mu}\, t^{\lambda\mu} - q^{\mu}_{;\mu} + q^{\lambda}\, a_{\lambda} \end{aligned}$$

and inserting here the decomposition (B.24) gives

$$T\,n\,\nabla_U s = \sigma_{\alpha\beta}\,t^{\alpha\beta} + \tfrac{1}{3}\,\theta\,t^{\alpha}_{\alpha} - q^{\alpha}_{;\alpha} + q^{\alpha}\,a_{\alpha}, \tag{B.37}$$

Finally, the relativistic Navier-Stokes equation is

$$h^{\alpha}_{\mu}\;T^{\beta}_{\alpha;\,\beta} = 0 \tag{B.38}$$

with $T^{\alpha\beta}$ given by (B.12), (B.16), and (B.35),

$$T = (\varrho + P)\,U \otimes U - P\,g + t + U \otimes q + q \otimes U. \tag{B.39}$$

# References

## Part I

### Modern Treatments of Differential Geometry for Physicists


1 W. Thirring: *A Course in Mathematical Physics,* Vol. 1, (Springer 1978); Vol. 2 (Springer 1979)
2 Y. Choquet-Bruhat, C. De Witt-Morette, M. Dillard-Bleick: *Analysis, Manifolds and Physics,* rev. ed. (North-Holland 1982)
3 G. von Westenholz: *Differential Forms in Mathematical Physics* (North-Holland 1978)
4 R. Abraham, J.E. Marsden: *Foundations of Mechanics,* 2nd ed. (Benjamin 1978)


### Short Selection of Mathematical Books


5 S. Kobayashi, K. Nomizu: *Foundations of Differential Geometry,* I, II (Interscience Publishers 1963/69)
6 Y. Matsushima: *Differentiable Manifolds* (Marcel Dekker, New York 1972)
7 R. Sulanke, P. Wintgen: *Differentialgeometrie und Faserbündel* (Birkhäuser 1972)
8 R.L. Bishop, R.J. Goldberg: *Tensor Analysis on Manifolds* (McMillan, New York 1968)


## Part II

### Classical Textbooks


9 W. Pauli: *Theory of Relativity* (Pergamon Press 1958)
10 H. Weyl: *Space-Time-Matter;* transl. by H.L. Brose, Nethuen, London 1922 (Springer 1970)
11 A.S. Eddington: *The Mathematical Theory of Relativity* (Chelsea Publishing Company 1975)


### Recent Books


12 L.D. Landau, E.M. Lifschitz: *The Classical Theory of Fields,* 4th rev. ed. (Addison-Wesley 1969)
13 S. Weinberg: *Gravitation and Cosmology* (Wiley & Sons 1972)
14 C.W. Misner, K.S. Thorne, J.A. Wheeler: *Gravitation* (Freeman 1973)
15 G. Ellis, S. Hawking: *The Large Scale Structure of Space-Time* (Cambridge University Press 1973)
16 R.U. Sexl, H.K. Urbantke: *Gravitation und Kosmologie.* 2. Aufl. (Bibliographisches Institut, Mannheim 1983)

17 R. Adler, M. Bazin, M. Schiffer: *Introduction to General Relativity,* 2nd ed. (McGraw-Hill 1975)
18 J.L. Synge; *Relativity, The General Theory* (North-Holland 1971)
19 W. Thirring: *A Course in Mathematical Physics,* Vol. 2, (Springer 1979)
20 J. Ehlers: *Survey of General Relativity Theory. In Relativity, Astrophysics and Cosmology,* ed. by W. Israel (Reidel 1973)

**Einstein Centenary Collections**

21 *General Relativity,* An Einstein centenary survey, ed. by S.W. Hawking, W. Israel (Cambridge University Press 1979)
22 *Einstein Commemorative Volume,* ed. by A. Held (Plenum 1980)

## Parts II and III: Quoted References

23 S. Deser: *General Relativity and Gravitation,* **1,** 9 (1970)
24 D. Lovelock: J. Math. Phys. **13,** 874 (1972)
25 L. Rosenfeld: Mem. Roy. Acad. Belg. Cl. Sci. **18,** No. 6 (1940)
26 E.B. Formalont, R.A. Sramek: Astrophys. J. **199,** 749 (1975); Phys. Rev. Lett. **36,** 1475 (1976); Comm. Astrophys. **7,** 19 (1977)
27 R.D. Reasenberg et al.: Astrophys. J. **234,** L 219 (1979)
28 J.B. Hartle: Phys. Reports **46,** 201 (1978)
29 J.M. Weisberg, J.H. Taylor: Gen. Rel. + Grav. **13,** 1 (1981); Astrophys. J. **253,** 908 (1982)
30 T.A. Weaver et al.: Astrophys. J. **225,** 1021 (1978)
31 S. Chandrasekhar: Am. Jour. Phys. **37,** 577 (1969); K.C. Wali: Physics Today, October 1982, p.33
32 K.S. Thorne: In *High Energy Astrophysics,* Lectures given at the Summer School at Les Houches (Gordon & Breach, New York) p. 259
33 V. Canuto: Ann. Rev. Astron. Ap. **12,** 167 (1974); Ann. Rev. Astron. Ap. **13,** 335 (1975)
34 G. Baym, C. Pethick: Ann. Rev. Nucl. Sci. **25,** 27 (1975); Ann. Rev. Astron. Ap. **17,** 415 (1979)
35 W.D. Arnett, R.L. Bowers: Ap. J. Suppl. **33,** 415 (1977)
36 B. Friedman, V.R. Pandharipande: Nucl. Phys. **A 361,** 502 (1981)
37 G. Baym et al.: Nucl. Phys. **A 175,** 225 (1971); Astrophy. J. **170,** 99 (1972)
38 N. Straumann: *Weak Interactions and Astrophysics,* Proceedings of the GIFT Seminar on Electro-Weak Interactions, Peniscola (Spain), May 1980, to be published
39 D.J. Helfand et al.: Nature **283,** 337 (1980)
40 H.Y. Chiu: *Stellar Physics* (Blaisdell 1968) Chap. 4
41 G. Glen, P. Sutherland: Astrophys. J. **239,** 671 (1980)
42 T. Takabuka: Progr. Theor. Phys. **48,** 1517 (1972)
43 G. Flowers: Astrophys. J. **180,** 911 (1973); **190,** 381 (1974)
44 C.G. Testa, M.A. Ruderman: Phys. Rev. **180,** 1227 (1969)
45 J.W. Negele, D. Vautherin: Nucl. Phys. **A 207,** 298 (1973)
46 B.L. Friman, O.V. Maxwell: Astrophys. J. **232,** 541 (1979)
47 G. Baym et al.: In *Mesons and Fields in Nuclei,* ed. by M. Rho, D. Wilkinson (North-Holland 1978)
48 S.O. Bäckmann, W. Weise: In *Mesons and Fields in Nuclei,* ed. by M. Rho, D. Wilkinson (North-Holland 1978)
49 O. Maxwell et al.: Astrophys. J. **216,** 77 (1977)
50 M.A. Ruderman et al.: Astrophys. J. **205,** 541 (1976)
51 Ch. Kindl, N. Straumann: Helv. Phys. Acta **54,** 214 (1981)
52 R.P. Kerr: Phys. Rev. Lett. **11,** 237 (1963)
53 E.T. Newman, A.I. Janis: J. Math. Phys. **6,** 915 (1965)

54 E.T. Newman et al.: J. Math. Phys. **6,** 918 (1965)
55 G.C. Debney et al.: J. Math. Phys. **10,** 1842 (1969)
56 R. Giacconi: The Einstein X-Ray Observatory, Scientific American, February 1980
57 J. Trümper et al.: Astrophys. J. (Lett.) **219,** L 105 (1978)
58 G. Boerner: Phys. Rep. **60,** 151 (1980)
59 J.N. Bahcall: Ann. Rev. Astron. Ap. **16,** 241 (1979)
60 W.H.G. Lewin, P.C. Joss: Space Science Rev. **28,** 3 (1981)
61 E. Flowers, N. Itoh: Astrophys. J. **206,** 218 (1976)
62 J. van Paradijs: Nature **274,** 650 (1978)
63 E.P.J. Van den Heuvel: *Enrico Fermi Summer School on Physics and Astrophysics of Neutron Stars and Black Holes,* Course 65 (North-Holland 1978)
64 A. Ashtekar, R.O. Hansen: J. Math. Phys. **19,** 549 (1978)
65 R. Schoen, S.T. Yau: Commun. Math. Phys. **65,** 45 (1976); Phys. Rev. Lett. **43,** 1457 (1979); Commun. Math. Phys. **79,** 231 (1981); Commun. Math. Phys. **79,** 47 (1981)
66 T. Parker, Ch. Tauber: Commun. Math. Phys. **84,** 223 (1982)
67 C.M. Will: *Theory and Experiment in Gravitational Physics* (Cambridge University Press 1981)
68 C.V. Vishveshwara: Phys. Rev. **D 1,** 2870 (1970)
69 F.J. Zerilli: Phys. Rev. Lett. **24,** 737 (1970)
70 R.A. Hulse, J.H. Taylor: Astrophys. J. **195,** L 51 (1975)
71 J.H. Taylor, J.M.Weisberg: Astrophys. J. **253,** 908 (1982)
72 R. Epstein: Astrophys. J. **216,** 92 (1977); Astrophys. J. **231,** 644 (1979); R. Blandford, S.A. Teukolsky: Astrophys. J. **205,** 580 (1976)
73 M. Schwarzschild: *Structure and Evolution of the Stars* (Princeton University Press 1958)
74 L.L. Smarr, R. Blandford: Astrophys. J. **207,** 574 (1976)
75 D.H. Roberts, A.R. Masters, W.D. Arnett: Astrophys. J. **203,** 196 (1976)
76 P. Crane, J.E. Nelson, J.A. Tyson: Nature **280,** 367 (1979); K.H. Elliott et al.: Mon. Not. Roy. Astr. Soc. **192,** 51 (1980)
77 S. Chandrasekhar: *An Introduction to the Study of Stellar Structure* (University of Chicago Press 1939)
78 F.J. Dyson, A. Lenard: J. Math. Phys. **8,** 423 (1967); J. Math. Phys. **9,** 698 (1968)
79 E.H. Lieb, W.E. Thirring: Phys. Rev. Lett. **35,** 687 (1975), see ibid. 1116 for errata
80 F.J. Dyson: J. Math. Phys. **8,** 1538 (1967)
81 P. Ehrenfest: *Collected Scientific Papers,* ed. by M.J. Klein (North-Holland, Amsterdam 1959) p. 617
82 E.H. Lieb: Rev. Mod. Phys. **48,** 553 (1976)
83 W.E. Thirring: *A Course in Mathematical Physics,* Vol. 4 (Springer 1982)
84 W.E. Thirring: In *Rigorous Atomic and Molecular Physics,* ed. by G. Velo, A.S. Wightman (Plenum Press 1981)
85 J.-M. Lévy-Leblond: J. Math. Phys. **10,** 806 (1969)
86 E.E. Salpeter: Astrophys. J. **134,** 669 (1961)
87 T. Hamada, E.E. Salpeter: Astrophys. J. **134,** 683 (1961)
88 L.D. Landau: Phys. Z. Sowjetunion **1,** 285 (1932)
89 V. Trimple: Rev. Mod. Phys. **54,** 1183 (1982); Rev. Mod. Phys. **55,** 511 (1983)
90 J.C. Wheeler: Rep. Progr. Phys. **44,** 85 (1981)
91 M.J. Rees, R.J. Stoneham (eds.): *Supernovae* (Reidel, Dordrecht 1982)
92 K. Nomoto: Astrophys. J. **253,** 798 (1982)
93 S.E. Woosley, T.A. Weaver: In *Nuclear Astrophysics,* ed. by C. Barnes, D. Clayton, and D. Schramm (Cambridge University, U.K. 1982)
94 K. Huang: *Quarks, Leptons and Gauge Fields* (World Scientific Publ. Co., Singapore 1982)
95 E. Leader, E. Predazzi: *Gauge Theories and the New Physics* (Cambridge University Press, Cambridge, U.K. 1982)

96 P. Goldreich, S. Weber: Astrophys. J. **238,** 991 (1980)
97 W. Hillebrandt: In *11th Texas Symposium on Relativistic Astrophysics,* Dec. 1982 (to appear)
98 W. Hillebrandt: Astron. Astrophys. **110,** L 3 (1982)
99 J. Kirk, J. Trümper: In *Accretion Driven X-ray Sources,* ed. by W.H.G. Lewin, E.P.J. van den Heuvel (Cambridge University Press, U.K. 1983)
100 S.I. Syrovatskii: Ann. Rev. Astron. Astrophys. **19,** 163 (1981)
101 A.G. Pacholcyk: *Radio Astrophysics* (Freeman 1970)
102 P. Goldreich, W.H. Julian: Astrophys. J. **157,** 869 (1969)
103 R.N. Manchester, J.H. Taylor: *Pulsars* (Freeman, San Francisco 1977)
104 W. Sieber, R. Wielebinski (eds.): *Pulsars,* IAU Symposium No. 95 (Reidel, Dordrecht 1981)
105 F.C. Michel: Rev. Mod. Phys. **54,** 1 (1982)
106 M. Abramowitz, I.A. Stegun: *Handbook of Mathematical Functions* (Washington: NBS 1964)
107 K. Nomoto, S. Tsuruta: Astrophys. J. **250,** L 19 (1981)
108 M.B. Richardson et al.: Astrophys. J. **255,** 624 (1982)
109 E.H. Gudmundsson: Thesis (University of Copenhagen 1981)
110 P.O. Mazur: J. Phys. A. Math. Gen. **15,** 3173 (1982)
111 S. Chandrasekhar: *The Mathematical Theory of Black Holes* (Oxford University Press 1983)
112 E. Müller, W. Hillebrandt: Astr. Astrophys. **80,** 147 (1981)
113 P.C. Joss, S.A. Rappaport: In *Accretion Driven X-ray Sources,* ed. by W.H.G. Lewin, E.P.J. van den Heuvel (Cambridge University Press, U.K. 1983)
114 S. Ayasli, P.C. Joss: Astrophys. J. **256,** 637 (1982)
115 P. Giannone, A. Weigert: Zs. f. Ap. **67,** 41 (1967)
116 C.T. Bolton: Astrophys. J. **200,** 269 (1975)
117 M. Oda: Space Sci. Rev. **20,** 757 (1977)
118 D.R. Gies, C.T. Bolton: Astrophys. J. **260,** 240 (1982)
119 B. Paczynski: Astron. Astrophys. **34,** 161 (1974)
120 B. Margon et al.: Astrophys. J. Lett. **185,** L 113 (1973); J. Bregman et al.: Astrophys. J. Lett. **185,** L 117 (1973)
121 H.A. Hill, R.T. Stebbins: Astrophys. J. **200,** 471 (1975)
122 R. Hellings: Talk given at the 10th Intern. Conference on General Relativity and Gravitation, Padova, July 1983; R. Hellings et al.: Phys. Rev. Lett. **51,** 1609 (1983); Phys. Rev. **D 28,** 1822 (1983)
123 H.A. Hill et al.: Phys. Rev. Lett. **49,** 1794 (1982)
124 M.J. Rees: Ann. N.Y. Acad. Sci. **302,** 613 (1977); Phys. Scripta **17,** 193 (1978); In *Origin of Cosmic Rays,* IAU Symposium 94, ed. by G. Setti, A. Wolfendale (D. Reidel, Dordrecht 1980)
125 A.C. Fabian, M.J. Rees: In *X-ray Astronomy,* ed. by W.A. Baity, L.E. Peterson (Pergamon 1979)
126 R.F. Mushotzky: In *11th Texas Symposium on Relativistic Astrophysics,* (Dec. 1982)
127 A.P. Lightman: Space Science Reviews **33,** 335 (1982)
128 H. Bondi: Mon. Not. Roy. Astron. Soc. **112,** 195 (1952)
129 I.D. Novikov, K.S. Thorne: In *Black Holes,* ed. by C. De Witt, B. De Witt (Gordon & Breach, New York 1973)
130 A.W. Gillman, R.F. Stellingwerf: Astrophys. J. **240,** 235 (1980)
131 W. Brinkmann: Astron. Astrophys. **85,** 146 (1980)
132 L.D. Landau, E. M. Lifshitz: *Quantum Electrodynamics* (Pergamon Press 1982)
133 W.J. Karzas, R. Latter: Astrophys. J. Suppl. **6,** 167 (1961)
134 S. Maxon: Phys. Rev. **A 5,** 1630 (1972)
135 J.R. Ipser, R.H. Price: Astrophys. J. **255,** 654 (1982); Astrophys. J. **267,** 371 (1983)
136 J. Schmid-Burgk: Astrophys. Space Science **56,** 191 (1978)
137 K.S. Thorne: Mon. Not. Roy. Astron. Soc. **194,** 439 (1981)

138 K.S. Thorne, R.A. Flammang, A.N. Zytkow: Mon. Not. Roy. Astron. Soc. **194,** 475 (1981)
139 L. Maraschi, R. Roasio, A. Treves: Astrophys. J. **253,** 312 (1982)
140 R.Z. Yahel: Astrophys. J. **252,** 356 (1982)
141 D. Freihoffer: Astron. Astrophys. **100,** 178 (1981)
142 C.F. von Weizsäcker: Z. Naturforsch. **3a,** 524 (1948)
143 R. Lüst: Z. Naturforsch. **7a,** 87 (1952)
144 K.H. Prendergast, G.R. Burbidge: Astrophys. J. Lett. **151,** L 83 (1968)
145 N.I. Shakura: Astron. Zh. **49,** 921 (1972)
146 J.E. Pringle, M. Rees: Astron. Astrophys. **21,** 1 (1972)
147 N.I. Shakura, R.A. Sunyaev: Astron. Astrophys. **24,** 337 (1973)
148 J.P. Cox, R.T. Giuli: *Principles of Stellar Structure,* Vols. I, II (Gordon & Breach, New York 1968)
149 R. Hoshi: Suppl. Progr. Theor. Phys. **70,** 181 (1981)
150 N. Straumann: Helv. Phys. Acta **49,** 269 (1976)
151 R.A. Sunyaev, J. Trümper: Nature **279,** 506 (1979)
152 P.L. Nolan, J.L. Matteson: Astrophys. J. **265,** 389 (1983)
153 P.L. Nolan et al.: Astrophys. J. **246,** 494 (1981)
154 A.A. Galeev, R. Rosner, G.S. Vaiana: Astrophys. J. **229,** 318 (1979)
155 V.M. Vasyliunas: Space Science Reviews **24,** 609 (1979)
156 P. Ghosh, F.K. Lamb: Astrophys. J. **232,** 259 (1979); Astrophys. J. **234,** 296 (1979)
157 F. Meyer, E. Meyer-Hofmeister: Astron. Astrophys. **106,** 34 (1982)
158 N.I. Shakura, R.A. Sunyaev: Mon. Not. Roy. Astron. Soc. **175,** 613 (1976)
159 D. Page, K.S. Thorne: Astrophys. J. **191,** 499 (1974)
160 M.A. Abramowicz, M. Jaroszynski, M. Sikora: Astron. Astrophys. **63,** 221 (1978)
161 M. Koslowski, M. Jaroszynski, M.A. Abramowicz: Astron. Astrophys. **63,** 209 (1978)
162 B. Paczynski, P. Wiita: Astron. Astrophys. **88,** 23 (1980)
163 B. Carter: In *Active Galactic Nuclei,* ed. by C. Hazard, S. Mitton (Cambridge University Press, U.K. 1979)
164 K.S. Thorne, D. Macdonald: Mon. Not. Roy. Astron. Soc. **198,** 339 (1982)
165 D. Macdonald, K.S. Thorne: Mon. Not. Roy. Astron. Soc. **198,** 345 (1982)
166 L.D. Landau, E.M. Lifschitz: *Fluid Mechanics* (Pergamon 1959)
167 P.A. Thompson: *Compressible-Fluid Dynamics* (McGraw-Hill 1972)
168 G.K. Batchelor: *An Introduction to Fluid Dynamics* (Cambridge University Press 1967)
169 T. Damour: In *Gravitational Radiation,* ed. by N. Deruelle, T. Piran (North-Holland 1983)
170 T. Damour: Talk given at the 10th Intern. Conference on General Relativity and Gravitation, Padova, July 1983
171 A.L. Fetter, J.D. Walecka: *Quantum Theory of Many-Particle Systems* (McGraw-Hill 1971)

# Subject-Index